Lecture Notes in Computer S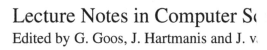

Edited by G. Goos, J. Hartmanis and J. v.

T0230204

Springer
Berlin
Heidelberg
New York
Barcelona
Hong Kong
London
Milan
Paris
Singapore
Tokyo

Egbert J.W. Boers et al. (Eds.)

Applications of Evolutionary Computing

EvoWorkshops 2001: EvoCOP, EvoFlight,
EvoIASP, EvoLearn, and EvoSTIM
Como, Italy, April 18-20, 2001
Proceedings

 Springer

Series Editors

Gerhard Goos, Karlsruhe University, Germany
Juris Hartmanis, Cornell University, NY, USA
Jan van Leeuwen, Utrecht University, The Netherlands

Main Volume Editor

Egbert J.W. Boers
Leiden University, Institute of Advanced Computer Science
Niels Bohrweg 1, 2333 CA Leiden, The Netherlands
E-mail: boers@liacs.nl

Cataloging-in-Publication Data applied for

Die Deutsche Bibliothek - CIP-Einheitsaufnahme

Applications of evolutionary computing : proceedings / EvoWorkshops
2001: EvoCOP ... Como, Italy, April 18 - 20, 2001. Egbert J. W. Boers
et al. (ed.). - Berlin ; Heidelberg ; New York ; Barcelona ; Hong Kong ;
London ; Milan ; Paris ; Singapore ; Tokyo : Springer, 2001
    (Lecture notes in computer science ; Vol. 2037)
    ISBN 3-540-41920-9

CR Subject Classification (1998): C.2, I.4, F.2-3, I.2, G.2, J.2, J.1, D.1

ISSN 0302-9743
ISBN 3-540-41920-9 Springer-Verlag Berlin Heidelberg New York

Springer-Verlag Berlin Heidelberg New York
a member of BertelsmannSpringer Science+Business Media GmbH

http://www.springer.de

© Springer-Verlag Berlin Heidelberg 2001
Printed in Germany

Typesetting: Camera-ready by author, data conversion by PTP Berlin, Stefan Sossna
Printed on acid-free paper   SPIN 10782476   06/3142   5 4 3 2 1 0

Volume Editors

Egbert J.W. Boers
Leiden Institute of Advanced
Computer Science
Leiden University
Niels Bohrweg 1
2333 CA Leiden, The Netherlands
Email: boers@liacs.nl

Jens Gottlieb
SAP AG
Neurottstrasse 16
69190 Walldorf, Germany
Email: jens.gottlieb@sap.com

Pier Luca Lanzi
Polytechnic of Milan
Piazza Leonardo da Vinci, 32
20133, Milan, Italy
Email: lanzi@elet.polimi.it

Robert E. Smith
The Intelligent Computer
Systems Centre
The University of The West of England
Coldharbour Lane
Frenchay, Bristol BS16 1QY
United Kingdom
Email: robert.smith@uwe.ac.uk

Stefano Cagnoni
Dept. of Computer Engineering
University of Parma
Parco Area delle Scienze 181/a
43100 Parma, Italy
Email: cagnoni@ce.unipr.it

Emma Hart
Napier University
School of Computing
219 Colinton Road
Edinburgh EH14 1DJ, UK
Email: emmah@dcs.napier.ac.uk

Günther R. Raidl
Algorithms and Data Structures Group
Institute of Computer Graphics
Vienna University of Technology
Favoritenstrasse 9-11/186
A-1040 Vienna, Austria
Email: raidl@ads.tuwien.ac.at

Harald Tijink
Data and Knowledge Systems Department
National Aerospace Laboratory NLR
P.O. Box 153
8300 AD Emmeloord, The Netherlands
Email: tijinkh@nlr.nl

Preface

Evolutionary Computation (EC) is a rapidly expanding field of computer science in which problem solving, optimization, and machine learning techniques inspired by genetics and natural selection are studied.

In recent years, a number of studies and results have been reported in the literature which have disclosed the potentials of EC techniques and shown their capability to solve hard problems in several domains.

This volume contains the proceedings of EvoWorkshops 2001, an event including the First European Workshop on Evolutionary Computation in Combinatorial Optimization (EvoCOP), the Second European Workshop on Evolutionary Aeronautics (EvoFlight), the Third European Workshop on Evolutionary Computation in Image Analysis and Signal Processing (EvoIASP), the First European Workshop on Evolutionary Learning (EvoLearn), and the Second European Workshop on Evolutionary Scheduling and Timetabling (EvoSTIM). These workshops were held in Como, Italy, on 18 and 19 April 2001, as part of EuroGP 2001, the Fourth European Conference on Genetic Programming.

EvoCOP focuses on applications of evolutionary algorithms and related heuristic search methods to various combinatorial optimization problems. It also covers general methodological aspects of such algorithms like operator analyses, search dynamics, fitness landscapes, and algorithmic comparisons, which are the driving force in gaining a better understanding of evolutionary search and hence support the design of effective evolutionary algorithms for combinatorial optimization problems of practical relevance.

EvoFlight is aimed at bringing together researchers and industrial parties to discuss the use of evolutionary computation in aerospace.

EvoIASP, held in 2001 for the third time, was the first event ever specifically dedicated to the applications of EC to image analysis and signal processing.

The aim of EvoLearn is to provide an opportunity for people interested in algorithms which "learn through evolution" to share ideas, discuss the current state of research, and to discuss the future directions of this particular area of Evolutionary Computation.

EvoSTIM presents the latest results in the fields of scheduling and timetabling, that are amongst the most successful applications of evolutionary techniques.

It was the aim of all workshops to give European and non-European researchers in these fields, as well as people from industry, an opportunity to present their latest research, discuss current developments and applications, besides fostering closer future interaction between members of all scientific communities that may benefit from the application of EC techniques.

EvoWorkshops 2001 were sponsored by EvoNet, the European Network of Excellence in Evolutionary Computation, as activities of EvoFlight, EvoIASP, Evo-Stim, the working groups on Evolutionary Aeronautics, on Evolutionary Image

Analysis and Signal Processing, on Evolutionary Scheduling and Timetabling of EvoNet, and of several other EvoNet members.

Fifty-two papers were accepted for publication out of 75 submissions, making EvoWorkshops 2001 the largest of the three events held since 1999. We are extremely grateful to all members of the program committee for their quick and thorough work.

April 2001

Egbert J.W. Boers
Stefano Cagnoni
Jens Gottlieb
Emma Hart
Pier Luca Lanzi
Günther R. Raidl
Robert E. Smith
Harald Tijink

Organization

EvoWorkshops 2001 was organized by EvoNet as part of EuroGP 2001.

1 Organizing Committee

EvoCOP co-chair:	Jens Gottlieb, SAP AG, Germany
EvoCOP co-chair:	Günther R. Raidl, Vienna University of Technology, Austria
EvoFlight co-chair:	Robert E. Smith, University of West of England, UK
EvoFlight co-chair:	Harald Tijink, NLR, The Netherlands
EvoIASP chair:	Stefano Cagnoni, University of Parma, Italy
EvoLearn chair:	Pier Luca Lanzi, Polytechnic of Milan, Italy
EvoSTIM chair:	Emma Hart, Napier University, Edinburgh, UK
EvoWorkshops chair:	Stefano Cagnoni, University of Parma, Italy
EuroGP co-chair:	Julian Miller, The University of Birmingham, UK
EuroGP co-chair:	Marco Tomassini, University of Lausanne, Switzerland
Local chair:	Pier Luca Lanzi, Polytechnic of Milan, Italy
Local chair:	Andrea G B Tettamanzi, Genetica srl, Italy

2 Program Committee

Giovanni Adorni, University of Genoa, Italy
Wolfgang Banzhaf, University of Dortmund, Germany
Egbert J.W. Boers, University of Leiden, The Netherlands
Alberto Broggi, University of Pavia, Italy
Larry Bull, University of West England, UK
Edmund Burke, University of Nottingham, UK
Stefano Cagnoni, University of Parma, Italy
Jie Cheng, J. D. Power & Associates, MI, USA
Ela Claridge, The University of Birmingham, UK
David Corne, University of Reading, UK
Carlos Cotta-Porras, University of Malaga, Spain
Peter Cowling, University of Nottingham, UK
Michiel de Jong, CWI, The Netherlands
Agoston E Eiben, Leiden University, The Netherlands
Terry Fogarty, Napier University, UK
David Fogel, Natural Selection, Inc., CA, USA
Jens Gottlieb, SAP AG, Germany
Jin-Kao Hao, University of Angers, France
Emma Hart, Napier University, UK
Daniel Howard, DERA, UK

Bryant Julstrom, St. Cloud State University, MN, USA
Dimitri Knjazew, SAP AG, Germany
Joshua Knowles, University of Reading, UK
Gabriele Kodydek, Vienna University of Technology, Austria
Mario Köppen, FhG IPK, Germany
Jozef Kratica, Serbian Academy of Sciences and Arts, Yugoslavia
Pier Luca Lanzi, Polytechnic of Milan, Italy
Yu Li, University of Picardie, France
Ivana Ljubic, Vienna University of Technology, Austria
Evelyne Lutton, INRIA, France
Elena Marchiori, Free University Amsterdam, The Netherlands
Dirk Mattfeld, University of Bremen, Germany
Zbigniew Michalewicz, University of North Carolina, NC, USA
Martin Middendorf, University of Karlsruhe, Germany
Julian Miller, The University of Birmingham, UK
Filippo Neri, University of Turin, Italy
Peter Nordin, Chalmers University of Technology, Sweden
Ben Paechter, Napier University, UK
Georgios I. Papadimitriou, Aristotle University, Greece
Riccardo Poli, The University of Birmingham, UK
Günther Raidl, Vienna University of Technology, Austria
Colin Reeves, Coventry University, UK
Geraldo Ribeiro Filho, UMC/INPE, Brazil
Peter Ross, Napier University, UK
Claudio Rossi, Ca' Foscari University of Venice, Italy
Franz Rothlauf, University of Bayreuth, Germany
Conor Ryan, University of Limerick, Ireland
Jim Smith, University of Western England, UK
Robert E. Smith, University of West of England, UK
Wolfgang Stolzmann, DaimlerChrysler AG, Germany
Thomas Stützle, Darmstadt University of Technology, Germany
Peter Swann, Rolls Royce plc, UK
Andrea G B Tettamanzi, Genetica srl, Italy
Harald Tijink, NLR, The Netherlands
Andy Tyrrell, University of York, UK
Christine Valenzuela, Cardiff University, UK
Marjan van den Akker, NLR, The Netherlands
Hans-Michael Voigt, GFaI - Center for Applied Computer Science, Germany

Sponsoring Institution

EvoNet, the Network of Excellence on Evolutionary Computing.

Table of Contents

Miscellaneous Applications

Assignment Problems

Analysis of Evolutionary Algorithms

Permutation Problems

EvoFlight Papers

EvoIASP Papers

EvoLearn Papers

EvoSTIM Papers

The Link and Node Biased Encoding Revisited: Bias and Adjustment of Parameters

Thomas Gaube and Franz Rothlauf*

Department of Information Systems (BWL VII)
University of Bayreuth / Germany
thomas.gaube@stud.uni-bayreuth.de, rothlauf@uni-bayreuth.de

Abstract. When using genetic and evolutionary algorithms (GEAs) for the optimal communication spanning tree problem, the design of a suitable tree network encoding is crucial for finding good solutions. The link and node biased (LNB) encoding represents the structure of a tree network using a weighted vector and allows the GEA to distinguish between the importance of the nodes and links in the network. This paper investigates whether the encoding is unbiased in the sense that all trees are equally represented, and how the parameters of the encoding influence the bias. If the optimal solution is underrepresented in the population, a reduction in the GEA performance is unavoidable. The investigation reveals that the commonly used simpler version of the encoding is biased towards star networks, and that the initial population is dominated by only a few individuals. The more costly link-and-node-biased encoding uses not only a node-specific bias, but also a link-specific bias. Similarly to the node-biased encoding, the link-and-node-biased encoding is also biased towards star networks, especially when using a low weighting for the link-specific bias. The results show that by increasing the link-specific bias, that the overall bias of the encoding is reduced. If researchers want to use the LNB encoding, and they are interested in having an unbiased representation, they should use higher values for the weight of the link-specific bias. Nevertheless, they should also be aware of the limitations of the LNB encoding when using it for encoding tree problems. The encoding could be a good choice for the optimal communication spanning tree problem as the optimal solutions tend to be more star-like. However, for general tree problems the encoding should be used carefully.

1 Introduction

The optimal communication spanning tree (OCST) problem [1] is defined to find a tree-structured communication network that connects all given nodes and satisfies their communication requirements for their minimum total cost. The number and positions of the network nodes are given a priori and the cost of the network is determined by the cost of the links.

Like other constrained spanning tree problems, the OCST problem is NP-hard [2, p. 207]. Thus, several genetic and evolutionary based algorithms (GEA) were proposed for solving the problem [3,4,5]. When using GEAs for the OCST

---

* Visiting researcher at the Illinois Genetic Algorithms Laboratory.

E.J.W. Boers et al. (Eds.) EvoWorkshop 2001, LNCS 2037, pp. 1–10, 2001.

problem and not applying genetic operators directly to the phenotypes [6], the design of a proper representation is demanding as there is a semantic gap between the structure of tree networks and common integer or bitstring representations. Recently researchers proposed encodings such as characteristic vectors [3,7,8,4], predecessor encodings [9], Prüfer numbers [5], random keys [10] and weighted encodings [11] for encoding the problem.

The characteristic vector indicates by a bitstring vector of length $n(n-1)/2$ if a link is used. This representation has high locality as small changes of the bitstring also result in small changes of the encoded tree. However, there are a lot of invalid solutions which makes using either repair operators, or specific mutation and recombination operators necessary.

When using the predecessor encoding, one node must be assigned to be the root of the network. The immediate predecessor on the path in the direction to the root is stored in a chromosome of length n. Therefore, invalid solution candidates can exist. Each element of the vector is of base n. As there are n choices for the root node, the encoding is redundant.

The Prüfer number encoding described in [12] encodes a tree with a string of length $n-2$ and each element of the string is of base n. However, although the encoding is compact and elegant, it suffers due to its weak locality. As a result, slight mutations of genes are followed by totally different network structures, and recombined offspring do not resemble the trees of their parents [13].

Rothlauf et. al. [10] extended the random key encoding [14] to tree network design problems. A vector of length $n(n-1)/2$ consisting of floating numbers ranging from zero to one was used as a chromosome. A decoding algorithm builds the tree based on the order of the values in the vector and skips edges which form cycles until the tree is complete.

The link and node biased (LNB) encoding is a representation from the class of weighted encodings and was developed by [15]. Additional encoding parameters are necessary to balance the importance of link and node weights. The encoding was proposed to overcome the problems of characteristic vectors, predecessor representations and Prüfer numbers. Later Abuali et. al. [16] compared different representations for probabilistic minimum spanning tree (PMST) problems and in some cases found the best solutions by using the LNB encoding. Raidl and Julstrom [11] observed for a similar weighted encoding which was used for the degree-constrained minimum spanning tree (d-MST) problem, solutions superior to those of several other optimization methods.

In this paper we want to investigate if there are any solution candidates preferred in the initial population, and how the setting of the two encoding parameters affects the bias of an arbitrary initial population. This is important because if the encoding prefers some solution candidates, a degradation of a GEA is often inescapable. To get rid of adjusting the encoding parameters, Palmer [15] presented in the original paper results, using only one of the two possible parameters. We want to investigate the limitations of this approach and show the dependency of the solution quality of the initial population when using both parameters.

The paper is structured as follows. In the following section we give a short description of the LNB encoding. In section 3 we give a theoretical reason for the problems with biased encoding by introducing the notion of building blocks (BB) and reviewing results from [17]. This is followed by an investigation into whether the LNB encoding using node weights only is biased, and how the individuals are represented in an initial population. In section 5 we investigate how the setting of the encoding parameters affects the bias of the encoding. The paper ends with concluding remarks.

2   A Short Description of the LNB Encoding

In this section we want to review the motivation and the resulting properties of the link-and-node-biased encoding as described in [15], and [18].

As the costs of a communication network strongly depend on the length of the links, network structures that prefer short distance links tend to have in general a higher fitness. It is useful to run more traffic over the nodes near the gravity center of an area, than over nodes at the edge of this area [19]. Thus, it is desirable to be able to characterize nodes to either be interior (some traffic only transits), or leaf nodes (all traffic terminates). As a result, the more important a link is, and the more transit traffic that crosses the node, the higher in general is the degree of the node. Nodes near the gravity center tend to have a higher degree than nodes at the edge. So the basic idea of the encoding is to encode the importance of a node. The more important the node is, the more traffic that should transit over this node.

When applying this idea to the OCST problem, the given distance matrix that defines the distances between any two nodes, is biased according to the importance of the nodes a link is connected to. If a node is not important, the modified distance matrix should increase the length of all links that are connected to this node.

The chromosome b holds the biases for each node, and has length n for an n node network. The values in the distance matrix d_{ij} are modified according to b using the weighting function

$$d'_{ij} = d_{ij} + p(b_i + b_j)d_{max}.$$

The bias b_i is a floating number between zero and one, d_{max} is the largest value in the distance matrix and p controls the influence of the biases. In the following we want to denote this approach as the node-biased encoding.

Using the bias-vector for encoding tree networks we get the encoded network structure by calculating the minimum spanning tree for the modified distance matrix. In the original work Prim's algorithm [20] was used. By running Prim's MST algorithm, nodes that are situated near other nodes will probably be interior nodes of high degree in the network. Nodes that are far away from the other nodes will probably be leaf nodes. Thus, the higher the bias of a node, the higher is the probability that it will be a leaf node. To finally get the tree's fitness the encoded network is evaluated by using the original distance matrix.

Palmer noticed in the original work that each bias b_i modifies a whole row and a whole column in the distance matrix. Thus, not all possible solution candidates can be encoded by this representation [15, pp. 66-67].

To overcome this problem he introduced in the second, extended version of the representation an additional link-bias. The chromosome holds biases not only for the n nodes but also for all possible $n(n-1)/2$ links and has overall length $l = n(n+1)/2$. Therefore, the weighting function for the elements in the distance matrix was extended to

$$d'_{ij} = d_{ij} + P_1 b_{ij} d_{max} + P_2 (b_i + b_j) d_{max}$$

with the link-specific bias b_{ij}. Using this representation the encoding could represent all possible trees. However, the string length is increased from $l = n$ to $l = n(n+1)/2$. In the following we want to denote this representation as the link-and-node-biased encoding.

Using the simple node-biased, or the more general link-and-node-biased encoding, makes it necessary to determine the value of one, respectively two, additional encoding parameters. In the original work from Palmer only results for the node-biased encoding and $p = 1$ are presented.

If the setting of the parameters could result in a biased representation of the individuals, a degradation of a GA is sometimes unavoidable as illustrated in the following section.

3 Unbiased Initial Populations and Building Block Supply

In this section we want to review the requirement for representations to be unbiased, and strengthen our investigation using work from [17]. Their model could be used for explaining why and how the quality of genetic search is affected by biased representations.

The equal distribution of the initial population is a desirable property of effective encodings [21]. Palmer formulated in his thesis necessary criteria for tree representations [15, pp. 39]:

> "It should be unbiased in the sense that all trees are equally represented; i.e., all trees should be represented by the same number of encodings. This property allows us to effectively select an unbiased starting population for the GA and gives the GA a fair chance of reaching all parts of the solution space."

Palmer recognized correctly that a widely usable encoding should be unbiased. For the LNB encoding he drew the conclusion at the end of his thesis that the

> "... new Link and Node Bias (LNB) encoding was shown to have all the desirable properties ..." [15, pp. 90]

including those to be unbiased. However, we will illustrate in the following sections that this claim is not true.

Using the notion of building blocks (BB) in the context of representations means that we want unbiased encodings to represent all BBs uniformly. The encoding should uniformly represent individuals containing high- and low-quality BBs. If the individuals are represented unbiased, then the BBs are represented unbiased, too. Biased encodings, however, overrepresent some specific building blocks in a randomly generated population.

The theoretical results from [17] could be used to explain why and how the quality of genetic search is affected by biased representations. They calculated the probability of failure for a GA as

$$\alpha = \exp\left(-x_0 \frac{2d}{\sigma_{BB}\sqrt{\pi m'}}\right),$$

where d is the signal difference between the best and second best BB, $m' = m - 1$ with m is the number of BB in the problem, σ_{BB}^2 is the BB variance and x_0 is the expected number of copies of the best BB in the randomly initialized population. The probability of GA failure goes with $O(\exp(-x_0))$.

The result from Harik et. al. tells us that when the number of copies of the best BB in the randomly initialized population is reduced, the probability of GA failure grows exponentially. Transferring these results to biased encodings, we recognize that encodings that overrepresent individuals, which consist of mainly low-quality BBs, result in an exponential decrease of GA solution quality.

However, in general, biased encodings do not always lead to a decrease in solution quality. If an encoding is biased towards individuals that are similar to the good solutions, the solution quality of a GA is increased exponentially. Therefore, researchers and practitioners should be careful with using biased encodings. If the good solutions are similar to the individuals the encoding is biased towards, it could be a good choice. However, if it is biased towards low quality solutions, a failure of the GA is inescapable.

4   The Node-Biased Encoding

It is known that the node-biased encoding is not capable of representing all possible network structures [15]. We want to investigate if the represented networks are encoded unbiased. We start with a distance matrix where all elements have the same value. This is followed by an investigation where the position of the nodes is chosen randomly.

4.1   All Links Have the Same Length

We assume that all values d_{ij} in the unbiased distance matrix are equal. Thus, the values in the biased distance matrix are determined by $\boldsymbol{b}$. We denote with b_l the lowest bias in $\boldsymbol{b}$. It is the bias for the lth node and all other biases are larger. Using this definition the modified length of each link connecting node i and j is always higher than the length of the link connecting either i and l or j and l:

$$d_{i,l} < d_{i,j} \text{ for } b_l = \min\{b_1, \ldots b_n\},$$

where $i, j, l \in \{0, \ldots, n\}$, $i \neq l$, $i \neq j$ and $l \neq j$. As the decoding algorithm chooses the shortest $n - 1$ links that do not create a cycle for creating the encoded network, the only structure that could be represented by the node-biased encoding is a star network with center l.

For an n node network the number of possible star networks is n, whereas the number of all possible networks is n^{n-2}. Thus, only a small fraction of networks could be represented by the node-biased encoding. However, at least the represented star networks are unbiased, as the elements of $\boldsymbol{b}$ are uniformly distributed in the initial population.

non-star	star with center			
	$l = 1$	$l = 2$	$l = 3$	$l = 4$
0%	25.01%	24.97%	24.92%	25.10%

Table 1. Average percentage of represented network types for a 4 node problem

We present an empirical verification of these results for a small 4 node problem in table 1. There are 16 possible networks, and 4 of them are stars with center l. For the experiments we created 1000 initial populations of size 1000. We see that it is not possible to create non-stars, and that the stars are represented uniformly. As a result, the node-biased representation is biased towards star networks if the distances between the nodes have the same value.

4.2   Random Length of Links

Now we assume that the nodes are placed randomly on a two-dimensional plane.

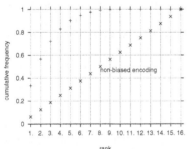

Fig. 1. Distribution of network types

For our investigation we randomly placed four nodes on a quadratic plane of size 1000 x 1000 and created randomly 1000 node-biased vectors. The distances between the nodes were calculated using the Euclidean distance. We performed 1000 experiments with different randomly located nodes.

If the representation would be unbiased each individual is created with probability $p = 1/n^{n-2} = 1/16 = 6.25\%$. However, our experiments revealed that star networks are strongly overrepresented. An average of 50.8% of all randomly created individuals are a star, whereas the portion of stars for all individuals is only $4/16 = 25\%$. Furthermore, the results show that all four stars are created uniformly. This means each of the four star networks is created with an average probability of about 12.7%

To investigate how the individuals are distributed in a randomly created population, we ordered the represented networks according to their frequency. In figure 1, we plot the cumulative frequency of the ordered number of copies an individual has in a randomly created population for a 4 node problem. If all individuals would be created with the same probability, the cumulative frequency would be linear over all possible individuals. However, for the node-biased encoding some individuals are created much more often. For our 4 node problem more than 90% of all individuals encode only five possible networks. In contrast to the overrepresentation of some individuals, some of the individuals are not represented at all. On average, 6 out of 16 possible networks are not represented at all when randomly creating 1000 individuals.

The simple node-biased encoding is biased towards star networks. A few individuals dominate a randomly created population and some solution candidates are impossible to be created. In the following section we want to investigate how the bias is influenced by the setting of its two parameters when using the link-and-node-biased encoding.

5   The Link-and-Node-Biased Encoding

Palmer [15] proposed the link-and-node-biased encoding using node and link biases as a way to overcome some of the problems with the node-biased (NB) encoding. In the following we investigate how the bias which we have noticed for the NB encoding, is influenced by the choice of the two parameters P_1 and P_2.

To investigate the bias of the LNB encoding, we randomly create a link-and-node-biased vector and measure the maximum number of links the individual has in common with one of the n stars. The more links an individual has in common with one of the stars, the more star-like it is. In figure 2 we present results for a randomly created 8, 16 and 32 node-problem. The average maximum number of links a randomly created link-and-node-biased individual has in common with a star is plotted over P_1 and P_2 and compared to an unbiased encoding. The number of links could vary from 2 (the individual represents a list network) to $n - 1$ (the individual represents a star network). The parameters P_1 and P_2 vary between 0 and 1, and we generated 2000 individuals for each parameter setting.

The results show that for all three problem instances that the bias of the LNB encoding strongly depends on P_2. For small values of P_2 the individuals are slightly biased towards non-star structures, whereas for large values of P_2 the individuals are strongly biased towards star networks. An increase of P_1 reduces the strong influence of P_2 on the bias of the encoding. For small P_2, the individuals are only slightly biased, and for large P_2 they are less biased towards stars.

The results show that with increasing P_2, a randomly created link-and-node-biased individual is more biased towards star networks. With increasing P_1, however, the population becomes less biased. In agreement with section 4 we see, that using only a node bias ($P_1 = 0$) leads to a strongly biased representation. To get a more unbiased representation large values should be used for P_1.

6   Summary and Conclusions

After a short review of tree network encodings we focus on the link and node biased encoding as described in [15]. We describe the encoding and illustrate that for setting the link-specific bias to zero we get the simplified node-biased

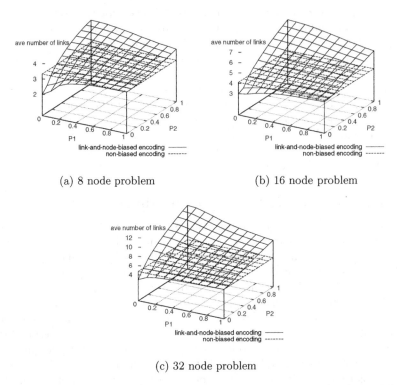

(a) 8 node problem   (b) 16 node problem

(c) 32 node problem

Fig. 2. Average maximum number of links a randomly generated individual has in common with a star. With increasing P_2 an individual is strongly biased towards star structures. Higher values for P_1 result in a more unbiased representation

encoding. Palmer stated as a necessary condition for encodings that it has to be unbiased in the sense that all individuals are equally represented. In section 3 we use the results from [17] to explain the effects of biased encodings on the solution quality of GAs. This is followed by an investigation into the bias of the node-biased encoding. Finally, we study how the bias of the link-and-node-biased encoding depends on the setting of both parameters.

We have seen that the simple node-biased encoding is biased towards star networks. Furthermore, a randomly created population is dominated only by a few individuals and some individuals could not be represented by the encoding at all. The investigation in the link-and-node-biased encoding reveals that the bias of the representation depends strongly on the used parameter setting. Using only a node-specific bias results, similar to the node-biased encoding, in a strong bias towards star networks. But fortunately the bias of the representation is decreased when using higher weights for the link-specific bias. Therefore, we strongly encourage users not to use only a node-specific, but also a link-specific bias, if they want to use the link-and-node-biased encoding, and if they are interested in having a more unbiased encoding.

Because optimal solutions for the optimal communication spanning tree problem often tend to be star-like, the LNB encoding could be a good choice for this problem. In general, however, the encoding has some serious problems, especially when using the simplified node-biased encoding. Researchers should therefore be careful when using this encoding for other problems because some network structures are not encoded at all, and a randomly generated individual is biased.

Acknowledgments. The authors would like to thank David E. Goldberg, Martin Pelikan, Kumara Sastri, Armin Heinzl, Clarissa von Hoyweghen and the anonymous reviewers of the paper for useful comments and help with the project.

Support for this work was provided by the Deutschen Forschungsgemeinschaft under SFB-DFG 1083. This work was also sponsored by the Air Force Office of Scientific Research, Air Force Materiel Command, USAF, under grant F49620-00-1-0163. Research funding for this work was also provided by the National Science Foundation under grant DMI-9908252. Support was also provided by a grant from the U. S. Army Research Laboratory under the Federated Laboratory Program, Cooperative Agreement DAAL01-96-2-0003. The U. S. Government is authorized to reproduce and distribute reprints for Government purposes notwithstanding any copyright notation thereon.

The views and conclusions contained herein are those of the authors and should not be interpreted as necessarily representing the official policies or endorsements, either expressed or implied, of the Air Force Office of Scientific Research, the National Science Foundation, the U. S. Army, or the U. S. Government.

References

1. T. C. Hu. Optimum communication spanning trees. *SIAM Journal on Computing*, 3(3):188–195, September 1974.
2. M. R. Garey and D. S. Johnson. *Computers and Intractability: A Guide to the Theory of NP-Completeness*. W. H. Freeman, New York, 1979.
3. L. Davis, D. Orvosh, A. Cox, and Y. Qiu. A genetic algorithm for survivable network design. In S. Forrest, editor, *Proceedings of the Fifth International Conference on Genetic Algorithms*, pages 408–415, San Mateo, CA, 1993. Morgan Kaufmann.
4. L. T. M. Berry, B. A. Murtagh, and S. J. Sugden. A genetic-based approach to tree network synthesis with cost constraints. In Hans Jürgen Zimmermann, editor, *Second European Congress on Intelligent Techniques and Soft Computing - EUFIT'94*, volume 2, pages 626–629, Promenade 9, D-52076 Aachen, 1994. Verlag der Augustinus Buchhandlung.
5. J. R. Kim and M. Gen. Genetic algorithm for solving bicriteria network topology design problem. In Peter J. Angeline, Zbyszek Michalewicz, Marc Schoenauer, Xin Yao, Ali Zalzala, and William Porto, editors, *Proceedings of the 1999 IEEE Congress on Evolutionary Computation*, pages 2272–2279. IEEE Press, 1999.
6. Y. Li and Y. Bouchebaba. A new genetic algorithm for the optimal communication spanning tree problem. In C. Fonlupt, J.-K. Hao, E. Lutton, E. Ronald, and M. Schoenauer, editors, *Proceedings of Artificial Evolution: Fifth European Conference*, page xx, Berlin, 1999. Springer.

7. K. S. Tang, K. F. Man, and K. T. Ko. Wireless LAN desing using hierarchical genetic algorithm. In T. Bäck, editor, *Proceedings of the Seventh International Conference on Genetic Algorithms*, pages 629–635, San Francisco, 1997. Morgan Kaufmann.

8. M. C. Sinclair. Minimum cost topology optimisation of the COST 239 European optical network. In D. W. Pearson, N. C. Steele, and R. F. Albrecht, editors, *Proceedings of the 1995 International Conference on Artificial Neural Nets and Genetic Algorithms*, pages 26–29, New York, 1995. Springer-Verlag.

9. M. Krishnamoorthy, A. T. Ernst, and Y. M. Sharaiha. Comparison of algorithms for the degree constrained minimum spanning tree. Tech. rep., CSIRO Mathematical and Information Sciences, Clayton, Australia, 1999.

10. F. Rothlauf, D. E. Goldberg, and A. Heinzl. Network random keys – a tree network representation scheme for genetic and evolutionary algorithms. Technical Report No. 8/2000, University of Bayreuth, Germany, 2000.

11. G. R. Raidl and B. A. Julstrom. A weighted coding in a genetic algorithm for the degree-constrained minimum spanning tree problem. In Janice Carroll, Ernesto Damiani, Hisham Haddad, and Dave Oppenheim, editors, *Proceedings of the 2000 ACM Symposium on Applied Computing*, pages 440–445. ACM Press, 2000.

12. H. Prüfer. Neuer Beweis eines Satzes ueber Permutationen. *Arch. Math. Phys.*, 27:742–744, 1918.

13. F. Rothlauf and D. E. Goldberg. Pruefernumbers and genetic algorithms: A lesson on how the low locality of an encoding can harm the performance of GAs. In Kalyanmoy Deb, Günther Rodolph, Xin Yao, and Hans-Paul Schwefel, editors, *Proceedings of the 2000 Parallel Problem Solving from Nature VI Conference*, pages 395–404. Springer, 2000.

14. J. C. Bean. Genetic algorithms and random keys for sequencing and optimization. *ORSA Journal on Computing*, 6(2):154–160, 1994.

15. C. C. Palmer. *An approach to a problem in network design using genetic algorithms.* unpublished PhD thesis, Polytechnic University, Troy, NY, 1994.

16. F. N. Abuali, R. L. Wainwright, and D. A. Schoenefeld. Determinant factorization: A new encoding scheme for spanning trees applied to the probabilistic minimum spanning tree problem. In L. Eschelman, editor, *Proceedings of the Sixth International Conference on Genetic Algorithms*, pages 470–477, San Francisco, CA, 1995. Morgan Kaufmann.

17. G. Harik, E. Cantú-Paz, D. E. Goldberg, and Brad L. Miller. The gambler's ruin problem, genetic algorithms, and the sizing of populations. *Evolutionary Computation*, 7(3):231–253, 1999.

18. C. C. Palmer and A. Kershenbaum. Representing trees in genetic algorithms. In *Proceedings of the First IEEE Conference on Evolutionary Computation*, volume 1, pages 379–384, Piscataway, NJ, 1994. IEEE Service Center.

19. A. Kershenbaum. *Telecommunications network design algorithms.* McGraw Hill, New York, 1993.

20. R. Prim. Shortest connection networks and some generalizations. *Bell System Technical Journal*, 36:1389–1401, 1957.

21. S. Ronald. Robust encodings in genetic algorithms: A survey of encoding issues. In *Proceedings of the Forth International Conference on Evolutionary Computation*, pages 43–48, Piscataway, NJ, 1997. IEEE.

An Effective Implementation of a Direct Spanning Tree Representation in GAs

Yu Li

LaRIA, Univ. de Picardie Jules Verne
33, Rue St. Leu, 80039 Amiens Cedex, France
fax: (33) 3 22 82 75 02
{yli@laria.u-picardie.fr}

Abstract. This paper presents an effective implementation based on predecessor vectors of a genetic algorithm using a direct tree representation. The main operations associated with crossovers and mutations can be achieved in $O(d)$ time, where d is the length of a path. Our approach can avoid usual drawbacks of the fixed linear representations, and provide a framework facilitating the incorporation of problem-specific knowledge into initialization and operators for constrained minimum spanning tree problems.

1 Introduction

Recently, many efforts have been spent to deal with constrained minimum spanning tree (MST) problems by using Genetic Algorithms (GA). These include the degree-constrained MST problem, the Optimal Communication Cost Spanning Tree (OCCST) problem, the capacitated-MST problem, etc [1]. The used GAs differ mainly in the way how spanning trees are represented. We can classify the tree representations in GAs into three categories.

The first is mainly based on the *fixed linear representations* (e.g. the Simple Genetic Algorithm of Goldberg [2]), such as, bit string, Predecessor vector, Prüfer number, etc [3, 4]. However, these linear representations often present some drawbacks: infeasible solutions, lack of locality, difficult to incorporate domain knowledge, etc.

The second is based on decoder techniques. A general decoder-based technique is the weight-coded approach [4, 5]. Another decoder-based technique consists of an integer vector which influences the order to connect vertices to the growing spanning tree by a modified version of Prim's algorithm [6].

Finally, the third category uses relatively simple and direct tree representations. In these approaches, crossovers and mutations directly manipulate trees. The advantage of direct tree representations is that they can avoid usual drawbacks of the fixed linear representations and facilitate the incorporation of problem-specific knowledge into initialization and operators. In [3, 7], two effective GAs directly working on trees are presented for the degree-constrained

E.J.W. Boers et al. (Eds.) EvoWorkshop 2001, LNCS 2037, pp. 11–19, 2001.

MST problem and the capacitated-MST problem. In [8], a direct tree representation together with variation genetic operators on trees are proposed and tested on the OCCST problem.

In [8], an adjacency list and edge lists [9] are used to implement genetic operators on trees. However, they are not effective for large instances, since the main operations of crossovers and mutations are in $O(n^2)$ time, where n is the number of vertices in a complete undirected graph. In this paper, we propose an effective implementation based on predecessor vectors. Some desirable properties of predecessor vectors allow the main operations of crossovers and mutations to be achieved in $O(d)$ time, where d is the length of a path.

The paper is organized as follows. Section 2 reviews crossovers and mutations based on a direct tree representation proposed in [8]. Section 3 discusses the distinction between representation and implementation in GAs and presents an effective implementation. Section 4 presents the first result of our implementation for the OCCST problem, and Section 5 concludes this paper.

2   Crossovers and Mutations Based on a Direct Tree Representation in [8]

Three types of crossovers manipulating trees have been proposed: *edge crossover*, *path crossover*, and *subtree crossover*. These crossovers are based on the exchange of edges.

edge crossover consists in randomly selecting some edges from two parent trees, and exchanging these edges between two parent trees to generate two offspring trees. To keep the spanning tree structure, the exchange proceeds as follows: first select an edge from parent 1 that is not already in parent 2, add the edge into parent 2, and finally delete another edge lying on the cycle introduced by the addition. *path crossover* and *subtree crossover* can be considered as the exchange of edges in paths or subtrees.

Figure 1 illustrates three examples of *edge crossover*, *path crossover* and *subtree crossover*.

Similar to the crossovers, three types of mutations have been proposed: *edge mutation*, *path mutation* and *subtree mutation*. The difference of the mutations from the crossovers is that edges, paths, and subtrees are selected not from a parent tree but rather from the original graph.

3   Effective Implementation Based on Predecessor Vectors

In our GA using the direct tree representation, crossovers and mutations are essentially additions and deletions of edges within a tree, together with searching for paths in a tree. It's important to design effective data structures and algorithms for these operations.

We observe that, if we use predecessor vectors as data structures to represent a rooted tree, we can generate a path between two vertices in $O(d)$ time, where

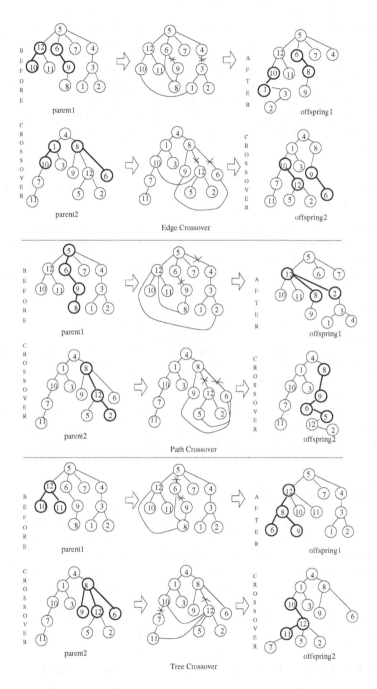

Fig. 1. Crossovers on trees

d is the length of the path. The main issue to use this data structure is how to maintain a tree rooted at some vertex during addition and deletion of edges.

Before presenting our implementation, we first discuss the distinction between representation and implementation in GAs. Then, we present the tree representation with a predecessor vector and the transformation of a tree rooted at a vertex into rooted at another vertex. Finally we propose an algorithm for adding and deleting an edge in a tree and an algorithm for generating a path between two vertices. We re-illustrate the example of *path crossover* of Figure 1 in Figure 5.

3.1 Distinction between Representation and Implementation in GAs

We emphasize the distinction between representation and implementation in GAs. In fact, this is a key issue in designing our GA. For a GA working on some representation, i.e. its operators directly manipulate this representation, there can be different implementations with different data structures. For example, in [8], an adjacency list and edge lists are used as data structures, here we can use predecessor vectors as data structures to implement the same GA.

This distinction has been mentioned in [10], however, it is not usually obvious in many GAs using the fixed linear representations. For example, in a GA using the predecessor encoding, the predecessor vector is not only used as data structure but also as representation, because operators directly manipulate it. However, in our GA, the predecessor vector is just used as data structure, and operators do not directly manipulate it.

This distinction can simplify the design of GAs. We first consider representation and concentrate on the design of initialization and genetic operators to adapt to a specific problem, then we consider effective data structures to implement the genetic operators.

3.2 Representing a Tree with a Predecessor Vector

Given a undirected connected graph G, the vertices in G are labeled with the number 1,2, ..., n. A *spanning tree* is a connected subgraph of G which covers all the vertices of G, but does not contains any cycle.

A spanning tree can be represented using a predecessor vector. It consists in designating the root r for the tree and then recording the predecessor of each node in the tree rooted at r in a vector *pred*. Let $pred[i] = j$ where j is the predecessor of i in T. As a convention, set $pred[r] = r$. Thus, every rooted tree T is represented by a vector of n numbers from 1 and n.

3.3 Transformation of a Tree Rooted at r to Rooted at r'

Since any vertex can be designated as the root, there are n rooted trees corresponding to a spanning tree. From the view of representing a tree, these n rooted trees are equivalent, since we can get the original tree from any one of them. In other word, it is not important to designate which vertex to be root.

The transformation of a tree rooted at r to rooted at r' can be achieved by simply reversing the linkage from r to r'. It just takes $O(d)$ time, where d is the length of the path from r to r'.

Figure 2 shows an example of transforming a tree rooted at 5 to another rooted at 2 and their corresponding predecessor vectors.

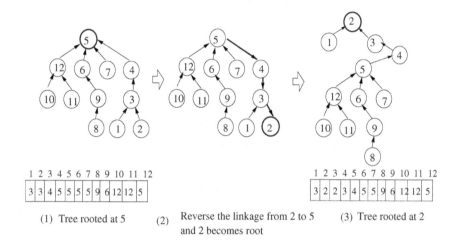

(1) Tree rooted at 5   (2) Reverse the linkage from 2 to 5 and 2 becomes root   (3) Tree rooted at 2

Fig. 2. Transformation of a tree rooted at 5 to another rooted at 2 and their corresponding predecessor vectors

3.4   Algorithm for Adding and Deleting an Edge

The main issue to use this data structure is how to maintain a tree rooted at some vertex when adding and deleting edges, because deleting an edge (u, v) will make u become another possible root, and adding an edge (i, j) can make i have two predecessors if i is not the root (Figure 3).

Based on the equivalence of spanning trees rooted at different vertices, we solve this problem as follows. For an edge (i, j) to be added, we first make i become the root by reversing the linkage from the root r to i, and search for an edge (u, v) which can be removed, e.g., an edge neither common to the two parents nor inserted recently. Then we delete (u, v) and add (i, j). Adding (i, j) makes i have a predecessor j and therefore i is not the root, and deleting (u, v) makes u become the root. This procedure leads to the same result as the operators presented in Section 2 which directly manipulate trees. This procedure is illustrated in Figure 3.

We describe the algorithm of adding and deleting an edge as follows.

Procedure addDeleteEdge($((i, j), pred)$)
```
if the edge (i, j) is not in the tree, then
    make i become a root by reversing the linkage from r to i
    search an edge (u, v) which can be removed on the path from j to i
    remove (u, v) from tree by pred[u] ← u, and u becomes a root
    add (i, j) to tree by pred[i] ← j
```

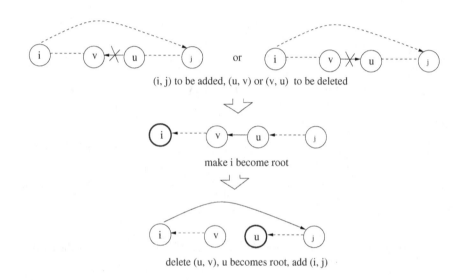

(i, j) to be added, (u, v) or (v, u) to be deleted

make i become root

delete (u, v), u becomes root, add (i, j)

Fig. 3. Illustration of *addDeleteEdge*()

3.5   Algorithm for Generating a Path in a Predecessor Vector

To generate the path from vertex i to vertex j, we construct in parallel a path from i and a path from j by following their predecessors, until they meet at a vertex k. The path from i meeting the path from j means that, the current vertex k to be put in the path from i exists in the path from j. The resulting path from i to j can be formed by concatenating the path from i to k and the path from k to j. This procedure is illustrated in Figure 4.

We give the algorithm for generating a path from i to j as follows.

Procedure getPath($i, j, pred$)
$$path_i \leftarrow \{i\}, \quad path_j \leftarrow \{j\}$$
$$(u_i, v_i) \leftarrow (i, pred[i])$$
$$(u_j, v_j) \leftarrow (j, pred[j])$$

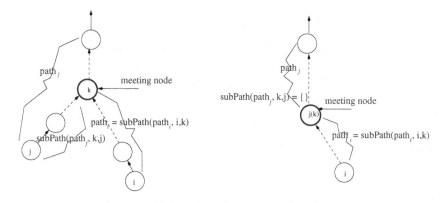

Fig. 4. Illustration of $getPath()$ for the case that $path_i$ meets $path_j$

```
while pathᵢ and pathⱼ do not meet at some vertex do
    if uᵢ is not root, then
        put vᵢ into pathᵢ
        if pathᵢ meets pathⱼ at vᵢ, then
            mark vᵢ with k by k ← vᵢ
        pass to the next edge by (uᵢ, vᵢ) ← (vᵢ, pred[vᵢ])
    if uⱼ is not root and two paths do not meet, then
        put vⱼ into pathⱼ
        if pathⱼ meets pathᵢ at vⱼ, then
            mark vⱼ with k by k ← vⱼ
        pass to the next edge by (uⱼ, vⱼ) ← (vⱼ, pred[vⱼ])
pathᵢⱼ ← subPath(pathᵢ, i, k) + subPath(pathⱼ, k, j)
```

3.6   Example of *Path Crossover* Based on This Implementation

We demonstrate the example of *path crossover* of Figure 1 based this implementation in Figure 5.

4   Application to the OCCST Problem

We re-implement the GA presented in [8] for the OCCST problem using the implementation algorithms above. First, our goal is just to test the feasibility of these algorithms. Table 1 gives a comparison about time used for instances in [8]. This comparison is very rough, for the experiments have not been finished. However, we notice an obvious improvement. Moreover, the larger the instance, the larger the improvement is.

The program is written in C++ and the tests are done on a 300 MHz AMD k6-2 PC under Linux. The algorithm is executed 5 times for each example. The parameters for the GA are set as below:

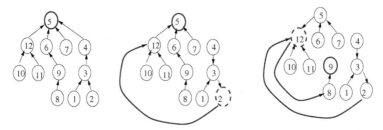

(1) the path 2-12-8 to be added   (2) make 2 become root, add (2,12),   (3) make 12 become root, add (12, 8),
into the tree rooted at 5   remove (5,4), and 5 become root   remove (9,6), and 9 become root

Fig. 5. Path Crossover in Figure 1 based on the predecessor vector implementation

- the crossover probability is 0.6;
- the mutation probability is 0.001;
- the population size is 200;
- the maximum generation is 50;
- the operators are *path crossover* and *path mutation*.

Table 1. Comparison of times on 4 instances for the OCCST problem

Instance (nb of vertices)	Old Implementation Best(time[s])	Avg(time[s])	New Implementation Best(time[s])	Avg(time[s])
12	5.83	6.06	11.09	11.51
24	73.01	76.34	60.86	63.55
35 (non-uniform distance)	340.64	348.12	160.01	164.35
35 (uniform distance)	277.58	284.19	109.78	112.67

5   Conclusions

This paper presents an effective implementation based on predecessor vectors of a GA using a direct tree representation. The main issue to use predecessor vectors as data structures is how to maintain a tree rooted at some vertex during addition and deletion of edges. Based on the equivalence of spanning trees rooted at different vertices, we propose an algorithm for adding and deleting an edge in $O(d)$ time, where d is the length of a path. We propose also an algorithm for generating a path between two vertices in $O(d)$ time.

In the future, we will exploit this representation and its implementation for constrainted MST problems. We will also extend this idea to other problems whose solutions are more complex, for example, a cycle, a graph, etc.

Acknowledgments. This work is partially supported by "le conseil Régional de Picardie" (Pôle modélisation, projet parallélisme).

We thank Chumin Li and Alain Cournier for their comments which helped to improve this paper.

References

1. Crecenzi P., Kann V. A compendium of NP optimization problems. Available online at http://www.nada.kth.se/theory/compendium/, Aug. 1998
2. Goldberg D. E. Genetic algorithms in search, optimization and machine learning. Addison-Wesley (Reading, Mass).
3. Raidl G. R. An Efficient Evolutionary Algorithm for the Degree-Constrained Minimum Spanning Tree Problem. Proc. of the 2000 IEEE Congress on Evolutionary Computation, San Diego, CA, pp. 104-111, July 2000.
4. Palmer C. C., Kershenbaum A. An approach to a problem in network design using genetic algorithms. Networks, vol. 26 (1995) 151-163.
5. Raidl G. R., Julstrom B. A. A Weighted Coding in a Genetic Algorithm for the Degree-Constrained Minimum Spanning Tree Problem. Proc. of the 15th ACM Symposium on Applied Computing, Como, Italy, pp. 440-445, March 2000.
6. 6. Knowles J., Corne D. A new evolutionary approach to the degree constrained minimum spanning tree problem. IEEE Transactions on Evolutionary Computation, Volume 4 number 2, pp. 125-134, July 2000.
7. Raidl G. R., Drexel C. A Predecessor Coding in an Evolutionary Algorithm for the Capacitated Minimum Spanning Tree Problem. Late-Breaking-Papers Proc. of the 2000 Genetic and Evolutionary Computation Conference, Las Vegas, NV, pp. 309-316, July 2000.
   bibitemli Li Y., Bouchebaba Y. A new genetic algorithm for the optical communication spanning tree problem. Proc. of Artifical Evolution 99, LNCS 1829, pp. 162-173, Dunkerque, France, 1999.
8. Gibbons A. Algorithmic graph theory. Cambridge University Press, New York.
9. Michalewicz Z. Genetic Algorithms + Data Structures = Evolutionnary Programs. Springer-Verlag, 3rd edition, 1996.

An Evolutionary Algorithm with Stochastic Hill-Climbing for the Edge-Biconnectivity Augmentation Problem

Ivana Ljubić and Günther R. Raidl

Institute for Computer Graphics and Algorithms, Vienna University of Technology,
Favoritenstraße 9–11/186, 1040 Vienna, Austria
{ljubic|raidl}@ads.tuwien.ac.at

Abstract. Augmenting an existing network with additional links to achieve higher robustness and survivability plays an important role in network design. We consider the problem of augmenting a network with links of minimum total cost in order to make it edge-biconnected, i.e. the failure of a single link will never disconnect any two nodes. A new evolutionary algorithm is proposed that works directly on the set of additional links of a candidate solution. Problem-specific initialization, recombination, and mutation operators use a stochastic hill-climbing procedure. With low computational effort, only locally optimal, feasible candidate solutions are produced. Experimental results are significantly better than those of a previous genetic algorithm concerning final solutions' qualities and especially execution times.

1 Introduction

All communication networks are designed for certain demands and requirements. In the course of time, traffic demands typically increase and the networks are often not as satisfying as at the beginning. Therefore, *the augmentation of networks* by additional links plays an important role in network design. In addition to the increase of band-width, an increase of robustness and survivability is often needed. A network can be made robust against failures in connections between two sites or against site failures. The costs of such augmentations should usually be as small as possible.

The robustness of a certain network, is in graph theory described by the *vertex* and *edge k-connectivity*. A connected undirected graph $G = (V, E)$ has edge (vertex) connectivity k if at least k edges (vertices) must be deleted to disconnect G. The removal of vertices hereby includes the removal of all adjacent edges. Therefore, a vertex k-connected network is always edge k-connected, but not necessarily vice versa.

The problem of augmenting a graph to become $k = 1$ connected is identical to the *minimum spanning tree* (MST) problem, which can be solved efficiently in polynomial time. However, in practical communication networks, a larger connectivity-level for higher reliability is often needed. From the other side, costs usually limit the connectivity level k to a small value.

E.J.W. Boers et al. (Eds.) EvoWorkshop 2001, LNCS 2037, pp. 20–29, 2001.

In this paper, we concentrate on edge-connectivity with $k = 2$. This problem is also called *edge biconnectivity augmentation problem* (E2AUG). Our goal is, therefore, to augment a given network with additional links of minimum total costs, to make it edge-biconnected.

Eswaran and Tarjan [1] showed that E2AUG is $\mathcal{NP}$-hard. The problem remains $\mathcal{NP}$-hard, even in the case when all connections have weights chosen from the set $\{1, 2\}$ only. Due to the hardness of the problem, it was addressed by heuristic methods including a hybrid genetic algorithm (HGA) [2]. In contrast to this previous HGA, we present here a new evolutionary algorithm (EA) based on a powerful preprocessing and a straight-forward edge-set representation. The recombination and mutation operators produce only feasible solution candidates and contain a local stochastic hill-climbing which removes redundant edges. That way, the EA searches the space of locally optimal solutions only.

The initialization, recombination, and mutation operators are specifically designed for the considered problem. Recombination and mutation preserve a great amount of parental structures, i.e. the locality is high. The average computational effort of recombination and mutation operators is low, which allows a fast execution on large graphs, too. Empirical results indicate the new EA is significantly better concerning the quality of final solutions, as well as the execution times when compared to the previous HGA and another iterative heuristic for the E2AUG.

The next section provides a mathematical definition of E2AUG and a summary of previous work related to the problem. In Section 3, an explanation of the preprocessing is given. Section 4 describes the EA with its stochastic local hill-climbing procedure in detail. An empirical comparison to the previous HGA is given in Section 5, and final conclusions are drawn in Section 6.

2   The Edge-Biconnectivity Augmentation Problem

Given are a connected, undirected graph $G = (V, E)$, and an additional set AUG of edges connecting nodes in V ($AUG \cap E = \emptyset$). Each edge $e \in AUG$ has associated costs $c(e) > 0$. The graph $G_A = (V, E \cup AUG)$ is edge-biconnected. The goal is to augment graph G using a subset S of edges from AUG with minimum total costs $c(S) = \sum_{e \in S} c(e)$, so that graph $G_S(V, E \cup S)$ is also edge-biconnected.

In graph G, an edge $e \in E$ is called a *bridge* if its deletion disconnects G. G_S must therefore not contain any bridges. That is why this problem is also called *bridge-connectivity augmentation problem*.

The problem has been stated the first time by Eswaran and Tarjan [1]. In this work a polynomial algorithm for E2AUG is given for the specific case when all edge costs are the same and graph G_A is complete. A survey on several related problems and approximation algorithms is given by Khuller [3].

In 1981, Frederickson and Jájá [4] proposed an approximative algorithm for E2AUG based on the following steps:

Firstly, all already biconnected components in G are shrinked into "super-nodes", whereby all self-loops are discarded; from multiple edges $e \in AUG$ connecting the same pair of nodes in the shrinked graph, only the cheapest edge is retained. In this way, the problem of augmenting a general connected graph G, can always be reduced to the problem of augmenting a spanning tree.

In the next step, the shrinked graph G is interpreted as a directed tree: a random root r is chosen and all edges in E are directed toward this root. After that, the algorithm searches for a minimum outgoing branching, i. e. a directed tree with paths from the root r to all other nodes (Gabow et al. [5]), using edges from the shrinked set AUG and E. A feasible set of augmenting edges $S \subset AUG$ can finally be derived from the set of edges included in this branching. It is proven that this algorithm determinates a solution with costs no greater than two times the costs of the optimal solution.

The time complexity of this algorithm has been improved by Khuller and Thurimella [6]. Recently, Zhu et al. [7,8] proposed an iterative scheme based on the branching algorithm. They provide a heuristic formula for measuring how good a certain augmenting edge in a determined outgoing branching can be. Using this formula, only one edge at a time, as one step of an iterative process, is fixed, and the edge's cost is set to zero. Then, a new minimum outgoing branching is derived, and the edge is fixed. This process continues until the evaluated branching has zero costs and a complete set S is obtained.

Ljubic et al. [2] proposed a hybrid genetic algorithm (called HGA) for E2AUG. This algorithm is based on a binary encoding, in which each bit corresponds to an edge in AUG, on standard uniform crossover, and on bit-flip mutation. Infeasible solutions are repaired using a greedy repair-algorithm in a Lamarckian way. This algorithm temporarily removes the bridges one by one from an infeasible solution and searches for the cheapest edge from AUG connecting the two separated components. For a better performance, the algorithm uses caching [9].

Although better results are obtained by HGA than by several previous approaches, this method has also disadvantages: The genetic code has length $|AUG|$, which is not efficient for larger complete or dense graphs. The required space and time to evaluate a solution is $O(|V|^2)$ in such cases, while the number of edges included in biconnectivity augmentation is always less then $|V|$, according to Mader's theorem [10]. Furthermore, many genetic codes created by recombination and/or mutation are mapped to one and the same phenotypic solution due to the repair operator. This effect endangers the GA to converge too quickly to suboptimal solutions.

In this paper, we present a new evolutionary approach which overcomes these disadvantages using a compact edge-set encoding, problem-based operators, and a local stochastic hill-climbing. This EA searches the space of local optima only. Such an approach belongs to a broader group of combinatorial optimization algorithms, called *local-search-based memetic algorithms* [11].

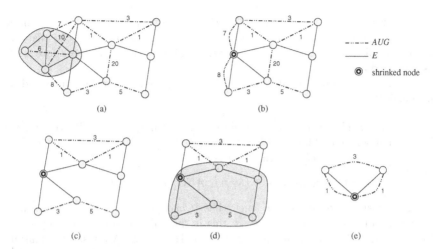

Fig. 1. An example for preprocessing: (a) given graph $G = (V, E)$ and set AUG, (b) after shrinking, (c) after elimination of edges, (d) after fixation of edges from AUG, (e) after another shrinking.

3 Preprocessing

Good preprocessing can reduce a problem's search space significantly. Our preprocessing is based on the next three steps which are illustrated in Fig. 1:

Shrinking: This reduction has been originally described by Frederickson and Jájá [4]. Edge-biconnected components of a graph are its maximal edge-biconnected subgraphs (maximal in the sense that no other node from the graph can be added, without violating biconnectivity). By this procedure, all edge-biconnected subgraphs are found and shrinked into "super-nodes". From the edges from AUG that connect the same components, only the cheapest ones are included and all others are discarded. Self-loops, i.e. augmenting edges connecting nodes of the same component, are always discarded. For each edge in the shrinked graph a reference back to the corresponding edge of the original problem is stored. After shrinking, the graph has always a tree structure, where all edges represent bridges of the initial configuration. Note that now, $G_S = (V, E \cup S)$ and $G_A = (V, E \cup AUG)$ can become multigraphs, since AUG and E are not necessarily disjoint anymore (see Fig. 1 (b)).

Edge elimination: Edge $e_0 \in E$ is *covered* by an edge $e = (u, v) \in AUG$ if e_0 lies on the tree path (in G) connecting nodes u and v. This procedure detects and removes each edge e_{in} from AUG, for which some other edge $e_{out} \in AUG$ exists, such that the tree-path (in G) covered by e_{in} is a subset of the tree path covered by e_{out} and e_{out} is cheaper then e_{in}. More formally, if $e_{in} = (u, v) \in AUG$, and:

$$Path(e) = \{e_0 \mid e_0 \in E \text{ and } e_0 \text{ is part of the path connecting } u \text{ and } v \text{ in } G\}$$

then, e_{in} is obsolete if:

$$\exists\, e_{out} = (s, t) \in AUG \text{ such that } Path(e_{in}) \subset Path(e_{out}) \land c(e_{in}) \geq c(e_{out}).$$

Frederickson and Jájá [4, pp. 276–277] proposed a dynamic programming algorithm for their branching based heuristic, that computes special distance values for all edges in AUG and identifies all obsolete edges in the above sense as a byproduct in $O(|V|^2)$ time.

Usually, the more star-like graph G is, the less edges can be eliminated by means of this procedure.

Edge fixation: This procedure identifies edges that must necessarily be included in the final solution. The sets $Cover(e_0)$, $e_0 \in E$, are the sets of all augmenting edges $e = (u, v) \in AUG$ such that e_0 lies on the (u, v)-path in G, i.e.

$$Cover(e_0) = \{e = (u, v) \in AUG \mid e_0 \in Path(e)\}. \qquad (1)$$

If there exists an edge $e_0 \in E$ such that $|Cover(e_0)| = 1$, then edge e must appear in all feasible solutions, since this is the only possibility to "cover" e_0. Edge e is therefore fixed by moving it from AUG to E. After such a fixation, a new edge-biconnected component is created, which can further be shrinked into a single new super-node. Since shrinking can enable further fixations, the process is repeated until no more edges from AUG can be fixed.

Note that the more sparse graph G_A is, the more edges can typically be fixed.

4   The Evolutionary Algorithm

Although preprocessing reduces the size of the problem, it is in general not able to solve the problem completely. Therefore, we apply the following evolutionary algorithm.

4.1   Edge-Set Encoding

Each candidate solution is directly represented by the set S of selected edges. For this purpose, we use a hashed array as data structure. Insertion as well as deletion of a single edge and the check whether an edge is contained or not take always constant time. Furthermore, the space needed to store each individual is $O(S)$, where $|S| < |V|$.

4.2   Stochastic Hill-Climbing

The central part of the new approach is the local stochastic hill-climbing procedure, incorporated in the initialization, crossover, and mutation operators. This procedure removes redundant edges from a given feasible solution in an indeterministic way until the solution becomes edge-minimal concerning the biconnectivity property. This algorithm checks each edge in S in random order if it can be removed without making the solution infeasible.

For each edge $e_0 \in E$ we determine the number of all edges from S that "cover" e_0 in an initial step:

$$n_{cov}(e_0) = |\{e \mid e \in S \text{ and } e_0 \in Path(e)\}|.$$

procedure • •••• ••• • •• • (**var** S):
begin
    $n_{cov}(e_0) \leftarrow 0$, for each $e_0 \in E$;
    for each $e \in S$ **do**
        for each $e_0 \in Path(e)$ **do**
            $n_{cov}(e_0) \leftarrow n_{cov}(e_0) + 1$;
    while not all edges $e \in S$ are processed
    select a yet unprocessed edge $e \in S$;
    if $n_{cov}(e_0) \geq 2$, $\forall e_0 \in Path(e)$
        $S \leftarrow S \setminus \{e\}$;
        for each $e_0 \in Path(e)$
            $n_{cov}(e_0) \leftarrow n_{cov}(e_0) - 1$;
end

Fig. 2. The EA's stochastic hill-climbing.

An edge $e \in S$ is then *redundant* and can be removed if:

$$\forall\, e_0 \in Path(e) \;:\; n_{cov}(e_0) \geq 2.$$

After each edge elimination from S, $n_{cov}(e_0)$ is updated accordingly.

The pseudo-code presented in Figure 2 shows the algorithm in more detail. During hill-climbing, each set $Path(e)$, $\forall e \in AUG$, can be determined in $O(|path(e)|)$ time, if a depth-first search started from an arbitrarily chosen root is performed and depth and parent informations are stored for each node [12]. Hence, the worst-case execution time needed for determining all $n_{cov}(e_0)$, $e_0 \in Path(e)$, is $O(|V||S|)$, but in average the whole algorithm runs in $O(|S| \log |V|)$ time.

4.3 Initialization

In order to create feasible initial solutions, we apply the described stochastic hill-climbing procedure to the whole set AUG of edges that can be used for augmentation, i.e. $S = AUG$. Due to the indeterminism of hill-climbing, generated solutions are in general different and enough initial diversity is provided.

4.4 Edge Crossover

When designing a suitable crossover operator, our main goal was to produce a new solution that inherits as many parental structures as possible in order to provide a high level of locality. This can be accomplished by setting a child's edge-set to the union of the parental edge-sets and applying local stochastic hill-climbing to it.

In this way, a new locally optimal solution is efficiently created out of the parental edges only. Nevertheless, even if the same two parents are again selected for mating, a different offspring is generated with high probability.

> **procedure** •••••••••• • •••••• •(var S):
> **begin**
>   **do** p_{mut} times:
>   choose $e \in S$ randomly;
> $S \leftarrow S \setminus \{e\}$;
>   **for each** $e_0 \in Path(e)$ **do**
> $n_{cov}(e_0) \leftarrow n_{cov}(e_0) - 1$;
>   **for each** $\{e_0 \mid e_0 \in Path(e) \wedge n_{cov}(e_0) = 0\}$ **do**
>   select e_1 from $Cover(e_0) \setminus \{e\}$ randomly;
> $S \leftarrow S \cup \{e_1\}$;
>   **for each** $e' \in Path(e_1)$ **do**
> $n_{cov}(e') \leftarrow n_{cov}(e') + 1$;
>   • ••••• ••• • •• •(S);
>   **end**

Fig. 3. Edge-delete mutation.

Meaningful building-blocks will be transmitted from parents to offsprings, and strong locality is provided. Since $|S| < |V|$ for each parental edge-set, the computational effort of edge crossover is only $O(|V| \log |V|)$.

4.5 Edge-Delete Mutation

The mutation operator's main purpose is to counteract premature convergence and to maintain enough diversity in the population by introducing new edges from $AUG \setminus S$. Edge-delete mutation replaces a randomly selected edge from S by one or more different, appropriate edges from $AUG \setminus S$, so that edge-biconnectivity is maintained. Since the offspring generated in this way is not necessarily edge-minimal anymore, local stochastic hill-climbing is finally applied again. The algorithm presented in Fig. 3 shows more details.

To be able to perform the mutation efficiently, we suggest to determine the set $Cover(e_0)$ as defined in (1) for each $e_0 \in E$ as a part of preprocessing. Then the mutation operator needs only $O(|V| \log |V|)$ time, too.

The parameter p_{mut} is the number of times an edge is substituted and therefore controls how strong mutation will actually change a certain solution.

4.6 Edge-Cost Based Heuristics

Usually, cheaper edges will appear more frequently in optimal solutions than expensive edges. Based on this observation, we include additional cost-based heuristic in the hill-climbing and mutation algorithms.

Heuristic in hill-climbing: During hill-climbing, the order of processing the edges in S is crucial. In each iteration, we bias the selection of the edge coming next towards more expensive edges by performing a tournament selection on all yet unprocessed edges in S. From a group of k_{impr} randomly chosen edges, the most expensive edge is selected.

Heuristic in edge-delete mutation: For mutation we select each replacement-edge (needed to cover a newly introduced bridge e_0) from $Cover(e_0) \setminus \{e\}$ by using a tournament selection. Now, cheaper edges need to be preferred; thus, the edge with the smallest costs is selected from a randomly chosen group of size k_{mut}.

4.7   General EA Properties

We used a steady-state evolutionary algorithm in which only one new solution is created by means of crossover and mutation. The new solution always replaces the worst solution with one exception: Only new solutions that are not dupli-cates of solutions already in population are accepted in order to maintain higher diversity. Parents are chosen using tournament selection with the group size k.

5   Experimental Results

In this section we present experimental results of the proposed EA with stochas-tic hill-climbing (EASHC) and the previous hybrid genetic algorithm (HGA) from [2].

Since the problem of augmenting a general connected graph G, can effectively be reduced to augmenting a tree (see Sec. 3), G is always a spanning tree in our test instances. Table 1 shows the main properties of the considered instances. Columns $|V_{pre}|$ and $|AUG_{pre}|$ indicate the numbers of nodes and the numbers of augmenting edges of G, respectively, after performing preprocessing. Instances A3 to N2 were created using a generator from Zhu et al. [8]. All graphs were randomly generated; column $c(e)$ shows the intervals from which costs for each $e \in AUG$ were chosen. Results for problem instances A3 to N2 were adopted from [2]. Instances R1 to E3 are new and larger than all previously tested ones. In contrast to the other instances, E1 to E3 are Euclidean problems in which nodes represent randomly placed points in the plane and edge costs correspond to the points' Euclidean distances.

It can further be observed that especially when G_A is sparse (as in instances A_3, B_1, B_6), the fixing of edges together with the new iterative shrinking and edge-elimination can dramatically reduce the problem size. On the other hand, if G_A is dense (instances N1 to E3), edge-fixation is not able to reduce the number of nodes, but the edge-elimination is more effective. Especially for larger problem instances, preprocessing times t_{pre} are neglectable in comparison to the EA's total execution times, see Table 2. Preprocessing was able to reduce the problem size $|AUG|$ significantly: in case of instance B1 to $\sim 1/7$, in average to about the half.

Suitable EA parameters were determined by extensive preliminary tests: pop-ulation size $|P| = 100$, group size for tournament selection of the EA $k = 5$, during recombination $k_{impr} = 5$, and during mutation $k_{mut} = 4$. The mutation rate is $p_{mut} = 5$. Each run was terminated when no new best solution was found within the last 100,000 created solutions.

Table 1. Properties of considered problem instances.

instance	$|V|$	$|AUG|$	$c(e)$ in	$|V_{pre}|$	$|AUG_{pre}|$	$t_{pre}[s]$
A3	40	29	[1,780]	12	13	0.1
B1	60	55	[1,1770]	8	4	0.1
B6	70	81	[1,2415]	31	39	0.2
N1	100	1104	[10,50]	100	687	0.6
N2	110	1161	[10,50]	110	734	0.6
R1	200	9715	[1,100]	200	3995	11.2
R2	200	9745	[5,100]	200	3702	10.9
E1	200	19701	Euclidean	200	4104	25.8
E2	300	11015	Euclidean	300	4462	31.5
E3	400	7621	Euclidean	400	4806	51.6

Table 2. Average results of HGA and EASHC.

inst.	best	HGA gap[%]	$\sigma(gap)$	eval	$t[s]$	EASHC gap[%]	$\sigma(gap)$	eval	$t[s]$
A3	6607.0	0.0	0.0	380	0.1	0.0	0.0	0	0.1
B1	15512.0	0.0	0.0	50	0.0	0.0	0.0	0	0.1
B6	19022.0	0.0	0.0	9400	4.5	0.0	0.0	7	0.2
N1	383.0	2.6	2.0	95350	230.4	0.5	0.4	3998	10.4
N2	429.0	2.3	1.3	120400	544.9	0.0	0.0	3793	11.3
R1	121.4	1.1	1.0	244325	12398.3	0.0	0.1	12410	135.3
R2	320.5	6.7	1.7	243085	11434.4	0.7	0.1	38912	218.5
E1	2873.8	12.8	5.6	236305	20740.1	1.0	0.9	34129	191.0
E2	9355.2	8.9	1.7	236480	22602.5	0.4	1.6	97764	731.0
E3	21329.1	8.4	2.0	246640	23970.4	0.5	1.3	113831	1451.4

Table 2 presents average results obtained from 100 runs/instance in case of EASHC, and 10 runs/instance in case of HGA. Column *best* shows each problem instance's best known solution. For both algorithms percentage *gaps* of the final solutions' average total costs to the best known values *best* and standard deviations of gaps $\sigma(gap)$ are given. *Evals* indicates the average number of evaluated solutions until the finally best solution had been obtained, and t is the corresponding execution time in seconds.

For all problem instances EASHC's final solutions are better than or at least equally good as those obtained by HGA. Nevertheless, EASHC needs in all cases significantly fewer iterations and was much faster. For instance E1, EASHC was more than 100 times faster.

6   Conclusion

The proposed EA has the following advantages: initialization, recombination, and mutation produce only feasible, locally optimal solution candidates, due to the local stochastic hill-climbing. The iterative process is computationally efficient since crossover and mutation run in $O(|V| \log |V|)$ time. Therefore, the approach scales well to larger problem instances.

The direct representation in combination with the proposed variation operators provides strong locality. In particular, crossover always generates new solutions out of inherited parental edges only. Other investigations indicate that the cost-based heuristics included in hill-climbing and mutation increase the performance of the proposed EA substantially. The proposed EA obtained better final solutions in dramatically shorter execution times than the previous hybrid genetic algorithm.

Future work will include a generalization of the approach for k-edge connectivity augmentation.

References

1. K. P. Eswaran and R. E. Tarjan. Augmentation problems. *SIAM Journal on Computing*, 5(4):653–665, 1976.
2. Ivana Ljubić, Günther R. Raidl, and Jozef Kratica. A hybrid GA for the edge-biconnectivity augmentation problem. In Kalyanmoy Deb, Günther Rudolph, Xin Yao, and Hans-Paul Schwefel, editors, *Proceedings of the 2000 Parallel Problem Solving from Nature VI Conference*, volume 1917 of *LNCS*, pages 641–650. Springer, 2000.
3. S. Khuller, B. Raghavachari, and N. Young. Low-degree spanning trees of small weight. *SIAM Journal of Computing*, 25(2):355–368, 1996.
4. G. N. Frederickson and J. Jájá. Approximation algorithms for several graph augmentation problems. *SIAM Journal on Computing*, 10(2):270–283, 1981.
5. H. N. Gabow, Z. Galil, T. Spencer, and R. E. Tarjan. Efficient algorithms for finding minimum spanning trees in undirected and directed graphs. *Combinatorica*, 6(2):109–122, 1986.
6. S. Khuller and R. Thurimella. Approximation algorithms for graph augmentation. *Journal of Algorithms*, 14(2):214–225, 1993.
7. A. Zhu. A uniform framework for approximating weighted connectivity problems. B.Sc. thesis, University of Maryland, MD, May 1999.
8. A. Zhu, S. Khuller, and B. Raghavachari. A uniform framework for approximating weighted connectivity problems. In *Proceedings of the 10th ACM-SIAM Symposium on Discrete Algorithms*, pages 937–938, 1999.
9. J. Kratica. Improving performances of the genetic algorithm by caching. *Computers and Artificial Intelligence*, 18(3):271–283, 1999.
10. W. Mader. Minimale n-fach kantenzusammenhängende Graphen. *Math. Ann.*, 191:21–28, 1971.
11. P. Moscato. Memetic algorithms: A short introduction. In D. Corne et al., editor, *New Ideas in Optimization*, pages 219–234. McGraw Hill, Berkshire, England, 1999.
12. R. E. Tarjan. Depth first search and linear graph algorithms. *SIAM Journal of Computing*, 1:146–160, 1972.

Application of GRASP to the Multiconstraint Knapsack Problem*

Pierre Chardaire, Geoff P. McKeown, and Jameel A. Maki

UEA Norwich, NR4 7TJ,
gpm@sys.uea.ac.uk

Abstract. A number of approaches based on GRASP are presented for the Multiconstraint Knapsack Problem. GRASP combines greedy construction of feasible solutions with local search. Results from applying our algorithms to standard test problems are presented and compared with results obtained by Chu and Beasley.

1 Introduction

In this paper, we consider the application of GRASP (*Greedy Randomized Adaptive Search Procedure*) to the multiconstraint knapsack problem (MKP). The paper is structured as follows. In the current section, we begin by giving a brief introduction to GRASP; we then introduce some notation for describing the MKP and briefly review other metaheuristic approaches for the MKP. In Section 2, we describe a number of GRASP approaches for the MKP. Our results are presented in Section 3.

GRASP was pioneered by Feo and Resende (see [1]), and is an iterative metaheuristic search process for solving optimization problems. Each iteration in a GRASP consists of a construction phase followed by a local search phase. The best solution generated during the iterative process is kept as the overall result (see figure 1). In a GRASP, we represent a solution as a subset of a given set of components, N, where N is specified as part of a problem instance, I. The idea is then to construct a solution component-by-component. If at some stage a partial solution, $S \subset N$, has been constructed, then the next component is selected from the set $F_S \subseteq N - S$ of unselected components whose inclusion is still under consideration. A solution is constructed using a greedy function, $g : 2^N \times N \to \mathbb{R} \cup \{\perp\}$. For $i \notin F_S$, $g(S, i) = \perp$. Otherwise, $g(S, i)$ is a measure of the "benefit" associated with the selection of $i \in F_S$. In a maximization (minimization) problem, $\perp$ (representing "undefined") could be $-\infty$ (∞). Thus, $g(S, \cdot)$ (which we abbreviate by g_S) denotes the greedy function given that a partial solution, S, has been constructed. At each stage of the construction process, the next component is picked at random from a *restricted candidate list (RCL)* containing a number of the best remaining candidate components.

---

* This work was supported in part by the UK Defence Research and Evaluation Agency, Malvern, England.

E.J.W. Boers et al. (Eds.) EvoWorkshop 2001, LNCS 2037, pp. 30–39, 2001.

$$
\begin{array}{l}
Input\_Problem\_Instance(I); \qquad //I = (\cdots, N, \cdots) \\
S^* \leftarrow Initialize\_Best(I); \\
\textbf{while not} \text{ finished } \textbf{do} \ \{ \\
\quad S \leftarrow Construct\_Solution(I, g, \alpha); \\
\quad S' \leftarrow Local\_Search(S); \\
\quad S^* \leftarrow Best\_Of(S', S^*); \ \} \\
Output \ S^*
\end{array}
$$

Fig. 1. Structure of a GRASP

The RCL is determined using g_S together with a parameter, α. In this paper, the RCL in a GRASP is defined to be the best $\lceil \alpha \, |F_S| \rceil$ candidate components, where $0 < \alpha \leq 1$. In this case, if α is fixed at the start of the algorithm, then at one extreme $(\alpha \to 0)$ the construction phase of a GRASP simply gives a greedy construction, whilst at the other extreme $(\alpha = 1)$ it delivers a random construction. Good choices of α will vary not only from one type of problem to another but often also between instances of a given problem. In this paper, we propose a scheme for determining suitable values of α dynamically for each problem instance. These values are selected by experimenting with randomly chosen values of α during a preliminary phase of the algorithm (the *warming period*) and retaining a small number of good values of α for use in subsequent iterations. From our experience on the test problems considered in this paper, suitable values of α tend to lie in the range $0 < \alpha \leq 0.25$.

When a solution, S, has been constructed, an iterative local search procedure is applied, which repeatedly updates S to an improved neighbouring solution, until there is no better neighbouring solution. GRASP has been applied to a wide range of combinatorial optimization problems (For numerous references to applications of GRASP, see the bibliography maintained by Resende at http://www.research.att.com/mgcr/doc/graspbib).

The multi-knapsack problem (*MKP*) can be formulated either as a maximization problem or as a minimization problem. A maximization instance of the *MKP* has the following form:

$$
MKP\_max \left\{
\begin{array}{ll}
\max px & (= Z) \\
\text{subject to } Rx \leq b & \\
x_j \in \{0,1\} \ j = 1, \ldots, n.
\end{array}
\right.
$$

Each of the entries in the $m \times n$ matrix $R = [r_{ij}]$ is a non-negative integer, as is each entry in the m-component right-hand-side vector, b. We assume that $p_j > 0$, and $r_{ij} \leq b_i < \sum_{j=1}^{n} r_{ij}$, for all $1 \leq i \leq m, 1 \leq j \leq n$. A minimization instance of the *MKP* has the following form:

$$
MKP\_min \left\{
\begin{array}{ll}
\min px & (= Z) \\
\text{subject to } Rx \geq b & \\
x_j \in \{0,1\} \ j = 1, \ldots, n.
\end{array}
\right.
$$

The *MKP* is applicable to a wide variety of resource allocation problems (see, for example, [2,3]).

We define $N = \{1, 2, \ldots, n\}$, $M = \{1, 2, \ldots, m\}$ such that $j \in N$ corresponds to the j-th 0-1 variable and $i \in M$ to the i-th constraint in an *MKP*. We then specify an instance of an *MKP* as a 6-tuple: $I = (opt, N, M, p, R, b)$, where $opt \in \{max, min\}$. Given a max-instance, I, of an *MKP*, we obtain a *complementary* min-instance, *comp(I)*, by complementing all the variables, i.e. setting $x = e - y$, where e denotes a vector with n components equal to 1. By doing this, the objective in a max-instance is transformed to $\max pe - py$. Since pe is a constant, this is equivalent to minimizing py.

Thus, $comp((max, N, M, p, R, b)) = (min, N, M, p, R, Re - b)$.

Similarly, $comp((min, N, M, p, R, b)) = (max, N, M, p, R, Re - b)$.

A number of researchers have applied metaheuristic approaches to the MKP (see, for example, [4,5]). Chu and Beasley's algorithm [5] appears to be the most successful GA to date for the MKP. In their algorithm, infeasible solutions are "repaired" using a greedy heuristic based on Pirkul's surrogate duality approach [6]. Because all of the previously used bench-mark problems for the MKP are solved in very short computing times using a modern MIP solver, Chu and Beasley introduced a new set of bench-mark problems in their paper. We have used these problems to test the GRASP approaches presented in this paper.

2   Application of GRASP to the Multi-knapsack Problem

We refer to our first GRASP for *MKP* as *GRASP_0*. This is a function of two arguments: an instance, I, of *MKP* and the user-defined parameter, α. The result of applying *GRASP_0* to a given pair of arguments is a subset of N corresponding to the set of variables assigned value 1 in the constructed solution. *GRASP_0* is applicable to both max and min instances of *MKP* and is defined as follows:

$$GRASP\_0(I, \alpha) = \begin{cases} G_0(I, \alpha), & \text{if } I.opt = max \\ N - G_0(comp(I), \alpha), & \text{otherwise.} \end{cases}$$

Here, the function G_0 is itself a GRASP but one restricted to just *MKP_max*. From above, if I is a min instance, then *comp(I)* is a max instance. The set returned when G_0 is applied to *comp(I)* corresponds to the set of variables assigned value 1 in a solution to *comp(I)*. The set of variables assigned 1 in the corresponding solution to the complementary problem, i.e. the original min instance, is then given by taking the complement of the set returned by G_0.

We begin by presenting G_0. Initially, all of the 0-1 variables are free with temporary value 0. We seek to construct a set of those variables to be assigned value 1. At each stage of the process, we add to the set under construction one of the "best" (w.r.t. the greedy heuristic being used) remaining free variables, but not necessarily the best variable itself. A solution is represented by $S \subseteq N$, corresponding to the set of variables assigned value 1 in that solution. Each construction starts from an empty set; thus, *Initialize_Best(I)* simply returns $\emptyset$. One approach for defining a greedy heuristic for *MKP_max* is to use the

```
S ← ∅;
(C_S, F_S) ← Update_CF(M, N, R, b);
if C_S = ∅ then return F_S;        // Delete for G_1
b' ← b;
while F_S ≠ ∅ do {
    l ← ⌈α |F_S|⌉;
    determine the elements θ_k ∈ F_S, k = 1, ..., l,
    giving rise to the l best benefits with respect to g_S;
    pick j ∈ {θ_1, θ_2, ..., θ_l} at random;
    S ← S ∪ {j};
    for each i ∈ C_S do
        b'_i ← b'_i − r_{ij};
    (C_S, F_S) ← Update_CF(C_S, F_S − {j}, R, b');
    if C_S = ∅ then {
        S ← S ∪ F_S;        // Delete for G_1
        F_S ← ∅; } }
return S
```

Fig. 2. The function $Construct\_Solution(I, g, \alpha)$ in G_0

profit coefficients, p, weighted in some fashion to take account of the knapsack constraints. A natural way of doing this is as follows. Let b' denote the residual resource vector at some stage of the construction process. Thus, if S repesents the current partial solution, then

$$b'_i = b_i - \sum_{j \in S} r_{ij}, \qquad i \in M.$$

Let $C_S \subseteq M$ be such that $i \in C_S$ if and only if the residual resource associated with the i-th constraint is still positive, i.e. $C_S = \{i \in M \mid b'_i > 0\}$. g_S is then defined by $g_S(j) = p_j / \sum_{i \in C_S} r'_{ij}$, for all $j \in F_S$, where $r'_{ij} = r_{ij}/b'_i$, for $i \in C_S, j \in F_S$. Free variables are ranked in *non-increasing order* of their g_S values. One alternative way of weighting the profit coefficients, and with which we have also experimented, is based on the idea of "pseudo utility ratios" introduced by Pirkul in [6]. The function $Construct\_Solution(I, g, \alpha)$ for G_0 is defined in figure 2. The function $Update\_CF$ (defined in figure 3), updates both the set of free variables (F) and the set of constraints still under consideration (C) for a given residual resource vector, b. If $F_S \neq \emptyset$ when no constraint remains under consideration, then for each $k \in F_S$, x_k may be set equal to 1. This is the purpose of the final **if** statement in the **while** loop of $Construct\_Solution$. We note that provided at least one left-hand-side coefficient, r_{ij}, in each of the given constraints is non-zero, then our use of $Update\_CF$ preserves as an invariant of $Construct\_Solution$ the fact that division by zero cannot occur when computing r'_{ij}.

When $GRASP\_0$ is applied to a min instance of MKP, we start from the all-zero solution of $comp(I)$, corresponding to the all-one solution of I in which each free variable has value 1. An alternative approach to starting each constructed solution of a max instance of *MKP* from the all-zero feasible solution, is to start each constructed solution from the infeasible all-one solution. For a min instance, this corresponds to starting each constructed solution from the infeasible all-zero solution. We refer to this alternative GRASP as *GRASP_1*, defined as follows:

$$GRASP\_1(I, \alpha) = \begin{cases} G_1(I, \alpha), & \text{if } I.opt = min \\ N - G_1(comp(I), \alpha), & \text{otherwise.} \end{cases}$$

Here, the function G_1 is a GRASP restricted to just *MKP_min*. The result of applying G_1 to a min instance is a subset of N corresponding to the set of variables assigned value 1 in the constructed solution. If I is a max instance, then *comp(I)* is a min instance. The result of applying G_1 to *comp(I)* specifies the set of variables assigned value 1 in a solution to *comp(I)*. The set of variables assigned 1 in the corresponding solution to the complementary problem, i.e. the original max instance, is then given by taking the complement of the set returned by G_1. Each iteration in G_1 seeks to construct a feasible solution of the given min instance starting from the all-zero infeasible solution. By selecting variables to be set to 1, we seek to construct a feasible solution. Constraint i is now removed from C_S as soon as it is satisfied by the current partial solution, S. This is indicated by $b'_i \leq 0$, where b'_i is defined as before but now represents the return on activity i still needed to achieve the minimum return, b_i. F_S now represents the set of those free variables whose selection would decrease infeasibility.

The main differences between G_1 and G_0 occur in the definition of the greedy heuristic and in the definition of *Update_CF*. The structure for *Construct_Solution* is the same for both GRASPs except for two statements that must be deleted for the G_1 case (see figure 2). For G_1, as soon as the current partial solution is feasible (corresponding to $C_S = \emptyset$), setting any further free variable to 1 would give a solution with a worse objective value. For G_0, on the other hand, if none of the remaining free variables can violate any constraint (corresponding to $C_S = \emptyset$), then setting each of these variables to 1 gives a solution with an improved objective value. The modified specification for *Update_CF* is given in figure 4. A natural greedy heuristic for G_1 is to rank the free variables still under consideration in *non-increasing* order of their g_S values, where

```
for each j ∈ F do
    if ∃i ∈ C such that r_ij > b_i then    // x_j cannot be set to 1
        F ← F − {j};
for each i ∈ C do
    if b_i = 0 then
        // r_ij = 0 for all j ∈ F, so constraint i is redundant
        C ← C − {i};
return (C, F)
```

Fig. 3. The function *Update_CF(C, F, R, b)*

```
for each i ∈ C do
    if b'_i ≤ 0 then  // constraint i is satisfied
        C ← C − {i};
for each j ∈ F do
    if r_{ij} = 0, ∀i ∈ C, then
        // setting x_j = 1 would not improve feasibility
        F ← F − {j};
return (C, F)
```

Fig. 4. The function $Update\_CF(C, F, R, b')$ for G_1

g_S is defined as for G_0 except that r'_{ij} is now defined by $r'_{ij} = \min\{r_{ij}/b'_i, 1\}$, for $i \in C_S, j \in F_S$.

With *GRASP_0*, each construction starts from the all-zero feasible solution whilst with *GRASP_1*, each construction starts from the all-one infeasible solution. We now consider a GRASP for *MKP_max* in which we start each construction from any given initial solution, S_0, which may be feasible or infeasible. Such an initial solution could, for example, be selected at random or according to some heuristic. If S_0 is feasible, we proceed as in *GRASP_0*, seeking to construct an improved feasible solution. Otherwise, we proceed as in *GRASP_1*, seeking to construct a feasible solution starting from the infeasible solution, S_0. A transformed specification for *Construct_Solution* that takes an initial solution, S_0, as a parameter and which covers both the feasible and infeasible cases is given in figure 5. Note that the all-zero *GRASP_0* and the all-one *GRASP_1* cases are both special cases of this more general algorithm. If the boolean flag *Gzero* is set, the *GRASP_0* approach is used to construct a solution, otherwise the *GRASP_1* approach is used. The transformed *Update_CF* function combines both the *GRASP_0* and the *GRASP_1* versions; the parameter *Gzero* determines which version should be applied. Each construction in an application of this GRASP starts from the same specified solution. Applying the GRASP a number of times allows the search to start in different areas of the search space.

For our experiments, we have used both 1-opt and 2-opt search strategies. In the former case, for each $j \in S$, we construct a new solution, S', using the *GRASP_0* approach starting from $S - \{j\}$ and for each $k \notin S$, we construct a new solution, S', using the *GRASP_1* approach starting from set $S \cup \{k\}$. In the 2-opt method, for each pair (j, k) such that $j \in S, k \notin S$, we start from $S - \{j\} \cup \{k\}$ and apply the general construction algorithm in figure 5 to this initial solution. A GRASP for *MKP_max* which uses a 2-opt local search is given in figure 6. Since I is always a max-instance, we supress the *opt* parameter of I. In our experiments, we have used the 1-opt search when a full 2-opt local search is too expensive.

We have also investigated the effect of using information obtained by solving linear programming relaxations to update the feasible candidate list, F_S. We present the idea in the context of the *Construct_Solution* function used in

```
Gzero ← is_Feasible(S, I);
if not  Gzero then {
    S ← N − S;
    b ← Re − b; }
for each i ∈ M do
    b'ᵢ ← bᵢ − Σⱼ∈S rᵢⱼ;
(C_S, F_S) ← Update_CF(Gzero, M, N − S, R, b');
if  Gzero and  C_S = ∅ then
    return  S ∪ F_S;
while  F_S ≠ ∅ do {
    l ← ⌈α|F_S|⌉;
    determine the elements θ_k ∈ F_S, k = 1,...,l,
    giving rise to the l best benefits with respect to g_S;
    pick j ∈ {θ₁, θ₂,...,θ_l} at random;
    S ← S ∪ {j};
    for each i ∈ C_S do
        b'ᵢ ← b'ᵢ − rᵢⱼ;
    (C_S, F_S) ← Update_CF(Gzero, C_S, F_S − {j}, R, b');
    if  C_S = ∅ then {
        if  Gzero then
            S ← S ∪ F_S;
        F_S ← ∅; } }
if  Gzero  then return  S else return  N − S;
```

Fig. 5. The function $Construct\_Solution(S, I, g, \alpha)$

$GRASP\_0$ (see figure 2). Each time round the **while** loop, we solve the LP relaxation corresponding to the integer program obtained from the given instance of MKP_max, I, by setting $x_j = 1$ for each $j \in S$. Let Z denote the optimal value of such an LP relaxation and let Z^* denote the value of the best solution so far found for I. If $Z \leq Z^*$, then we can terminate the current construction, since it cannot lead to a better solution than our current best. Otherwise, we consider each $j \in F_S$ which has an integer value (0 or 1) in the optimal solution of the current LP relaxation. If $x_j = 0$, we can decide whether or not to fix x_j to 0 permanently in the solution we are constructing. To do this, we use the reduced cost, d_j say, corresponding to x_j. $d_j (\leq 0)$ represents the minimum change in the objective value if the value of x_j changes from 0 to 1 (irrespective of any changes made to the values of other variables). Thus, if $Z + d_j \leq Z^*$, then changing the value of x_j from 0 to 1 cannot lead to a better solution than our current best solution, so we can remove j from F_S. Similarly, if $x_j = 1$ in the optimal solution of the current LP relaxation, then if $Z - d_j \leq Z^*$, there is no point in setting $x_j = 0$. Hence, we remove j from F_S and add it to S. A consequence of permanently fixing one or more variables to 1 is that there may be a number of fractional-valued variables which cannot be set to 1 in any feasible integer solution in the resulting sub-space. If this is the case, all such fractional-valued variables are fixed to zero and the resulting reduced LP relaxation is solved. If

```
Input_Problem_Instance(I);          // I = (N, M, p, R, b)
S₀ = Input_Solution(I);
S* ← S₀;
while not  finished do {
    S ← Construct_Solution(S₀, I, g, α);
    // Perfom Local_Search(S):
    S' ← S;
    for each  (j, k) ∈ S × (N − S) do {
        S'' ← S − {j} ∪ {k};
        S'' ← Construct_Solution(S'', I, g, α);
        S' = Best_Of(S', S''); }
    S* = Best_Of(S', S*); }
Output S*
```

Fig. 6. A GRASP for *MKP_max*

necesssary, this process is repeated. Finally, when picking an element from F_S, we select only from amongst those elements whose corresponding variable is currently fractional-valued. The underlying greedy heuristic that we use is defined by $g_S(j) = p_j\beta_j$, where β_j is the value of a fractional-valued variable, x_j.

3   Results for MKP Test Problems

We give results for three versions of our general GRASP for *MKP_max*. Version 1 corresponds to *GRASP_0*, in which we start each construction from the all-zero solution and version 2 corresponds to *GRASP_1*, in which we start each construction from the all-one solution. Version 3 (referred to as *GRASP_X*), the constuction is based on the linear programming approach described above. Each run of the algorithms we have implemented consists of a warming period, during which suitable values of α are determined, followed by a post-warming period during which only these values of α are used. For each iteration of the warming period, a value of α is selected at random from the interval $\left[\frac{3}{2n}, 0.25\right]$ where n is the size of the problem instance. The choice of left-hand limit in this range ensures that the RCL will always have length at least 2, thereby excluding the pure greedy case.

We have tested our algorithms with the 27 problem sets available at the OR-Library maintained by Beasley (see http://mscmga.ms.ic.ac.uk/info.html). There are 10 instances per problem set. Each problem set is characterized by a number, m, of constraints, a number, n, of variables and a *tightness ratio*, $0 \le t \le 1$. The vector b of an MKP instance with tightness ratio t is $(b_i = t \sum_j R_{ij})_{i=1,n}$. In other words the closer to 0 the tightness ratio the more constrained the instance. In our experiments, the maximum total number of iterations is set to 200 for all of the $n = 100$ and the $n = 250$ problem instances, and to 50 for all of the $n = 500$ problem instances; the number of iterations in the warming period is set to 70, when $n = 100$ or $n = 250$, and to 20 when $n = 500$; the algorithms

Table 1. Comparison between *GRASP_0*, *GRASP_X* and *CBGA*

Instance			Percentage Relative Error, E						CPU					
			GRASP_0			*GRASP_X*			*GRASP_0*		*GRASP_X*		*GA_CHU*	
m	n	t	ave.	min	max	ave.	min	max	best	total	best	total	best	total
5	100	0.25	0.06	0.00	0.27	0.10	0.00	0.32	57	139	26	138	10	346
5	100	0.5	0.02	0.00	0.12	0.05	0.00	0.12	54	178	62	199	24	347
5	100	0.75	0.00	0.00	0.01	0.05	0.00	0.15	24	128	29	143	27	362
5	250	0.25	0.09	-0.04	0.16	0.14	0.08	0.25	1349	2458	1152	2640	51	682
5	250	0.5	0.03	0.00	0.09	0.04	0.01	0.06	1576	3263	1162	2978	277	709
5	250	0.75	0.01	0.00	0.02	0.02	0.00	0.07	750	2237	816	2190	196	763
5	500	0.25	0.18	0.15	0.24	0.51	0.27	0.95	34	44	44	64	265	1272
5	500	0.5	0.09	0.03	0.15	0.36	0.09	0.62	37	65	42	96	391	1346
5	500	0.75	0.05	0.03	0.07	0.27	0.11	0.51	44	82	98	151	386	1413
10	100	0.25	0.47	0.00	0.90	0.21	0.00	0.55	103	214	41	166	98	384
10	100	0.5	0.17	0.00	0.40	0.16	0.00	0.49	114	295	122	287	97	419
10	100	0.75	0.04	0.00	0.14	0.04	0.00	0.15	86	243	62	206	17	463
10	250	0.25	0.22	0.06	0.41	0.17	0.00	0.34	1583	3287	1303	3710	359	871
10	250	0.5	0.11	-0.01	0.24	0.04	-0.03	0.10	2510	4809	1749	5085	342	932
10	250	0.75	0.04	0.00	0.08	0.03	0.00	0.06	1749	1708	1027	3433	128	1011
10	500	0.25	0.37	0.23	0.62	0.55	0.04	0.92	36	57	41	94	703	1505
10	500	0.5	0.18	0.07	0.31	0.30	0.09	0.43	40	86	75	167	562	1729
10	500	0.75	0.13	0.08	0.18	0.18	0.06	0.30	33	97	139	231	938	1932
30	100	0.25	0.70	0.19	1.32	0.34	0.00	0.86	258	606	168	573	177	605
30	100	0.5	0.29	0.00	0.70	0.08	-0.17	0.31	326	948	380	927	118	782
30	100	0.75	0.07	0.00	0.23	0.02	0.00	0.08	184	698	346	850	90	904
30	250	0.25	0.93	0.61	1.44	0.21	0.00	0.54	1710	2733	1283	2581	583	1500
30	250	0.5	0.48	0.33	0.73	0.10	0.02	0.26	1531	3782	1912	3472	902	1980
30	250	0.75	0.18	0.08	0.36	0.07	0.01	0.12	1500	2703	1011	2504	1059	2441
30	500	0.25	1.20	0.50	1.81	0.35	0.13	0.85	63	97	241	533	1127	2438
30	500	0.5	0.57	0.31	0.82	0.31	0.15	0.52	104	169	450	877	1122	3199
30	500	0.75	0.33	0.22	0.45	0.18	0.02	0.39	100	196	823	1192	1903	3888

stop after 100 iterations without improvement of the best solution when $n = 100$ or $n = 250$ and after 30 such iterations when $n = 500$. When $n = 100$ or 250, the 2-opt heuristic is used during improvement phases, but only the 1-opt heuristic is used when $n = 500$. In table 1, we display results for each data set for *GRASP_0* and *GRASP_X*, with results from Chu & Beasley's GA (*CBGA*) included for comparison. The results for *GRASP_1* are of a similar quality to those obtained for *GRASP_0* and are not given in this paper. The percentage relative error, E, is defined to be $100(z_{\mathrm{GRASP}} - z_{\mathrm{GA}}) / \max(z_{\mathrm{GRASP}}, z_{\mathrm{GA}})$, where z_{GRASP} and z_{GA} denote the best solution values found by the GRASP and by *CBGA*, respectively. The columns labelled "ave." "min", "max" give the average, minimum and maximum values, respectively, of E over the 10 instances in the data set. CPU times are averaged over the instances of each data set and are given for Chu & Beasley's machine (approximatively 4.7 times slower than a Dec 500/400). "total" gives the average total execution time of the algorithm whereas "best" gives the average CPU time to find the best solution returned by the algorithm.

For data sets with $(m, n) = (5, 100)$ and $(m, n) = (10, 100)$ all but one of the solutions found by $CBGA$ were proven to be optimal using an integer programming solver. For both versions of our GRASP, the average gap for each of the $n = 100$ data sets is a fraction of 1%. $GRASP\_X$ obtains results as good as those for $CBGA$ in one third of the $n = 100$ problem instances and obtains one result which is better than the result for $CBGA$. Total computing times are generally slightly better for GRASP than for $CBGA$ on the $n = 100$ problem sets, although $CBGA$ usually obtains its best solution more quickly. In general, GRASP performs less well on the $n = 250$ problem sets. However, for both versions of the GRASP, the maximum gap for each of these data sets is nearly always much less than 1%. Furthermore, $GRASP\_0$ obtains two results, and $GRASP\_X$ one result, better than the corresponding results for $CBGA$. For these problem sets, the 2-opt heuristic leads to expensive computing times for GRASP. For the $n = 500$ data sets, only a single pass of our 1-opt local search was performed. For $GRASP\_0$, this results in methods 20 to 30 times faster than $CBGA$, and even $GRASP\_X$ is 10 to 20 times faster than $CBGA$ for all but the $m = 30$ instances, for which $GRASP\_X$ is still approximately 4 or 5 times faster.

4   Conclusions

The quality of the results obtained by both versions of our GRASP is encouraging. The maximum gap is less than 1% for all of the 270 instances for $GRASP\_X$ and for all but one of the data sets for $GRASP\_0$. Using the more general GRASP of figure 5, which can start from any initial solution may lead to even better results.

References

1. Feo, T.A., Resende, M.G.C.: Greedy randomized adaptive search procedures. J. of Global Optimization **6** (1995) 109–133
2. Weingarter, H. M.: Mathematical programming and the analysis of capital budgeting problems. Markham Publishing (1967) Chicago
3. Shih, W.: A Branch and Bound Method for the Multiconstraint Zero-One Knapsack Problem. J. of the Operational Res. Soc. **30** (1979) 369–378
4. Cotta, C., Troya, J. M.: A hybrid genetic algorithm for the 0-1 multiple knapsack problem. In: Artificial neural nets and genetic algorithms 3, eds. Smith, G. D., Steele, N. C., Albrecht, R. F., Springer-Verlag (1998) 251–255
5. Chu, P. C., Beasley, J. E.: A genetic algorithm for the multidimensional knapsack problem. J. of Heuristics **4** (1998) 63–86
6. Pirkul, H.: A Heuristic Search Procedure for the Multiconstraint Zero-One Knapsack Problem. Naval Research Logistics **34** (1987) 161–172

Path Tracing in Genetic Algorithms Applied to the Multiconstrained Knapsack Problem

Jens Levenhagen[1], Andreas Bortfeldt[2], and Hermann Gehring[2]

[1] Astrium GmbH, 88039 Friedrichshafen, Germany
[2] Dept. of Business Informatics, University of Hagen, 58084 Hagen, Germany

Abstract. This contribution investigates the usefulness of F. Glover's path tracing concept within a Genetic Algorithm context for the solution of the multiconstrained knapsack problem (MKP). A state of the art GA is therefore extended by a path tracing component and the Chu/Beasley MKP benchmark problems are used for numerical tests.

1 Introduction

The multiconstrained knapsack problem (MKP) is a well-known problem in combinatorial optimization and is defined as follows:

$$\text{maximize } Z = \sum_{j=1}^{n} p_j \cdot x_j$$

$$\text{subject to} \quad \sum_{j=1}^{n} w_{ij} \cdot x_j \leq b_i \; , \; w_{ij} \geq 0 \; , \; b_i \geq 0 \; , \tag{1}$$

$$x_j \in \{0,1\}$$

with $i = 1, \ldots, m$ and $j = 1, \ldots, n$. Here, the x_j denote specific items, w_{ij} denote their weights (or sizes), p_j their profits and b_i the knapsack capacities. It is worth mentioning that each item x_j appears either in all knapsacks or in none of them.

This kind of problem frequently occurs within the context of decision making for project realization: assume that n possible projects have to share scarce project factors like money, time and staff. Therefore a sensible project selection has to be carried out which maximizes the overall profit (referring to an arbitrary quantifiable cost variable) while at the same time all the factor constraints are met. In that case the x_j can be identified as possible projects and the w_{ij} as their factor consumptions. Because one single constraint is assigned to each project factor, the b_i represent the factor resources [1]. Further applications of the MKP are cutting stock problems or cargo loading problems [2].

Being a combinatorial optimization problem the MKP can be solved by exact algorithms or by heuristics. As it is NP-hard in general[1] the application of heuristics is generally recommendable.

---

[1] The MKP is pseudo-polynomial if weights and profits are bounded.

E.J.W. Boers et al. (Eds.) EvoWorkshop 2001, LNCS 2037, pp. 40–49, 2001.

During the last few years several powerful Genetic Algorithms (GA) were developed for obtaining near-optimal solutions of the MKP. Chu and Beasley found their GA superior to all former approaches and defined a collection of 270 difficult benchmark problems[2] [2]. The GA by Raidl kept the overall concept of the Chu and Beasley algorithm but differed by the specific realization of MKP dependent operators [1]. An empirical comparison between these GAs was carried out by Gottlieb. Moreover, he introduced additional initialization concepts which lead to considerable performance improvements of both GAs [3].

Path tracing (PT) and *embedded local search* (ELS) represent two recent approaches for GA hybridization. PT was introduced by Glover as a generalization of Scatter Search [4]. Genetic Algorithms with ELS represent a possible realization of Memetic Algorithms (MA) which were founded by Moscato [5] and, among others, investigated by Merz and Freisleben [6],[7]. A hybrid concept of both PT and ELS was discussed by Reeves and Yamada in [8],[9].

Since at least PT is a relatively new optimization technique it was our first aim to show how a GA for the MKP can be extended by both PT and ELS. Moreover, the paper investigates whether this extension leads to a significant performance improvement or, on the contrary, whether the underlying fitness landscape ([7]) might give an explanation on why the approach does not work. We proceed as follows: in section 2 a short introduction to the GAs by Chu/Beasley/Gottlieb and Raidl/Gottlieb is given. Section 3 describes the essential elements of PT and ELS whereas section 4 deals with the extension of the RGGA by PT and ELS. Finally, numerical test results for this extended GA are presented in section 5 while section 6 concludes the paper.

2   Two State of the Art Genetic Algorithms for the Multiconstrained Knapsack Problem

Genetic algorithms rank with the most popular heuristics for the solution of combinatorial optimization problems and adapt the transmittance and selection mechanisms of natural evolution; see for example [10],[11],[12]. In many cases the combination of a GA with another optimization scheme – as a certain way of hybridization – leads to a significant performance improvement. This is due to the increased exploitation or exploration properties of the additional scheme which will support or complement standard genetic operators like *Crossover* or *Mutation*. Because our aim was to determine whether PT and ELS are sensible GA hybridization concepts for the solution of the MKP, either the Chu/Beasley/Gottlieb-GA (CBGGA) or the Raidl/Gottlieb-GA (RGGA) seemed reasonable as a powerful starting-point for the implementation. The essentials of both GAs will be briefly put together in the following.

MKP *independent* features of both GAs are stated in Table 1. Herein, P_c denotes the crossover probability, P_m the mutation probability and N the population size. The chromosome length equals the number of items n, the allele (0 or 1) of a specific gene j corresponds to the value of x_j, and the *bit-flip operator* performs a 0/1 or 1/0 allele change on a gene.

----

[2] *http://mscmga.ms.ic.ac.uk/info.html*

Table 1. MKP independent GA features

GA feature	Realization
• Encoding	• binary
• Selection	• tournament (2 chromosomes)
• Crossover	• uniform crossover
• Mutation	• bit-flip operator
• Population concept	• steady-state
	• no duplicates
• Design parameter set $\mathcal{D}$	• $P_c = 0.9/1.0, P_m = 0.01, N = 100$

MKP *dependent* features account for a specific property of all knapsack problems which refers to search space reduction as follows. The whole search space $\mathcal{S}$, given by $\{0,1\}^n$, can be partitioned into subspaces $\mathcal{F}$ and $\mathcal{I}$ which comprise the *feasible* and *infeasible* MKP solutions, respectively. A feasible solution is defined as a chromosome that fulfills all problem constraints whereas an infeasible solution is a chromosome that violates at least one of them. $\mathcal{F}$ and $\mathcal{I}$ are separated by the *boundary* (cf. [3])

$$\mathcal{B} = \{C \in \mathcal{F} \mid C \prec \tilde{C} \text{ implies } \tilde{C} \in \mathcal{I}\}, \qquad (2)$$

where C is a boundary chromosome and no 0/1 allele exchange can be performed on it without violating at least one constraint. Hence C represents a *local optimum* of the MKP with respect to a bit-flip induced neighbourhood $\mathcal{N}(C,1)$ (see section 3.1) and $\mathcal{B}$ is build up by the totality of local optima. Obviously it is reasonable to bias the search towards $\mathcal{B}$ as the global optimum is necessarily also a local one.

Both GA contain the following concepts for this bias search procedure. A heuristic *repair operator* is applied to infeasible solutions $C \in \mathcal{I}$ (produced by mutation and/or crossover) which makes them feasible ($C^* \in \mathcal{F}$). A succeeding heuristic local optimization yields the desired boundary solutions ($C^{**} \in \mathcal{B}$). Moreover, the initialization procedure is also adapted to yield only boundary chromosomes [2],[3],[1]. With regard to our objective CBGGA and RGGA show comparable performance properties and could both be used as a starting-point for the PT/ELS implementation. While CBGGA achieves a slightly better overall performance [3], RGGA has significant convergence advantages [1]. The latter point caused us to choose the RGGA because a great number of numerical tests for PT/ELS was to be carried out.

3 Path Tracing and Embedded Local Search

3.1 Path Tracing

In the past it was observed as a specific property of several problems that the local optima concentrate within a relatively small part of $\mathcal{S}$, the so-called *big valley* [7],[8], which also contains the global optimum. Unfortunaly until today no formal characterization of this big valley can be given. Hence, its existance can only serve as an intuitive guideline for the development of new exploitation operators which make use of this problem property.

Providing that a big valley really exists a relatively high weight should be admitted to the exploitation, i.e. the search should be focused towards the big valley region. A systematic and promising way of doing this is the application of path tracing (PT).

Path tracing is based on the consideration that the probability of finding good new solutions C_{p_k} on an interpolating path between two existing big valley solutions C_1, C_2 is very high. So for its application a promising interpolating path scheme has first to be determined. In case of binary encoding the bit-flip operator automatically induces the *Hamming metric*

$$d_{\mathrm{H}}(C_j, C_k) = \sum_{i=1}^{n} |C_j(i) - C_k(i)| \tag{3}$$

which equals the number of different alleles between two chromosomes C_j and C_k. $d_{12}! = d_{\mathrm{H}}(C_1, C_2)!$ distinct paths exist between C_1 and C_2 thus forcing the path tracing concept to choose a promising one out of this possibly very large number. Since only information on the starting point C_1 and the end point C_2 is available at the beginning, a stepwise procedure is required. Therefore, we introduce the 1-neighbourhood of chromosomes C_i:

$$\mathcal{N}(C_i, 1) = \{C_j \mid d_{\mathrm{H}}(C_i, C_j) = 1\} . \tag{4}$$

Starting with C_1, each step consists of a propagation to a promising 1-neighbourhood chromosome which shortens the distance to C_2 by 1. There are exactly $d_{12} - 1$ steps to be made, see Figure 1. To prevent high population con-

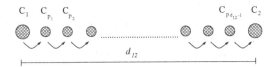

Fig. 1. Path tracing

densation PT can be extended by the introduction of an *extrapolation* concept. A probability P_e is determined by which the actual distance to C_2 is increased by 1. The expected number of steps then raises from $d_{12} - 1$ to

$$E(n_S) = \frac{d_{12}}{1 - 2 \cdot P_e}, \quad \text{with} \quad 0 < P_e < 0.5 \tag{5}$$

and the given validity range of P_e assures that the PT algorithm terminates within finite time. However, P_e should be chosen considerably smaller, i.e. 0.2 or 0.3. Additionally a tabu list concept is necessary to exclude an exact repersuing of the already traced path [14].

3.2   Embedded Local Search

Chromosomes created by PT are very promising if a big valley exists but they are boundary solutions only by coincidence. Therefore, an additional local optimization procedure can be applied to path chromosomes C_{p_k} to fulfill the desired boundary property. For this purpose the ELS concept is very well suited as it can be initiated at any time for an arbitrary actual chromosome and is able to perform many direction changes before reaching the local optimum. If, on the other hand, ELS was initiated at arbitrary points of S without the coarse predirecting of PT it would in most cases only find weak local optima. So both concepts supplement each other and a combined strategy is recommended.

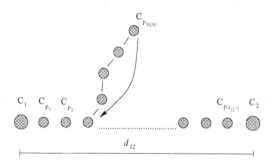

Fig. 2. Path tracing with embedded local search

The stepwise *direction selection principle* plays an important role for ELS. It can be realized in two ways [8]:

- Selection of the *fittest* 1-neighbourhood chromosome if it dominates the actual ELS solution. This is called the principle of *steepest fitness ascent*. If there is no superior neighbour the actual solution represents a local optimum and ELS is terminated.
- Selection of the *first* identified 1-neighbourhood chromosome which dominates the actual ELS solution (*principle of next fitness ascent*). The termination condition is the same as before.

Figure 2 shows an initiation of ELS at path chromosome C_{p_3}. After its termination at the local optimum $C_{p_3(LS)}$, PT is continued at C_{p_3} and a steady-state-like replacement strategy is applied: if $C_{p_3(LS)}$ dominates the worst population member C_{worst}, the former replaces the latter.

4   Extension of the Raidl/Gottlieb GA by Path Tracing and Embedded Local Search

Figure 3 shows the extended RGGA algorithm, further denoted as ERGGA. Herein, P_{pt} denotes the initialization probability of PT and N_{actual} the evaluated number of non-duplicate chromosomes. It is further assumed that the fitness function equals the MKP cost function: $F = Z$. Skipping line 08 yields the original RGGA.

```
01: initialize population;
02: set $N_{actual} = N$;
03: evaluate population;
04: store fittest population member $C_{best}$;
05: WHILE $N_{actual} < N_{max}$ DO
06:     perform tournament selection;
07:     perform uniform crossover;
08:     perform path tracing with probability $P_{pt}$;
09:     perform mutation;
10:     apply repair operator and local optimization;
11:     IF new chromosome $C_{new}$ is no duplicate THEN
12:         evaluate $C_{new}$;
13:         IF $F(C_{new}) > F(C_{worst})$ THEN replace $C_{worst}$ by $C_{new}$; ENDIF;
14:         IF $F(C_{new}) > F(C_{best})$ THEN set $C_{best} = C_{new}$; ENDIF;
15:         $N_{actual} := N_{actual} + 1$;
16:     ENDIF;
17: ENDWHILE;
18: return $C_{best}$, $F(C_{best})$;
```

Fig. 3. Algorithm of the extended Raidl/Gottlieb-GA (ERGGA)

The PT (Figure 4) and ELS (Figure 5) realizations were done according to sections 3.1 and 3.2, respectively. In PT, after initializing both the tabu list and C_p with C_1 the sets $\mathcal{G}_{eq}$ and $\mathcal{G}_{dif}$ of genes with equal and different alleles in C_p and C_2 are determined. These sets are ordered according to [1]. Then a bit-flip operation is applied to a gene $G \in \mathcal{G}_{dif}$ (interpolation) or $G \in \mathcal{G}_{eq}$ (extrapolation) of C_p which yields C_p^+. Hereafter the RGGA repair operator ([1]) removes a possible constraint violation. To prevent an artificial extrapolation induced by the repair procedure it is only applied to genes with a 1-allele for both C_p^+ and C_2. The resulting new chromosome C_p^{++} is afterwards subjected to a duplicate and tabu list check. If the result is negative C_p is set equal to C_p^{++} and the tabu list is updated. Then ELS is initiated on C_p with probability P_{els}. If the result of the check is positive the algorithm continues with a bit-flip on $\mathcal{G}_{dif} \setminus G$ or $\mathcal{G}_{eq} \setminus G$. Finally, PT terminates either under condition $d_{p2} = 1$ or if no chromosome C_p^{++} can be found.

In ELS, $C_{p(LS)}$ is initialized with C_p. Then a subset $\mathcal{N}^+(C_{p(LS)}, 1)$ of $\mathcal{N}(C_{p(LS)}, 1)$ is determined which comprises all chromosomes with the following properties: 1) no duplicates, 2) no constraint violation, 3) not in tabu list and 4) superior to $C_{p(LS)}$. If $\mathcal{N}^+(C_{p(LS)}, 1)$ is empty $C_{p(LS)}$ represents a local optimum and the loop terminates. If not, $C_{p(LS)}$ is set equal to the fittest chromosome in $\mathcal{N}^+(C_{p(LS)}, 1)$ and the loop is repeated. Finally, $C_{p(LS)}$ is compared to both the worst and fittest population member and possibly replaces one or both of them.

```
01: initialize C_p and tabu list with C_1;
02: WHILE d_H(C_p, C_2) > 1 DO
03:     set Path_Successor_Found = FALSE;
04:     REPEAT
05:         determine G_eq and G_dif;
06:         get C_p^+ by applying bit-flip operator to G ∈ G_dif of C_p (interpol.) OR
            get C_p^+ by applying bit-flip operator to G ∈ G_eq of C_p (extrapol.);
07:         get C_p^++ by applying repair operator to C_p^+;
08:         IF C_p^++ is no duplicate AND C_p^++ ∉ tabu list THEN
09:             set Path_Successor_Found = TRUE;
10:             set C_p = C_p^++;
11:             insert C_p into tabu list;
12:             perform embedded local search with probability P_els;
13:         ELSE
14:             remove G from G_eq or from G_dif;
15:         ENDIF;
16:     UNTIL Path_Successor_Found = TRUE OR (G_eq = {} AND G_dif = {});
17: ENDWHILE;
18: return to RGGA scheduling;
```

Fig. 4. Path tracing (PT) algorithm

```
01: initialize C_p(LS) with C_p;
02: REPEAT
03:     identify N^+(C_p(LS), 1) as promising subset of N(C_p(LS), 1);
04:     IF N^+(C_p(LS), 1) ≠ {} THEN
05:         determine C_p(LS) as best chromosome in N^+(C_p(LS), 1);
06:     ENDIF;
07: UNTIL N^+(C_p(LS), 1) = {};
08: IF F(C_p(LS)) > F(C_worst) THEN replace C_worst by C_p(LS) ENDIF;
09: IF F(C_p(LS)) > F(C_best) THEN set C_best = C_p(LS) ENDIF;
10: return to PT scheduling;
```

Fig. 5. Embedded local search (ELS) algorithm (steepest descent)

5   Numerical Tests

The RGGA/ERRGA as depicted in figure 3 was implemented in Fortran 77 as an extension of a standard GA by *D.L.Carroll (University of Illinois)*[3]. 90 Chu/Beasley MKP benchmark problems with $n = 100$ and $m \in \{5, 10, 30\}$, $\alpha \in \{0.25, 0.50, 0.75\}$ acted as test instances, where n denotes the number of variables, m the number of constraints and α the MKP specific *tightness ratio* [2]. Preliminary investigations were executed and showed that a run-time of 240 seconds on a Pentium III processor (450 MHz) was long enough to cancel

---

[3] *http://www.staff.uiuc.edu/∼carroll/ga.html*

out randomizing effects. Moreover, the high convergence rate of RGGA leads to very good results even for this modest run-time [14]. Altogether 810 test runs were performed to identify the following optimal design parameter set for the ERGGA:

$$\mathcal{D}^*_{\mathrm{opt}} = \{P_{\mathrm{c}} = 0.9, P_{\mathrm{m}} = 0.01, P_{\mathrm{pt}} = 0.005, P_{\mathrm{e}} = 0, P_{\mathrm{els}} = 0.5\} \ . \tag{6}$$

Moreover, the next fitness descent strategy proved superior to the steepest descent strategy.

After identification of both $\mathcal{D}^*_{\mathrm{opt}}$ and the direction selection principle the ERGGA was to challenge RGGA in extended tests on 9 test instances. These comprised all possible m/α combinations and the GA runs terminated after 10^6 non-duplicate individuals as proposed by Chu/Beasley and Raidl. Columns 4 and 5 of table 2 show the achieved cost function values Z which differ only in one case. By comparing these results to those of Chu/Beasley (CBGA, column 3) one observes equivalence in 7 and 8 of 9 instances, respectively. Finally, columns 6 and 7 contain the number of evaluated non-duplicate chromosomes that were necessary to reach the Z values: the RGGA seems to have some considerable advantages for higher values of m and α.

Table 2. Performance test results of RGGA and ERGGA

m	α	cost function value Z			# Chromosomes	
		CBGA	RGGA	ERGGA	RGGA	ERGGA
5	0.25	24216	24216	24216	79943	31965
	0.50	42545	42545	42545	559963	537109
	0.75	61520	61520	61520	22194	48313
10	0.25	21875	21821	21875	14780	386739
	0.50	41395	41395	41395	250957	174295
	0.75	56377	56377	56377	24488	70963
30	0.25	20754	20754	20754	249596	286284
	0.50	41304	41304	41304	142058	189556
	0.75	58884	58842	58842	946	3649

In summary, appreciable performance differences between ERGGA and RGGA could not be identified for what reason the application of ERGGA is *not recommendable* for a MKP. Mainly responsable for this result seems to be the absence of a big valley structure which is assumed to be necessary for a successful use of the PT concept. The existence or non-existence of a big valley structure was determined graphically by means of *fitness-distance-plots* (FDP). Following Merz and Freisleben [7], the existence of a big valley can be assumed if the FDP shows an approximate linear connection between Hamming distances $d_{\mathrm{H}}(C_{\mathrm{opt}}, C_i)$ (x-axis) and fitness differences $\Delta F = F(C_{\mathrm{opt}}) - F(C_i)$ (y-axis), where C_{opt} denotes the (known) global optimum and C_i denotes an arbitrary local optimum. Figure 6 shows a typical FDP for a MKP and unfortunately indicates that *MKP do not contain a big valley structure.*[4] This seems at least to be true for the tested MKP instances and with respect to the used search space metric, for which the Hamming distance was chosen.

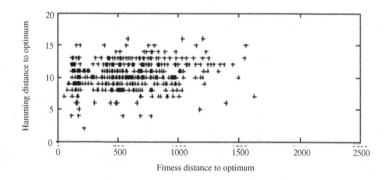

Fig. 6. Fitness distance plot

6   Conclusions

The numerical results of our particular PT/ELS realization, together with the graphical analysis, may confirm the assumption that an application of PT/ELS should be restricted to problem instances which show a big valley structure. Because a formal characterization of that structure can not be given today, an empirical identification of *problem classes* or *subclasses* with this property should be considered for future work. We strongly believe that applying a PT/ELS concept to these problems will lead to a considerable GA performance improvement.

The authors wish to thank the three anonymous referees for their helpful comments and suggestions.

References

1. Raidl, G.R., *An Improved Genetic Algorithm for the Multiconstrained 0-1 Knapsack Problem*, Proceedings of the 5th IEEE International Conference on Evolutionary Computation, pp. 207-211, 1998
2. Chu, P.C. and Beasley, J.E., *A Genetic Algorithm for the Multidimensional Knapsack Problem*, Journal of Heuristics, 4 : pp. 63-86, 1998

---

[4] C_{opt} was known a-priori for the selected instance of figure 6.

3. Gottlieb, J., *On the Effectivity of Evolutionary Algorithms for the Multidimensional Knapsack Problem*, in: Fonlupt, C. et.al.(Eds.): Proceedings of Artificial Evolution, pp. 23-37, Lecture Notes in Computer Science, Vol. 1829, Springer, 2000

4. Glover, F., *Scatter Search and Path Relinking*, in: Corne, D. et.al.(Eds.): New Ideas in Optimization, pp. 297-316, McGraw-Hill, 1999

5. Moscato, P., *Memetic Algorithms: A Short Introduction*, in: Corne, D. et.al.(Eds.): New Ideas in Optimization, pp. 219-234, McGraw-Hill, 1999

6. Merz, P. und Freisleben, B., *Genetic Local Search for the TSP: New Results*, Proceedings of the 1997 IEEE International Conference on Evolutionary Computation, pp. 159-164, IEEE Press, 1997

7. Merz, P. and Freisleben, B., *Fitness Landscapes and Memetic Algorithm Design*, in: Corne, D. et.al.(Eds.): New Ideas in Optimization, pp. 245-260, McGraw-Hill, 1999

8. Reeves, C.R. and Yamada, T., *Embedded Path Tracing and Neighbourhood Search Techniques*, in: Miettinen,K. et.al.(Eds.): Evolutionary Algorithms in Engineering and Computer Science, pp. 95-111, Wiley, 1998

9. Reeves, C.R. and Yamada, T., *Goal-Oriented Path Tracing Methods*, in: Corne, D. et.al.(Eds.): New Ideas in Optimization, pp. 341-356, McGraw-Hill, 1999

10. Falkenauer, E., *Genetic Algorithms and Grouping Problems*, Wiley, 1998

11. Goldberg, D.E., *Genetic Algorithms in Search, Optimization and Machine Learning*, Addison-Wesley, 1989

12. Holland, J.H., *Adaption in Artificial Systems*, The University of Michigan Press, 1975

13. Reeves, C.R. (Ed.), *Modern Heuristic Techniques for Combinatorial Problems*, Blackwell Scientific Publications, 1993

14. Levenhagen, J., *Ein genetischer Algorithmus mit Pfadverfolgung zur Loesung des mehrfach restringierten Rucksackproblems*, Diploma thesis, University of Hagen, 2000

On the Feasibility Problem of Penalty-Based Evolutionary Algorithms for Knapsack Problems

Jens Gottlieb

SAP AG
Neurottstr. 16, 69190 Walldorf, Germany
jens.gottlieb@sap.com

Abstract. Constrained optimization problems can be tackled by evolutionary algorithms using penalty functions to guide the search towards feasibility. The core of such approaches is the design of adequate penalty functions. All authors, who designed penalties for knapsack problems, recognized the feasibility problem, i.e. the final population contains unfeasible solutions only. In contrast to previous work, this paper explains the origin of the feasibility problem. Using the concept of fitness segments, a computationally easy analysis of the fitness landscape is suggested. We investigate the effects of the initialization routine, and derive guidelines that ensure resolving the feasibility problem. A new penalty function is proposed that reliably leads to a final population containing feasible solutions, independently of the initialization method employed.

1 Introduction

Penalty-based evolutionary algorithms employ a fitness function

$$fit(x) = f(x) - penalty(x)$$

based on the original objective function f, which is assumed to be maximized here, and an additional penalty term. A solution x that violates some given constraint is called unfeasible and "punished" by a value $penalty(x) > 0$, while a feasible solution x is assigned $penalty(x) = 0$. Unfeasible solutions can be discarded from search by large penalties, or they can contribute to the search if graded penalty terms are used. Although some early guidelines were proposed by Richardson et al. [1], who suggested using pessimistic estimations of the distance from feasibility, the actual degree of pessimism which should be used is still unclear, and penalty-based evolutionary algorithms frequently suffer from the *feasibility problem*: they terminate with a completely unfeasible population.

Researchers dealing with penalty-based approaches for knapsack problems recognized the feasibility problem, in particular for highly constrained problems. A penalty function measuring the relative degree of the constraint violation was proposed by Khuri and Batarekh for the unidimensional knapsack problem [2] and adapted to the multidimensional knapsack problem by Thiel and Voss, who reported this approach as having problems in finding any feasible solution [3].

E.J.W. Boers et al. (Eds.) EvoWorkshop 2001, LNCS 2037, pp. 50–59, 2001.
© Springer-Verlag Berlin Heidelberg 2001

Michalewicz and Arabas employed several penalty functions for the unidimensional knapsack problem, which are based on the amount of constraint violation, and observed all approaches failing to find any feasible solution for highly restrictive instances [4]. This was also recognized by Olsen, who suggested penalties depending on the actual amount of constraint violation but also reported weak penalties preventing the search from finding any feasible solution for restrictive problems [5]. Recently, Hinterding compared some penalties of different strength and remarked that only unfeasible solutions were produced for the smallest strength for the unidimensional knapsack problem considered [6]. Often, the employed initialization routine produces only unfeasible solutions, which can prevent an EA from locating the feasible region [3,7]. Therefore the choice of an appropriate initialization routine is crucial, too. However, well-designed penalties should also be able to guide an unfeasible population towards the feasible region of the search space.

There are other constraint-handling techniques like repair algorithms or decoders. Here we focus on penalty functions only and refer to [8] for an in-depth comparison between those techniques and the penalty-based approach. Our goal is to evaluate the sensitivity of penalty functions to the initialization method, and to derive properties that guarantee solving the feasibility problem.

We consider the multidimensional knapsack problem and its most important properties in section 2, and present five penalty functions in section 3. A static analysis of the induced fitness landscapes is given in section 4, and the average Hamming weight dynamics of evolutionary search are investigated in section 5. Conclusions are given in section 6.

2   The Multidimensional Knapsack Problem

The *multidimensional knapsack problem (MKP)* is an NP-complete combinatorial optimization problem with a wide area of applications like capital budgeting, cargo loading and project selection [9,10]. It is stated as

$$\text{maximize} \quad \sum_{j \in J} p_j x_j \tag{1}$$

$$\text{subject to} \quad \sum_{j \in J} r_{ij} x_j \le c_i, \quad i \in I \tag{2}$$

$$x_j \in \{0, 1\}, \qquad j \in J \tag{3}$$

with $I = \{1, \dots, m\}$ and $J = \{1, \dots, n\}$ denoting the sets of resources and items, respectively. Each item j has a profit $p_j > 0$ and a resource demand $r_{ij} \ge 0$ of resource i, which is limited by its capacity $c_i > 0$. The goal is to determine a set of items with maximum profit, which does not exceed the resource capacities. In this paper, we assume $r_{ij} > 0$ for all i and j without loss of generality [8].

The *search space* $S = \{0, 1\}^n$ is partitioned into the *feasible region*

$$F = \{x \in S \mid \sum_{j \in J} r_{ij} x_j \le c_i \text{ for all } i \in I\},$$

and the *unfeasible region* $U = S \setminus F$. We assume the most natural neighbourhood for the MKP, which is defined by Hamming distance 1, and define the boundary of the feasible region as the set of those feasible solutions, for which all neighbours with higher Hamming weight are unfeasible. The boundary is denoted by B and it is equivalent to the set of all local optima [8]. Figure 1 illustrates a search space partitioned into F and U, and the corresponding boundary B. Note that the degree of feasibility decreases with increasing Hamming weight of the solution candidates, due to the structure of the constraint.

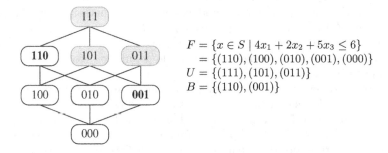

$$F = \{x \in S \mid 4x_1 + 2x_2 + 5x_3 \le 6\}$$
$$= \{(110), (100), (010), (001), (000)\}$$
$$U = \{(111), (101), (011)\}$$
$$B = \{(110), (001)\}$$

Fig. 1. Example for a unidimensional knapsack problem with $S = \{0,1\}^3$

3   Penalty Functions

Penalty functions can be classified by their severity with respect to unfeasible solutions. While some penalties allow unfeasible solutions to be evaluated as being superior to some feasible solutions, strict penalties lead to favouring all feasible solutions to all unfeasible solutions. The most obvious penalty function to achieve such severity is

$$pINF(x) = \infty, \quad x \in U,$$

which is sometimes called *death penalty* because all unfeasible individuals $x \in U$ reveive the worst possible evaluation $-\infty$, drastically restricting their chance for reproduction. Another variant also severely punishing unfeasible individuals is

$$pOFF(x) = 1 + \sum_{j \in J} p_j, \quad x \in U,$$

which is based on an offset term exceeding the sum of all profits. As this value is larger than the maximum objective value, all feasible individuals achieve a higher fitness than all unfeasible solutions; this basic idea has been suggested by Khuri et al. for the subset sum problem [11]. Note that higher offset terms could also be used, but they yield exactly the same overall performance when rank-based selection schemes are used. The main difference to *pINF* is that unfeasible individuals can be distinguished with respect to the fitness function. However, this distinction is solely based on the objective function, i.e. the amount of constraint violation is not considered.

While the previous functions severely penalize all unfeasible individuals, the following penalties allow unfeasible solutions to get a better overall evaluation than some feasible solutions. Given the maximum profit coefficient $p_{max} = \max\{p_j \mid j \in J\}$ and $NVC(x) = |\{i \in I \mid \sum_{j \in J} r_{ij} x_j > c_i\}|$ as the number of violated constraints of solution $x \in S$, we consider

$$pKBH(x) = p_{max} \cdot NVC(x), \quad x \in U ,$$

which has been proposed by Khuri et al. [7]. The basic idea is to estimate the distance from feasibility by the number of violated constraints, and to correlate the penalty to the objective function by the factor p_{max}. Since many other authors also used this function [12,13,14,15], it is viewed as standard penalty function for the MKP.

Defining $CV(x, i) = \max(0, \sum_{j \in J} r_{ij} x_j - c_i)$ as the amount of constraint violation for constraint $i \in I$ and $x \in S$, x is feasible if and only if $CV(x, i) = 0$ for all constraints $i \in I$. Hoff et al. proposed the function

$$pHLM(x) = \sqrt{\sum_{i \in I} CV(x, i)} , \quad x \in U,$$

based on the sum of all constraint violations [16,17]. Thus, unfeasible individuals with a higher distance from feasibility are assigned a higher penalty. However, this function is not correlated to the objective function, which causes problems as we shall see later.

Assuming $r_{min} = \min\{r_{ij} \mid i \in I, j \in J\}$ as the minimum resource consumption, we propose the new function

$$pCOR(x) = \frac{p_{max} + 1}{r_{min}} \cdot \max\{CV(x, i) \mid i \in I\}, \quad x \in U .$$

The factor $(p_{max} + 1)/r_{min}$ is used to correlate the objective function based on profits with the resource demands that exceed the capacity constraints. This factor is based on a pessimistic estimation of the profit that would get lost if items were removed from the knapsack in order to obtain feasibility. Thus, the penalty term increase is higher than the objective function increase for moves away from feasibility.

4   Analysis of the Fitness Landscape

This section analyzes penalty functions concerning the fitness landscapes they induce. We suggest the concept of fitness segments in section 4.1, which leads to identifying perfect fitness landscapes in section 4.2 and characteristic properties of the penalty functions in section 4.3.

4.1 Characterizing Fitness Landscapes by Fitness Segments

The search space $S = \{0,1\}^n$ forms a complete lattice with infimum $(0,\dots,0)$ and supremum $(1,\dots,1)$, which correspond to the solutions with maximum and minimum feasibility, respectively. Each penalty function induces a fitness landscape defined by S, the neighbourhood and the fitness function, and the behaviour of evolutionary search is mainly affected by the structure of this fitness landscape, if the employed operators induce a neighbourhood similar to the one we supposed. A thorough analysis of the complete fitness landscape is difficult due to the huge size of S, hence we suggest focusing on representative parts only. Rather than sampling some points of the search space in a purely random fashion, we analyze fitness segments that consist of a specific set of solutions.

A sequence $(x^0, x^1, \dots, x^n)$ of distinct solutions $x^i \in S$ is called a *segment of S* if x^i has Hamming weight i for all i, and x^i and x^{i+1} are neighbours for $i \in \{0,\dots,n-1\}$. Obviously, $x^0 = (0,\dots,0)$ and $x^n = (1,\dots,1)$ hold for each segment, and two successive solutions of a segment are identical except for one bit. There exist $n!$ segments of the given search space S since each segment corresponds to a path of length n from the infimum to the supremum of S. Some index $k \in \{0,\dots,n\}$ exists for each segment, such that the solutions $x^0,\dots,x^k$ are feasible, while $x^{k+1},\dots,x^n$ are not. The index k indicates the transition from feasibility to unfeasibility, which is critical for penalty functions. The segment $(000,001,011,111)$ in figure 1 has index $k=1$.

The fitness values $(fit(x^0), fit(x^1), \dots, fit(x^n))$ of a segment $(x^0, x^1, \dots, x^n)$ form a sequence which is termed *fitness segment*. Each element of a fitness segment is identified by the Hamming weight of the corresponding solution. A fitness segment describes the slope of one slice of the whole (high-dimensional) fitness landscape. The complete search space being represented by all its segments, the fitness landscape is characterized by all its fitness segments. The slope of a fitness segment enables us to draw conclusions about the basic structure of the complete landscape, and can be described by concepts like hills and valleys, which are intuitively used to characterize 2-dimensional fitness landscapes.

4.2 The Bias Towards the Boundary of the Feasible Region

A reasonable penalty function should be able to guide unfeasible populations towards feasibility, since the evolutionary search otherwise might converge in the unfeasible region. Thus, the fitness function must be strictly monotonic decreasing in the unfeasible region. Speaking in terms of landscape concepts,

1. there must be a big valley beyond the boundary of the feasible region,
2. this valley must not contain plateaus or peaks which would induce new local optima in the unfeasible region, and
3. the valley must be descending rather than ascending to prevent the supremum from becoming a new local optimum.

Figure 2 shows a fitness segment of a "perfect" penalty function in an idealized fashion, where the feasible part of the segment is shaded grey. On the one hand,

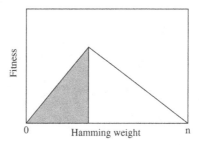

Fig. 2. Idealized fitness segment for a "perfect" penalty function

feasible populations are driven towards the boundary of the feasible region due to the objective function. On the other hand, the slope of the segment's unfeasible part leads unfeasible populations towards feasibility and hence towards the boundary. We call a penalty function *biased towards the boundary of the feasible region* if the slopes of all fitness segments exhibit the same basic tendency as visualized in figure 2. Such penalty functions do not suffer from any feasibility problem since unfeasible populations are reliably guided towards the boundary.

4.3   A Comparison of the Penalty Functions

Figure 3 shows representative fitness segments of the first instance with $m = 30$ and $n = 500$ from Chu's benchmark suite[1] for the penalty functions introduced in section 3. A simple technique was used to implement the infinite penalty term of *pINF*: we employed the penalized fitness value -1 for unfeasible individuals, exhibiting an equivalent fitness landscape for rank-based parent selection.

The functions *pINF* and *pOFF* induce an extremely wide and deep valley beyond the boundary of the feasible region. However, these valleys are problematic since *pOFF* introduces the supremum as new local optimum with a very large basin of attraction, and in case of *pINF* the valley is in fact a very large plateau containing many new local optima.

The functions *pKBH* and *pHLM* cause a monotonic increase of the fitness values for sufficiently high Hamming weights and thus introduce the supremum as new local optimum. While *pKBH* also introduces new local optima near the boundary of the feasible region, *pHLM* induces a fitness landscape with the supremum as only local optimum. These functions generally do not preserve the local optimality of the boundary, and they fail to introduce a sufficiently deep and wide valley to separate the boundary from unfeasibility.

Resembling the perfect slope shown in figure 2, the function *pCOR* prevents the supremum from becoming locally optimal and yields a separation of the boundary from the unfeasible region by a wide and deep valley that is descending. Thus, *pCOR* perfectly achieves the desired properties and is therefore expected to resolve the feasibility problem.

---

[1] http://mscmga.ms.ic.ac.uk/jeb/orlib/mknapinfo.html

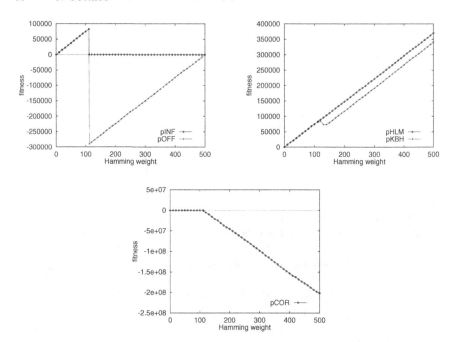

Fig. 3. Slopes of representative fitness segments for the penalty functions

5   Analysis of Average Hamming Weight Dynamics

While the previous section provides a static analysis of fitness landscapes, this section empirically confirms that the structure of fitness segments reliably predicts the actual dynamics of evolutionary algorithms with standard operators. We consider an EA with bit string representation, population size 100, uniform crossover applied with probability $p_c = 0.7$, standard bit mutation (flipping each bit with probability $p_m = 1/n$), parent selection via tournaments of size 2, steady-state replacement (deleting the worst) and duplicate elimination. The EA generates 1 000 000 non-duplicate solutions and its initialization method is parameterized by $b \in [0, 1]$ such that the expected Hamming weight of an initialized solution is $b \cdot n$. Varying the parameter b, we can analyze the effects of initialization and check whether the EA is sensitive to the initial population.

The actual neighbourhood structure induced by the EA's variation operators differs from the neighbourhood we used in the previous sections. We perceive the mutation operator's neighbourhood – i.e. the set of solutions that are reachable by one mutation – as basic neighbourhood, which is frequently enlarged by the use of crossover. The mutation operator can generate any solution in the search space, but in expectation it changes one bit and its neighbourhood therefore resembles the natural neighbourhood of Hamming distance 1. Thus, it is reasonable to rely on the approximation of the variation operators' neighbourhood by the neighbourhood of Hamming distance 1.

The population's average Hamming weight describes its location in the search space, and hence an evolutionary search process can be perceived as a trajectory of the average Hamming weights associated to the population sequence. For the considered initialization routine, the first population is located at Hamming weight level $b \cdot n$. Figure 4 shows the dynamics during the first 10 000 generations for the functions $pINF$, $pOFF$, $pHLM$ and $pCOR$, the initialization parameters $b \in \{0.05, 0.20, 0.35, 0.50, 0.65, 0.80, 0.95\}$ and the MKP instance considered in section 4.3: it consists of $n = 500$ items and its boundary is located approximately at Hamming weight level 120.

In the case of $pINF$ and $pOFF$ the population is reliably guided towards the boundary of the feasible region, if the initial population contains feasible solutions. However, $pINF$ causes a random search at Hamming weight level $b \cdot n$ if the initial population is completely unfeasible and too far away from the boundary. This is due to the large plateau induced by $pINF$. The function $pOFF$ causes convergence towards the supremum for sufficiently high values of b, i.e. the population is guided away from feasibility. These results suggest using $pINF$ and $pOFF$ with an initialization parameter b which ensures that the initial population contains at least one feasible solution.

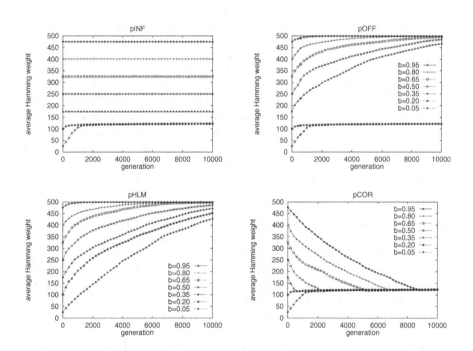

Fig. 4. Average Hamming weight dynamics for selected penalties and parameters b

The function *pHLM* illustrates the feasibility problem in an impressive manner since the population is always guided towards maximum unfeasibility, independently of the initial population's average Hamming weight. Thus, unlike for *pINF* and *pOFF*, even the choice of appropriate initialization methods does not overcome the feasibility problem. Therefore we recommend not to use this function in the knapsack domain. Although the fitness landscape induced by *pKBH* is somehow more advantageous than *pHLM*, it also leads to convergence in the unfeasible region in most cases since its fitness landscape contains new local optima in the unfeasible region [8]. Thus, its use should also be avoided.

The function *pCOR* achieves the perfect behaviour: any population, be it feasible or unfeasible, is reliably guided towards the boundary of the feasible region because all local optima are located there. This function is insensitive to the initial population, making it a very robust choice conerning the feasibility issue.

6   Conclusion

The feasibility problem is a serious issue in penalty-based evolutionary algorithms since it prevents effective search, due to convergence in the unfeasible region of the search space. This problem was recognized and investigated empirically by several authors, but previous work was mainly concerned about the outcome of the search, i.e. the final population.

In this paper we considered the multidimensional knapsack problem and five penalty functions with different characteristic properties, originating from different design principles like dominance of feasibility or degree of unfeasibility. We presented a static analysis of the fitness landscapes induced by these penalties. Our analysis is based on fitness segments, which can be determined easily and allow to draw meaningful conclusions about the whole fitness landscape. Using the notion of fitness segments, certain properties of penalty functions were derived that guarantee to solve the feasibility problem. Empirical experiments concerning the average Hamming weight dynamics of the penalty functions confirmed our hypothesis that a wide, deep, and descending valley in the fitness landscape should separate the boundary of the feasible region from unfeasibility.

Our results imply guidelines for successful penalty-based evolutionary search applied to knapsack problems: the feasibility problem is resolved if (i) the penalty function is biased towards the boundary of the feasible region, or (ii) the initial population contains at least one feasible solution and unfeasible solutions are assigned a lower fitness than all feasible solutions. This guideline is not only valid for knapsack problems, but also for all covering and packing problems [8] since their local optima lie in the boundary of the feasible region, too. Thus, the basic results also apply for many other well-known problems like set covering, subset sum, independent set, set packing and various knapsack problems.

Acknowledgement. The author would like to thank the referees for their helpful comments.

References

1. J. T. Richardson, M. R. Palmer, G. Liepins, and M. Hilliard. Some Guidelines for Genetic Algorithms with Penalty Functions. In *Proc. 3rd International Conference on Genetic Algorithms*, 191 - 197, Morgan Kaufmann, 1989

2. S. Khuri and A. Batarekh. Heuristics for the Integer Knapsack Problem. In *Proc. 10th International Computer Science Conference*, 161 - 172, Santiago, Chile, 1990

3. J. Thiel and S. Voss. Some Experiences on Solving Multiconstraint Zero-One Knapsack Problems with Genetic Algorithms. *INFOR*, Volume 32, No. 4, 226 - 242, 1994

4. Z. Michalewicz and J. Arabas. Genetic Algorithms for the 0/1 Knapsack Problem. In *Proc. 8th International Symposium on Methodologies for Intelligent Systems*, 134 - 143, Springer, 1994

5. A. L. Olsen. Penalty functions and the knapsack problem. In *Proc. 1st IEEE Conference on Evolutionary Computation*, 554 - 558, 1994

6. R. Hinterding. Representation, Constraint Satisfaction and the Knapsack Problem. In *Proc. Congress on Evolutionary Computation*, 1286 - 1292, Washington DC, 1999

7. S. Khuri, T. Bäck, and J. Heitkötter. The Zero/One Multiple Knapsack Problem and Genetic Algorithms. In *Proc. ACM Symposium on Applied Computation*, 188 - 193, ACM Press, 1994

8. J. Gottlieb. *Evolutionary Algorithms for Constrained Optimization Problems*. Dissertation, Technical University of Clausthal, 1999. Shaker, Aachen, 2000

9. P. C. Chu and J. E. Beasley. A Genetic Algorithm for the Multidimensional Knapsack Problem. *Journal of Heuristics*, Volume 4, No. 1, 63 - 86, 1998

10. S. Martello and P. Toth. *Knapsack Problems*. John Wiley & Sons, 1990

11. S. Khuri, T. Bäck, and J. Heitkötter. An Evolutionary Approach to Combinatorial Optimization Problems. In *Proc. 22nd Annual ACM Computer Science Conference*, 66 - 73, ACM Press, New York, 1994

12. F. Corno, M. Sonza Reorda, and G. Squillero. The Selfish Gene Algorithm: A New Evolutionary Optimization Strategy. In *Proc. 13th Annual ACM Symposium on Applied Computing*. 1998

13. H. A. Mayer. ptGAs - Genetic Algorithms Evolving Noncoding Segments by Means of Promoter/Terminator Sequences. *Evolutionary Computation*, Volume 6, No. 4, 361 - 386, 1998

14. G. Rudolph and J. Sprave. A Cellular Genetic Algorithm with Self-Adjusting Acceptance Threshold. In *Proc. 1st IEE/IEEE International Conference on Genetic Algorithms in Engineering Systems: Innovations and Applications*, 365 - 372, IEE, London, 1995

15. G. Rudolph and J. Sprave. Significance of Locality and Selection Pressure in the Grand Deluge Evolutionary Algorithm. In *Proc. 4th Conference on Parallel Problem Solving from Nature*, 686 - 695, Springer, 1996

16. A. Hoff, A. Løkketangen, and I. Mittet. Genetic Algorithms for 0/1 Multidimensional Knapsack Problems. In *Proc. Norsk Informatikk Konferanse*, 1996

17. A. Løkketangen. A Comparison of a Genetic Algorithm and a Tabu Search Method for 0/1 Multidimensional Knapsack Problems. In *Proc. Nordic Operations Research Conference*, 1995

Coloured Ant System and Local Search to Design Local Telecommunication Networks

Roberto Cordone and Francesco Maffioli

DEI - Politecnico di Milano
{roberto.cordone,francesco.maffioli}@polimi.it

Abstract. This work combines local search with a variant of the *Ant System* recently proposed for partitioning problems with cardinality constraints. The *Coloured Ant System* replaces the classical concept of *trail* with p trails of different "colours", representing the assignment of an element to one of the classes in the partition. We apply the method with promising results to the design of local telecommunication networks. The combination of the *Coloured Ant System* with local search yields much better results than the two approaches alone.

1 Introduction

The design of local telecommunication networks (e. g. cable television companies providing internet access [1]) often requires to link demand nodes to a wide area network through a number of concentrator devices taking over the traffic flow. The capacity of the concentrators is limited and the demands form tree-shaped subnetworks, as a higher connectivity is unjustified at this level. The *Weight Constrained Graph Tree Partition Problem* (*WC-GTPP*) models the situation by an undirected graph $G(V, E)$ of n vertices and m edges. Given a cost function $c : E \rightarrow \mathbb{N}$ on the edges and a weight function $w : V \rightarrow \mathbb{N}$ on the vertices, determine a spanning forest $F(V, X)$ of p trees $T_r(U_r, X_r)$, such that the weight of each tree falls in a given range $[W^-; W^+]$ and F has minimum cost.

This paper adapts the *Ant System* to the *WC-GTPP* and combines it to local search. Instead of a single undifferentiated trail, the ants release trails of p different colours, according to the tree each vertex is assigned to. Hence the name of *Coloured Ant System* (*CAS*) [2]. The algorithm relaxes the weight bound, penalizing its violations with a factor tuned by feedback. The *CAS* interacts with *Swinging Forest*, a local search heuristic based on operations such as vertex exchanges, tree splittings and tree removals. At each iteration, the *CAS* submits the best current solution to local search before updating the trails. The two approaches take advantage of each other: the *CAS* provides a smart initialization to local search; local search provides a smart trail update to the *CAS*.

Sect. 2 briefly surveys the literature on local network design problems, and gives some references on the *Ant System* and its variants. Sect. 3 describes the *CAS* algorithm. Sect. 4 describes *Swinging Forest* and its interactions with the *CAS*. The last section presents some computational results and conclusions.

E.J.W. Boers et al. (Eds.) EvoWorkshop 2001, LNCS 2037, pp. 60–69, 2001.

2   Survey

The field of local telecommunication networks (servers, cable TV companies, etc. . .) poses a number of practical problems in which secondary devices must be connected to a given number of primary ones, minimizing the total connection cost and keeping a balance between the traffic of the disjoint subnetworks. Similar trade-offs between cost and balance occur in electric or radio broadcasting networks, but also in electoral districting and Cluster Analysis.

Though fairly general, the *WC-GTPP* is a good model for these cases. It is strongly $\mathcal{NP}$-hard, by direct reduction from *SAT*, and in general not approximable [3], though a $(2p-1)$ algorithm exists if the number of vertices in each tree is given and the triangle inequality holds [4]. Similar problems have been deeply studied. In the *Capacitated Minimum Spanning Tree* problem (*CMST*) all secondary devices are linked to a single primary one, with an upper bound on the traffic in each branch [5]. The lower weight constrained minimum spanning forest problem bounds from below the weight of the trees and does not fix their number. These problems are strongly $\mathcal{NP}$-hard, but approximable [6,7].

The *Ant System* has been first proposed for the *Travelling Salesman Problem* [8], spreading to other fields in Combinatorial Optimization, such as *Graph Colouring* [9] and the *Quadratic Assignment Problem* [10]. Several variants propose effective ways to update the trail function during the execution: the *Max-Min Ant System* limits it into a given interval [11], the *Ant-Q* approach updates it by reinforcement learning mechanisms [12], the *Ant Colony System* by moving averages [13]. The probability distribution of choices can be based on the rank of the alternatives, instead of their absolute value [14]. In the field of routing problems, the *Ant System* combined with local search has brought to strong improvements upon the two approaches applied singularly [15].

3   Coloured Ant System

The *Ant System* involves a population of agents (*ants*), who build in parallel greedy solutions to a problem. Their behaviour is not trivially greedy: they partly take random choices biased by a *trail function*, which is shaped by the results of previous runs. To reformulate the natural analogy for our problem, an agent is a caste of ants divided into p colonies. Each colony builds a hive on a different vertex and takes possession of a connected subgraph from it. The colony patrols a network of edges, whose cost evaluates the necessary effort, to control food sources located on the vertices, whose amount is measured by the weights. The global effort must be minimum and the amount of food owned by each colony neither too small nor too large. Each colony marks the vertices with a trail of a typical "colour". Season after season, they retire into the hives and resume colonizing the graph, driven by the costs, the weights and the residual trails. When the ants find better solutions, they resettle their hives accordingly. Several independent castes operate on the graph, sharing the p hives and the trail information. Ants of different colonies can occupy the same vertex, if they belong to different castes; otherwise, they cannot.

Algorithm 1 starts by selecting p hives. Then, it repeats a main loop for $I_{\max}$ iterations or $T_{\max}$ seconds since the beginning, or I_{best} iterations since the discovery of the best known solution. Each time, the agents build S solutions, annexing the vertices to the hives one by one. Solutions violating the weight bound are admitted, but their cost is increased by a penalization factor π. As the basic step assigns a vertex v to a tree T_r, the trails are released on the vertices and their colours refer to the trees. The algorithm updates the trail function τ_{rv} in multiple points, both strengthening and weakening it.

Algorithm 1.   CAS(G, c, w, W, p)
$\{v_1, \ldots, v_p\} := StartingHives(G, c, p);$ { Choose p starting hives }
$\tau_{rv} := \tau_0 \quad \forall r, \forall v;$ { Distribute p uniform trails }

$i := 1; \; i^* := 1; \; c^* := +\infty;$
While $i \le I_{\max}$ **and** $ElapsedTime() \le T_{\max}$ **and** $i \le i^* + I_{\text{best}}$ **do**
    { Build S solutions in parallel }
    $U_r^{(s)} := \{v_r\} \quad \forall r, \forall s;$ { The ants retire }
    While $\exists s : \left(V \setminus \cup_r U_r^{(s)}\right) \ne \emptyset$ **do**   { While unassigned vertices exist... }
        For $s := 1$ **to** S **do**
            $(\tilde{v}, \tilde{r}) := BestAnnexion(G, c, w, W, p, \pi, U_r^{(s)});$ { ... take a choice... }
            $U_{\tilde{r}}^{(s)} := U_{\tilde{r}}^{(s)} \cup \{\tilde{v}\};$ { ... perform it... }
            $\tau_{\tilde{r}\tilde{v}} := (1 - \rho)\tau_{\tilde{r}\tilde{v}} + \rho\tau_0;$ { ... and deter repetitions }
        EndFor
    EndWhile

    { Determine the best current solution }
    $F := BestSolution(G, c, w, W, p, \pi, U_r^{(s)});$
    $\tau_{rv} := (1 - \rho)\tau_{rv} + \rho/c_F \pi_F \quad \forall r, \forall v \in U_r^{(\tilde{s})};$ { Intensify the trails }

    { Update the best known solution and the hives }
    If $c_F < c^*$ **then** $c^* := c_F; \; i^* := i; \; v_r := RootVertex(U_r) \quad \forall r;$
    $UpdatePenalization(\pi);$
    $i := i + 1;$
EndWhile;

3.1   The Greedy Heuristic

The basic greedy heuristic adapts Prim's algorithm for the *Minimum Spanning Forest* problem [16]: find the unassigned vertex closest to a tree and assign it, repeatedly, until the solution spans the whole graph. Besides the edge costs, the algorithm takes into account the violations of the weight bound and the trails. Instead of minimizing the connection cost $\bar{c}_{rv}$ between vertex v and colony r

$$\bar{c}_{rv} = \min_{u \in U_r} c_{uv} \qquad (1)$$

it maximizes a *request factor* f_{rv}

$$f_{rv} = \frac{\tau_{rv}}{\bar{c}_{rv}\pi_{rv}} \tag{2}$$

where π_{rv} is a penalization factor and τ_{rv} is the trail function.

The heaviest tree which could be thus obtained spans the whole graph minus the $p - 1$ lightest vertices. Let w_M be its weight; $w_m = \min_{v \in V} w_v$ is the weight of the lightest tree achievable. Therefore, the penalization factor:

$$\pi_{rv} = 1 + \pi \max\left(\frac{w_{U_r} + w_v - W^+}{w_M - W^+}, \frac{W^- - w_{U_r} - w_v}{W^- - w_m}, 0\right) \tag{3}$$

ranges from 1 for a feasible assignment to $1 + \pi$ for the most unbalanced.

Each time, the algorithm chooses at random whether to obey a

- *deterministic strategy*: perform the assignment with the highest value of f_{rv}
- *stochastic strategy*: choose at random vertex v and tree T_r with a probability proportional to f_{rv}: $p_{rv} = f_{rv} / \sum_{s=1}^{p} \sum_{u \in V} f_{su}$

The former is adopted with probability q (*deterministic factor*), the latter with probability $1 - q$. Higher values of q favour choices with stronger request factors.

3.2 The Penalization Factor

Most of the time we find out that, if the optimal solution is far from the frontier of the feasible space, the greedy heuristic finds it easily. However, in the most interesting cases the optimal solution is close to unfeasibility and the penalization factor becomes relevant. We use the following scheme, which seems reasonable: when the current S solutions are prevailingly feasible, π decreases to drive the search to the unfeasible region; when they are unfeasible, π increases. If S_f current solutions are feasible,

$$\pi := \pi \, 2^{(S-2S_f)/S} \,. \tag{4}$$

3.3 The Trail Function

The trail function saves information from previous runs in order to repeat the good choices and avoid the wrong ones. Its effective management can be interpreted in terms of the *diversification* and *intensification* principles. On one hand, it is profitable to avoid sticking in already explored regions. On the other hand, it may be profitable to explore more thoroughly the regions close to good solutions, hoping to find better ones by slight changes. A correct balance of these complementary strategies is a key issue to create an efficient heuristic.

At the beginning, the ants release a uniform small trail τ_0 of each colour on each vertex, as a sort of "ground value". While the literature derives τ_0 from the cost of a heuristic solution [13], we employ the cost of the minimum spanning forest, c_{MSF}, which usually has the same order of magnitude:

$$\tau_0 = \frac{1}{nc_{MSF}} \,. \tag{5}$$

Diversifying the Trail Function. During the greedy heuristic, after an agent assigns vertex v to tree r, it draws the associated trail closer to the ground level:

$$\tau_{rv} = (1 - \rho)\, \tau_{rv} + \rho \tau_0 \ . \tag{6}$$

The other agents consider this assignment less attractive and prefer different solutions. Equation (6) describes the "evaporation" of the trail: the *oblivion factor* ρ tunes the strength of greediness versus memory.

Intensifying the Trail Function. At the end of the greedy heuristic, the best performing agent increases the trails corresponding to its solution $\tilde{F}$. This favours the same choices in the following iterations, leading to similar solutions. The better is $\tilde{F}$, the stronger is the effect:

$$\tau_{rv} = (1 - \rho)\, \tau_{rv} + \rho \ \frac{1}{c_{\tilde{F}} \pi_{\tilde{F}}} \ . \tag{7}$$

3.4   The Root Choice

The position of the roots influences remarkably the final result. At first, we choose them so as to cover the graph uniformly. Given a seed root, the second root is the farthest vertex in the graph, the third is the vertex with the highest total distance from the first two, and so on. Then, alternatively, we apply the greedy heuristic with no random choices and a uniform trail, and we move the roots to the centroids of the trees obtained. The process ends when the roots stabilize or repeat cyclically, and it is performed n times, using each vertex as the seed. The best root assignment overall initializes the *CAS*. Every time the best known solution changes, the centroids of its trees become the new roots.

4   The *Swinging Forest* Procedure

Local search is based on the concept of *neighbourhood*, a function associating to each feasible solution F a subset $\mathcal{N}(F)$ of "neighbour solutions". Commonly, this is the set of solutions obtained applying to F a given family of manipulations (*moves*). Local search procedures start from a current solution, assume one of its neighbours as the *incumbent solution*, and replace the former with it.

Algorithm 2 outlines the behaviour of *Swinging Forest*. First, the *Exchange* procedure improves the starting solution F moving vertices from tree to tree, one by one or in pairs. This is the merging of two related neighbourhoods: $\mathcal{N}_1(F)$ includes the feasible forests whose vertices belong, all but one, to the same trees as in F; $\mathcal{N}_2(F)$ includes those whose vertices belong, all but two, to the same trees. We adopt a *steepest descent* strategy, choosing the best solution in the whole neighbourhood as the incumbent. The procedure ends when no neighbour solution is better than the current one. When the weight bound severely limits these moves, the approach is not very effective.

Then, *Swinging Forest* splits one tree and reoptimizes the solution on and on, until the number of trees p_F reaches p_M. The aim is to redistribute the vertices in spite of the weight bound, thanks to the slack capacity of the new trees. Then, the algorithm tries to remove each tree displacing its vertices into the other ones. When it has processed all trees, *Exchange* reoptimizes the solution. This phase ends when the number of trees reaches p_m. Then, the algorithm splits some trees to retrieve p. If the cost has decreased, a new period starts. If it has not or it is no longer possible to obtain p trees, the algorithm terminates.

Algorithm 2.   SwingingForest(G, c, w, W, p, F)
$c^* := \infty$;
While $p_F = p$ **and** $c_F < c^*$ **do**
    $F := Exchange(G, c, w, W, F)$;   { Steepest descent }
    If $c_F < c^*$ **then** $F^* := F$; $c^* := c$;

    While $p_F < p_M$ **do**   { Increase the number of trees }
        $F := TreeSplitting(G, c, w, W, F)$;
        $F := Exchange(G, c, w, W, F)$;
    EndFor;

    $Stop := False$;
    While $p_F > p_m$ **and** $Stop = False$ **do** { Reduce the number of trees }
        $F' := TreeRemoval(G, c, w, W, F)$;
        If $p_{F'} = p_F$ **and** $c_{F'} \geq c_F$
            then $Stop := True$;
            else $F := Exchange(G, c, w, W, F')$;
    EndFor;

    While $p_F < p$ **do**   { Retrieve p trees }
        $F := TreeSplitting(G, c, w, W, F)$;
        $F := Exchange(G, c, w, W, F)$;
    EndFor;
EndWhile;
Return F^*;

Vertex Exchanges. An exhaustive search in $\mathcal{N}_1(F) \cup \mathcal{N}_2(F)$ takes $O\left(n^2 p^2 \gamma(m, n)\right)$ time, where $\gamma(m, n)$ is the time to evaluate the minimum spanning forest after a move. For the sake of efficiency, we update the spanning forest as in [17], instead of recomputing it from scratch. Moreover, a move is often available along a sequence of steps. Keeping a list of the best feasible moves on each couple or triplet of trees, one can perform the best move in the list, cancel those involving the trees modified and evaluate them again. The other moves need not be updated.

Tree Splitting. The algorithm increases the number of trees in the current forest by removing the most expensive edge. This splits a tree in two, and at least one of the resulting subtrees (possibly both) has a large amount of weight slack. Thus, the new neighbourhood is larger.

Tree Removal. The procedure lists the vertices of a tree, from the leaves up to the root, as this seems the most natural way to "prune" a tree. It performs the best feasible transfer of the first vertex, if any exists. Then, it considers the second vertex, and so on. In the end, if the tree is empty or the solution has improved, the new solution replaces the current one. Otherwise, the original solution is retrieved. To make the removal easier, the trees are processed in increasing order of cardinality and, if ties occur, of weight.

4.1   The *Coloured Ant System* and *Swinging Forest*

Local search is effective in finding good solutions, the *Ant System* in managing intensification and diversification: we believe that they can gain much from each other. Our algorithm applies local search to the best solution found at each iteration of the *CAS* before using it to update the trails. This provides a smarter intensification of the trail, which can drive the *CAS* better. Conversely, the best solution found by the *CAS* is a smarter starting point for *Swinging Forest*. To limit the computational burden, we simply explore the $\mathcal{N}_1$ or the $\mathcal{N}_1 \cup \mathcal{N}_2$ neighbourhood at each step, and apply *Swinging Forest* only in the end, on the best solution overall.

5   Computational Results

We led an experimental campaign on the *CAS* and *Swinging Forest*, both as stand-alone algorithms and combined. They were run on a Pentium II 450 MHz with a Linux operating system. Benchmark instances for the *WC-GTPP* were not available in the literature: we built them by adapting the *CMST* instances of the OR-Library [18]. For each original instance, we set two values for the number of trees p, so as to obtain tightly and loosely constrained instances. The *TC* and *TE* problems have 41 or 81 vertices, homogeneous weights and Euclidean costs. Three upper bounds limit the weight of each tree: $W^+ = 3$, 5 and 10 for $n = 41$, $W^+ = 5$, 10 and 20 for $n = 81$. The *CM* problems do not satisfy the triangle inequality and have non uniform weights. We consider two problem sizes: $n = 50$ and $n = 100$. There are three weight upper bounds: $W^+ = 200$, 400 and 800. Each combination of family, size, weight bound and number of trees corresponds to 5 instances, which leads to a total of 180.

5.1   Parameter Settings

In our experiments, the *CAS* always runs n times, since the maximum number of iterations $I_{\max}$ is set to n, the maximum number of non improving iterations I_{best} and the maximum time $T_{\max}$ to infinite. The number of agents S is set to n. The penalization coefficient is at first $\pi_0 = 100$, but in few steps it shifts to the correct range, usually stabilizing on a floating behaviour rather than a fixed value. In the end, the minimum and maximum number of trees used by *Swinging Forest* are set, respectively, to $p_{\text{m}} = \lfloor p/1.1 \rfloor$ and $p_{\text{M}} = \lceil 1.1p \rceil$.

Table 1. Results of the *CAS* with different parameter settings.

Class	q	$\rho = 0.2$ Gap	(max.)	$\rho = 0.3$ Gap	(max.)	$\rho = 0.4$ Gap	(max.)
TC40	0.25	8.2%	25.9%	**8.1%**	25.9%	8.7%	25.9%
	0.5	8.5%	27.4%	8.3%	27.4%	9.1%	27.4%
TC80	0.25	4.0%	13.8%	3.8%	13.8%	4.1%	13.8%
	0.5	3.5%	12.9%	**3.3%**	12.9%	3.7%	12.9%
TE40	0.25	9.9%	27.7%	9.6%	27.7%	10.6%	27.7%
	0.5	**9.3%**	25.6%	9.6%	25.6%	10.4%	25.6%
TE80	0.25	8.9%	61.3%	8.8%	54.8%	10.9%	56.4%
	0.5	7.9%	41.3%	**7.6%**	42.4%	9.4%	47.3%
CM50	0.25	30.4%	194.5%	32.3%	159.6%	31.8%	172.2%
	0.5	31.7%	148.9%	**29.9%**	140.0%	38.8%	230.8%
CM100	0.25	37.8%	102.4%	37.1%	104.0%	41.4%	87.1%
	0.5	**34.2%**	79.0%	34.3%	91.9%	38.4%	92.7%
Total	0.25	16.5%	194.5%	16.6%	159.6%	17.9%	172.2%
	0.5	15.8%	148.9%	15.5%	140.0%	18.3%	230.8%

5.2 Experience on the *CAS* Parameters

Table 1 sums up our experience on the deterministic factor q and the oblivion factor ρ. After experimenting with various settings, we consider two values for q: 0.5 (half of the time the choice is taken at random proportionally to the request factor, the other half it is greedy) and 0.25 (randomness prevails). As for ρ, we compare a longer ($\rho = 0.2$) a medium ($\rho = 0.3$) and a shorter memory ($\rho = 0.4$). The first column reports the problem classes. For each one, the results are structured into two rows (corresponding to $q = 0.25$ and $q = 0.5$) and three columns ($\rho = 0.2$, $\rho = 0.3$ and $\rho = 0.4$). We report both the average and the maximum gap of the solution with respect to the minimum spanning forest lower bound. The setting with $q = 0.5$ and $\rho = 0.3$ performs consistently better and appears to be more stable. This is confirmed by the number of best results obtained: $q = 0.25$ determines 61, 63 and 55 best solutions out of 180, respectively for $\rho = 0.2$, 0.3 and 0.4; $q = 0.5$ determines 78, 116 and 70. It is rather likely that, when $\rho = 0.2$ bad starting solutions influence too strongly the algorithm. Roughly speaking, this influence falls under 5% after 14 diversifying updates, since $(1 - \rho)^{14} \approx 0.047$ (see (6)). When $\rho = 0.3$, 9 updates are enough.

5.3 A Comparison between Algorithms

Table 2 compares the results of the algorithms described, namely the greedy heuristic run deterministically with an infinite penalization factor, the *CAS* with $q = 0.5$ and $\rho = 0.3$, *Swinging Forest* initialized with the five best solutions provided by the greedy heuristic (*SF(5)*) and the *CAS* followed by *Swinging Forest* applied on the best known solution (*CAS + SF*). Each column reports the average and the maximum gap and the execution time T_{tot} in seconds. For

Table 2. The combination of *CAS* and *Swinging Forest* is better than either method.

Class	Greedy			CAS			SF(5)			CAS+SF		
	Gap	(max.)	T_{tot}	Gap	(max.)	T_{tot}	Gap	(max.)	T_{tot}	Gap	(max.)	T_{tot}
TC40	11.5%	34.8%	0.1	8.3%	27.4%	3.1	6.1%	17.8%	14.0	4.0%	12.6%	6.1
TC80	4.3%	11.7%	0.8	3.3%	12.9%	69.4	1.6%	9.5%	116.5	1.5%	5.0%	105.3
TE40	11.9%	31.9%	0.1	9.6%	25.6%	3.1	6.7%	18.5%	15.5	4.0%	12.0%	6.6
TE80	10.3%	28.0%	0.7	7.6%	42.4%	71.1	3.7%	15.3%	282.6	3.0%	11.9%	123.9
CM50	20.0%	171.6%	0.2	29.9%	140.0%	7.1	16.5%	73.6%	59.2	5.1%	15.7%	22.9
CM100	39.4%	72.4%	3.0	34.3%	91.9%	224.7	14.7%	29.8%	1208.3	13.3%	28.5%	477.6
Total	16.2%	171.6%	0.8	15.5%	140.0%	63.1	8.2%	73.6%	282.7	5.2%	28.5%	123.7

Table 3. The interaction between the *CAS* and *Swinging Forest*.

Class	CAS+SF			CAS(N1)+SF			CAS(N2)+SF		
	Gap	(max.)	T_{tot}	Gap	(max.)	T_{tot}	Gap	(max.)	T_{tot}
TC40	4.0%	12.6%	6.1	3.9%	12.6%	6.2	2.8%	7.4%	42.9
TC80	1.5%	5.0%	105.3	1.4%	5.0%	106.3	1.0%	4.3%	786.87
TE40	4.0%	12.0%	6.6	3.6%	10.0%	6.9	2.7%	6.8%	43.8
TE80	3.0%	11.9%	123.9	2.8%	7.7%	127.8	2.2%	10.6%	995.6
CM50	5.1%	15.7%	22.9	6.0%	19.2%	21.0	4.5%	11.8%	115.2
CM100	13.3%	28.5%	477.6	15.5%	26.5%	461.4	10.6%	21.8%	2502.9
Total	5.2%	28.5%	123.7	5.5%	26.5%	121.6	3.9%	21.8%	747.9

the *CAS*, the time required to find the best solution is often much lower, since the stopping conditions have not yet been optimized. The first three algorithms do not dominate each other. The greedy heuristic is very fast, but unstable: in 7 cases out of 180 it could not find a feasible solution (the better result on class *CM50* is deceiving). If restarted, *Swinging Forest* performs better than the *CAS*, but in a much longer time. By contrast, their combination gives the best results, and the most stable, in much lower time. We may conclude that the *CAS* is at least an effective initialization procedure.

5.4 Interaction between *CAS* and *Swinging Forest*

Table 3 considers a more strict combination of the *CAS* with local search. Algorithm *CAS+SF* simply applies *Swinging Forest* to the best solution found by the *CAS*. Algorithms *CAS(N1)+SF* and *CAS(N2)+SF* also improve the best solution found at each iteration of the *CAS* by a steepest descent exploration of, respectively, $\mathcal{N}_1$ and $\mathcal{N}_1 \cup \mathcal{N}_2$. Exploring $\mathcal{N}_1$ seems to bring little advantage, and even some loss on the *CM50* and *CM100* classes . By contrast, the exploration of $\mathcal{N}_1 \cup \mathcal{N}_2$ remarkably improves the quality of the results, as well as their robustness. The computational time, though much longer, is comparable to a multiple start (column *SF(5)* in Table 2). We may conclude that the *CAS* and local search gain much indeed by a strict interaction.

References

1. R. Patterson, E. Rolland, and H. Pirkul. A memory adaptive reasoning technique for solving the capacitated minimum spanning tree problem. Working paper, University of California, Riverside, September 4th, 1998.

2. R. Cordone and F. Maffioli. A coloured ant system approach to graph tree partition. In *Proceedings of the ANTS' 2000 Conference*, Brussels, Belgium, September, 2000.

3. R. Cordone and F. Maffioli. On graph tree partition problems. In *Proceedings of EURO XVII*, Budapest, Hungary, July 16-19th, 2000.

4. N. Guttmann-Beck and R. Hassin. Approximation algorithms for minimum tree partition. *Discrete Applied Mathematics*, 87(1–3):117–137, October 1st, 1998.

5. A. Amberg, W. Domschke, and S. Voß. Capacitated minimum spanning trees: Algorithms using intelligent search. *Combinatorial Optimization: Theory and Practice*, 1:9–39, 1996.

6. K. Altinkemer and B. Gavish. Heuristics with constant error guarantees for the design of tree networks. *Management Science*, 32:331—341, 1988.

7. C. Imielińska, B. Kalantari, and L. Khachiyan. A greedy heuristic for a minimum weight forest problem. *Operations Research Letters*, 14:65–71, September 1993.

8. M. Dorigo, V. Maniezzo, and A. Colorni. The Ant System: Optimization by a colony of cooperating agents. *IEEE Transactions on Systems, Man, and Cybernetics Part B: Cybernetics*, 26(1):29–41, 1996.

9. D. Costa and A. Hertz. Ants can colour graphs. *Journal of Operational Research Society*, 48:295–305, 1997.

10. V. Maniezzo and A. Colorni. The ant system applied to the quadratic assignment problem. *IEEE Transactions on Knowledge and Data Engineering*, 1999.

11. T. Stützle and H. Hoos. The $\mathcal{MAX}$–$\mathcal{MIN}$ Ant System and local search for the traveling salesman problem. In T. Bäck, Z. Michalewicz, and X. Yao, editors, *Proceedings of The IEEE Conference on Evolutionary Computation, IEEE World Congress on Computational Intelligence*, pages 309–314, Piscataway, NJ, 1997. IEEE Press.

12. M. Dorigo and L. M. Gambardella. A study of some properties of Ant-Q. *Lecture Notes in Computer Science*, 1141:656–665, 1996.

13. M. Dorigo and L. M. Gambardella. Ant colony system: A cooperative learning approach to the traveling salesman problem. *IEEE Transactions on Evolutionary Computation*, 1(1):53–66, April 1997.

14. B. Bullnheimer, R. F. Hartl, and C. Strauss. A new rank based version of the ant system: A computational study. *Central European Journal for Operations Research and Economics*, 7(1):25–38, 1999.

15. L. M. Gambardella, E. Taillard, and G. Agazzi. Ant colonies for vehicle routing problems. In D. Corne, M. Dorigo, and F. Glover, editors, *New Ideas in Optimization*. McGraw–Hill, 1999.

16. R. C. Prim. Shortest connection networks and some generalizations. *Bell System Technical Journal*, 36:1389, 1957.

17. M. Gendreau, J.-F. Larochelle, and B. Sansò. A tabu search heuristic for the Steiner tree problem. *Networks*, 34(2):162–172, September 1999.

18. J. E. Beasley. OR–Library. http://mscmga.ms.ic.ac.uk/info.html, 1999.

Cooperative Ant Colonies for Optimizing Resource Allocation in Transportation

Karl Doerner, Richard F. Hartl, and Marc Reimann

Institute of Management Science, University of Vienna, Brünnerstrasse 72,
A-1210 Vienna, Austria
{karl.doerner, richard.hartl, marc.reimann}@univie.ac.at
http://www.bwl.univie.ac.at/bwl/prod/index.html

Abstract. In this paper we propose an ACO approach, where two
colonies of ants aim to optimize total costs in a transportation network. This main objective consists of two sub goals, namely fleet size
minimization and minimization of the vehicle movement costs, which
are conflicting for some regions of the solution space. Thus, our two ant
colonies optimize one of these sub-goals each and communicate information concerning solution quality. Our results show the potential of the
proposed method.

1 Introduction

In the last decade, a new meta-heuristic called Ant Colony Optimization (ACO)
has attracted increasing attention, as a tool to solve various hard combinatorial
optimization problems (cf. e.g. [1], [2], [3], [4], [5]). It is based on research done in
the early nineties by Dorigo et al. (see e. g. [6], [7], [8]) on the Ant System, which
was inspired by the behavior of real ant colonies searching for food. Information
concerning the quality of food sources is communicated between the members of
the colony via an aromatic essence called pheromone. Over time this information
will lead to the reinforcement of some paths, which lead to rich food sources,
while other paths will not be used anymore.

In the context of combinatorial optimization problems this mechanism is implemented as an adaptive memory which, together with a local heuristic function
called visibility, guides the search of the artificial ants through the solution space.
Thus, the artificial ants base their decisions on their own rule of thumb and on
the experience of the colony as a whole. The objective values correspond to the
quality of the discovered food. A convergence proof for a generalized Ant System
Algorithm is provided in [9].

In order to be able to successfully implement such an ACO algorithm for a
given combinatorial optimization problem, problem specific knowledge is necessary to identify an appropriate rule of thumb to guide the search of the ants to
promising regions of the search space. However, for most problems more than
one good heuristic exists, and the quality of each one generally depends on the
current problem constellation. Furthermore, for problems with multiple goals
available heuristics which seek to optimize the objective function with respect

E.J.W. Boers et al. (Eds.) EvoWorkshop 2001, LNCS 2037, pp. 70–79, 2001.

to one goal can perform rather poor with respect to some other goals. Thus, such problems are normally solved using either a lexicographic approach or an objective function, which sums up the values associated with each goal. The former approach is based on a ranking of the goals with respect to their importance, in the latter approach each goal can be assigned a weight before the sum is taken.

The aim of this paper is to overcome these problems and develop a method which finds comprehensive solutions for problems with multiple objectives. We propose an algorithm, where two colonies of ants aim to optimize total costs in a transportation network. These costs consist of fixed costs associated with the fleet size and variable vehicle movement costs. In general, minimal fleet sizes will not cause minimal vehicle movement costs, these two goals are rather conflicting. Thus, our two ant colonies optimize one goal each and communicate information about good solutions in order to enhance the general solution quality. However, due to the size of the costs the main goal is to minimize the fleet size required. Thus, in our approach we have a master population which optimizes the fleet size. This 'master' population is supported by a 'slave' population which optimizes empty vehicle movements and communicates outstanding solutions to the 'master' population. A similar approach for a related problem, the Vehicle Routing Problem with Time Windows, was proposed in [10]. In that approach one colony optimizes fleet size, while the other one minimizes vehicle movements. However, there the populations communicate only if one population improves the global best solution. Thus, both colonies consider fleet size as the main goal. On the contrary, in our approach, the 'slave' population does not consider fleet size, but rather aims to truly minimize the empty vehicle movements.

The remainder of this paper is organized as follows. In section 2 we describe the problem we consider. Our new approach for handling multiple objectives is proposed in section 3. In section 4 we present our numerical results. We close in section 5 with some final remarks and an outlook on future research.

2   Description of the Problem

In this paper we aim to solve a problem, where full truckloads have to be transported between a number of locations in a network. Each location acts as a depot for the fleet of homogeneous trucks, available to perform the service. Due to the fact that we consider only full truckloads each truck goes directly from the pickup location to the delivery location of a shipment. Upon delivery it is available to perform another task. However, each truck has to return to its home depot after two periods. This is due to legal restrictions. The planning horizon, i.e. the time for which shipment information is available, is eight periods. Thus, each truck can be used repeatedly within the planning horizon. Finally, the shipments can not be delivered at any time, but only during pre-specified time windows.

In practice, logistics service providers are generally faced with such a problem. They have to satisfy customer orders, which require the delivery of goods between pickup and delivery locations. In general, if these orders are of small size and distances are large, they are not transported directly from their source to their destination but via the locations of the service provider. Such a situation is depicted in Figure 1. An order which has to be delivered from customer a to customer b is first shipped to the distribution center i associated with customer a. There, together with other orders requiring transportation to the same region, it is consolidated to a full truckload and delivered to the receiving distribution center j, from where it is finally delivered to customer b. While the local transportation is generally performed with small trucks, such a problem is treated in [11], the long distance movements between distribution centers are performed by larger trucks. The problem considered in our paper is depicted with bold lines in Figure 1.

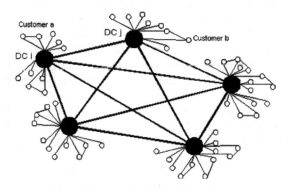

Fig. 1. Structure of the distribution network

It is obvious that the utilization of these large trucks is very important for the service provider. Thus, empty vehicle movements have to be minimized. Apart from that, the necessary fleet size should be as small as possible to keep the fixed costs low. Therefore, the overall objective is to minimize total costs, consisting of fixed costs associated with the utilized vehicle fleet and variable costs associated with the total distance traveled by all the trucks. Thus, we have to minimize fleet size and empty vehicle movements simultaneously. Note, that these two goals may not be equally important. In a situation, where a fleet of vehicles and a number of drivers are available fleet size may be less important, especially if unused trucks or drivers could not be utilized for other services. However, in our case we look at a situation, where utilization of trucks or drivers for alternate services is possible, and thus fleet size minimization is the main goal.

As stated above, minimizing empty vehicle and minimizing fleet size may be antagonistic. For example, empty vehicle movements can be kept small if trucks wait at their locations for nearby pickups. However, these waiting times have a negative impact on vehicle utilization. This in turn might increase the fleet size necessary to satisfy all transportation requests. Given this tradeoff between the goals we will now propose our new approach developed for this multi-objective problem.

3   Cooperative Ant Colonies to Handle Multiple Objectives

In our algorithm two objectives are optimized simultaneously by coordinating the activities of two ant colonies. The ants of the larger master population optimize the main objective function. The ants of the smaller slave population optimize a minor goal and 'inject' their knowledge into the pheromone information of the main population. The two ant colonies are coordinated by the procedure CooperativeACO.

Table 1. The CooperativeACO procedure

---

procedure CooperativeACO
    for $i := 1$ **to** $max\_Iterations$
      ACO (η^{master}; Γ^{master}; Λ^{master});
      $solution\_vectors\_slave\_pop :=$
      ACO (η^{slave}; Γ^{slave}; Λ^{slave});
      PheromoneInjection to the pheromone information of the master population

---

The procedure CooperativeACO, described in Table 1 initializes the two ant colonies (master population and slave population), handles the communication of the two populations via one global pheromone information and controls the termination of the algorithm. In detail, for a number of $max\_Iterations$ the procedure ACO is executed for the 'master' population as well as the 'slave' population. Both algorithms are called with the parameter settings for the priority rule (η^{master}, η^{slave}), population size (Γ^{master}, Γ^{slave}), and the number of best ants (Λ^{master}, Λ^{slave}) for the pheromone update. After the run of the two ACO Procedures the PheromoneInjection for the master population is executed by using the results ($solution\_vectors\_slave\_pop$), produced by the ACO procedure with the parameter settings for the 'slave' population. The resulting pheromone information is replicated in the local pheromone information of the 'master' population.

Let us now turn to a detailed description of the ACO procedure (subsection 3.1), where we briefly describe the two basic ACO phases, namely the construction of a feasible solution and the trail update for the global pheromone information (in section 3.2).

3.1   The Ant Colony Algorithm

The master and the slave population use the same ACO algorithm to construct feasible solutions. Starting at time $t = 0$ a truck is sequentially filled with orders until the end of the planning horizon T is reached, or no more order assignment is feasible. At this point another vehicle is brought into use, t is set to $t = 0$ and the order assignment is continued. This procedure is repeated until all orders are assigned.

For the selection of orders that have not yet been assigned to trucks, two aspects are taken into account: how promising the choice of that order is in general, and how good the choice of that order was in previous iterations of the algorithm. The first information is the visibility, the second is stored in the pheromone information.

The proposed ant system can be described by the algorithm given in Table 2.

Table 2. The ACO procedure

```
procedure ACO (η; Γ; Λ)
    Initialization of the ACO;
    set a number of ants on each depot;
    for Ant := 1 to Γ
        while not all orders are assigned
            initialize a new truck;
            t = 0;
            select a home base for the truck;
            while ∃ ηᵢⱼ(t) > 0      ∀i, j ∈ J
                select an order using formula (3);
                update t;
            evaluate the objective function;
        for λ := 1 to Λ
            improve the solution using the post optimization procedure
            evaluate the objective function;
        update local pheromone information;
    return solution_vectors;
```

In the initialization phase, Γ ants are generated and each depot is assigned the same number of ants. Then the two basic phases - construction of tours and trail update - are executed for a given number of iterations. To improve the solution quality a post optimization procedure will be applied, which seeks to improve a solution by finding the optimal depot for each truck given the orders assigned. For each truck, all possible depot assignments are considered and the one yielding least costs is chosen. Note however, that the sequence of the orders assigned to each vehicle must not be changed.

Visibility. Let J denote the set of orders and D denote the set of depots. The visibility information is stored in a matrix η, each element in the matrix is denoted by $\eta_{ij}(t)$, where $\eta_{ij}(t)$ is positive, if and only if the assignment of order j after order i is feasible. An assignment of order j is feasible, if the order can be scheduled on the current vehicle without violation of its time window. Hence, it is clear that η depends on the time. Note that in each iteration only the row associated with the order assigned in the previous iteration has to be evaluated. The actual value of the visibility of order j depends on the priority rule incorporated in the algorithm. Based on this information we can define the set $\Omega_i(t) = \{j \in J : j$ is an order feasible to assign$\}$.

It is obvious that the choice of the priority rule substantially influences the solution quality. In our problem at hand, we want to minimize total costs, that means a minimization of both empty vehicle movements, as well as number of trucks required. Therefore, the master population of the Cooperative ACO uses a priority rule, which leads to good solutions with respect to total costs. This priority rule takes into consideration the minimization of the empty vehicle movements only insufficiently. Thus, we introduce a slave population, which uses a priority rule suitable to minimize this goal. The relevant information discovered by the ants of the slave population is inserted in the global pheromone information and can be used by the ants of the master population.

Visibility for the master population. The priority rule for the master population is:

$$\eta_{ij}^{master}(t) = \begin{cases} e^{-4\cdot(EDD_j + 2\cdot EPST_j(i,t))} & \text{if } j \in \Omega_i(t) \\ 0 & \text{otherwise} \end{cases} \quad \forall j \in J, \qquad (1)$$

This priority rule aims to maximize truck utilization by avoiding waiting times. It takes into account the due dates (EDD), as well as the earliest possible pickup times $(EPST)$ of the orders. While the EDD measure exactly represents the due dates, the $EPST$ measure takes into account waiting times and connecting empty vehicle movements. A more detailed description of the priority rule can be found in [12].

Visibility for the slave population. The priority rule for the slave population is:

$$\eta_{ij}^{slave}(t) = \begin{cases} e^{-16\cdot DIST(i,j)} & \text{if } j \in \Omega_i(t) \\ 0 & \text{otherwise} \end{cases} \quad \forall j \in J. \qquad (2)$$

This priority rule is solely based on the distance traveled to get from the delivery location of the last customer assigned (i) to the pickup location of customer j, $DIST(i,j)$. It is obvious, that this priority rule is well suited for the minimization of empty vehicle movements.

Decision rule. Given the visibility and pheromone information, a feasible order j is selected to be visited immediately after order or depot i according to a random-proportional rule that can be stated as follows:

$$\mathcal{P}_{ij}^{\xi}(t) = \begin{cases} \dfrac{[\tau_{ij}^{\xi}]^{\alpha}[\eta_{ij}^{\xi}(t)]^{\beta}}{\sum_{h \in \Omega_i(t)}[\tau_{ih}^{\xi}]^{\alpha}[\eta_{ih}^{\xi}(t)]^{\beta}} & \text{if } j \in \Omega_i(t) \\ \\ 0 & \text{otherwise,} \end{cases} \quad \forall j \in J, \xi \in \{master, slave\}. \quad (3)$$

This probability distribution is biased by the parameters α and β that determine the relative influence of the trails and the visibility, respectively. τ_{ij}, represents the current pheromone information, i.e. the value τ_{ij} represents the pheromone information of assigning order j immediately after order i.

3.2   Pheromone Information

Pheromone Update. After the two ant populations have constructed a feasible solution, the global pheromone trails are updated. We use a pheromone update procedure, where only a number of the best ants, ranked according to solution quality, contribute to the pheromone trails. Such a procedure was proposed in [1]. The update rule is as follows:

$$\tau_{ij}^{\xi} = \rho \cdot \tau_{ij}^{\xi} + \sum_{\lambda=1}^{\Lambda^{\xi}} \Delta\tau_{ij}^{\xi,\lambda} \text{ , where } \xi \in \{master, slave\} \quad \forall i,j \in J, \qquad (4)$$

where ρ is the trail persistence (with $0 \leq \rho \leq 1$). Only the Λ^{master} best ants of the master population and the Λ^{slave} best ants of the slave population update the pheromone information. If an order j was performed immediately after an order i in the solution of the λ-th best ant of Λ^{master} or of Λ^{slave} the pheromone trail is increased by a quantity $\Delta\tau_{ij}^{\lambda}$. This update quantity can be represented as

$$\Delta\tau_{ij}^{\xi,\lambda} = \begin{cases} 1 - \frac{\lambda-1}{\Lambda^{\xi}} \text{ if } 1 \leq \lambda \leq \Lambda^{\xi}, \text{where } \xi \in \{master, slave\} \\ \\ 0 \quad\quad \text{otherwise} \end{cases} \quad \forall i,j \in J. \quad (5)$$

Pheromone Injection. After both ACO procedures have been executed the pheromone injection is performed in order to communicate good solutions from the slave to the master population. Formally this can be written as

$$\tau_{ij}^{master} = \tau_{ij}^{master} + \sum_{\lambda=1}^{\Lambda^{slave}} \Delta\tau_{ij}^{slave,\lambda}. \qquad (6)$$

4   Numerical Analysis

In this section we will present the results of our numerical analysis. We generated a set of test problems with a network of 8 distribution centers, 512 transportation orders and a planning horizon of 8 periods. Given these settings, the problems differ with respect to the average time window lengths associated with the orders. These average time window lengths were varied between 1 and 8, 1 meaning that every order has to be delivered within one period, while 8 means that no time window restrictions have to be respected. The global objective, which is also the objective function for the 'master' population is given by

$$TC = 20 \cdot FS + 1 \cdot MC,$$

where TC denotes the total costs, FS is the fleet size and MC denotes the vehicle movement costs. An interpretation of the weights used in this objective function can be found in [12]. The objective function for the 'slave' population is given by

$$VC = 1 \cdot MC,$$

where VC denotes the variable vehicle movement costs. In this case the population does not take the fleet size into account when evaluating solutions. The following parameter setting was chosen for the Ant System:

$$\alpha = 1, \ \beta = 1, \ \rho = 0.5, \ \tau_0 = 0.1, \ \Lambda^{master} = 8, \ \Lambda^{slave} = 1, \ \Gamma^{master} = 128,$$
$$\Gamma^{slave} = 32 \text{ and } maxIterations = 30.$$

Let us now turn to the analysis of our cooperative ant colonies. Note, that the results presented in this section are based on averages over 40 runs for each problem. In order to show the effects of the use of the memory in general, and information sharing in particular we compare three different cases.

- Case 1: 4800 randomized greedy solutions
- Case 2: 1 master population of 160 ants, no slave population
- Case 3: Cooperative ant colonies: 1 master population with 128 ants, 1 slave population with 32 ants, slave population reports good solutions to master population

Case 1 represents a guided stochastic search where the solutions are built using only the heuristic information given in 1. The solutions are completely independent of each other as no memory is utilized. Case 2 represents a basic ACO algorithm where one population of ants searches the solution space. Each ant in the colony utilizes the same heuristic information, which in our case is the one presented for the master population in the last section. The ants communicate via the pheromone information. Finally, Case 3 represents our new approach which is based on a cooperative system of two ant colonies as presented in the last section. In Table 3 these three cases are compared with respect to empty vehicle movements and fleet sizes. By comparing Cases 2 and 3 with Case 1 it can be clearly seen, that the use of a memory leads to significantly better results than a simple guided stochastic search.

Let us now turn to the much more interesting question whether information sharing between populations is useful. The results shown in the table suggest that our Cooperative Ant Colony approach outperforms the basic Ant System. Except for one case (the problem with an average time window length of 4 periods) our new approach always finds lower total costs than the more traditional Ant System. If we split these costs up, we can analyze these results with respect to fleet sizes and empty vehicle movements. Let us first consider empty vehicle movements. Table 3 shows, that our new approach always finds less empty vehicle movements. The average improvement in empty vehicle movements is 2.55%. Thus, there is always a positive influence of information sharing. If we take a look at the fleet sizes, we see that apart from one problem (the problem with an average time window length of 4 periods) this improvement of the vehicle movements is achieved without detrimental effects on the fleet size. On the contrary, for three problems, those with average time windows of 1, 6 and 7 periods respectively, our new approach even improves the necessary fleet size. Furthermore, for two problems, namely the problems with time window lengths of 1 and 7 periods respectively, the best results found with our new approach utilize one truck less than the best solutions associated with the basic Ant System. Furthermore, the variability in the results of our new approach is generally smaller than the variability in the results of the basic Ant approach.

Note, that while the improvements seem to be rather small, the presented results are based on first simulations. Thus, we strongly expect that parameter fine tuning or slight modifications could possibly lead to an increase in these improvements.

Table 3. Solution comparison between several advanced Ant System algorithms

Length of Time Windows	Case 1			Case 2			Case 3		
	a	b	c	a	b	c	a	b	c
1	809.389	62.093	29.95	750.658	42.362	28	748.775	41.479	27.95
2	742.504	53.708	27.025	680.264	31.968	25	679.969	31.673	25
3	777.839	56.044	28.675	701.506	33.210	26	700.657	32.361	26
4	744.139	55.844	27	672.297	32.001	24.6	674.806	30.510	24.8
5	719.449	51.153	26	632.908	24.612	23	632.306	24.010	23
6	707.477	51.181	25.4	619.145	24.349	22.325	616.455	23.659	22.225
7	698.046	49.250	25.025	609.003	20.707	22	607.961	20.165	21.975
8	695.278	47.483	24.975	586.872	18.576	21	586.442	18.146	21

a: Total costs   b: Empty vehicle movements   c: Fleet size

5   Conclusions and Future Research

In this paper we have proposed an ACO approach, where two ant colonies cooperatively solve a multi-objective transportation problem. The objective function was to minimize total costs, consisting of fixed costs for the utilization of the fleet and variable costs for the transportation movements. In our algorithm a large population of ants solves the problem with respect to the fleet size costs, as these costs are the major component in the objective function. This large population is supported by a small population, which aims to minimize total vehicle movement costs and communicates good solutions to the larger master population.

Our results can be viewed as a proof of concept for the proposed method. We showed, that the communication of the two ant colonies improved the solution quality. While these improvements are not very large, they highlight the potential of communication between colonies with different problem solving approaches.

Future research will deal with improvements of this concept, as well as other communication mechanisms which enhance the performance of agent based optimization algorithms.

Acknowledgments. The authors are grateful for financial support from the Austrian Science Foundation (FWF) under grant SFB #010 'Adaptive Information Systems and Modeling in Economics and Management Science' and from the Oesterreichische Nationalbank (OENB) under grant #8630.

References

1. Bullnheimer, B., Hartl, R. F. and Strauss, Ch.: An improved ant system algorithm for the vehicle routing problem. Annals of Operations Research **89** (1999) 319–328
2. Costa, D. and Hertz, A.: Ants can colour graphs. Journal of the Operational Research Society **48**(3) (1997) 295–305
3. Dorigo, M. and Gambardella, L. M.: Ant Colony System: A cooperative learning approach to the Travelling Salesman Problem. IEEE Transactions on Evolutionary Computation **1**(1) (1997) 53–66
4. Dorigo, M., Di Caro, G. and Gambardella, L. M.: Ant Algorithms for Discrete Optimization. Artificial Life **5**(2) (1999) 137-172
5. Stützle, T. and Dorigo, M.: ACO Algorithms for the Quadratic Assignment Problem. In: Corne, D., Dorigo, M. and Glover, F. (Eds.): New Ideas in Optimization. Mc Graw-Hill, London (1999)
6. Colorni, A., Dorigo, M. and Maniezzo, V.: Distributed Optimization by Ant Colonies. In: Varela, F. and Bourgine, P. (Eds.): Proc. Europ. Conf. Artificial Life. Elsevier, Amsterdam (1991)
7. Dorigo, M.: Optimization, Learning and Natural Algorithms. Doctoral Dissertation. Politecnico di Milano, Italy (1992)
8. Dorigo, M., Maniezzo, V. and Colorni, A.: Ant System: Optimization by a Colony of Cooperating Agents. IEEE Transactions on Systems, Man and Cybernetics **26**(1) (1996) 29–41
9. Gutjahr, W. J.: A graph-based Ant System and its convergence. Future Generation Computing Systems. **16** (2000) 873–888
10. Gambardella, L. M., Taillard, E. and Agazzi, G.: MACS-VRPTW: A Multiple Ant Colony System for Vehicle Routing Problems with Time Windows. In: Corne, D., Dorigo, M. and Glover, F. (Eds.): New Ideas in Optimization. McGraw-Hill, London (1999)
11. Irnich, St.: A Multi-Depot Pickup and Delivery Problem with a Single Hub and Heterogeneous Vehicles. European Journal of Operational Research **122**(2) (2000) 310-328
12. Doerner, K. F., Gronalt, M., Hartl, R.F., and Reimann, M.: Optimizing Pickup and Delivery Operations in a Hub Network with Ant Systems. POM Working Paper 07/2000

An ANTS Algorithm for Optimizing the Materialization of Fragmented Views in Data Warehouses: Preliminary Results

Vittorio Maniezzo[1], Antonella Carbonaro[1],
Matteo Golfarelli[2], and Stefano Rizzi[2]

[1] Department of Computer Science, University of Bologna, Italy
{maniezzo, carbonar}@csr.unibo.it
[2] DEIS, University of Bologna, Italy
{mgolfarelli, srizzi}@deis.unibo.it

Abstract. The materialization of fragmented views in data warehouses has the objective of improving the system response time for a given workload. It represents a combinatorial optimization problem arising in the logical design of data warehouses which has so far received little attention from the optimization community. This paper describes the application of a metaheuristic approach, namely the ANTS approach, to this problem. In particular, we propose an integer programming formulation of the problem, derive an efficient lower bound and embed it in an ANTS algorithm. Preliminary computational results, obtained on the well-known TPC-D benchmark, are presented.

1 Introduction

Data warehouses are enjoying increasing market success being foremost systems for companies willing to improve the support given to decision processes and data analysis procedures. A data warehouse enables the executives to retrieve summary data, derived by "cleaning" and integrating those present in operational information systems; primary issues are flexible query interface and fast query response.

The design of a data warehouse starts with the identification of relevant data in the company information system; these data must be integrated, reorganized in a multidimensional fashion and possibly aggregated in order to be of effective use. After that, conceptual, logical and physical design phases are encompassed. The problem addressed in this paper belongs to logical design and has the objective of minimizing the query response time by reducing the number of disk pages to be accessed. This may be obtained by defining appropriate tables of aggregated data (*views*) and by including in them only the data which are actually requested by some query.

Storing aggregated data obviously leads to redundancy, thus generating a trade-off between effectiveness and amount of memory to be allocated. The algorithm presented in this paper directly addresses the optimization of this trade-off, when working on real-world large-scale data repositories. To the best of our

E.J.W. Boers et al. (Eds.) EvoWorkshop 2001, LNCS 2037, pp. 80–89, 2001.

knowledge, no algorithmic solution has been presented so far in the literature for the problem of interest, which goes under different names, such as the problem of materializing fragmented views, vertical fragmentation problem or vertical partitioning problem. The problem has been described in [1], where no optimization algorithm is proposed. In [2], a related problem is described, aimed at building data indices to enhance performance in parallel implementations of data warehouses. In [3] the problem is formalized and a branch-and-bound approach is devised.

This paper is structured as follows. In Section 2 we introduce the necessary background on data warehouses with reference to logical design. In Section 3 we define the vertical fragmentation problem (VFP) and propose a mathematical formalization of VFP and a possible linear relaxation of the formulation leading to an effective polynomial-time lower bound. Section 4 reviews the ANTS approach and describes the adaptation of ANTS to the VFP, while Section 5 presents preliminary computational results obtained on the TPC-D benchmark. Finally, Section 6 concludes the paper.

2   Background

The description of a data warehouse must start from the adoption of a suitable data modeling language. The most widely accepted one is the multidimensional modeling technique [4], which denotes data by means of an n-dimensional (hyper)cube, where each *dimension* corresponds to a characteristic of the data. Each element of the cube is usually characterized by quantitative attributes, called *measures*, which are computed from the operational information system. Furthermore, each dimension is related to a set of *attributes* defining a hierarchy of aggregation levels. Elements of the cube can then be aggregated along these hierarchies, in order to retrieve summary values for measures.

For example, a 3-dimensional cube with dimensions Store, Product and Date might represent the sales in a chain store; the measures could be Quantity and Revenue. In this case, for each product, each element of the cube would measure the quantity sold in one store in one day and the corresponding revenue. An interesting aggregation could be that computing the total monthly revenue for each category of products.

The design of a data warehouse goes thorough successive phases [5], among which are conceptual design, logical design and physical design. The objective of logical design, relevant for this paper, is the minimization of query response time. This is obtained by pre-defining a set of queries, called *workload*, that the system is likely to have to answer more often. This is possible on the one hand, because the user typically knows in advance which kind of data analysis will be carried out more often for decisional or statistical purposes, and on the other hand, because a substantial amount of queries are aimed at extracting summary data to fill standard reports. Actually, taking into account all possible queries is computationally infeasible, but it is possible to identify a reduced set of significant and frequent queries which are considered to be representative of the actual workload.

More formally, a cube f is a 4-tuple $< Patt(f), Meas(f), Attr(f), R >$, where:

- $Patt(f)$ is a set of *dimensions*;
- $Meas(f)$ is a set of *measures*;
- $Attr(f)$ is a set of *attributes* (being the dimensions particular attributes, we have $Patt(f) \subseteq Attr(f)$);
- R is a set of functional dependencies $a_i \rightarrow a_j$ defined between pairs of attributes in $Attr(f)$, where $a_i \rightarrow a_j$ denotes both the case in which a_i directly determines a_j and the case in which a_i transitively determines a_j.

Obviously every attribute which is not a dimension itself must be derivable from a dimension, that is $\forall a_j \in Attr(f) \setminus Patt(f) \ (\exists a_i \in Patt(f); a_i \rightarrow a_j)$.

In the TPC-D benchmark [6], which consists of a database of orders issued to a company, one of the cubes of interest represents order line items; it is named *LineItem* and is defined by:

$Patt(LineItem) = \{$Part, Supplier, Order, ShipDate, ShipMode, ReturnFlag, ReceiptDate, CommitDate, Status$\}$,

$Meas(LineItem) = \{$UnitPrice, Qty, ExtPrice, Discount, DiscPrice, Charge, Tax$\}$

and by attributes with specific functional dependencies.

Given a cube f, an *aggregation pattern* (or simply a *pattern*) on f is a set p, $p \subseteq Attr(f)$, such that no functional dependency exists between any pair of attributes in p: $\forall a_i \in p (\not\exists a_j \in p; a_i \rightarrow a_j)$.

With reference to the *LineItem* cube, examples of patterns are $Patt(f)$, $\{$Part, OMonth, SNation$\}$, $\{$Brand, Type$\}$, $\{\}$.

Let p_i and p_j be two patterns $(p_i \neq p_j)$; p_i is *coarser* than p_j $(p_i < p_j)$ if every element in p_i is either also in p_j or is functionally dependent on some element in p_j. For example, $\{$Brand, CRegion$\} < \{$Brand, Customer, Supplier$\}$.

A query q on a cube f is characterized by the pattern $Patt(q)$ on which data must be aggregated and by the measures $Meas(q)$ required in output. Part of the queries the user formulates may require comparing measures taken from distinct cubes; in the OLAP terminology, these are called *drill-across queries*. Intuitively, a drill-across query can be formulated on cubes sharing one or more attributes. Consider for instance the *PartSupplier* cube characterized by:

$$Patt(PartSupplier) = \{$Part, Supplier, Date$\},$$
$$Meas(PartSupplier) = \{$AvailQty, SupplyCost$\}$$

A possible drill-across query is the one comparing the total available quantity and the total quantity sold for each part, characterized by $Patt(q) = \{$Part$\}$, $Meas(q) = \{$AvailQty, Qty$\}$.

Given a cube f, each pattern on f determines a possible view to be materialized. Given a workload expressed as a set of queries, we will call *candidate views* those being potentially useful to reduce the workload execution cost [7]. Let $Cand(f)$ be the set of the candidate views for cube f; each $v \in Cand(f)$ is defined by its pattern $Patt(v)$. For each cube f, the view at pattern $Patt(f)$ is always a candidate. We will denote with P the set of patterns of all the candidate views on all the cubes involved in the workload.

3   The Vertical Fragmentation Problem

In presence of a memory constraint, which requires to use a maximum memory size, only a subset of the candidate views can be actually materialized. Thus, several techniques have been proposed to select the subset to be materialized in order to optimize the response to the workload (e.g, [8], [9]). All the approaches in the literature store, for each view $v \in Cand(f)$, all the measures in $Meas(f)$. In this paper, we evaluate how the solution can be further optimized by materializing views in *fragments* including measures requested together by at least one query. In fact, some queries on f may require a subset of $Meas(f)$; thus, it may be worth materializing fragments including only a subset of $Meas(f)$ (*partitioning*). On the other hand, the access costs for drill-across queries may be decreased by materializing fragments which include measures taken from different cubes (*unification*).

With the term *fragmentation* we denote both partitioning and unification of views. The approach we propose in this paper is aimed at determining an optimal set of fragments to materialize from the candidate views.

A fragment v is useful to solve query q iff $Patt(q) \leq Patt(v)$ and $Meas(q) \cap Meas(v) \neq \emptyset$. If several fragments are necessary to retrieve all the measures in $Meas(q)$, they must be aggregated on $Patt(q)$ and then joined.

In order to specify objective and constraints of fragmentation, some further notation must be introduced.

Given a cube f and a workload Q, it is possible to partition the measures $Meas(f)$ into subsets (*minterms*) such that all the measures in a minterm are requested together by at least one query in Q and do not appear separately in any other query in Q. We call *terms* the sets of measures obtained as the union of any combination of minterms, even from different cubes (of course, all minterms are also terms). We denote with T the set of all terms.

The fragmentation problem can now be modeled over a *fragmentation array* $\Xi = [x_{ijk}]$, which is a tridimensional array of 0-1 binary variables whose dimensions correspond to the queries $q_i \in Q$, to the patterns $p_j \in P$ and to the terms $t_k \in T$, respectively. Each cell of the array corresponds to a fragment candidate to materialization; setting $x_{ijk} = 1$ means stating that query q_i will be answered accessing (also) the fragment defined by the measures in t_k and pattern p_j.

A value assignment for variables x_{ijk} is feasible if:

1. for every query, each measure required is obtained by one and only one fragment;
2. for every pattern, each measure is contained in one and only one fragment.

The objective function to minimize is based on the number of disk pages to access in order to satisfy the workload.

3.1   Mathematical Formulation

Problem VFP can be formulated as follows. Let $\mathcal{Q}$ be the index set of the queries in the workload and $\mathcal{P}$ the index set of the patterns in P. For every query $i \in \mathcal{Q}$, $\mathcal{P}_i$ denotes the subset of $\mathcal{P}$ containing the indices of all patterns p_j which are useful to solve query q_i $(Patt(q_i) \leq p_j)$ and for which $p_j = Patt(v)$, where v is a candidate view for at least a cube f involved in query i, i.e., $v \in Cand(f)$.

The index set $\mathcal{T}$ contains the indices of the terms in T; we will further denote by $\mathcal{T}_i$ the subset of indices of the terms which contains at least one measure in $Meas(q_i)$ $(i \in \mathcal{Q})$.

Problem VFP asks to minimize the workload execution cost, subject to a number of constraints. The cost is computed as the sum of the costs c_{ijk} of obtaining, for each query $i \in \mathcal{Q}$, the relevant term $k \in \mathcal{T}_i$ from pattern $j \in \mathcal{P}_i$.

Let x_{ijk} be a 0-1 variable which is equal to 1 if and only if query i is executed on pattern j to get the term k. Let y_{jk} be a 0-1 variable which is equal to 1 if and only if the pattern j is used to get the term k, in which case an amount b_{jk} of disk space out of the maximum available space amount B is needed. Problem VFP is then as follows.

$$(VFP) \quad z(VFP) = Min \quad \sum_{i \in \mathcal{Q}} \sum_{j \in \mathcal{P}_i} \sum_{k \in \mathcal{T}_i} c_{ijk} x_{ijk} \tag{1}$$

$$s.t. \quad \sum_{j \in \mathcal{P}_i} \sum_{k \in \mathcal{T}_i} x_{ijk} = 1 \quad i \in \mathcal{Q} \tag{2}$$

$$\sum_{k \in \mathcal{T}} y_{jk} \leq 1 \quad j \in \mathcal{P} \tag{3}$$

$$x_{ijk} \leq y_{jk} \quad i \in \mathcal{Q}, j \in \mathcal{P}, k \in \mathcal{T}_i \tag{4}$$

$$\sum_{\substack{j \in \mathcal{P} \\ k \in \mathcal{T}}} b_{jk} y_{jk} \leq B \tag{5}$$

$$x_{ijk} \in \{0, 1\} \quad i \in \mathcal{Q}, j \in \mathcal{P}, k \in \mathcal{T} \tag{6}$$

$$y_{jk} \in \{0, 1\} \quad j \in \mathcal{P}, k \in \mathcal{T} \tag{7}$$

Equations (2) impose that each measure specified in a query must be obtained by one and only one pattern (thus, implicitly, that each query in the workload must be satisfied); inequalities (3) require that, in each pattern, a measure can belong to only one term; inequalities (4) link the x and y variables and inequality (5) is the memory knapsack constraint. Finally, constraints (6) and (7) are the integrality constraints.

By a linear relaxation of integrality constraints we get problem LVFP whose optimal solution value $z(LVFP)$ constitutes a lower bound to $z(VFP)$.

Remark . Consider a problem VFP' which is obtained from VFP by fixing to 1 some x_{ijk} variable. Since fixing a x_{ijk} entails fixing to one the corresponding y_{jk} variable due to constraints (4), and since fixing to 1 a x_{ijk} entails fixing to 0 all variables appearing with it in the relevant constraints (2), it follows that VFP' is a subproblem of VFP having less variables (all the fixed ones can be disposed of) and less constraints (all those defined only over the expunged variables). Moreover, the value of B in constraint (5) is decreased by the amount corresponding to the sum of the b_{jk} of the y_{jk} variables fixed to 1. The partial solution PS will have a cost $z(PS)$ due to the fixed x_{ijk} variables. A *lower bound* to the cost of the best solution S containing PS is obviously $z'(S) = z(PS) + z(VFP')$. It is easy to notice that $z(VFP')$ can be approximated from below by adding the optimal values of the dual variables associated to the constraints of problem LVFP maintained in problem VFP'.

4   The ANTS Metaheuristic

ANTS [10] is a technique to be framed within the Ant Colony Optimization (ACO) class, whose first member called Ant System was initially proposed by Colorni, Dorigo and Maniezzo [11]. The main underlying idea of all ACO algorithms is that of parallelizing search over several constructive computational threads, all based on a dynamic memory structure incorporating information on the effectiveness of previously obtained results and in which the behavior of each single agent is inspired by the behavior of real ants.

The collective behavior emerging from the interaction of the different search threads has proved effective in solving combinatorial optimization problems.

An *ant* is defined to be a simple computational agent, which iteratively constructs a solution for the problem to solve. Partial problem solutions are seen as *states*; each ant *moves* from a state ι to another one ψ, corresponding to a more complete partial solution. At each step σ, each ant k computes a set $A_k^\sigma(\iota)$ of feasible expansions to its current state, and moves to one of these according to a probability distribution specified as follows.

For ant k, the probability $p_{\iota\psi}^k$ of moving from state ι to state ψ depends on the combination of two values:

1. the attractiveness $\eta_{\iota\psi}$ of the move, as computed by some heuristic indicating the *a priori* desirability of that move;
2. the trail level $\tau_{\iota\psi}$ of the move, indicating how proficient it has been in the past to make that particular move: it represents therefore an *a posteriori* indication of the desirability of that move.

In ANTS, the *attractiveness* of a move is estimated by means of lower bounds (upper bounds in case of maximization problems) to the cost of the completion of a partial solution. In fact, if a state ι corresponds to a partial problem solution it is possible to compute a lower bound to the cost of a complete solution containing ι. Therefore, for each feasible move $(\iota\psi)$, it is possible to compute the lower bound to the cost of a complete solution containing ψ: the lower the bound the better the move. The use of LP bounds is a very effective and straightforward general policy, whenever tight such bounds have been identified for the problem to solve.

Trails are updated when all ants have completed a solution, increasing or decreasing the level of trails corresponding to moves that were part of "good" or "bad" solutions, respectively.

The specific formula for defining the probability distribution of moving from a state to another one makes use of a set $tabu_k$ which indicates a problem-dependent set of infeasible moves for ant k. Different authors use different formulae, but according to the ANTS approach [10] probabilities are computed as follows: $p_{i\psi}^k$ is equal to 0 for all moves which are infeasible (i.e., they are in the tabu list), otherwise it is computed by means of formula (8), where α is a user-defined parameter $(0 \leq \alpha \leq 1)$.

$$p_{i\psi}^k = \frac{\alpha \cdot \tau_{i\psi} + (1-\alpha) \cdot \eta_{i\psi}}{\Sigma_{(iv) \notin tabu_k} (\alpha \cdot \tau_{iv} + (1-\alpha) \cdot \eta_{iv})} \tag{8}$$

Parameter α defines the relative importance of trail with respect to attractiveness. After each iteration t of the algorithm, that is when all ants have completed a solution, trails are updated following formula (9).

$$\tau_{i\psi}(t) = \tau_{i\psi}(t-1) + \Delta\tau_{i\psi} \tag{9}$$

where $\Delta\tau_{i\psi}$ represents the sum of the contributions of all ants that used move $(i\psi)$ to construct their solution. The ants' contributions are proportional to the quality of the achieved solutions , i.e., the better an ant solution, the higher will be the trail contribution added to the moves it used.

In ANTS, the trail updating procedure evaluates each solution against the last k ones globally constructed by ANTS. As soon as k solutions are available, their moving average $\overline{z}$ is computed; each new solution z_{curr} is compared with $\overline{z}$ (and then used to compute the new moving average value). If z_{curr} is lower than $\overline{z}$, the trail level of the last solution's moves is increased, otherwise it is decreased. Formula (10) specifies how this is implemented:

$$\Delta\tau_{i\psi} = \tau_0 \cdot (1 - \frac{z_{curr} - LB}{\overline{z} - LB}) \tag{10}$$

where $\overline{z}$ is the average of the last k solutions and LB is a lower bound to the optimal problem solution cost.

Based on the described elements, the ANTS metaheuristic is the following.

The metaheuristic just introduced must be specified to make it an algorithm, that is a heuristic procedure for problem VFP. The only element to define is the lower bound to use as an estimate of the attractiveness of a move.

The lower bound used was the linear time approximation of the dual solution introduced in remark 1. That is, at every step we compute the cost of the partial solution so far constructed and we remove from the mathematical representation of the problem all constraints of type (2) and (4) which are saturated by the incumbent solution and all variables which cannot belong to any feasible solution due to those already fixed. The lower bound is obtained as the sum of all dual variables associated with the remaining constraints, with the values computed in the optimal solution of the linear relaxation of the whole problem.

5   Computational Results

The ANTS algorithm described in Section 4 has been coded in Microsoft Visual C++ and run on a Pentium III, 733 MHz machine working under Windows 98. As a linear programming solver, in order to compute the lower bounds we used CPLEX 6.6. The test set has been obtained from the TPC-D benchmark [6],

ANTS algorithm

1. (Initialization)
   Compute a (linear) lower bound LB to the problem to solve.
   Initialize $\tau_{\iota\psi}$, $\forall(\iota, \psi)$.

2. (Construction)
   For each ant k **do**
      repeat
         compute $\eta_{\iota\psi}$, $\forall(\iota, \psi)$, as a lower bound to the cost of a complete
         solution containing ψ.
         choose the state to move to, with probability given by (8).
         append the chosen move to the k-th ant's set $tabu_k$.
      until ant k has completed its solution.
   end for.

3. (Trail update)
   For each ant move $(\iota\psi)$ **do**
      compute $\Delta\tau_{\iota\psi}$.
      update the trail matrix by means of (9) and (10).
   end for.

4. (Terminating condition)
   If not(end-test) go to step 2.

Fig. 1. Pseudo code for the ANTS algorithm

which is a standard in the data warehousing field. The benchmark is defined on a 1 Gb sized database composed by 3 star schemes and contains data about items sold by a company. Since the standard TPC-D contains only 17 queries, to generate more challenging instances we added 13 queries structurally similar to the already present ones, as already proposed in [3]. The set of candidate views is obtained by means of the approach proposed in [7]. ANTS allowed us to select the subset of the fragments to be materialized, optimizing the query response time by reducing the number of disk pages to be accessed under the given memory constraint. In order to evaluate the algorithm effectiveness, we have defined a number of instances, all derived from the TPC-D by randomly selecting a progressively greater subset of the 40 queries.

Experimental runs are still under way. Table 1 shows the preliminary results obtained so far. The table columns show:
− the problem name (*prob*), where the number indicates how many queries were used to build the instance;
− the number of constraints (m);
− the number of variables (n);
− the number of constraints which can be removed by a specific preprocessing routine (*reduct*);
− the lower bound $z(LVFP)$ (*lvfp*);
− the cpu time to compute $z(LVFP)$ ($t\ lvfp$);

– the percentual deviation from $z(LVFP)$ of the upper bound (zub);
– the cpu time to compute the ANTS upper bound $(t\ zub)$.

Problems VFP3 and VFP5 are small-sized problems that we used to fine tune the algorithm elements. For all other problem dimensions we present three instances, which were obtained by randomly selecting the specified number of queries out of the possible 40. Notice the high variability of difficulty deriving from different query sets.

On the small instances, ANTS was able to identify the optimal solution, as testified by the fact that the lower bound has a cost equal to that of the best solution found by ANTS. On bigger instances, the distance between z(LVFP) and the best solution cost found by ANTS increases with the problem size: on those instances more CPU time than the 30 minutes allowed in this test is needed to get good quality results.

Table 1. ANTS results on the set of VFP test problems

prob	m	n	reduct	lvfp	t lvfp	zub	t zub
VFP3	94	76	20	50475.0	0.05	0.00	7.34
VFP5	729	704	34	281816.2	0.22	2.18	1138.22
VFP10A	4780	5338	245	65190.0	0.66	0.00	568.37
VFP10B	360	358	5	33976.0	0.06	0.00	1367.23
VFP10C	592	512	97	116463.0	0.11	0.00	4.23
VFP15A	4962	5528	242	67971.0	0.82	0.12	1046.24
VFP15B	2365	2544	108	132775.7	0.39	0.72	384.38
VFP15C	6410	7118	317	286063.2	4.67	2.03	183.59
VFP20A	9466	10397	405	128094.4	10.77	3.28	428.39
VFP20B	7745	8285	326	165426.2	2.41	5.92	1634.53
VFP20C	36854	40108	1358	186915.7	100.46	10.77	823.48
VFP25A	25855	28121	874	173367.4	10.99	6.57	1772.16
VFP25B	8182	8766	314	169133.6	2.47	19.27	1003.80
VFP25C	57964	62733	1965	145471.3	254.90	33.98	204.55
VFP30A	44758	48133	1339	237247.4	118.36	22.00	1538.23
VFP30B	62973	67484	1735	208801.4	343.29	35.76	704.86
VFP30C	75489	81026	2206	178171.3	625.93	43.88	839.11

6   Conclusions

In this paper we presented a preliminary report about the use of a state-of-the-art metaheuristic approach for optimizing the materialization of fragmented views in data warehouses. We have shown how the problem is amenable to mathematical programming formalization and how efficient lower bounds can be derived.

The lower bound has been embedded in an ANTS framework, obtaining a heuristic approach for solving the materialization (or vertical fragmentation) problem. Preliminary computational results on standard problem test sets from the literature confirm that the proposed approach is promising.

However, further work need to be accomplished. Specifically, implementation details, such as the use of efficient data structures and the definition of a well-tuned local optimization procedure, still have to be defined and possible different mathematical formulations should be studied.

References

1. D. Munneke, K. Wahlstrom, and M. Mohania. Fragmentation of multidimensional databases. In *Proc. 10th Australasian Database Conf.*, pages 153-64, 1999.
2. A. Datta, B. Moon, and H. Thomas. A case for parallelism in data warehousing and OLAP. In *Proc. IEEE First Int. Workshop on Data Warehouse Design and OLAP Technology*, 1998.
3. M. Golfarelli, D. Maio, and Rizzi S. Applying vertical fragmentation techniques in logical design of multidimensional databases. In *Proceedings of 2nd International Conference on Data Warehousing and Knowledge Discovery (DaWaK 2000)*, pages 11-23, 2000.
4. R. Kimball. *The data warehouse toolkit.* John Wiley & Sons, 1996.
5. M. Golfarelli and S. Rizzi. Designing the data warehouse: key steps and crucial issues. *J. of Computer Science and Information Management*, 2(3), 1999.
6. F. Raab, editor. *TPC Benchmark(tm) D (Decision Support), Proposed Revision 1.0.* Transaction Processing Performance Council, San Jose, 1995.
7. E. Baralis, S. Paraboschi, and E. Teniente. Materialized view selection in multidimensional database. In *Proc. 23rd VLDB*, pages 156-65, 1997.
8. V. Harinarayan, A. Rajaraman, and J. Ullman. Implementing Data Cubes Efficiently. In *Proc. ACM Sigmod Conf.*, 1996.
9. J. Yang, K. Karlaplem, and Q. Li. Algorithms for Materialized View Design in Data Warehousing Environments. In *Proc. 23rd VLDB*, pages 136-45, 1997.
10. V. Maniezzo. Exact and approximate nondeterministic tree-search procedures for the quadratic assignment problem. *INFORMS J. on Computing*, 11(4):358-69, 1999.
11. A. Colorni, M. Dorigo, and V. Maniezzo. Distributed optimization by ant colonies. In *Proc. ECAL'91, European Conference on Artificial Life.* Elsevier, 1991.

A Genetic Algorithm for the Group–Technology Problem

Ingo Meents

IBM Deutschland Speichersysteme GmbH
Hechtsheimer Strasse 2, 55131 Mainz, Germany
`meents@de.ibm.com`

Abstract. The design and production planning of cellular manufacturing systems requires the decomposition of a company's manufacturing assets into cells. The set of machines has to be partitioned into machine-groups and the products have to be partitioned into part-families. Finding the machine-groups and their corresponding part-families leads to the combinatorial problem of simultaneously partitioning those two sets with respect to technological requirements represented by the part-machine incidence matrix. This article presents a new solution approach based on a grouping genetic algorithm enhanced by a heuristic motivated by cluster analysis methods.

1 Introduction

The basic idea of group technology (GT) is to decompose a manufacturing system into smaller subsystems. The subject of the analysis are machines, products (parts), and the part-machine incidence, i.e. the allocation of products to machines. The decomposition process creates subsets of machines (machine-groups) each of which is responsible for the production of a certain subset (part-family) of the products. A machine-group and its corresponding part-family are grouped together in a cell. The problem is to find the cells of a manufacturing system so that the dependencies between different cells are as small as possible.

The implementation of cellular manufacturing systems is supposed to avoid some of the disadvantages of job shop and continuous flow production systems. If a cell contains all machines needed for the production of its part-family, only the raw materials and finished goods have to be transported into or from the cell, respectively. Thus there is less traffic with easier routings on the shop floor in contrast to job shop systems. In comparison with continuous flow lines cellular production systems offer more flexibility and are more likely to compensate disturbances in the production process. In general, cellular manufacturing systems are supposed to offer advantages with respect to lead time, costs, and quality.

Section 2 of the article presents a mathematical formulation of the GT problem followed by a short literature review with special emphasis on modern approaches. A new genetic algorithm (GA) and an associated local search heuristic are presented in Sect. 3. The GA allows to partition the sets of machines and products simultaneously and is able to find the inherent number of cells. A performance analysis of the GA is given in Sect. 4.

E.J.W. Boers et al. (Eds.) EvoWorkshop 2001, LNCS 2037, pp. 90–99, 2001.

2   Problem Formulation

The relationship between the machines and products is modeled by the binary machine–part incidence matrix A. Let m be the number of machines, n the number of products, $M = \{1, \ldots, m\}$ the index set of the machines, $N = \{1, \ldots, n\}$ the index set of the products, and $I\!B = \{0, 1\}$ the set of boolean values. Then the entries of the incidence matrix $A = (a_{i,j})_{i \in M, j \in N} \in I\!B^{m \times n}$ are defined as $a_{ij} = 1$ if machine i processes product j and $a_{ij} = 0$ otherwise. If the machines and products are numbered arbitrarily, the machine-groups and part-families are not directly visible as shown by the sample matrix[1] A in (1). A shows that machine 1 processes parts 2, 4, and 5, machine 2 processes parts 1 and 3, etc.

$$
A = \begin{array}{c} \\ \end{array}
\begin{matrix} p_1\ p_2\ p_3\ p_4\ p_5 \\ \left(\begin{matrix} & 1 & & 1 & 1 \\ 1 & & 1 & & \\ & & 1 & & 1 \\ 1 & & & 1 & \end{matrix} \right) \begin{matrix} m_1 \\ m_2 \\ m_3 \\ m_4 \end{matrix} \end{matrix}
\qquad
A' = \begin{matrix} p_1\ p_3\ p_2\ p_4\ p_5 \\ \left(\begin{matrix} 1 & 1 & & & \\ 1 & 1 & & & \\ \hline & & 1 & 1 & 1 \\ & & 1 & & 1 \end{matrix} \right) \begin{matrix} m_2 \\ m_4 \\ m_1 \\ m_3 \end{matrix} \end{matrix}
\qquad (1)
$$

Under certain conditions the rows and columns of the matrix can be permuted so that cells become visible. These permutations do not change the incidences described by the matrix A if the machine and part indices are permuted together with their rows and columns. Matrix A' in (1) visualizes the blocks almost completely filled with ones along the main diagonal of the matrix. The machine-groups $G_1 = \{m_2, m_4\}$ and $G_2 = \{m_1, m_3\}$ and the part families $F_1 = \{p_1, p_3\}$ and $F_2 = \{p_2, p_4, p_5\}$ form the cells $C_1 = \{G_1, F_1\}$ and $C_2 = \{G_2, F_2\}$.

   In general, matrices do not always allow to permute their columns to form disjunct blocks along the main diagonal. For example, if matrix A' in (1) had another entry indicating that product p_5 has also to be processed on machine m_2, product p_5 would belong to machine-groups G_1 and G_2. Product p_5 is the critical part and is called *bottleneck part*. Machines m_2 and m_1 are called *bottleneck machines* as they are needed by a part belonging to more than one cell.

   The GT problem requires partitioning two sets of machines and products under the restriction that the number of the partitions in both sets have to be equal and that the inter-cell dependencies are minimized. Note that for every pair of partitions of machines and parts a permuted matrix can be derived so that the cells represented by the partitions are arranged along the main diagonal of the matrix.

   The first ideas in the area of GT were presented by Burbidge in the early 1970s. Since then various solution approaches including cluster analysis, sorting algorithms, graph theory, and mathematical programming have been presented. [1] offers a survey on different solution techniques. More recent work on the application of stochastic search algorithms can be found in [2] (simulated annealing) and [3,4,5,6] (genetic algorithms). [7] provides a review on modern solution

---

[1] For clarity, the entries with $a_{ij} = 0$ are shown as blanks.

techniques. An in-depth discussion with special emphasis on genetic algorithms for the GT problem is [8]. [9] covers simple[2] partitioning problems; one of the industrial application examples deals with part-families.

3   Solution Approach Using a Genetic Algorithm

Genetic algorithms are heuristic search methods based on the evolutionary principle of the "survival of the fittest". A population of chromosomes (solutions) evolves over time by repeated phases of variation (recombination, mutation, inversion), evaluation, and selection (usually biased towards those chromosomes with higher fitness) [10].

In order to solve the GT problem by a GA the goal function, a coding for the chromosomes, and the corresponding genetic operators have to be defined among others. Two requirements are important: Firstly, the GA finds the inherent number of cells, i.e. the number of cells is variable. Secondly, the machines and parts are to be partitioned simultaneously.

Moreover, problem-independent techniques like selection procedures and termination criteria have to be specified.

3.1   Goal Function

In order to evaluate different chromosomes of a population a measure is needed that is at least partially ordered. Many papers on GT use a measure called *grouping efficiency*, for example [11]. Defining[3]

$$\alpha = \text{number of block elements with } a_{ij} = 0$$
$$\beta = \text{number of block elements with } a_{ij} = 1$$
$$\gamma = \text{number of non-block elements with } a_{ij} = 0$$
$$\delta = \text{number of non-block elements with } a_{ij} = 1$$

the measure that has to be minimized is calculated by

$$\eta = q \frac{\alpha}{\alpha + \beta} + (1 - q) \frac{\delta}{\gamma + \delta}, \qquad \text{where } 0 \leq q \leq 1 \ . \tag{2}$$

Thus η reflects the homogenity of the cells, i.e. it decreases with a decreasing number of zeroes inside the blocks along the main diagonal and with a decreasing number of ones outside those blocks. The weight factor q determines whether the utilization within the cell or the avoidance of bottlenecks is more important. Small values of q lead to few large inhomogeneous blocks whereas values close to 1 lead to many small homogeneous blocks. From the definition follows that $0 \leq \eta \leq 1$, but normally the grouping efficiency has an upper bound strictly less than 1. This upper bound is dependent on the incidence matrix.

---

[2] The term *simple* refers to partitioning problems with one set of objects (which still are NP-hard) in contrast to the two sets of the GT problem.

[3] Note that α and β as well as γ and δ are swapped in contrast to the original definition in order to get a minimization problem.

The sample matrix A' in (1) yields $\alpha = 1, \beta = 9, \gamma = 10, \delta = 0$ and by setting $q = 0.5$ the grouping efficiency sums up to $\eta = 0.05$. To get strictly positive goal function values in a larger range of values $\eta' = 1000 \cdot (1 + \eta)$ is used for the GA.

3.2 Chromosome Representation

One of the main topics in the design of a GA for a problem is a suitable encoding of a problem solution in a chromosome. Different possibilities of coding a solution to the GT problem are group-number encoding, permutation lists with delimiters, and permutation lists without delimiters but with a greedy decoding heuristic and the grouping GA. For an in-depth discussion of these representations for simple partitioning problems see [10], and the extensions for the GT problem can be found in [8].

The coding for the GT problem can be derived from the codings for simple partitioning problems. However, special requirements for the GT problem — the simultaneous partitioning of two object sets and the variable number of partitions — have to be accounted for. Combining all machines and parts into one large set leads to problems if a group is assigned some machines but no products. This case arises especially if both the difference between m and n and the number of groups k are large. In this case the block diagonal matrix is degenerate, see matrix A_1 in (3). Moreover, some solutions may contain machines that do not have an associated product within a block (or the other way round). In this case $a_{ij} = 0$ for a row i or a column j, e. g. matrices A_2 and A_3 in (3).

$$A_1 = \begin{pmatrix} \boxed{\begin{matrix} 1 & 1 \\ 1 & 1 \end{matrix}} & & \\ & \begin{matrix} 1 & 0 & 1 \\ 0 & 0 & 0 \\ & \boxed{1} \end{matrix} & \cdots \\ & & \ddots \end{pmatrix}, A_2 = \begin{pmatrix} \ddots & & \\ & \boxed{\begin{matrix} 1 & 0 & 1 \\ 0 & 0 & 0 \\ 1 & 1 & 1 \end{matrix}} & \\ & & \ddots \end{pmatrix}, A_3 = \begin{pmatrix} \ddots & & \\ & \boxed{\begin{matrix} 1 & 0 & 1 \\ 1 & 0 & 0 \\ 1 & 0 & 1 \end{matrix}} & \\ & & \ddots \end{pmatrix}. \quad (3)$$

Degenerate solutions like these can be handled by typical methods applied in constraint optimization with genetic algorithms, for example

- rejection of bad chromosomes and repetitive application of the genetic operators until a non-degenerate chromosome is found
- penalty terms in the goal function of the genetic search
- genetic operators and/or repair algorithms maintaining feasibility.

The number of degenerate solutions is dependent on the problem parameters m, n and the inherent number of cells k. In unfavorable circumstances the number of degenerate solutions is quite large so that the few non-degenerate solutions will dominate the population and finally lead to premature convergence. For the same reason the repetitive recombination — until non-degenerate solutions are found — is too expensive. The algorithm presented here is supposed to search the space of feasible solutions excluding degenerate solutions. This is achieved by the special operators described in Sect. 3.3.

The straight-forward encoding for simple partitioning problems is group number encoding (GNE). The chromosomes are vectors of length n where n denotes the number of objects to group. Each element takes an integer value from the range $1, \ldots, k$. The j-th element of the vector c, $c_j \in \{1, \ldots, k\}$, shows that object j belongs to group K_{c_j}, i.e.

$$\begin{bmatrix} c_1 \ c_2 \ \ldots \ c_n \end{bmatrix} \Leftrightarrow O_i \in K_{c_i} \forall i = 1, \ldots, n \ .$$

For example, the chromosome $C_1 = \begin{bmatrix} 1 \ 2 \ 2 \ 3 \ 3 \ 1 \end{bmatrix}$ represents the partition $P = \{\{O_1, O_6\}, \{O_2, O_3\}, \{O_4, O_5\}\}$.

In order to partition machines and products simultaneously both sets of objects have to be added to the chromosome. The problems left are to assure that the same number of groups is formed for both parts and that degenerate solutions are avoided.

The main problem of GNE with 1 or 2-point-crossover is that the partitions can be destroyed arbitrarily by the genetic operators. The approach described in [12] allows for complete partitions getting inherited by following the idea that the building blocks of a grouping GA are the groups themselves and not just single objects. The genetic operators work on the groups instead of on the objects.

For the grouping GA a chromosome consists of two parts, the object-part and the group-part. The object-part is the group-number-encoding of the objects. The group-part of the chromosome contains the identifiers of the groups used in the object-part. For partitions with a varying number of groups the group-part is of variable length. The genetic operators work on the group-part, whereas the object-part is adjusted as needed to keep track of the objects' group membership.

In addition to the partitioning algorithms described in [12] the GT problem requires the partition of two different sets of objects, thus the chromosome contains two fixed-length parts and a variable-length part. The chromosome

$$C = [\ \overbrace{1\ 2\ 1\ 3\ 3\ 2}^{\text{machines}} : \overbrace{3\ 3\ 1\ 2\ 3\ 3\ 1\ 1\ 2}^{\text{products}} \ \underbrace{}_{\text{object-part}} \ :: \ \underbrace{1\ 2\ 3}_{\text{group-part}} \ ]$$

represents a partition with the three groups $1, 2, 3$ shown in the group part. The object–part is interpreted according to the group–number–encoding: O_i belongs to partition $(C)_i = c_i$. By using the same numbers in the group and product-part, machine-groups and product-families are assigned automatically to each other so that the production cells can be derived directly from the chromosome. Thus the chromosome C represents the three production cells $\{m_1, m_3, p_3, p_7, p_8\}$, $\{m_2, m_6, p_4, p_9\}$, and $\{m_4, m_5, p_1, p_2, p_5, p_6\}$.

3.3   Genetic Operators

Crossover. The crossover operator for the GT problem is based on the ideas described in [12]. The two chromosomes P_1 and P_2 are crossed to produce the offspring C_1 and C_2.

Firstly, two crossing sites $r_1 \neq r_2$ are chosen in the group-part of the chromosome, i.e. $r_1, r_2 \in \{0, \ldots, \min\{l_{G_1}, l_{G_2}\}\}$ where l_{G_i} denotes the length of the group-part of P_i. For example, let $r_1 = 0$ and $r_2 = 2$

$$P_1 = \left[\, 1\,2\,1\,3\,3\,2 : 3\,3\,1\,2\,3\,3\,1\,1\,2 :: |\,1\,2\,|\,3\,\right]$$
$$P_2 = \left[\, \underline{0}\,\underline{1}\,\underline{0}\,\underline{3}\,\underline{1}\,\underline{3} : \underline{0}\,\underline{1}\,\underline{0}\,\underline{2}\,\underline{3}\,\underline{1}\,\underline{1}\,\underline{0}\,\underline{2} :: |\,\underline{0}\,\underline{1}\,|\,\underline{2}\,\underline{3}\,\right] \ .$$

Secondly, the group-part of P_1 is copied into the group-part of C_1. Then the group-part of P_2 between r_1 and r_2, is inserted in the group-part of C_1 after position r_2.[4]

Thirdly, the object-part of C_1 is constructed on the basis of the alleles of P_1 and P_2: First of all, the group-numbers of the inserted group-part are assigned to the objects as in P_2. If there are any unassigned objects left they get the group numbers as assigned in P_1. This leads to the chromosome

$$C_1 = \left[\, \underline{0}\,\underline{1}\,\underline{0}\,3\,\underline{1}\,2 : \underline{0}\,\underline{1}\,\underline{0}\,2\,3\,\underline{1}\,\underline{1}\,\underline{0}\,2 :: 1\,2\,\underline{0}\,\underline{1}\,3\,\right] \ .$$

Underlined alleles of C_1 are taken from P_1, the others from P_2. The construction of the object-part of C_1 encounters the following cases:

1. A group g of P_1 is overwritten completely. In this case group g has to be removed from the group-part of C_1 (group 1 in the example).
2. A group g of P_1 is not modified. In this case nothing has to be done.
3. A group g contains only machines but no products. This is a degenerate solution, the machines of group g are assigned randomly to other groups.
4. A group contains only products but no machines. This is also a degenerate solution that has to be repaired by assigning its products to other groups.
5. A group g of P_1 has lost some of their machines and/or products. In this case the group either remains part of the solution with probability p or is destroyed completely with probability $(1 - p)$ where p is the ratio of the groups' original and current sizes,

$$p = \frac{\text{number of machines and products of group } g \text{ in } P_1}{\text{number of machines and products of group } g \text{ in } C_1} \ .$$

The second offspring C_2 is constructed in the same way, with the roles of P_1 und P_2 swapped. Without the corrections in step 5 the crossover operator tends to generate too many degenerate solutions and the number of classes in the offspring increases in comparison with its parents.

Mutation. The three mutation operators work on the group part. Again, the object-part is adjusted as needed. Each operator might generate degenerate solutions that can be corrected by the same mechanism applied after the crossover operator.

---

[4] Note that the groups 1 from P_1 and $\underline{1}$ from P_2 are considered to be different groups.

The *Mutation Distribute* operator selects one of the groups in the group part at random and assigns each of its objects to a new group. As an assignment to the selected group is allowed, the operator either reduces the number of groups by one or keeps it constant, consider group 3, for example:

$$\left[1\,2\,1\,3\,3\,2:3\,3\,1\,2\,3\,3\,1\,1\,2::1\,2\,3\right] \rightarrow \left[1\,2\,1\,3\,2\,2:1\,3\,1\,2\,2\,3\,1\,1\,2::1\,2\,3\right] \ .$$

The *Mutation Combine* operator randomly selects two different groups g_1 and g_2 and combines them into one by substitution of all alleles g_1 by g_2 and removal of g_1 from the group-part. This reduces the number of groups by one. For example, the groups 1 and 2 are combined:

$$\left[1\,2\,1\,3\,3\,2:3\,3\,1\,2\,3\,3\,1\,1\,2::1\,2\,3\right] \rightarrow \left[2\,2\,2\,3\,3\,2:3\,3\,2\,2\,3\,3\,2\,2\,2::2\,3\right] \ .$$

The *Mutation Divide* operator introduces a new group to the chromosome by selecting a group at random and moving each of its objects to the new group with a probability of $p = 1/2$. In the example, group 3 is divided:

$$\left[1\,2\,1\,3\,3\,2:3\,3\,1\,2\,3\,3\,1\,1\,2::1\,2\,3\right] \rightarrow \left[1\,2\,1\,3\,4\,2:3\,3\,1\,2\,4\,4\,1\,1\,2::1\,2\,3\,4\right].$$

Inversion. The inversion operator is supposed to shorten good schemas so that they do not get disrupted by genetic recombination and thus can be inherited completely. Inversion selects a part of the group-part and inverses the order of its groups, as shown in the following example:

$$\left[1\,2\,4\,4\,3\,3\,1\,2\,1\,1:1\,2\,3\,4\right] \longrightarrow \left[1\,2\,4\,4\,3\,3\,1\,2\,1\,1:\underline{3\,2\,1}\,4\right] \ .$$

The object-part of a chromosome remains unchanged.

3.4   Heuristic

A mechanism to improve the performance of a GA is to include further heuristics for local search. This section presents a heuristic motivated by cluster analysis techniques for the GT problem.

Hierarchical cluster analysis methods are based on similarity coefficients that allow to construct dendrograms showing which objects to cluster. Many cluster analysis approaches to the GT problem apply the Jacard-Similarity-Coefficient [13]. The similarity s_{ij} of machines i and j is calculated as

$$s_{ij} = \frac{\sum_{k=1}^{n}(a_{ik} \wedge a_{jk})}{\sum_{k=1}^{n}(a_{ik} \vee a_{jk})} \quad i, j = 1, \ldots, n \ . \tag{4}$$

The numerator in (4) is the number of products to be manufactured on machines i and j. The denominator is the number of products to be processed on machines i or j. For any i and j holds that $s_{ij} \in [0, 1]$. Moreover, $s_{ij} > 0$, unless $a_{ik} = a_{jk} = 0$ for all $k = 1, \ldots, n$. In this case the matrix has got two rows with zeros only. But machines without any incidence can be ignored. For the GA presented in this article two matrices of Jacard-Similarity-Coefficients are calculated, one for machines and — by analogy — one for products, $S^M = (s_{ij}^M), i, j = 1, \ldots, m$ and $S^P = (s_{ij}^P), i, j = 1, \ldots, n$, respectively.

Any time a machine i has to be assigned to a new group during the optimization, the group of a randomly selected machine $j \in S_i$ is taken, where

$$S_i = \{j | s_{ij}^M \geq s_{ik}^M \quad \forall k = 1, \ldots, m, \quad i \neq j\} \ . \tag{5}$$

S_i is the set of machines most similar to machine i with respect to the similarity coefficient. In the same way, parts can be assigned to the group of the part most similar to it. If $\max_{j=1,\ldots,n} s_{ij} = 0$ for a machine i, then there are no machines similar to machine i and machine i is assigned randomly to a group.

Cases 3, 4, and 5 of the crossover operation and all of the mutation operators require random assignments of objects to groups. Instead of these random assignments the heuristic can be used.

This heuristic only allows to assign machines to machines or products to products. A heuristic that assigns machines to products is more difficult. A simple approach could assign a product only to a group where it is processed or to the group in which it has the most process steps. But as this is dependent on a special partition of the machines, it requires a re-calculation for any solution thus introducing a new quite expensive step. In contrast, the similarity matrices are static and need only be calculated once in the initialization phase of the GA. More complex heuristics could try an assignment of machines to products. But this is exactly the problem the GA is supposed to solve.

4   Results

A large number of experiments showed how to set the parameters to achieve the best GA performance on several test problems taken from literature. The GA applied is the modified GA [10] with a population size of $n + m$ (number of machines + number of products), a selection rate of $\frac{2}{3}(m + n)$, and elitist strategy. The crossing rate is set to 0.6 and the mutation rates for the mutation operators *distribute*, *combine*, and *divide* are 0.14, 0.04, and 0.04, respectively. Linear scaling is applied with a factor of 1.9. For each problem 20 optimization runs are started. The coefficient of the goal function is set to $q = 0.25$. The GA is stopped at a maximum of 2000 iterations or earlier if 300 successive generations do not find any improvements. The groups of the initial population are randomly assigned, degenerate solutions are repaired as needed.

The six test problems and the partitions for the reference solutions are taken from [14,15,16,16,15,17], respectively. It is a selection of small and large matrices that have a more or less perfect block diagonal form. Table 1 shows the results. The columns are the number of the test problem, the problem size $m + n$, the value of the best known solution, the minimum, average, and maximum minimum of the 20 runs, the average deviation from the best known solution, the average number of generations, the number of runs the optimum was found, the total time, and the average time to convergence, i.e. the average of the total time subtracted by the time for 300 generations to check for convergence. The experiments were carried out on a Pentium 90 personal computer. According to the table the GA with the heuristic performs very well.

The average deviations from the optimum show that the GA finds good solutions for all of the six problems. For two problems (4,5) the optimum solution is found in every single run and for problems (2,3) better solutions are found[5].

Table 1. Selected numerical results

#	Size	Ref.	Min. min	Av. min	Max. min	Av. dev.	Av. #gen	# opt.	Ti. tot.	Ti. con.
1	140	1027.7	1027.24	1032.11	1050.84	0,424	471,95	10	88:08	1:36
2	59	1133.22	1121,33	1127,17	1130,92	-0,538	560,6	20	27:46	0:39
3	55	1062.86	1061.4	1062.78	1062.86	-0,007	357,1	20	13:08	0:07
4	28	1062,50	1062,50	1062,50	1062,50	0	325,1	20	2:10	0:01
5	25	1020,00	1020,00	1020,00	1020,00	0	319.05	20	1:47	0:00.3
6	16	1146.88	1146,88	1148,97	1151,53	0,18	358,7	11	0:42	0:00.3

Problem 1 has the maximum difference between the minimum and the maximum minimum, but the average minimum is quite close to the minimum minimum. This shows that runs not finding optimal solutions are not too far off.

The average number of generations increases with the problem size except for problem 6. This is due to the higher grouping efficiency, i.e. the cells are less heterogeneous and thus more difficult to find. Note that although problem 1 has minimum grouping efficiency, it leads to the worst (in comparison) GA performance, probably due to the large problem size. For small problems optimum solutions are found within the the first generations.

5 Conclusion

This article presented a new GA algorithm for the GT problem and an associated local search heuristic. Both the sets of machines and products are partitioned simultaneously and the algorithm allows to find the inherent number of cells. Experiments proved a good performance on a number of problems taken from GT literature. The performance is dependent on the structure of the incidence matrix. The less homogeneous the blocks are, the less often the GA is able to find the optimum. But, nevertheless, the optimum was found, just the number of times finding it decreased.

The solution presented showed to be suitable for the design of cellular manufacturing systems on the basis of grouping efficiency. Future research will concentrate on more complex goal functions and constraints that reflect other important characteristics of manufacturing systems like varying product mixes and demands over different periods, processing times and sequences, the variation of utilizations within the cells, and distances on the shop floor. As this mainly influences the goal function, the extended grouping GA with its special operators is expected solve these problems satisfactorily as well.

Acknowledgements. The author likes to thank Prof. Th. Hanschke, Technical University of Clausthal, Dr. J. Gottlieb, SAP AG, and the anonymous referees for their suggestions to improve the first version of the paper.

---

[5] Note that the optimum solution is dependent on the factor q in the goal function.

References

1. S. S. Heragu. Group technology and cellular manufacturing. *IEEE Transactions on Systems, Man, and Cybernetics*, 24(2):203–215, February 1994.

2. W. H. Chen and B. Srivastava. Simulated annealing procedures for forming machine cells in group technology. *European Journal of Operational Reasearch*, 74:100–111, 1994.

3. Y. Gupta, M. Gupta, A. Kumar, and C. Sundaram. A genetic algorithm-based approach to cell composition and layout design problems. *International Journal of Production Research*, 34(2):447–482, 1996.

4. J.A. Joines, C.T. Culbreth, and R.E. King. Manufacturing cell design: an integer programming model employing genetic algorithms. *IIE Transactions*, 28(1), 1996.

5. V. Venugopal and T. T. Narendran. A genetic algorithm approach to the machine-component grouping problem with multiple objectives. *Computers and Industrial Engineering*, 22(4):469–480, 1992.

6. C. Zaho and Z. Wu. A genetic algorithm for manufacturing cell formation with multiple routes and multiple objectives. *International Journal of Production Research*, 38(2):385–395, 2000.

7. S. A. Mansouri, S. M. Moattar Husseini, and S. T. Newman. A review of modern approaches to multi-criteria cell design. *International Journal of Production Research*, 38(5):1201–1218, 2000.

8. I. Meents. Genetic algorithms for the group technology problem. Master's thesis, Technical University of Clausthal, 1997. (in German).

9. E. Falkenauer. *Genetic Algorithms and Grouping Problems*. John Wiley & Sons Ltd., Baffins Lane, Chichester, West Sussex PO19 1UD, England, 1998.

10. Z. Michalewicz. *Genetic Algorithms + Data Structures = evolution programs*. Springer, Berlin, Heidelberg, New York, 3. edition, 1996.

11. M. P. Chandrasekharan and R. Rajagopalan. An ideal seed non–hierarchical clustering algorithm for cellular manufacturing. *International Journal of Production Research*, 24(2):451–464, 1986.

12. E. Falkenauer. A new representation and operator for genetic algorithms applied to grouping problems. *Evolutionary Computation*, 2(2):123–144, 1994.

13. S. M. Taboun, S. Sankaran, and S. Bhole. Comparison and evaluation of similarity measures in group technology. *Computers and Ind. Eng.*, 20(3):343–353, 1991.

14. M. P. Chandrasekharan and R. Rajagopalan. ZODIAC - An algorithm for concurrent formation of part-families and machine-cells. *International Journal of Production Research*, 25(6):835–85, 1987.

15. A. Ballakur and H. J. Steudel. A within-cell utilization based heuristic for designing cellular manufacturing systems. *International Journal of Production Research*, 25(5):639–665, 1987.

16. M. P. Chandrasekharan and R. Rajagopalan. MODROC: An extension of rank order clustering for group technology. *International Journal of Production Research*, 24(5):1221–1233, 1986.

17. W. S. Chow and O. Hawaleshka. An efficient algorithm for solving the machine chaining problem in cellular manufacturing. *Computers and Industrial Engineering*, 22(1):95–100, 1992.

Generation of Optimal Unit Distance Codes for Rotary Encoders through Simulated Evolution

Stefano Gregori[1], Roberto Rossi[1], Guido Torelli[1], and Valentino Liberali[2]

[1] Università degli Studi di Pavia, Dipartimento di Elettronica
Via Ferrata 1, 27100 Pavia, Italy
s.gregori@ele.unipv.it, r.rossi@ele.unipv.it, g.torelli@ele.unipv.it
[2] Università degli Studi di Milano
Dipartimento di Tecnologie dell'Informazione
Via Bramante 65, 26013 Crema, Italy
vliberali@crema.unimi.it

Abstract. An evolutionary algorithm is used to generate unit distance codes for absolute rotary encoders. The target is to obtain a code suitable for disk size reduction, or for resolution increase, thus overcoming the limitations of conventional Gray codes. Obtained results show that simulated evolution can produce codes suitable for this purpose.

1 Introduction

Evolutionary algorithms, built on the key concept of Darwinian evolution [1], are a broad class of optimization methods which could provide innovative solutions to problems heigh computational complexity [2], when exhaustive approaches become unfeasible.

In this paper, simulated evolution is used to generate unit distance codes for rotary encoders having better properties than the conventional Gray code. As shown in Sect. 2, the widely used Gray code limits the size reduction of the rotating part and the maximum achievable resolution. To overcome this constraint, the evolutionary algorithm described in Sect. 3 has been implemented. The obtained results, shown in Sect. 4, demonstrate that the evolutionary approach can provide a suitable solution, being able to manage computationally complex problems in a limited amount of time.

2 The Rotary Encoder System

2.1 Mechanical Assembly

An absolute rotary encoder is an angular position transducer, generally used to monitor the rotation of mechanical parts [3]. It is made up of a glass disk, referred to as code disk, with a suitable number of concentric traces, each having alternated opaque and transparent zones, whose patterns allow the absolute angular position of the disk to be determined. All traces are illuminated by a

E.J.W. Boers et al. (Eds.) EvoWorkshop 2001, LNCS 2037, pp. 100–109, 2001.

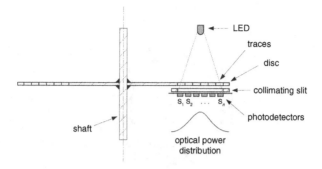

Fig. 1. Mechanical assembly of an absolute rotary encoder.

light-emitting diode (LED) from one side of the disk. The light beam is modulated by the traces and, after passing through the collimating slit (a narrow radial slit which limits the width of the light beam), is collected by a photodetector array on the opposite side of the disk (one photodetector is associated to each trace). Each photodetector generates an electric signal, which is then amplified and converted into a digital waveform. The set of digital waveforms coming from the photodetector array contains the disk angular position information. The number of coding bits (n), which gives the encoder resolution, is equal to the number of disk traces, whose pattern, read by the photodetector array, identifies, univocally, the angular position of the disk.

Figure 1 shows the mechanical assembly of an absolute rotary encoder. It is apparent that the impinging optical power is not the same for all the elements of the photodetector array. The light power in the radial direction can be approximated as a Gaussian distribution, with the maximum corresponding to the central photodetector. Furthermore, the signals produced by different photodetectors are affected by substantial variations of the photogenerated current due to the type and ageing of the LED and temperature drifts. Moreover, due to causes such as the geometry of the disk traces, of the photodetectors, and of the collimating slit, when the disk rotates, the current signal provided by each photodetector turns out to have a trapezoidal shape.

In modern encoders specific electronic readout channels are designed with the aim to provide thermal stabilization and prevent performance degradation [4]. Moreover, a calibration procedure is required to achieve adequate alignment of the LED, the disk code, and the photodetectors. Digital calibration circuits allow a reduction in calibration time. However, due to the increasing demand for small-dimension high-resolution encoders, the alignment procedure is still critical, especially for encoders with a large number of traces.

2.2   Codes for Absolute Rotary Encoders

In practice, the use of the natural binary code to define the pattern of opaque and transparent zones on the traces of the code disk, entails problems such as

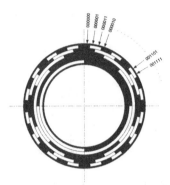

Fig. 2. Example of a code disk with Gray code. The disk has six traces, which are read by the photodetector array from the innermost to the outermost one in the radial direction. In all read words, an opaque (black) zone corresponds to a bit '0' and a transparent (white) zone corresponds to a bit '1'.

spurious transients in correspondence of the positions of the code disk where more than one bit has to switch at the same time. Indeed, in real systems these bits will not switch at the same time, thus causing a readout error during the transitions. For example, when passing from the codeword 011111 (decimal 31) to the codeword 100000 (decimal 32), six bits should change over at the same time. In the case of non-perfect alignment, the six bits do not switch simultaneously, thus causing a readout error during the transition.

To avoid spurious reading during transitions, unit distance codes are used for the code disks. In these codes, words associated with adjacent angular positions differ by only one bit: therefore, a single bit has to be switched during any transition. The most widely used unit distance code in encoder disks is the conventional Gray code (Fig. 2).

2.3   Design Constraints

The minimum angular step measured with an n-bit encoder is given by $\alpha = 2\pi/2^n$. To obtain high accuracy, it is necessary to use code disks with a high number of traces (n). On the other hand, technical specifications force the manufacturer to make the code disk radius small, so as to limit the size and the weight of the system as well as the moment of inertia added by the encoder to the coupled drive shaft.

In the case of a Gray coded disk, the main design constraint regards the length of the opaque and transparent zones of the outermost trace. Indeed, this length should be larger than the width of the collimating slit, so as to allow the photodetector to be fully illuminated when a transparent zone passes above it or completely darkened when an opaque zone passes above it.

Any opaque and transparent zone in the outermost trace when using a Gray code corresponds to a sequence '00' and '11', respectively, along the trace itself.

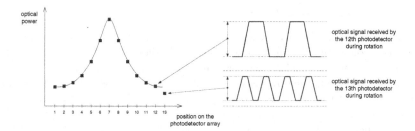

Fig. 3. Typical measured example of optical power distribution over the photodetector array. The optical power reduction in the outermost trace (number 13) derives from the use of a collimating slit narrower than the opaque and transparent zones of the outermost trace.

The corresponding angle is $\beta_G = 2\alpha$, that is two times the minimum angular step, because the sequence contains two equal bits ('00' or '11'). Since the length of any opaque or transparent zone must be larger than the width of the collimating slit c, the following relation must hold

$$2 \, \sin\left(\frac{\beta_G}{2}\right) r_n \cong \beta_G \, r_n \geq c \tag{1}$$

where r_n is the radius of the outermost trace. Therefore, the constraint on the outermost trace radius is given by

$$r_n \geq \frac{c}{\beta_G} = \frac{c}{2\alpha} = c\,\frac{2^n}{4\pi} \tag{2}$$

Considering c fixed, to increase the number of bits by one entails to double the radius of the outermost trace, which is, in practice, the radius of the whole code disk. For this reason, when it is important to reduce the dimensions of the encoder, the radius of the outermost trace is generally chosen so that the corresponding photodetector is never fully illuminated nor fully darkened, thus worsening the reading conditions of the ensuing signal.

An example of this situation is a commercial 13-bit encoder, whose disk code has a radius of 21 mm. In this case, the length of the opaque and transparent zones of the outermost trace is $2\pi \cdot 21/2^{12}$ mm $= 31$ μm, that is smaller than the width of the collimating slit (45 μm). Fig. 3 shows the optical power distribution over the photodetector array. As the light source is placed on the axis corresponding to the central photodetector (number 7), this receives the maximum optical power, while the outermost photodetector (number 13) receives the minimum optical power, which is further reduced due to the small dimensions of the transparent zones of the corresponding trace. The signal produced by this photodetector is more difficult to be correctly read because the photodetector is never fully illuminated nor fully darkened.

2.4   Non-conventional Codes

A way to overcome the above constraint set by the conventional Gray code, is to build non-conventional unit distance codes that allow us to have longer opaque and transparent zones along the outermost trace (the length of the zones in the inner traces must be obviously kept not smaller than the outermost trace, in terms of angular width). This is possible by increasing the length of the sequences of consecutive '0's and of '1's, respectively, along the outermost trace. In this way, it is possible to increase the number of traces with no need for increasing the disk radius, or, alternatively, to reduce the radius for any given condition of photodetector array dimensions and illumination.

If q is the minimum length of a sequence along the traces in a unit distance code, the corresponding angle is $\beta_E = q\alpha$, while in a Gray code it is $\beta_G = 2\alpha$ (as $q = 2$). Assuming the photodetector array length is $\Delta r = r_n - r_1$, if the minimum-length sequences are contained in m traces (with $m \leq n$) the constraint on the radius of the outermost trace is

$$r_n \geq \frac{c}{q\alpha} + \frac{m-1}{n-1}\Delta r \tag{3}$$

In a Gray code, this constraint is stated in equation (2). In this case, the new code is better than the Gray code if the following condition is verified:

$$\frac{m-1}{n-1}\Delta r < \frac{c}{\alpha}\left(\frac{1}{2} - \frac{1}{q}\right) \tag{4}$$

When reducing the disk radius, a further limit on the minimum radius has to be considered in order to avoid readout errors. This limit depends on the error margin in transforming the trapezoidal waveform given by the photodetector into a digital signal. This limit can be expressed as

$$\arcsin\left(\frac{c}{r_1}\frac{V_{EM}}{V_{SA}}\right) < \alpha \tag{5}$$

where V_{SA} is the peak-to-peak signal amplitude, V_{EM} is the error margin and r_1 is the radius of the innermost trace.

Provided that this condition is satisfied, the only drawback of non-conventional codes is the need for a decoding logic to obtain the natural binary code. However, this task does not introduce any significant delay in processing, and can be easily carried out with a lookup table.

3   The Evolutionary Algorithm

The search for an optimal non-conventional code becomes very difficult as the number of traces n increases. To overcome this limitation, an evolutionary algorithm has been implemented to find unit distance codes. The proposed algorithm uses variable population size, generation by mutation, and rank-based selection; its main aspects are described in the following subsections.

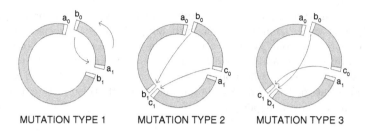

Fig. 4. The three possible mutations of a code: (1) reversal of a portion of the code; (2) extraction and insertion in a different position; (3) extraction, reversal and insertion in a different position.

3.1   Encoding

The genetic encoding is simply the sequence of the 2^n words which represent all the sectors of the disk (each sector is the radial portion of the disk having the minimum measurable angular width α). For an encoder, a "valid" code must satisfy the following conditions:

1. **completeness**: the code must be complete, i.e. it must contain all the 2^n words, without repeating any of them;
2. **unit distance**: two adjacent words must differ by the value of one and only one bit;
3. **cyclicity**: the last word must have unit distance from the first one.

Condition 3 (cyclicity) is required for a rotary encoder only; in the case of a linear position encoder, this condition is not necessary.

3.2   Mutation

In order to satisfy all the constraints enumerated in the previous subsection, new codes are generated by mutation only. Mutation consists in generating a new code (the child) from a previous one (the parent) by means of one of the following operations:

1. reversing the order of a portion of the genetic code;
2. extracting a portion of the genetic code and inserting it in a different position;
3. extracting a portion of the genetic code and inserting it in a different position after reversal.

The three mutation operators are sketched in Fig. 4. After generation, both the parent and the child are included in the new population.

The cut points (indicated as a_0-b_0, a_1-c_0, and b_1-c_1 in Fig. 4) are chosen so that the resulting code still satisfies the three conditions stated above (completeness, unit distance, and cyclicity). The search for the good cut points is carried out in an exaustive way on all the positions of the considered code.

None of the three above mutation rules is guaranteed to work for any possible unit distance code. To maximize the probability of success of such an operation, this step is carried out as follows. One of the three allowed types of mutation is randomly selected and applied to the code. If it is possible, then a child is obtained and added to the population. Otherwise, one of the two remaining mutation strategies is randomly chosen and tried. If still no result is possible, the last mutation rule is performed.

3.3   Fitness

The fitness function is a measure of the minimum sequence length. To distinguish between different codes with the same minimum length q, an additional term has been added, to favour codes with a lower number of minimum length sequences.

The fitness f is defined as:

$$f = q + \frac{1}{2} - \frac{N_q}{2^n} \tag{6}$$

where N_q is the number of q-length sequences.

The conventional Gray code has a fitness $f_G = 2$.

3.4   Selection

The population is seeded with a single individual: the Gray code described in Sect. 2.

New codes are generated by randomly picking existing codes and applying one of the three allowed mutation types according to the strategy described above. A linear rank-based selection is performed when the population size exceeds a predefined value N_{max}. Selection is carried out by assigning a survival probability to each code in the population. The best code is assigned a probability of 1, while all the others are given lower probabilities according to their rank in such a way that, on average, the best N individuals are most likely to survive. The maximum population size has been set to $N_{max} = 2N$.

From our experiments, we noticed that a population size N in the range from 50 to 100 maintains a good genetic variability without slowing down the algorithm excessively.

4   Experiments and Results

4.1   Results

The algorithm described in the previous section has been used to search unit distance codes for an encoder disk with 6, 7, and 8 bits.

Fig. 5 shows the fitness evolution as a function of the number of individuals generated. It is worth remarking that, from (6), the integer part of the fitness

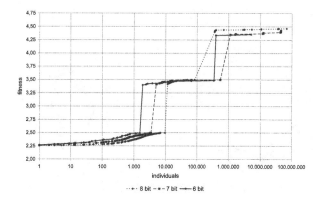

Fig. 5. Evolution of the fitness for three disk codes with 6, 7, and 8 bits, respectively.

corresponds to the minimum sequence length, and the sharp steps in the curves indicate that the minimum length has been improved.

It is also observed that the algorithm produces similar behaviours of the fitness for the three values considered ($n = 6$, 7, and 8), despite the exponential increase of the number of codewords.

An example of this kind of code, obtained using an evolutionary algorithm, is illustrated in Fig. 6. While in the conventional Gray code the sequences of consecutive '0's or '1's along the traces have a minimum length of 2, in the non-conventional code the minimum length is 4. In this case, the angle corresponding to the minimum sequence is doubled: for the Gray code $\beta_G = 2\alpha$, while for the non-conventional unit distance code $\beta_E = 4\alpha$.

However, while for the Gray code the constraint is on the radius of the outermost trace, for this new code the constraint is on the radius of the most internal trace containing a minimum-length sequence (in this case, the second innermost trace). Assuming the photodetector array length is $\Delta r = r_6 - r_1$, according to (3), the constraint on the radius of the outermost trace is

$$r_6 > c\frac{2^6}{8\pi} + \frac{4}{5}\Delta r \qquad (7)$$

For any given unity distance code, a code with the same minimum length and a higher number of bits can be easily obtained using the basic building procedure for conventional Gray codes. This consists in copying the considered code with 2^n words, flipping it and adding a row containing a sequence of 2^n '0' and 2^n '1' as shown in Fig. 7.

For example, starting from the 6-bit code in Fig. 6, a 13-bit code with minimum length 4 can be obtained. The code disk obtainable with this unit distance code can be designed with the radius of the innermost trace $r_1 = 6.05$ mm and the radius of the outermost trace $r_2 = 13.25$ mm, while for the 13-bit Gray coded disk $r_{G1} = 13.3$ mm and $r_{G2} = 20.5$ mm, in both cases assuming to have the same array length $\Delta r = 7.2$ mm and the same width of the collimating slit $c = 45$ μm.

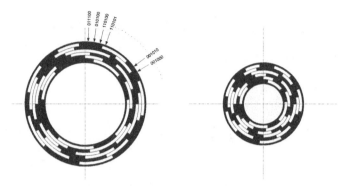

Fig. 6. Example of unit distance code obtained with an evolutionary algorithm. The code disk on the left is a 6-trace disk with the same radius of the Gray disk in Fig. 2; the code disk on the right has the same code, but its radius is reduced so as to have minimum opaque or transparent zones length equal to that of the Gray encoded disk in Fig. 2.

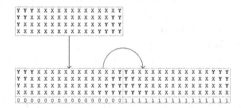

Fig. 7. Procedure generate a unity distance code with $n + 1$ bits starting from a unity distance code with n bits.

4.2   Discussion

The obtained results have been compared with the results obtained with a local search heuristic method. In this case we used a fixed population with a single individual. This individual generate a child through mutation, who generates a grandchild, who generates a great-grandchild. Selection is then applied to choose the individual with the best fitness and discard the other three. With this method neither with 6, 7 or 8 bit codes we were able to get codes with minimum length 4, within 10^8 generations.

Fig. 8 shows the fitness evolution as a function of the number of individuals generated using a local search heuristic method compared with the one generated with the evolutionary algorithm previously described. From this figure, we conclude that the evolutionary algorithm is capable to perform better exploration of the design space, since it has a better chance to escape from local maxima of the fitness function, while the heuristic search has an intrinsically limited capacity of avoiding local maxima.

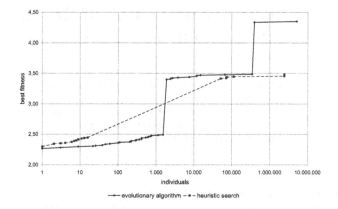

Fig. 8. Evolution of the fitness for two disk codes with 6 bits using heuristic search and evolutionary algorithm.

5   Conclusion

This paper has described an evolutionary approach to the generation of optimal unit distance codes for rotary encoders. Results demonstrate that unit distance codes better than the conventional Gray code can be obtained through simulated evolution. The increase of the minimum sequence length can be exploited either to reduce the encoder size (maintaining the same resolution), or to increase resolution (with the same disk size).

Acknowledgments. The authors would like to thank the reviewers for their very useful comments and suggestions.

References

1. C. Darwin. *On the Origin of Species by Means of Natural Selection.* John Murray, London, UK, 1859.
2. T. Bäck. *Evolutionary Algorithms in Theory and Practice.* Oxford University Press, Oxford, UK, 1996.
3. R. Ohba. *Intelligent Sensor Technology.* John Wiley & Sons, Chichester, UK, 1992.
4. D. Maschera, A. Simoni, L. Gonzo, M. Gottardi, S. Gregori, V. Liberali, and G. Torelli, "CMOS front-end for optical rotary encoders", *Proc. 7th Int. Conf. on Electronics, Circuits, and Systems*, Kaslik, Lebanon, Dec. 2000, pp. 891-894.

On the Efficient Construction of Rectangular Grids from Given Data Points

Jan Poland, Kosmas Knödler, and Andreas Zell

Universität Tübingen, WSI-RA, Sand 1, D - 72076 Tübingen, Germany
poland@informatik.uni-tuebingen.de,
http://www-ra.informatik.uni-tuebingen.de

Abstract. Many combinatorial optimization problems provide their data in an input space with a given dimension. Genetic algorithms for those problems can benefit by using this natural dimension for the encoding of the individuals rather than a traditional one-dimensional bit string. This is true in particular if each data point of the problem corresponds to a bit or a group of bits of the chromosome. We develop different methods for constructing a rectangular grid of near-optimal dimension for given data points, providing a natural encoding of the individuals. Our algorithms are tested with some large TSP instances.

1 Introduction

Many combinatorial optimization problems provide their data in an input space with a given dimension d. For example, the classical traveling salesman problem (TSP) consists of n points in the two-dimensional plane. An efficient representation of a tour for solving the TSP with a genetic algorithm is the adjacency coding (see [1], [2] or [3]): A tour is defined by its adjacency matrix or, equivalently, by a list containing the successor for each city. This *locus*-based representation has shown to be much more appropriate for use with genetic algorithms than a time-based representation containing a permutation of the cities. It is characterized by the fact that each city corresponds to a certain part of the chromosome.

When using a locus-based representation together with a standard N-point crossover operator that does not permute the chromosomes, the arrangement of the points in the representation is important. For example in the case of the TSP, arranging the cities in the order given by a rough approximation of the TSP results in a better performance of the genetic algorithm than using an arbitrary arrangement, as shown in [1]. This is due to the fact that the crossover operator can exploit neighbourhood relations of the preordered cities.

However, since the original problem is two-dimensional, a two-dimensional representation as suggested by Bui and Moon ([4]) would be more natural. They present a d-dimensional N-point crossover operator that applies to d-dimensional grids instead of one-dimensional strings. If this idea is employed for the TSP with a locus-based representation, the cities have to be allocated to the grid points. This is a preliminary task that has to be completed before the start of the genetic

E.J.W. Boers et al. (Eds.) EvoWorkshop 2001, LNCS 2037, pp. 110–120, 2001.
© Springer-Verlag Berlin Heidelberg 2001

algorithm and should not consume too much time. On the other hand, a good arrangement may increase the performance of the genetic algorithm.

Clearly, the grid encoding is not the only possibility for a d-dimensional representation, one could, for example, try a graph-based encoding. But the grid encoding has the advantage of a very simple and efficient crossover procedure and is therefore very convenient for a genetic algorithm that uses crossover as an important operator.

Hence, the problem considered here can be defined as follows. Given are n data points $(x_1, \ldots, x_n)$ in the d-dimensional space. Find an appropriate grid size $(g_1, \ldots, g_d)$ and an unambiguous (injective) allocation $\imath : \{1 \ldots n\} \mapsto \{1 \ldots g_1\} \otimes \{1 \ldots g_2\} \otimes \ldots \otimes \{1 \ldots g_d\}$, such that the allocation is good in the following sense: Points that are close to each other should be mapped to neighbouring grid points, while distant points should be mapped to distant nodes in the grid. Thus, the transformation to the grid throws away as little neighbourhood information as possible, and one can hope that a genetic algorithm is able to use this information for faster convergence. Note that the injectivity of $\imath$ implies $g_1 \cdot g_2 \cdot \ldots \cdot g_d \geq n$. On the other hand, this product should be as small as possible, since each unused grid point means unused space in the chromosomes and therefore poorer performance of the genetic algorithm.

In [5], B. Moon and C. Kim study a similar problem, the two-dimensional embedding of graphs. Their assumptions are weaker, since they exploit only the adjacency information (i.e. distance in the TSP case). In the contrary, we will exploit the locus information.

2   Calculating the Grid Size

We start with a brief discussion of different approaches. Basically, there are two possible ways: One can either fix the grid size before allocating the points, or one can determine the grid size while arranging the points. Suppose we want to do the latter. An algorithm could perform a depth first search, in analogy to the graph embedding algorithm in [5]. In each step, the grid is extended by an additional point. But the decision where to extend the grid and in which direction is difficult in general, since it depends crucially on points that are not yet processed. Furthermore, if $n < g_1 \cdot g_2 \cdot \ldots \cdot g_d$, then unused grid points have to be inserted at some places. Suboptimal decisions that result in an ineffective allocation are almost certain, which carries the need for repairing algorithms. Hence, constructing the grid without previously fixing the grid size can become very expensive, in particular for large data sets (for example $n \approx 1000$).

For a concrete example, consider the points displayed in Fig. 1 (a). A step by step arrangement may lead to the situation shown in Fig. 1 (b). It would be relatively easy to repair this arrangement by inserting two empty positions, thus arriving at Fig. 1 (c). In contrast, all modifications that could result in the optimal allocation Fig. 1 (d) are quite expensive. Another situation: If point 6 is placed to the right of point 5 instead of below, the result is Fig. 1 (e) or, if point

112   J. Poland, K. Knödler, and A. Zell

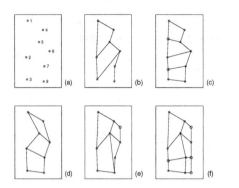

Fig. 1. (a) Eight data points, (b) - (f) their allocation to different grids, (d) is optimal

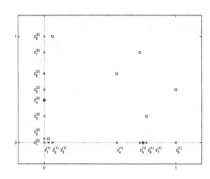

Fig. 2. The points have been normalized and projected to the coordinate axis

7 is choosen to be below point 5, Fig. 1 (f). Note that for the last arrangement four additional empty positions have to be inserted, which is not optimal at all.

In general, we are interested in a reasonable minimum of grid points. This ensures a minimum of unused space for the individuals, while Fig. 1 suggests that the neighbourhood relation is fairly well preserved by different arrangements (each of the arrangements (c) through (f) would pass this criterion). Therefore, we pursue the other way and fix the grid size before allocating the points.

We assume that the data points cover a region that has approximately the shape of a d-dimensional hypercube with balanced aspect ratio, i.e. that the data set is not poorly scaled. Otherwise, the scaling information should be exploited for calculating the grid size.

Consider again Fig. 1 (a). We are looking for a method that calculates the grid size for this set of points. A 2×4 grid would be optimal, as we can see in the figure, a 3×3 grid would be tolerable, too.

To this aim, we first normalize the points to $[0,1]^d$ and name the resulting points $(y_1, y_2, \ldots, y_n)$. Then we project the points to each of the d coordinate axis, obtaining d vectors of length n $\{(y_1, \ldots, y_n)^{(k)}, k = 1 \ldots d\}$, see Fig. 2. Sorting $(y_1, \ldots, y_n)^{(k)}$ for each $k \in \{1 \ldots d\}$ *separately* in ascending order yields $(z_1, \ldots, z_n)^{(k)}$, $k \in \{1 \ldots d\}$, having $z_1^{(k)} = 0$ and $z_n^{(k)} = 1$ for each k. We define

$$r_k = \frac{1}{\sum_{j=1}^{n-1}(z_{j+1}^{(k)} - z_j^{(k)})^2} + 1 \quad \text{for each} \quad k \in \{1 \ldots d\}.$$

This number r_k estimates the requested number of points for each dimension k.

To motivate this statement, consider more closely the projections onto the x-axis in Fig. 2. We have eight different points $z_1^{(1)}, \ldots, z_8^{(1)}$, but they are not equidistantly distributed. Instead, $z_1^{(1)}$, $z_2^{(1)}$ and $z_3^{(1)}$ form a cluster, and the same is true for $z_5^{(1)}$, $z_6^{(1)}$ and $z_7^{(1)}$. Thus one should expect $r_1 = 4$ rather than $r_1 = 8$.

Consider n sorted points $z_1, \ldots, z_n \in [0, 1]$ with $z_1 = 0$ and $z_n = 1$. Suppose that each z_j is one of the $m \leq n$ points $\{\frac{i}{m-1} \; : \; 0 \leq i \leq m - 1\}$, which are *equidistantly* located. If $m < n$, then at least two points coincide. Let X be a random variable with uniform distribution on $[0, 1]$ and define $\delta(\xi)$ as the distance of ξ to the closest point z_j:

$$\delta(\xi) = \min_{1 \leq j \leq n} |z_j - \xi| \quad \text{for} \quad \xi \in [0, 1].$$

Then the expectation of $\delta(X)$ can be computed:

$$
\begin{aligned}
E(\delta(X)) = \int_0^1 \delta(\xi)d\xi &= \sum_{j=1}^{n-1} \int_{z_j}^{z_j+1} \min_{i \in \{j, j+1\}} |z_i - \xi| \, d\xi \\
&= \sum_{j=1}^{n-1} \frac{1}{4}(z_{j+1} - z_j)^2 = \sum_{j=1}^{m-1} \frac{1}{4(m-1)^2} = \frac{1}{4(m-1)}.
\end{aligned}
$$

Thus, we can reconstruct the number of distinct points m by means of $E(\delta(X))$:

$$m = \frac{1}{4 \cdot E(\delta(X))} + 1 = \frac{1}{\sum_{j=1}^{n-1}(z_{j+1} - z_j)^2} + 1.$$

This is the formula stated above. Note that this expression is continuous in each z_j, so if you take a point z_j that coincides with its successor and move it a little to the left, the left-hand side changes only a little, too. Hence a cluster that is for example formed by $z_1^{(k)}$, $z_2^{(k)}$ and $z_3^{(k)}$ in Fig. 2 yields almost the same value as a cluster of three coinciding points. This property enables the formula to be a good estimate for the number of different points. In our example (Fig. 2) the calculations yield $r_1 = 4.1045$ and $r_2 = 7.0624$.

Now one notes that the product of the requested grid sizes is in general much larger than n: We have $r_1 \cdot r_2 = 28.9880$ and $n = 8$ in our example. A simple reflexion explains this fact: If n points are equidistantly placed on the line through $(0, 0)$ and $(1, 1)$, the result will be $r_1 = n$ and $r_2 = n$, hence $r_1 \cdot r_2 = n^2$.

At this stage, we recall our assumption that the data set is not poorly scaled. Thus, each dimension can be treated in the same way, and we set

$$s_k = r_k \cdot \sqrt[d]{\frac{n}{r_1 \cdot r_2 \cdot \ldots \cdot r_d}} \quad \text{for each} \quad k \in \{1, \ldots, d\}$$

and obtain the desired grid size for each dimension having $s_1 \cdot s_2 \cdot \ldots \cdot s_d = n$, while the relations are preserved.

Unfortunately, $s_1, \ldots, s_d$ are no integer numbers yet. We could obtain integers by setting $g_k = \lceil s_k \rceil$. But since we are interested in a grid as small as possible, this is not a good choice. Thus, we simply try each reasonable combination $g_1, \ldots, g_d$. Since the dimension d is normally quite small, this is no drawback for the performance. For large d (about $d \geq 15$), a different method has to be used instead.

In order to select the best grid size, we define

$$q_1 = \max_{1 \leq k \leq d} |\log(\frac{g_k}{s_k})| \quad \text{and} \quad q_2 = \log(\frac{g_1 \cdot \ldots \cdot g_d}{n}).$$

The ratio q_1 can be considered as the bias of the desired size relations, while q_2 is the factor by which the grid is too large. We define the optimal grid size as the vector $(g_1, \ldots, g_d)$ for which $\max\{q_1, q_2\}$ attains its minimum while $g_1 \cdot \ldots \cdot g_d \geq n$. The number of grid points $g_1 \cdot \ldots \cdot g_d$ will be denoted by n_{grid} in the sequel. In our example (Fig. 2), the algorithm found the 2×4 grid to be optimal.

We point out that this method for calculating the grid size is most appropriate when the projections to the coordinate axis form significant clusters, otherwise the algorithm will produce a nearly quadratic grid. This is a desirable behaviour for data sets that are reasonably scaled and oriented. If the orientation is not appropriate, i.e. not roughly parallel to the coordinate axis, a transformation using the eigenvectors of the covariance matrix (see Section 4) can fix that problem. The algorithm is not appropriate for exploiting the scaling information of a poorly scaled data set, but this is normally a simple task which can be done by multiplying the requested grid sizes r_k with the corresponding scaling factors.

3   Allocation of the Points

Once the grid size has been computed, the allocation of the points is performed by a simple heuristic, similar to the heuristic that is often used for the TSP. We define the n_{grid} grid points $\gamma_1, \ldots, \gamma_{n_{grid}}$ as an arbitrary enumeration of the set

$$\left\{ \left(\frac{1}{2} + \frac{i_1 - 1}{g_1}, \ldots, \frac{1}{2} + \frac{i_d - 1}{g_d} \right) \quad : \quad 1 \leq i_1 \leq g_1, \ldots, 1 \leq i_d \leq g_d \right\}.$$

Then, we start with a random allocation of the points, see Fig. 3 (a). Each step of the heuristic selects randomly two edges, i.e. two existing connections between a point and a grid point, and exchanges the allocation if this reduces the sum of the distances. Instead of two edges, the heuristic can select one edge and one unused grid point (they exist if $n_{grid} > n$) and exchange the allocation if the distance is reduced. The heuristic aborts after $5000 \cdot n_{grid}$ steps maximum. Fig. 3 illustrates the function of the heuristic in our example. Note that the result (Fig. 3 (b)) is the optimal arrangement (Fig. 1 (d)).

Instead of the distances, the heuristic can also minimize the *square* distances. This results in a slightly different behaviour and is a little faster, since no square root has to be computed. If the standard grid defined above does not yield satisfactory results, a different grid as described in the next section can improve the performance.

4   Data Preprocessing and Self Organizing Maps

If the shape that is covered by the data points (x_j) differs from a (scaled) hypercube, both the calculation of the grid size and the heuristic that uses a rectangular grid (γ_i) may yield poor results. In many cases, a simple rotation of the data

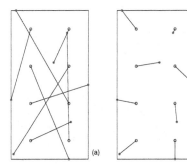

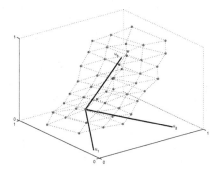

Fig. 3. Allocation of the points to the grid: (a) random initialization, (b) final result of the heuristic

Fig. 4. A data set in $\mathbb{R}^3$ and the coordinate system formed by the eigenvectors of the covariance matrix

is sufficient to fix this problem. To this aim, we apply a very efficient statistical method that makes use of the covariance matrix of the data. Let $\bar{x} = \frac{1}{n}\sum_{j=1}^{n} x_j$ and $X = ((x_1 - \bar{x}) \ \ldots \ (x_n - \bar{x}))$. Since XX' is symmetric, its eigenvectors $u_1, \ldots, u_d$ form an orthogonal basis. We set $U = (u_1 \ \ldots \ u_d)$ and $y_j = U'x_j$ for each $1 \leq j \leq n$. This transformation yields a data set that is oriented mainly parallel to the coordinate axis (Fig. 4).

If the data set has a more complex shape e.g with a curvature, the use of a Self Organizing Map (SOM) may be appropriate. This is a class of Neural Networks introduced by Kohonen, see [6] or [7] for details. Self Organizing Maps provide a powerful tool for classifying large sets of points in the d-dimensional space with regard to neighbourhood relations.

We take a rectangular SOM with a highly unbalanced aspect ratio and only few codebook vectors. Thus, the SOM can reproduce the curvature of the data, when it is trained with the points (x_j), see Fig. 5 (a). In the resulting map, we can define a coordinate system for each codebook vector (in the figure, this is shown for the third point from the left). With this information, we can "straighten" the SOM and obtain a transformation to a nearly rectangular grid, while the neighbourhood relations are preserved.

A grid defined by a SOM can be also very appropriate as a base for the heuristic, instead of the standard grid. For this aim, we build a SOM of size $g_1 \times \ldots \times g_d$ and train it with the data points (x_j), see Fig. 5 (b). The resulting codebook vectors define the grid (γ_i), which is rectangular and contains the neighbourhood information, but is adapted to the data shape. For such a grid, the heuristic that minimizes the square distances has shown to be best suitable.

5   Practical Tests

The single steps have been described by now. The complete algorithm for constructing a rectangular grid from given data points can be summarized as follows.

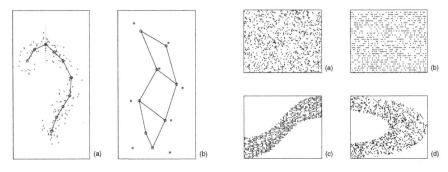

Fig. 5. Different applications of a SOM: (a) a SOM for data preprocessing, (b) the SOM defines the grid

Fig. 6. The test data sets, each defines 1000 cities for a TSP

Algorithm

1. *Data preprocessing.* Choose either no preprocessing, rotation according to the eigenvectors of the covariance matrix, or transformation using a SOM.
2. *Grid size calculation.*
3. *Grid definition.* Use either a standard grid or define the grid by a SOM.
4. *Choice of the heuristic.* Choose the sum of the distances or the sum of the square distances to be minimized.

The algorithm is implemented in the MATLAB environment. This implies that the performance is lower than, for example, C-code would be. This is tolerable since there are no time-consuming loops, except for the heuristic, which is therefore directly coded in C. For the SOM features, we use the SOM toolbox, see [8]. The genetic algorithm is the MATLAB implementation presented in [9].

For testing our algorithm, we used four rather large traveling salesman instances with 1000 cities each. The cities are shown in Fig 6. The first data set (a) simply consists of randomly placed points. The second (b) is defined by randomly placed points aligned to a grid. The third data set (c) is the second transformed to a sigmoid curve, and the last (d) is the first transformed to a half circle.

To estimate the performance of the grid allocations, there is a direct and an indirect way. The former consists in defining a measure for the preservation of neighbourhood relations under the grid transformation. To this aim, consider the distance matrix D^x defined by $D_{ij}^x = \|x_i - x_j\|$ $(1 \leq i, j \leq n)$ and the distance matrix D^γ that contains the distances after transformation to a standard rectangular grid. D^x is normalized such that $\sum_i D_{ij} = 1$ for each j. Moreover, we define a weight matrix W to be inverse proportional to the square of the distances and normalized: $W_{ij} = c \cdot (D_{ij}^x)^{-2}$ for $i \neq j$ and $W_{ij} = 0$ for $i = j$ and $\sum_i W_{ij} = 1$ for each j. Then, a measure M_j for the preservation of neighbourhood relations for one point x_j is defined as the weighted mean of all distance differences that arise from the grid transformation. Thereby the grid distances

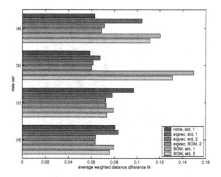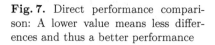

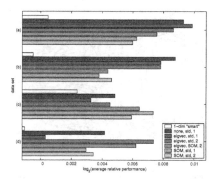

Fig. 7. Direct performance comparison: A lower value means less differences and thus a better performance

Fig. 8. Indirect performance comparison: A higher value means a shorter tour and therefore a better performance

must be scaled in order to achieve the best fit to the distances. As an explicit formula, this gives

$$M_j = \min_{s \in \mathbb{R}^+} \sum_{i=1}^{n} W_{ij} \cdot |D_{ij}^x - s \cdot D_{ij}^\gamma| \quad \forall \, 1 \le j \le n.$$

The measure M for the overall preservation of the neighbourhood relations is defined as the mean of all M_j.

Fig. 7 shows this evaluation of the grid transformation for the four test data sets and several allocation strategies with different preprocessing, grid definition and heuristic. Each combination corresponds to a color in the figure which is explained in the legend by a comma separated list: The first entry is the preprocessing method (none, eigenvectors or SOM), the second the grid definition (standard or SOM) and the third is the heuristic (1 for absolute distances and 2 for square distances). We observe that the evaluation for the grid allocation varies not very much, except for a larger value for the SOM preprocessing for the first two data sets. This is not unexpected since these data sets cover already a quadratic area, thus a rectangular SOM is likely to generate a worse arrangement.

The indirect evaluation of a grid arrangement is given in terms of the shortest city tour computed by a genetic algorithm using the grid arrangement for encoding. As a mutation operator, we employ the well known TSP-heuristic (see e.g. [10]) that randomly performs edge exchange and single point insertion. We use the same combinations for preprocessing, grid definition and heuristic as above. In addition, for reference purpose, we try a "smart" one-dimensional arrangement according to a run of the TSP-heuristic as suggested in [1] as well as an arbitrary one-dimensional arrangement. For each setting, 20 GA runs have been performed, from which we take the averages. The following GA settings

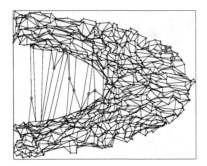

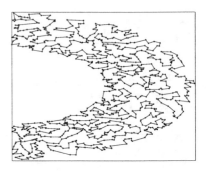

Fig. 9. Grid defined by a SOM with eigenvector preprocessing and quadratic heuristic for data set (d)

Fig. 10. The optimal path found with the coding shown in Fig. 9

were used: population size $\mu = 40$, number of generations $t_{max} = 100$, nonlinear ranking selection ($q = 7$), crossover and mutation probablity $p_{cross} = 0.5$ and $p_{mut} = 0.1$. Fig. 8 shows the resulting indirect performances relative to the average performance of the arbitrary one-dimensional arrangement. We use logarithmic (base 2) scale, where a positive value means a shorter average tour length.

The performance differences are minimal (less than 1%), but reproducible. The two-dimensional arrangement is always better than both the arbitrary and the smart one-dimensional arrangement. These relations remain valid also in earlier generations of the GA run, e.g. after $t = 30$, 50, or 80 generations. Thus, our coding implies a faster GA convergence. It is interesting to observe that the smart one-dimensional arrangement is sometimes worse than the reference. Further, there is no outstanding correlation between the direct and the indirect performance.

Fig. 9 and Fig. 10 show examples of a grid gained with a SOM and the resulting optimal path for the last test data set.

The time complexity of each of the steps is clearly linear in the number of points n. In the practical experiments with $n = 1000$ cities, the time for the first step ranges from 0 (no preprocessing) over 0.01 sec (eigenvector transformation) up to 0.8 sec (SOM). The second step (calculation of the grid size) needs about 0.01 sec. The standard grid definition takes no measurable time, while the SOM grid calculation costs 99 sec. This is due to the fact that a large amount of code-book vectors have to be trained. Nevertheless, even this is not much compared to the running time of the genetic algorithm which is in this case about 48 min. The last step, the heuristic, takes 2.3 sec for minimizing the distance sum and 1.1 sec for minimizing the square distance sum. In any case, the overall algorithm takes little time in relation to the following genetic algorithm.

6   Conclusions

The presented methods allow the efficient arrangement of a given set of data points in a rectangular grid under preservation of the neighbourhood relations. This can increase the performance of genetic algorithms using a locus-based representation, when each data point corresponds to a part of the chromosome.

The algorithms developed in this paper can be applied also in completely different situations. For example, to define the grid size of a SOM, the ideas from Section 2 can be used.

The natural encoding results in a small performance improvement for the TSP. Tested with other problems, the improvements gained by a d-dimensional encoding were similarly measurable, but small. On the other hand, there is no obvious relation between the direct performance measure (neighbourhood preservation) and the genetic performance. These facts raise some questions: Is there a direct performance measure that is correlated to the indirect performance? Are there (d-dimensional) encoding schemes that yield more improvement? What are characteristics of an optimal encoding scheme for a locus-based representation?

Acknowledgments. We thank Alexander Mitterer, Thomas Fleischhauer and Frank Zuber-Goos for helpful discussions. This research has been supported by the BMBF (grant no. 01 IB 805 A/1).

References

1. T. N. Bui and B. R. Moon. A new genetic approach for the traveling salesman problem. In *International Conference on Evolutionary Computation*, pages 7–12, 1994.
2. J. Grefenstette, R. Gopal, B. Rosmaita, and D. Van Gucht. Genetic algorithms for the travelling salesman problem. In *Proceedings of the first International Conference on Genetic Algorithms and Application*, pages 160–168, 1985.
3. A. Homaifar, S. Guan, and G. E. Liepins. A new approach on the traveling salesman problem by genetic algorithms. In *5th International Conference on Genetic Algorithms*, pages 460–466, 1993.
4. T. N. Bui and B. R. Moon. On multidimensional encoding/crossover. In *6th International Conference on Genetic Algorithms*, pages 49–56, 1995.
5. B. R. Moon and C. K. Kim. A two-dimensional embedding of graphs for genetic algorithms. In *7th International Conference on Genetic Algorithms*, pages 204–211, 1997.
6. T. Kohonen. *Self-Organization and Associative Memory*. Springer-Verlag, 3rd edition, 1989.
7. A. Zell. *Simulation neuronaler Netze*. Addison-Wesley, Bonn, 1994.
8. J Vesanto, J. Himberg, E. Alhoniemi, and J. Parhankangas. SOM Toolbox for Matlab 5. Technical Report A57, Helsinki University of Technology, http://www.cis.hut.fi/projects/somtoolbox/, April 2000.
9. J. Poland and K. Knödler et al. A genetic algorithm with variable alphabet coding for a new NP-complete problem from application. Preprint, 2000.
10. K. Knödler, J. Poland, A. Mitterer, and A. Zell. Optimizing data measurements at test beds using multi-step genetic algorithms. Preprint, 2000.

An Evolutionary Annealing Approach to Graph Coloring

Dimitris A. Fotakis[1], Spiridon D. Likothanassis[1], and Stamatis K. Stefanakos[2]

[1] Computer Engineering & Informatics Department,
University of Patras, 26500 Greece
{fotakis, likothan}@cti.gr
[2] Department of Computer Science, ETH Zurich,
CH-8092, Zurich, Switzerland
stefanak@iiic.ethz.ch

Abstract. This paper presents a new heuristic algorithm for the graph coloring problem based on a combination of genetic algorithms and simulated annealing. Our algorithm exploits a novel crossover operator for graph coloring. Moreover, we investigate various ways in which simulated annealing can be used to enhance the performance of an evolutionary algorithm. Experiments performed on various collections of instances have justified the potential of this approach. We also discuss some possible enhancements and directions for further research.

1 Introduction

In this work, we consider one of the most extensively studied NP-hard problems, the graph coloring problem (GCP). In the GCP, one wishes to color the vertices of a given graph with a minimum number of colors, such that no two adjacent vertices receive the same color. A coloring which uses k colors is called a k-coloring and can be regarded as the partition of the vertices of the graph to distinct color classes. The minimum number of colors (or the minimum color classes) needed to color a specific graph G, is called the chromatic number $\chi(G)$. Given a specific coloring, if two adjacent vertices have the same color we say that these vertices are in conflict and call the connecting edge between them, a conflicting edge. For an excellent bibliographic survey on the GCP, see [1].

Given the NP-hardness of the GCP, it is natural to use heuristic methods in order to tackle it. The simplest heuristics work on a given permutation of the vertices and color each one sequentially, with the minimal possible color. Other heuristics, like RLF [2] and Dsatur [3], try to generate the permutation of the vertices dynamically, as they proceed with the coloring.

In the case where a better solution is needed, more time-consuming methods have to be applied. Simulated Annealing (SA) [4] and Tabu Search (TS) [5] have been applied successfully to the GCP [6,7,8,9]. The basic idea behind TS and SA is to allow some uphill moves during the search process, hoping that the algorithm will be able to escape local minima.

E.J.W. Boers et al. (Eds.) EvoWorkshop 2001, LNCS 2037, pp. 120–129, 2001.

The first extensive study of the application of Genetic Algorithms (GAs) [10] to the GCP was made by Fleurent and Ferland [11][1]. Their work showed that GAs that use standard crossover operators are not able to compete on the GCP with heuristic algorithms based on TS and SA. On the other hand by maintaining a population of possible solutions and using a meaningful problem-specific crossover, together with a powerful local search method they were able to improve on the solutions obtained by local search, both in terms of solution quality and time. Since then, considerable work has been done in applying hybrid algorithms (i.e. algorithms combining two heuristics in order to improve the efficiency of both elements of the algorithm) to the GCP. Most of the proposed algorithms [13,14,15] use either TS or a simple steepest descent algorithm as the local search ingredient and utilize a specialized crossover in order to get the best results.

To the best of our knowledge, Simulated Annealing has never been used in a hybrid Genetic Algorithm for the GCP. A hybrid scheme like that seems very promising: by maintaining a population of solutions we can investigate a large part of the search space and move within promising areas of it by the use of suitable genetic operators. On the other hand SA gives us the opportunity to exploit (improve) quickly any good solutions that we might come up with and also serves as a fine mutation operator in order to move to new solutions and therefore keep the population diversity in high levels. In this paper we present a new hybrid genetic algorithm for the GCP which uses simulated annealing in order to enhance the search process and also a new crossover specialized for the graph coloring problem called Maximum Independent Set Crossover (MISC).

In the next section we present the generic form of our evolutionary annealing algorithm. In section 3 we present the GCP-specific details of our implementation. Then, we present the results we obtained on various test instances (section 4). Finally, in the last section we give some conclusions and ideas for future work.

2   The Generic Evolutionary Annealing Algorithm

In this section we present the Evolutionary Annealing (EVA) algorithm. We do not concentrate on any problem specific details but only present the building blocks of the algorithm in their generic form. We will deal with the GCP-specific details and the genetic operators in the next section.

The basic idea behind EVA is to maintain a population of solutions and to evolve them using genetic operators such as mutation and crossover. Using simulated annealing we can enhance the search process by two ways: first, we use simulated annealing to exploit quickly any good solutions that we might come upon with. This is achieved by starting the annealing process at low temperatures. On the other hand, when the process is stuck, we can use simulated annealing in order to generate different solutions using the existing ones; this is achieved by starting the cooling process from high temperatures and thus allowing many uphill moves. The main EVA procedure is given below:

---

[1] There exists also the earlier work of Davis [12], but he did not consider the combination of GAs with local search.

The main EVA algorithm
% *pop_size*: population size
% *p_mut*: mutation probability
% *off_size*: number of offsprings generated by crossover

> Initialize the population P with *pop_size* solutions
> **Repeat**
> > **Repeat**
> > > $P = $ evolve(P, *off_size*, *p_mut*)
> >
> > **Until** term_cond_GA $=$ TRUE
> > Choose a solution $s \in P$
> > $s_{new} = $ anneal(s)
> > Add s_{new} to the population and delete s
>
> **Until** term_cond_EVA $=$ TRUE
> Return the best solution found.

The procedure *evolve* implements a one-generation evolution process which is repeated in order to obtain a complete evolution process. The model used here is a variation of the $(\mu + \lambda)$ evolutionary algorithm. In our case, the offsprings are generated only by crossover and are included in the enlarged population, while the solutions that are selected for mutation are replaced directly (without competition) by the new ones. The exact procedure is presented here:

Algorithm evolve(P, *off_size*, *p_mut*)
% P: Population of pop_size solutions
% $f()$: cost function

> Evaluate the population P
> **Repeat**
> > Choose two solutions $s_1, s_2 \in P$
> > $s_c = $ crossover(s_1, s_2)
> > $s_c = $ anneal(s_c)
> > Compute $f(s_c)$
> > Add s_c to P
>
> **Until** iterations $=$ *off_size*
> $\forall s \in P$
> > with probability *p_mut1*
> > > $s = $ mutate(s)
> > > $s = $ anneal(s)
> >
> > with probability *p_mut2*
> > > $s = $ anneal(s)
>
> Discard the worst solutions from P until there
> are *pop_size* solutions left
> Return the new population

The basic idea behind simulated annealing is that the search process can escape local minima if we allow some uphill moves. Starting with an initial solution $s = s_0$ we choose at random a neighbor of s. If that move improves the cost function we perform it; otherwise we move to that solution with probability

$p = e^{-\Delta/t}$, where Δ is the deterioration in the cost function. The parameter t, which is called *temperature*, is used in order to control the acceptance ratio: In the beginning, the temperature is set to a high value, and therefore a move to a worse is solution is often; as the search goes on, the temperature is reduced by a factor a and thus the probability of a downhill move decreases. The initial temperature together with the factor a and the number of trials performed in each iteration is called the *cooling schedule*. The procedure *anneal* which is presented below, implements a full annealing cycle.

Algorithm anneal(s_0)
% s_0: initial solution
% $f()$: cost function
% $N(s)$: the neighbors of s
% We consider a minimization problem

> Compute an initial temperature t_0 depending on $f(s_0)$
> Let $t_0 = t_f \cdot t_0 + t_p$ where t_f, t_p are user-defined parameters
> **Repeat**
>> **Repeat**
>>> Compute the size of the neighborhood $N = |N(s_0)|$
>>> Let nrep= $N \cdot S_f$, where S_f is a user-defined parameter
>>> Randomly choose $s \in N(s_0)$
>>> $\Delta = f(s) - f(s_0)$
>>> **If** $\Delta < 0$
>>>> Let $s_0 = s$
>>> **Else**
>>>> With probability $e^{-\Delta/t_0}$ put $s_0 = s$
>>> **Until** iterations = nrep
>>> Let $t_0 = a \cdot t_0$
> **Until** term_cond_SA = TRUE
> Return s_0

Our implementation is similar to a typical SA algorithm with the exception that the neighborhood size is re-computed for every solution and therefore varies from time to time (in most implementations that we are aware of, the neighborhood size is computed once and remains the same during the whole optimization cycle). This feature makes a difference if we choose a neighborhood that reduces its size over the optimization process. As we shall see in the next section, that is the case in our implementation for the GCP.

The annealing procedure turns out to be a very important feature of our algorithm, as it is used after every crossover or mutation. The parameters that need to be defined by the user for the *anneal* procedure are namely t_f, t_p, s_f, a. These parameters are used so that we shall be able to control better the cooling schedule of the annealing process.

An interesting characteristic of our implementation is that the user is able to define different values for these parameters, according to which component of the algorithm calls the procedure *anneal*. For example, we are able to define a different cooling schedule for the annealing process that takes place in the initialization phase instead of the one performed after each crossover. This is very crucial for the efficiency of the algorithm, since it facilitates the full exploitation of the "cooling" process: In one extreme, we may use annealing as a random walk in the search space by using very high temperatures and therefore increase the diversity of the population when it becomes too low. It is well known that maintaining the diversity of the population in high levels is very important in GAs, because otherwise the crossover is not able to produce different solutions and that leads to premature convergence. On the other hand, we can see annealing as a greedy local search (using zero temperatures), which could be handy when we are dealing with good solutions (e.g. with 1 or 2 conflicts) that we want to improve.

3   GCP-Specific Details of Our Implementation

In order to find a proper coloring of a given graph $G(V, E)$, one usually starts out by trying to color the graph with a large enough number of colors k. Upon succeeding, the number of colors is reduced by one and a new attempt is made. Our evolutionary annealing algorithm (EVA) seeks out proper k-colorings and stops when it finds one.

Initialization. The first generation of solutions is created by a modified RLF algorithm, which uses at maximum k colors and assigns randomly those vertices that cannot be colored, within that range. Before the solutions are entered into the population, they are improved for a number of iterations by simulated annealing.

Search space and cost function. A solution (i.e. a not necessarily proper k-coloring) is a partition of the vertices of the graph in k color classes: $s = \{C_1, C_2, \ldots, C_k\}$. In order to evaluate such a solution we just count the number of conflicting edges. Thus, the cost function in both the *anneal* and *evolve* processes is

$$f(s) = |\ \{(u, v) \in E \mid u \in C_i \text{ and } v \in C_i\ ,\ 1 \leq i \leq k\}\ |.$$

Obviously, in order to find a proper k-coloring we need to minimize $f()$ and seek for a zero-cost solution.

Neighborhood operator. A solution s_2 is a neighbor of s_1 if the two solutions differ in a single location of a vertex v in the coloring classes, and additionally v is in conflict with another vertex. In order to obtain such a neighbor from a given solution s, we first choose at random a conflicting edge and then choose with equal probability one of its adjacent vertices. Finally, we move that vertex to another color class.

It is obvious that this neighborhood is not symmetric because a valid coloring would have no neighbors; apparently that is not a problem, for if we ever find a legal coloring there is no point in continuing any further with the search process. Another thing to be noted here is that not all the solutions have the same neighborhood size. As we have mentioned earlier, in order to speed up the search process, our annealing algorithm recomputes the neighborhood size for every solution considered, thus spending more time (making more trials) when starting off with a solution that has many conflicts.

Crossover operator. The crossover operator used in our implementation is called Maximum Independent Set Crossover (MISC), as it is based on the notion of maximizing the independent sets that already appear in the parent solutions, in a way similar to the RLF algorithm. Our crossover works on two parent solutions $s_1 = \{ C_1^1, \ldots, C_k^1 \}, s_2 = \{ C_1^2, \ldots, C_k^2 \}$ and produces an offspring solution $s = \{ C_1, \ldots, C_k \}$ as follows: In each step the chromatic class C_i of the offspring solution is constructed by setting $C_i = I_j^m$, where I_j^m is the largest independent subset of a chromatic class of the solutions s_m, $m = 1, 2$ (we check all the classes in order to find the largest independent subset). We consider two sets V, U: V contains the uncolored vertices, while U contains at step the vertices that can't be colored at that moment i.e. the vertices that are in conflict with the color class C_i that is being constructed. Initially V contains all the vertices of the graph and $U = \emptyset$. After setting $C_i = I_j^m$ we delete all the vertices of C_i from V and add to U all the vertices that are adjacent to a member of C_i. The vertices of $V - U$ are then added to C_i and are deleted from V. Finally we set $U = \emptyset$, delete the vertices of C_i from the solutions s_1, s_2 and repeat the process until $V = \emptyset$. The offspring is then improved for a number of iterations by the *anneal* procedure.

Mutation operator. The mutation operator used tries to give to each conflicting vertex v a feasible color. If that's not possible (meaning that its adjacent vertices have been already assigned to all k color classes), we then assign color $i \in [1, k]$ with probability proportional to

$$p(i) = d(v) - \phi(v, i),$$

where $d(v)$ is the degree of the vertex v and $\phi(v, i)$ is the number of the vertices adjacent to v that own the color i. As in the crossover operator, the outcome of the mutation is then improved by the *anneal* procedure. We hope that this mutation, will move the conflicts to another part of the graph when the search is stuck and thus in the long term we will be able to improve the solution.

4   Experimental Results

In this section we present the results we were able to obtain using the EVA algorithm on various test instances and compare them with other algorithms. We performed experiments on the following instances:

- Three random graphs ($G_{125, 0.5}$, $G_{250, 0.5}$, $G_{500, 0.5}$) with unknown chromatic number, which we generated using the random graph generators of LEDA [16].
- Twelve Leighton graphs with 450 vertices and chromatic numbers 5,15 and 25, which were taken from the 2nd DIMACS challenge benchmarks [17].
- Two application graphs, *school* and *school_nsh*, also used in [18].

We are interested in these particular instances because they have been largely studied in the literature and thus constitute a good reference for comparisons. Besides that, some of the Leighton graphs have proved to be very difficult. The application graphs, which were generated from class scheduling problems, are used in order to find out the performance and therefore the usefulness of EVA on such problems.

In order to evaluate the algorithm's performance we use the criterion of the quality of the best solution found, that is the number of colors k used. For the Leighton and the application graphs we check whether the algorithm is able to solve the instance optimally. If optimum cannot be found, we increase the number of colors used, until we find a feasible coloring. For the three random graphs, due to the fact that their chromatic number is unknown, we perform multiple runs for various values of k and measure the ratio of successful runs (meaning these that come up with a legal coloring).

Table 1. Results for random graphs per 5 runs.

Instance	k	succ (fail)	avg. t (sec)
$G_{125,0.5}$	16	(5)	-
	17	4 (1)	512.49
	18	5	115.51
	19	5	13.26
$G_{250,0.5}$	27	(5)	-
	28	1 (4)	50864
	29	2 (3)	7867.23
	30	5	2746.69
$G_{500,0.5}$	50	(5)	-
	51	1 (4)	140043
	52	3 (2)	67560
	53	5	36136

We compare our results with those given for three well known algorithms:

- DCNS. The algorithm of C. Morgenestern [8], based on Simulated Annealing. It features a distributed implementation with various child-processes coloring in parallel a population of possible solutions.
- PGC. This algorithm of G. Lewandowski and A. Condon [18] , is a parallel implementation based on DCNS and Johnson's XRLF [7].
- FF. This is a hybrid GA-TS algorithm by Fleurent and Ferland [11]. Their given results do not base on one method only: For the most difficult instances they first partially color the graph with an Independent Set Extraction method based on XRLF and then color the remainder graph with their algorithm.

In table 1 we present the results obtained by EVA on the three random graphs. In table 2 we give the best results that were obtained for all the instances and the parameters we used. Our algorithm is able to solve all but two Leighton

Table 2. Best results obtained by EVA and parameters used.

Instance	k	$|P|$	offsize	p_{mut1}	p_{mut2}	t (sec)
le450_5a	5	8	2	0.5	0.7	366
le450_5b	5	8	2	0.5	0.7	146
le450_5c	5	1	-	-	-	3
le450_5d	5	1	-	-	-	3.2
le450_15a	15	10	3	0.2	0.5	4010
le450_15b	15	8	3	0.2	0.1	481
le450_15c	15	10	3	0.2	0.2	73392
le450_15d	15	14	4	0.2	0.2	38755
le450_25a	25	1	-	-	-	2
le450_25b	25	1	-	-	-	2.1
le450_25c	27	8	3	0.2	0.1	224
le450_25d	27	1	0	0.2	0.1	18.07
$G_{125,0.5}$	17	6	2	0.2	0.2	411
$G_{250,0.5}$	28	14	2	0.1	0.1	50864
$G_{500,0.5}$	51	15	2	0.1	0.1	140043
school	14	8	3	0.3	0.5	12
school_nsh	14	8	3	0.3	0.3	29

graphs optimally. Let us note that four Leighton graphs (le450_5c, le450_5d, le450_25a, le450_25b) are easily solved in the initialization phase by the RLF algorithm. For the two graphs that were not solved optimally (le450_25c, le450_25d), the small execution times required to solve them with 27 colors, suggest that they should be solved with 26 colors, given sufficient running times. That was not the case though, as we were not able to find legal colorings with 26 colors for very large running times.

The two scheduling graphs constitute no problem as they are solved easily in very small running times. As is noted in [18], where more experiments with such graphs were conducted, several applications, like register allocation and scheduling, provide graphs which are easy to color. Our algorithm beats their parallel implementation (using 32 processors) in terms of running time and thus we believe that EVA is well suited for solving problems arising from such applications.

We present the comparative results in table 3. Only Morgenstern's algorithm is able to solve all Leighton graphs optimally. Fleurent and Ferland manage to color the two harder Leighton instances using one color less than our algorithm. For the smaller random graphs, EVA gives similar results with the other algorithms and is only out-beaten on the $G_{500,0.5}$ graph. In order to find a 51-coloring of $G_{500,0.5}$ we spent nearly 40 hours of computation time, while DCNS finds 49-colorings in half the time. Let us note here, that the three algorithms used for comparative purposes are among the best presented in the literature and each one uses a different technique in order to scale up well with the problem instance. Our algorithm does not use a distributed or parallel implementation, nor a pre-coloring technique that reduces the size of the instance, as the one used in FF.

Table 3. Comparative results ("-" means that the algorithm was not tested on that instance).

Instance	$\chi(G)$	DCNS	PGC	FF	EVA
le450_5a	5	-	5	5	5
le450_5b	5	-	5	5	5
le450_5c	5	-	5	5	5
le450_5d	5	-	5	5	5
le450_15a	15	-	15	15	15
le450_15b	15	-	15	15	15
le450_15c	15	15	16	15	15
le450_15d	15	15	16	15	15
le450_25a	25	-	25	25	25
le450_25b	25	-	25	25	25
le450_25c	25	25	27	26	27
le450_25d	25	25	27	26	27
$G_{125,0.5}$	-	17	17	-	17
$G_{250,0.5}$	-	28	29	-	28
$G_{500,0.5}$	-	49	52	49	51
school	14	-	14	-	14
school_nsh	14	-	14	-	14

Note that we do not take into consideration the running times of the algorithms; the only criterion used for the comparison is the number of colors needed to legally color a given graph. Comparing the execution times of the four algorithms is a difficult task, as each algorithm is run on a different machine. Our experiments carried over in a SUN Sparc 10 workstation, that was also accessible to other users.

5   Conclusions and Future Work

In this paper, we introduced a new hybrid algorithm for graph coloring combining a genetic algorithm and simulated annealing. Our algorithm maintains a population of possible solutions in order to investigate a large part of the search space and uses simulated annealing to improve quickly the best solutions found and produce new solutions in different parts of the search space. We also presented a new crossover for graph coloring, called Maximum Independent Set Crossover which builds the color classes of the offspring by expanding the color classes of the parents in an RLF fashion.

We conducted experiments on three classes of instances: random graphs, Leighton graphs, and application graphs. The experiments showed that our algorithm gives excellent results in terms of solution quality and speed for small graphs, but is not able to scale up its performance well as the instance size increases. This can be partly overcome by a parallel or distributed implementation which is not difficult due to the nature of our algorithm. Also, pre-coloring techniques can be used for large instances as presented in [11].

It is our opinion that the combination of genetic algorithms and simulated annealing is a very promising technique that we have slightly investigated in this work and a lot remain to be done.

References

1. P.M Pardalos, T. Mavridou, and J. Xue. The graph coloring problem: A bibliographic survey. In D.-Z. Du and P.M. Pardalos, editors, *Handbook of Combinatorial Optimization, Vol. 2*, pages 331–395. Kluwer Academic Publishers, 1998.
2. F.T. Leighton. A graph colouring algorithm for large scheduling problems. *Journal of Research of the National Bureau of Standards*, 84(6):489–503, 1979.
3. D. Brélaz. New methods to color the vertices of a graph. *Communications of the ACM*, 22(4):251–256, April 1979.
4. E. Aarts and J. Korst. *Simulated Annealing and Boltzmann Machines: A Stochastic Approach to Combinatorial and Neural Computing*. Interscience series in discrete mathematics and optimization. John Wiley & Sons, New York, 1989.
5. F. Glover. Future paths for integer programming and links to artificial intelligence. *Computers and Operations Research*, 13:533–549, 1986.
6. M. Chams, A. Hertz, and D. de Werra. Some experiments with simulated annealing for coloring graphs. *European Journal of Operational Research*, 32(2):260–266, 1987.
7. D.S. Johnson, C.R. Aragon, L.A. McGeoch, and C. Schevon. Optimization by simulated annealing: An experimental evaluation; part II, graph coloring and number partitioning. *Operations Research*, 39(3):378–406, May-June 1991.
8. C.A. Morgenstern. Distributed coloration neighborhood search. In [17], pages 335–357. American Mathematical Society, 1996.
9. A. Hertz and D. de Werra. Using tabu search techniques for graph coloring. *Computing*, 39(4):345–351, 1987.
10. Z. Michalewicz. *Genetic Algorithms + Data Structures = Evolution Programs*. Springer-Verlag, NY, 1992.
11. C. Fleurent and J.A. Ferland. Genetic and hybrid algorithms for graph coloring. *Annals of Operations Research*, pages 437–461, 1995.
12. L. Davis. Order-based genetic algorithms and the graph coloring problem. In *Handbook of Genetic Algorithms*. Van Nostrand Reinhold, 1991.
13. D. Costa, A. Hertz, and O. Dubuis. Embedding a sequential procedure within an evolutionary algorithm for coloring problems. *Journal of Heuristics*, 1:105–128, 1995.
14. P. Galinier and J.K. Hao. Hybrid evolutionary algorithms for graph coloring. *Journal of Combinatorial Optimization*, 1998.
15. R. Dorne and J.K. Hao. A new genetic local search algorithm for graph coloring, 1998.
16. K. Mehlhorn and S. Naher. *LEDA: A platform for combinatorial and geometric computing*. Cambridge University Press, 1999.
17. D.S. Johnson and M.A. Trick, editors. *Cliques, Coloring, and Satisfiability: Second DIMACS Implementation Challenge*, volume 26 of *DIMACS Series in Discrete Mathematics and Theoretical Computer Science*. American Mathematical Society, 1996. contains many articles on cliques and coloring.
18. G. Lewandowski and A. Condon. Experiments with parallel graph coloring heuristics and applications of graph coloring. In [17], pages 309–334. American Mathematical Society, 1996.

A Constructive Evolutionary Approach to School Timetabling

Geraldo Ribeiro Filho[1] and Luiz Antonio Nogueira Lorena[2]

[1] UMC/INPE - Av Francisco Rodrigues Filho, 399
08773-380 – Mogi das Cruzes – SP Brazil
Phone: 55-11-4791-3743
geraldo@lac.inpe.br

[2] LAC/INPE - Caixa Postal 515
12.201-970 São José dos Campos – SP - Brazil
Phone: 55-12-345-6553
lorena@lac.inpe.br

Abstract. This work presents a constructive approach to the process of fixing a sequence of meetings between teachers and students in a prefixed period of time, satisfying a set of constraints of various types, known as school timetabling problem. The problem is modeled as a bi-objective problem used as a basis to construct feasible assignments of teachers to classes on specified timeslots. A new representation for the timetabling problem is presented. Pairs of teachers and classes are used to form conflict-free clusters for each timeslot. Teacher preferences and the process of avoiding undesirable waiting times between classes are explicitly considered as additional objectives. Computational results over real test problems are presented.

1 Introduction

The timetabling problem consists in fixing a sequence of meetings between teachers and students in a prefixed period of time (typically a week), satisfying a set of constraints of various types. A large number of variants of the timetabling problem have been proposed in the literature, which differ from each other based on the type of institution involved (university or high school) and distinct constraints. A typical timetable instance requires several days of work for a manual solution[1].

Several techniques have been developed to automatically solve the problem[2, 3]. We therefore see algorithms based on integer programming[4], network flow, and others. In addition, the problem has also been tackled by reducing it to a well-studied problem: *graph coloring*[5]. More recently, some approaches based on search techniques appeared in the literature[6]; among others, we have *simulated annealing*[7], *tabu search*[8] and *genetic algorithms*[9, 10, 11].

We consider in this paper a problem known as *school timetabling*: the weekly scheduling for all the classes of a high school, avoiding teachers meeting two classes in the same time, and vice versa. Our main objective was to help administrative staff of public schools in Brazil. The particular characteristics observed for Brazilian public schools are:

E.J.W. Boers et al. (Eds.) EvoWorkshop 2001, LNCS 2037, pp. 130-139, 2001.
© Springer-Verlag Berlin Heidelberg 2001

- Full use of available rooms;
- Closed timetabling – at any timeslot all rooms are occupied;
- Usual timeslot conflicts of classes and teachers; and
- Soft constraints for teachers – preferences to some determined timeslots and in general avoiding the waiting timeslots (windows).

Genetic Algorithms (GA) are very well known, having several applications to general optimization and combinatorial optimization problems[12]. GA is based on the controlled evolution of a structured population, and is considered as an *evolutionary algorithm*[13]. The basis of a GA are the recombination operators and the schema formation and propagation over generations. This work presents an application of a Constructive Genetic Algorithm to school timetabling problems.

The Constructive Genetic Algorithm (CGA) is a recently developed approach of Lorena and Furtado [14] that provides some new features to GA, such as a population formed only by schemata, recombination among schemata, dynamic population size, mutation in complete structures, and the possibility of using heuristics in schemata and/or structure representation. Schemata do not consider all the problem data. The schemata are recombined, and they can produce new schemata or structures. New schemata are evaluated and can be added to the population if they satisfy an evolution test. Structures can result from recombination of schemata or complementing of good schemata. A mutation process is applied to structures and the best structure generated so far is kept in the process.

In this work, the school timetabling problem is considered as a clustering problem to be solved using the CGA. Our CGA application presents various new features compared to others GA applications to school timetabling. They include a specific representation for clustering problems, specialized recombination and local search mutation.

2 CGA Modeling

The CGA is proposed to address the problem of evaluating schemata and structures in a common basis. While in other evolutionary algorithms evaluation of individuals is based on a single function (the fitness function), in the CGA this process relies on two functions, mapping the space of structures and schemata onto $\mathfrak{R}_+$.

2.1 Representation

Considering p timeslots in a week, and respecting the lecture requirements of each class, we can form all possible pairs of (teacher, class), which should be implemented in the p timeslots. Let n be the total of possible pairs.

The soft constraints for teachers are considered implicitly encoded in the representation. The set of teachers is partitioned on three levels, according the number of classes and overall time dedicated to the school. All the teachers are asked to identify undesirable timeslots (preference constraints) conformable with their number of classes per week.

Pairs (teacher, class) are represented by binary columns. For example, considering 4 teachers and 5 classes, the column corresponding to the pair (2,3) is

$$
a =
\begin{array}{ll}
0 & \\
1 & \leftarrow \text{teacher 2} \\
0 & \\
0 & \\
-- & \\
0 & \\
0 & \\
1 & \leftarrow \text{class 3} \\
0 & \\
0 &
\end{array}
$$

The CGA works over a population of schemata (strings) formed by n symbols, one for each column. For example: $s = $ (#,0,0,0,#,0,1,#,1,0,0,0,1,#,#,0,#,0,1,0,0,0,1,#), is a possible schema. There are three possible symbols:

1   →   the corresponding column is a *seed* to form a cluster
            (there is always exactly p seeds inside each schema or structure);

0   →   the corresponding column is assigned to a cluster; and

→   the column is considered temporarily out of the problem.

The dissimilarity between two columns is then calculated to non-seed columns and all the other columns assigned to a cluster. The result is used to identify the cluster to which non-seed columns will be assigned. The dissimilarity measure between two columns is given by:

$$
d_{jk} = 1 - \frac{\sum_i \left| a_i^k - a_i^j \right|}{\sum_i \left(a_i^k + a_i^j \right)} \tag{1}
$$

where:

a_i^k is the value (zero or one) on position i at column k, and

a_i^j is the value (zero or one) on position i at column j.

To find out the cluster to which a non-seed column will be assigned,

- columns are ordered according to the teacher level and the number of preference constraints,
- we take the seed column that is most dissimilar,

the columns (the non-seed and the chosen seed) are merged into a single one (simple binary OR operation – see *figure 1* for an example) that becomes a new seed column. The process then continues until all non-seed columns are assigned to a cluster.

After columns to clusters assignments, exactly p clusters $C_1(s), C_2(s), ..., C_p(s)$ are identified, corresponding to the p available timeslots.

0	**0**	0
1	**1**	0
0	**1**	1
0	**0**	0
--	**--**	--
0	**1**	1
0	**0**	0
1	**1**	0
0	**0**	0
0	**0**	0

Fig. 1. Merging two strings.

2.2 Modeling

Let X be the set of all structures and schemata that can be generated by the *0-1-#* string representation of *section 2.1.*, and consider two functions f and g, defined as $f : X \rightarrow {}_+$ and $g : X \rightarrow {}_+$ such that $f(s_i) \le g(s_i)$, for all $s_i \in X$. We define the double fitness evaluation of a structure or schema s_i, due to functions f and g, as *fg-fitness*.

The *CGA* optimization problem implements the *fg-fitness* with the following two objectives:

(*interval minimization*)	Search for $s_i \in X$ of minimal $\{g(s_i) - f(s_i)\}$, and
(*g maximization*)	Search for $s_i \in X$ of maximal $g(s_i)$.

Considering the schema representation, the *fg-fitness* evaluation increases as the number of labels # decreases, and therefore, structures have higher *fg-fitness* evaluation than schemata. To attain these purposes, a problem to be solved using the CGA is modeled as the following *Bi-objective Optimization Problem* (BOP):

$$Min \quad \{g(s_i) - f(s_i)\}$$
$$Max \quad g(s_i)$$

$$\text{subj. to} \quad g(s_i) \ge f(s_i) \quad \forall s_i \in X$$

Functions f and g must be properly identified to represent optimization objectives of the problem at issue. For each schema $s_i \in X$, exactly p clusters $C_1(s_i), C_2(s_i), ..., C_p(s_i)$ are identified. Functions g and f are defined by

$$g(s_i) = \sum_{j=1}^{p} \left[\left(|C_j(s_i)| - 1 \right) |C_j(s_i)| \right] / 2 \text{ and } f(s_i) = g(s_i) - \sum_{j=1}^{p} \left[conflicts(C_j(s_i)) \right].$$

Considering graphs formed by vertices as columns and the edges as possible conflicts between columns (clashes of teachers or classes), function $g(s_i)$ can be interpreted as the total number of possible conflicts in p complete graphs of size

$\left|C_j(s_i)\right|$. Function $f(s_i)$ decreases this number by the true number of conflicts on the clusters $C_j(s_i)$. When $f(s_i) = g(s_i)$ the p clusters $C_j(s_i)$ are free of conflicts (a possible feasible solution).

3 The Evolution Process

The *BOP* defined above is not directly considered as the set X is not completely known. Instead we consider an evolution process to attain the objectives (*interval minimization* and *g maximization*) of the *BOP*. At the beginning of the process, the following two *expected values* are given to these objectives: a non-negative real number $g_{max} > Max_{s_i \times} g(s_i)$, that is an upper bound to $g(s_i)$, for each $s_i \in$ X; and the interval length $d\,g_{max}$, obtained from g_{max} using a real number $0 < d \le 1$.

Let $g_{max} = mult.p.\dfrac{(n/p-1).\,n/p}{2}$. This upper bound is obtained by dividing the number of vertices n in p clusters with approximately the same number of elements (the expression n/p gives the large integer smaller than n/p), and the same procedure used for $g(s_i)$ is applied, where the positive factor *mult* is considered to certify that $g_{max} > Max_{s_i \, X} g(s_i)$.

The evolution process is then conducted considering an adaptive rejection threshold, which contemplates both objectives in BOP. Given a parameter $\alpha \ge 0$, the expression

$$g(s_i) - f(s_i) \ge d\,g_{max} - \alpha.d[g_{max} - g(s_i)] \tag{2}$$

presents a condition for rejection of a schema or structure s_i from the current population.

The right hand side of (2) is the threshold, composed of the expected value to the interval minimization $d\,g_{max}$, and the measure $[g_{max} - g(s_i)]$, that shows the difference of $g(s_i)$ and g_{max} evaluations. For $\alpha = 0$, the expression (2) is equivalent to comparing the interval length obtained by s_i and the expected length $d\,g_{max}$. Schemata or structures are discarded if expression (2) is satisfied. When $\alpha > 0$, schemata have higher possibility of being discarded than structures, as structures present, in general, smaller differences $[g_{max} - g(s_i)]$ than schemata.

The *evolution parameter* α is related to time in the evolution process. Considering that the good schemata need to be preserved for recombination, α starts from 0, and then increases slowly, in small time intervals, from generation to generation. The population at the evolution time α, denoted by P_α, is dynamic in size according to the value of the adaptive parameter α, and can be eventually emptied during the process.

The parameter α is isolated in expression (2), yielding the following expression and the corresponding *rank* to s_i: $\alpha \geq \dfrac{dg_{\max} - [g(s_i) - f(s_i)]}{d[g_{\max} - g(s_i)]} = \delta(s_i)$

At the time they are created, structures and/or schemata receive their corresponding rank value $\delta(s_i)$. This rank is then compared to the current evolution parameter α. At the moment a structure or schema is created, it is then possible to have some figure of its survivability. The higher the value of $\delta(s_i)$, and better is the structure or schema to the BOP, and they also have more surviving and recombination time.

3.1 Selection, Recombination, and Mutation

Functions f and g defined in *section 2.2.* drives the evolution process to reach feasible solutions (structures free of conflicts), but the soft constraints are not directly considered. The selection will consider explicitly the soft constraints. Define a new measure $d(s_i) = \dfrac{[d_1(s_i) + w_{pref}.d_2(s_i) + w_{window}.d_3(s_i)]}{1 + w_{pref} + w_{window}}$, where

$d_j(s_i) = [g_j(s_i) - f_j(s_i)]/g_j(s_i)$, j = 1,2,3; w_{pref} is the preference constraint weight, w_{window} is the window constraint weight, and $g_1(s_i) = g(s_i)$, $f_1(s_i) = f(s_i)$ (as defined in section 2.2), $g_2(s_i) = $ number of columns, $f_2(s_i) = g_2(s_i) -$ number of columns with preference in conflict, $g_3(s_i) = $ number of columns, and $f_3(s_i) = g_3(s_i) -$ number of windows.

The population is kept in a non-decreasing order according to the following key: $\Delta(s_i) = (1 + d(s_i))/(n - n_\#)$, where $n_\#$ is the number of # labels in s_i. Schemata with small $n_\#$ and/or presenting small $d(s_i)$ are better and appear in first order positions.

The method used for selection takes one schema from the n first positions in the population (*base*) and the second schema from the whole population (*guide*). Before recombination, the first schema is complemented to generate a structure representing a feasible solution (all #'s are replaced by 0's). A mutation process is applied to this complete structure and it is compared to the best solution found so far, which is kept throughout the process. The recombination merges information from both selected schemata, but preserves the number of labels 1 (number of timeslots) in the new generated schema.

At each generation, exactly n new individuals are created by recombination. If a new individual is a schema, it is inserted into the population; otherwise the new individual is a structure, the mutation process is applied, and it is compared to the best solution found so far.

Recombination

---

if $s_{base}(j) = s_{guide}(j)$ then $s_{new}(j) \leftarrow s_{base}(j)$

if $s_{guide}(j) = \#$ then $s_{new}(j) \leftarrow s_{base}(j)$

if $s_{base}(j) = \#$ or 0 and $s_{guide}(j) = 1$ then

$\qquad s_{new}(j) \leftarrow 1$ and $s_{new}(i) \leftarrow 0$ for some $s_{new}(i) = 1$

if $s_{base}(j) = 1$ and $s_{guide}(j) = 0$ then

$\qquad s_{new}(j) \leftarrow 0$ and $s_{new}(i) \leftarrow 1$ for some $s_{new}(i) = 0$

The mutation process has three parts. The purpose of the first two parts is to repair infeasible solutions eventually produced by recombination. The third part maximizes soft constraints satisfaction.

The mutation process can be described as: 1) Class feasibility - for each cluster, while there are repeated classes in this cluster, find in other clusters a missing class on this one, and swap the columns; 2) Teacher feasibility - for each cluster, while there are repeated teachers in this cluster, find in other clusters a missing teacher on this one, for the same class, and swap the columns; and 3) Teacher preference improvement - make columns ordered according to teachers level and number of constraints, and for each column, if the teacher is constrained in the present cluster, find in other clusters a unconstrained teacher which is missing on this cluster and is feasible to swap, and swap the columns.

4 Computational Tests

The computational tests consider four instances, corresponding to two typical Brazilian high schools. Three periods were considered for the *Gabriel* school, respectively, *morning*, *afternoon* and *evening*, and only one period for the *Massaro* school.

When the tests were performed, the schools activities were already begun. The data used in the tests were taken from feasible solutions given by the schools administrative staff. As the teachers precedence levels and preference timeslots were unknown to us, this information were artificially generated. The set of teachers was partitioned into three levels, according to the number of classes and overall time dedicated to the school: teachers giving classes in less than 50% of the all timeslots in the week were considered at level three, between 50% and 75% were considered at level two, and those giving classes in more than 75% of the timeslots, had the precedence level considered one. Teachers in level one precedes the others and so on. The teachers undesirable timeslots (preference constraints) were artificially identified considering their number of classes per week and the real solution manually obtained by the schools administrative staff.

Tables 1 to 4 show the results for weights (w_{pref} and w_{window}) varying on the set {0, 0.5, 1}. Three runs were made for each weight combination and the average results are reported in the tables lines. The first column resumes the data: number of teachers, classes, timeslots and preference constraints (total and particularized for the teachers in level one). The other columns show the weight values, percentage of preferences attendance (total and for teachers in level one) and number of windows (total and for the teachers in level one) at the best schedule obtained.

It can be seen in the tables that the weights have direct influence on the soft constraints attendance. The percentages of attendance for teacher preferences and final number of windows are comparable to those obtained by manual schedule, aiming the possibility of being in future an important component of administrative school tools.

All tests were made considering the initial population composed of 100 schemata, generated randomly, and considering for each schema, 20% of it's positions filled with zeros, exactly p with ones and #'s in all remaining positions. For each algorithm run the maximum number of generations was set to 60, and 30 new schemata or structures were created at each generation. For each selection, the base schema was taken from the best 33% individuals of the population and the guide schema was taken from the whole population. Computational times reported correspond to a Pentium II 266 MHz machine.

Comparison between the computer generated solutions and real manually obtained solutions was not considered because of the lack of information like teachers preferences timeslots and teachers precedence level, which in practice can be very subjective.

5 Conclusion

The school timetabling problem is very challenging for public schools in Brazil. Several days of work are normally employed to manually solve these problems. We have proposed in this paper a constructive evolutionary approach to school timetabling problems. It considers the usual feasibility problem of teachers and classes allocation avoiding conflicts, and also some soft constraints, like teacher preferences and to avoid waiting times.

The problem was considered as a clustering problem, and adapted to the application of a recently proposed Constructive Genetic Algorithm (CGA). The CGA has been successfully applied to other clustering problems[14]. The weights used at the selection phase may extend the CGA to the class of multicriteria algorithms. The mutation process was highly specialized to this problem. Some algorithm parameter tuning can give even better results. Computational tests with real world instances was promising and the algorithm may result on a useful tool for Brazilian high schools.

Acknowledgments. The second author acknowledges Fundação para o Amparo a Pesquisa no Estado de S. Paulo - FAPESP (proc. 96/04585-6 and 99/06954-7) for partial financial support. The authors acknowledge the referees for their useful suggestions and comments.

Table 1. Results for Gabriel – morning.

Gabriel morning	W_{pref}	W_{window}	% prefer.	% prefer. (1)	Number of Windows	Number of Windows (1)	Times (sec.)
	0	0	89.39	83.33	55.00	12.33	718.67
	0	0.5	88.33	74.24	33.33	7.33	625.33
30 teachers	0	1	89.85	80.30	33.33	8.67	599.33
17 classes	0.5	0	93.18	83.33	43.00	8.67	687.33
5x5 Timeslots	0.5	0.5	91.52	81.82	36.00	7.33	632.00
220 pref.	0.5	1	90.91	81.82	37.00	10.00	601.33
22 pref. (1)	1	0	93.18	81.82	42.67	11.33	681.00
	1	0.5	92.12	87.88	35.67	7.33	628.00
	1	1	92.88	83.33	36.67	9.67	594.67

Table 2. Results for Gabriel – afternoon.

Gabriel Afternoon	W_{pref}	W_{window}	% prefer.	% prefer.(1)	Number of Windows	Number of Windows (1)	Times (sec.)
	0	0	92.75	75.76	48.67	6.00	840.33
	0	0.5	92.31	69.70	32.00	3.33	740.00
38 teachers	0	1	93.28	75.76	34.33	3.00	692.00
17 classes	0.5	0	94.52	75.76	49.67	3.33	758.67
5x5 timeslots	0.5	0.5	93.72	83.33	38.67	4.00	687.67
377 pref.	0.5	1	94.16	81.82	35.67	3.00	668.67
22 pref.(1)	1	0	95.05	84.85	51.33	4.33	732.00
	1	0.5	94.16	80.30	41.67	4.33	679.00
	1	1	93.37	77.27	35.67	4.33	648.67

Table 3. Results for Gabriel – evening.

Gabriel evening	W_{pref}	W_{window}	% prefer.	% prefer. (1)	Number of Windows	Number of Windows (1)	Times (sec.)
	0	0	88.17	75.31	25.00	4.67	574.33
	0	0.5	88.17	76.54	12.33	2.67	518.67
38 teachers	0	1	88.69	79.01	13.00	1.67	503.33
17 classes	0.5	0	90.59	77.78	22.67	3.33	486.33
5x4 timeslots	0.5	0.5	90.24	87.65	15.33	2.00	478.00
386 pref.	0.5	1	89.55	82.72	13.33	2.67	480.33
27 pref.(1)	1	0	90.85	76.54	26.67	2.33	451.00
	1	0.5	90.59	77.78	16.33	3.00	444.67
	1	1	89.90	83.95	16.33	3.00	446.33

Table 4. Results for Massaro.

Massaro	W_{pref}	W_{window}	% prefer.	% prefer. (1)	Number of windows	Number of windows (1)	Times (sec.)
	0	0	85.79	66.67	11.33	2.33	182.00
	0	0.5	88.80	86.67	4.67	0.33	169.67
18 teachers	0	1	89.89	76.67	4.00	1.67	163.00
11 classes	0.5	0	93.44	86.67	7.00	1.33	163.67
5x4 timeslots	0.5	0.5	92.62	93.33	4.00	0.67	159.00
122 pref.	0.5	1	93.17	96.67	6.33	1.00	160.00
10 pref. (1)	1	0	93.72	90.00	7.67	1.67	157.33
	1	0.5	93.44	86.67	5.33	1.00	158.33
	1	1	92.90	83.33	6.00	2.33	158.00

References

1. de Werra D.: An introduction to timetabling. European Journal of Operations Research. 19 (1985) 151–162.
2. Carter M. W.; Laporte G.: Recent developments in practical course timetabling. In Carter M. W.; Burke E. K. (eds.) Lecture Notes in Computer Science 1408. Springer-Verlag, Berlin (1998) 3–19.
3. Schaerf, A.: A survey of automated timetabling. Artificial Intelligence Review. n.13 (1999) 87-127.
4. Tripathy, A.: School timetabling : A case in large binary integer linear programming. Management Science, 30 (12) (1984) 1473-1489.
5. Neufeld, G. A.; Tartar, J.: Graph coloring conditions for existence of the solution to the timetabling problem. Communications of the ACM. n. 17 v.8 (1974).
6. Coloni A.; Dorigo, M.; Maniezzo, V.: Metaheuristics for high school timetabling. Computational Optimization and Applications. n. 9 (1998) 275-298.
7. Dowsland K. A.: Simulated annealing solutions for multi-objective scheduling and timetabling. In Modern Heuristic Search Methods. Wiley, Chichester, England (1996) 155–166.
8. Hertz A.: Tabu search for large scale timetabling problems. European Journal of Operations Research, 54 (1991) 39–47.
9. Burke E. K.; Elliman D. G.; Weare R. F.: A hybrid genetic algorithm for highly constrained timetabling problems. In Larry J. Eshelman (ed.) Genetic Algorithms: Proceedings of the 6th In-ternation Conference, San Francisco. Morgan Kaufmann. (1995) 605–610.
10. Burke E. K.; Newall J. P.; Weare R. F.: Initialization strategies and diversity in evolutionary timetabling. Evolutionary Computation 6(1) (1998) 81–103.
11. Corne D.; Ross P.; Fang H.: Fast practical evolutionary time-tabling. In Fogarty T.C. (ed.) Lecture Notes in Computer Science 865. Springer-Verlag, Berlin (1994) 250–263.
12. Holland, J.H.: Adaptation in natural and artificial systems. MIT Press (1975) 11-147.
13. Michalewicz Z.: Genetic Algorithms + Data Structures = Evolution Programs. 3rd edn. Springer Verlag, Berlin Heidelberg New York (1996).
14. Lorena, L. A N. and Furtado, J. C.: Constructive Genetic Algorithms for Clustering Problems. Evolutionary Computation - to appear (2000). Available from http://www.lac.inpe.br/~lorena/cga/cga_clus.PDF.

A Co-evolutionist Meta-heuristic for the Assignment of the Frequencies in Cellular Networks

Benjamin Weinberg[1], Vincent Bachelet[2], and El-Ghazali Talbi[1]

[1] USTL-LIFL-UPRESA 8022 CNRS,
59655 Villeneuve d'Ascq CEDEX, France
{Weinberg, Talbi}@lifl.fr
[2] CREGI-FUCaM B 7000 Mons, Belgium bachelet@fucam.ac.be

Abstract. This paper presents a new approach, the COSEARCH approach, for solving the Problem of Assigning Frequencies (FAP) on antennas of a cellular telecommunication network. The COSEARCH approach is a co-evolutionist method in which complementary meta-heuristics, such as genetic algorithm (GA) or tabu search (TS), cooperate in parallel via an adaptive memory (AM). We introduce an original encoding and two new cross-over operators suited to FAP. COSEARCH for the FAP is compared with other studies and its efficiency is revealed on both medium and large instances.

Keywords: frequency assignment problem, graph colouring problem, meta-heuristics, co-evolution, hybrid.

1 Introduction

In this paper, we consider the so-called Frequency Assignment Problem (FAP) for cellular telecommunication networks. It consists in the allocation of a set of frequencies on all antennas[1] of the network. Usually, radio resources are controlled by the government, so the telecommunication companies have to cope with expensive narrow ranges of frequencies. Moreover, the needs of consumers are growing exponentially. Thus, efficient assignments of frequencies on the network are required to exploit the radio resource in an (almost) optimal manner. Hence, industrials have to reuse the same frequencies on several antennas, unless they do not interfere. According to the propagation of waves, the signal power is decreasing with the distance; thus, equal frequencies should be assigned to distant stations to limit the interferences. This assignment problem can be formulated as a set-T-colouring problem, which is a $\mathcal{NP}$-hard problem. Several meta-heuristics have been proposed for solving this colouring problem: ant colonies system based on the greedy algorithm DSATUR [12], tabu search (TS) [9,3,4], genetic algorithms (GA) [8,11,10], and hybrid algorithms [6,7,5].

---

[1] Also called Base Transceiver Stations (BTS)

E.J.W. Boers et al. (Eds.) EvoWorkshop 2001, LNCS 2037, pp. 140–149, 2001.
© Springer-Verlag Berlin Heidelberg 2001

In the remainder of this article, we first present the FAP, followed by the encoding of a configuration and the operators: especially two new crossover operators, well-suited to the FAP. Then, the co-evolutionist meta-heuristic COSEARCH we have designed, is specified. We give the idiosyncrasies of the different agents which co-evolve. Finally, the results of COSEARCH are compared to other studies.

2   Frequency Assignment Problem

We use the formulation of the FAP as a set-T-colouring problem (STCP). This problem can be viewed as a generalised version of the well known graph colouring problem [9]. An instance of the STCP can be stated as follows. Given:

- the number of stations of the network $N \in \mathbb{N}$
- the set of stations $V = [1, N] \cap \mathbb{N}$
- the separation constants between two antennas $A : V \times V \longrightarrow \mathbb{N}$
$$v_i, v_j \longmapsto a_{ij}$$
- the number of required frequencies per antenna $T : V \longrightarrow \mathbb{N}$,
$$v \longmapsto T(v)$$

Find a configuration or a mapping $C : V \longrightarrow 2^{\mathbb{N}}$ such that $\forall v \in V, |C(v)| = T(v)$. Such a mapping fits with an assignment of a set of frequencies to each antenna of the network. For any configuration, the size of the interval of frequencies used, called *spectrum span* is stated as:

$$\chi(C) = \max_{v \in V} [\max_{f \in C(v)} f] - \min_{v \in V} [\min_{f \in C(v)} f]$$

The problem consists in the search of the minimal spectrum span configuration does not violate the separation constraints. These constraints are formulated as follows:

- frequencies assigned to different stations v and w must be separated by at least $A(v, w)$,

$$\forall v, w \in V, \ v \neq w \qquad \forall k \in C(v) \qquad \forall l \in C(w) \qquad |k - l| \geq A(v, w)$$

- frequencies assigned to the same station v must be separated by at least $A(v, v)$.

$$\forall v \in V, \qquad \forall k \in C(v), \qquad \forall l \in C(v), \qquad k \neq l, \qquad |k - l| \geq A(v, v)$$

3   Encoding and Operators

In the mobile telecommunication networks, the instances are strongly structured: the transceivers are grouped by stations and the stations are grouped by sites (one site is equipped with 1 to 3 stations). Hence, we propose a natural encoding of a configuration of the network which integrates this particular structure (see Figure1). A configuration is encoded by an array indexed by the stations. For each station, the set of frequencies is encoded as a sorted array. This sort is used because different assignments of the same set of frequencies to the same station are equivalent with regard to the problem, because transceivers of a same station are not differentiated.

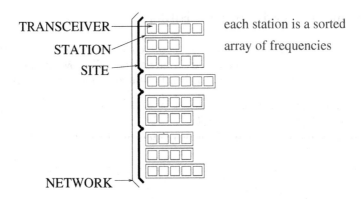

Fig. 1. Encoding of a configuration: an array of sorted array of frequencies.

We defined several operators for moving in the search space. The neighbour-hood operator, we have chosen, is the usual one: change one frequency assignment. This operator induces a notion of distance between configurations: the distance between two configurations is defined as the minimum number of applications of the neighbourhood operator to move from one configuration to the other one.

$$d(C_1, C_2) = card(\{f \mid \forall i \in V \ f \in C_1(i), \ f \notin C_2(i)\})$$

This distance is used in COSEARCH to guide the search. Aside the neighbour-hood operator, we use two complementary cross-overs. These new operators, we have designed, are well suited to the FAP. They preserve the structure of the problem, and thus do not act as blind operators. The first one, called SX for site cross-over, is a coarse grain cross-over. It works on site level. It realizes a two points cross-over on the whole configuration but does not separate neither the frequencies of a stations nor the stations of a site (see Figure 2). The second cross-over, FX for frequency cross-over, is a fine grained one. It shuffles the frequencies of stations. It requires two stages: the first stage acts on each set of frequencies of each station. This step works as a one point cross-over and generates two new sets. But one of these sets may be unavailable (unsorted), hence it is always discarded (even if, they are well sorted). At the second stage, two new configurations are built by mixing the stations. One cutting point is defined.

Parents   *Offsprings*

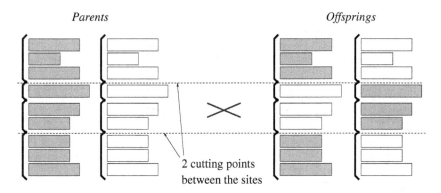

2 cutting points
between the sites

Fig. 2. Site cross-over (SX): a 2 points cross-over on site level.

Child 1 inherits, for station before the cutting point, of corresponding shuffled set of frequencies and inherits, for stations after the cutting points, of set of frequencies of parent 1. The opposite is done for child 2 (see Figure 3). The two cross-over operators, SX and FX, operate in complementary manners. SX does not split configurations of sites: complete sites are exchanged. Children may inherit well assigned sites from the parent configurations but no improvement into a site is done. On the contrary, with FX, assignments of the parents are crossed for each station separately, and the assignments on stations are improved.

To evaluate of a configuration, the cost function we use is stated as the number of violated constraints:

$$f(C) = \sum_{\substack{i,j \in V \\ f_{i,k} \in C(i) \\ f_{j,l} \in C(j)}} CI(f_{i,k}, f_{j,l}) + \sum_{\substack{i \in V, \\ f_{i,k} \in C(i) \\ f_{i,l} \in C(i) \\ k \neq l}} CO(f_{i,k}, f_{i,l})$$

where CI is the number of violated constraints for distant stations:

$$CI(f_{i,k}, f_{j,l}) = \begin{cases} 1 \text{ if } |f_{i,k} - f_{j,k}| < T_{ij} \\ 0 \quad\quad \text{else} \end{cases}$$

and CO is the number of violated constraints inside for a station:

$$CO(f_{i,k}, f_{i,l}) = \begin{cases} 1 \text{ if } |f_{i,k} - f_{i,k}| < T_{ii} \\ 0 \quad\quad \text{else} \end{cases}$$

4   COSEARCH for Solving the FAP

The meta-heuristic COSEARCH is based on three complementary agents. Theses agents evolve in parallel and cooperate through an adaptive memory (AM). COSEARCH was first applied to the Quadratic Assignment Problem [1]. As a

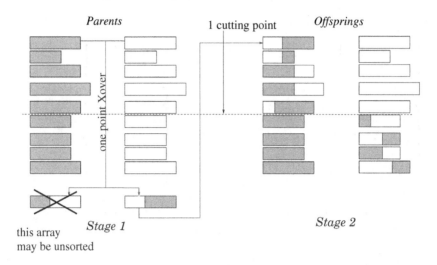

Fig. 3. Frequency cross-over (FX) shuffle the frequencies on each station.

key point for efficiency, the heuristic COSEARCH is intended to balance the exploration and the exploitation of the search space. The exploration is useful to guarantee a good approximation of the global optimum. The exploitation of a region of the space is useful because the efficient configurations are usually gathered in few regions, depending on the structure of the instance to be tackled.

The three complementary agents are:

- the searching agent SA, a neighbourhood-based method which improves an initial configuration;
- the diversifying agent DA, which provides new configurations in unexplored areas of the search space;
- the intensifying agent IA, which provides new configurations in already visited promising areas.

During the search, all agents refers to the AM which is accumulating the knowledge of the search space (see Figure 4). This AM essentially provides informations about the explored regions and the promising regions. It collects the set of visited configurations (or an approximation of this set), and the set of interesting configurations. In the COSEARCH approach, an interesting configuration corresponds to an efficient configuration and/or configuration being in a region that has not been explored. The exploration mechanism is based on a cycling strategy of diversification. The SA starts from initial the configurations of the AM, computes, and then provides back the set of visited configurations to the AM (explored regions). The DA refers to the explored regions and yields back configurations in unexplored regions. On the other side, in the intensifying cycle, the search agent starts from initial configurations, computes, and send efficient

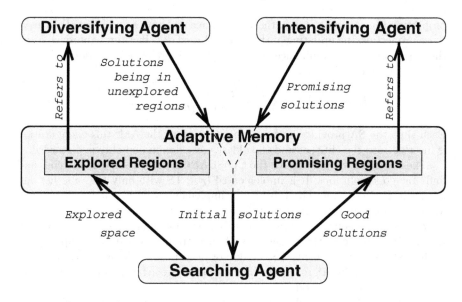

Fig. 4. Scheme of the co-evolutionist meta-heuristic COSEARCH.

configurations to the AM (promising regions). The AM collects the most interesting configurations. Afterward, the IA refers to the AM and provides new configurations in the good regions toward the SA. The balance of this two cycles is a key point of the robustness and the efficiency of the meta-heuristic COSEARCH. This balance is effected by the alternation of the cycles.

In the following of this section, we present the specifications of the meta-heuristic COSEARCH for tackling the FAP. We give first the details of the AM, then the different agents are detailed. In the AM, the storing of all explored configurations is approximated by a frequency matrix M. M is a $N \times F$ matrix, where N is the number of stations[2] and F is the current used spectrum span. In M, element $m_{f,s}$ is the number of assignments of the frequency f to the station s during the previous searches. We use an approximation of the explored regions because the complete information about the explored regions is too heavy in terms of memory. To evaluate the distance of a configuration from explored regions, we use the function g defined by:

$$g : \{\text{configurations}\} \longrightarrow \mathbb{N}$$
$$C \longmapsto \sum_{s,f \in C(s)} m_{s,f}$$

The greater $g(C)$ is, the closer c is to the already visited portion of the space.

The part of the AM concerned with the promising regions collects a limited number of "interesting" configurations called elites. In this set, elites have two

---

[2] The number of transceivers is much bigger.

properties: they are scattered in the search space and have a good cost. To achieve these properties, the updating of the elites is based on a bicriteria approach using the notion of Pareto dominance [13]. We define the relationship *"dominate"* on the set of elite $\mathcal{E}$ as follows:

$$c1 \; domines \; c2 \iff \left\{ \begin{array}{c} d(C1, \mathcal{E} \setminus \{C2\}) \geq d(C2, \mathcal{E} \setminus \{C1\}) \\ and \; f(C1) \leq f(C2) \end{array} \right.$$

Each time the SA provides its best configuration to the AM, this configuration is inserted into the elite set and the oldest most dominated elite is discarded.

The SA is the main agent for the CO-SEARCH architecture. It searches and finds the good solutions. It is implemented by a TS with only the tabu list and the aspiration criterion The tabu list is implemented by a matrix[3] $N \times F$ [5]. Moreover, we use a heuristic to reduce the neighbourhood: during the search of the best neighbour, the transceivers that are not interfered are not checked. Indeed, the change of the frequencies of such transceivers do not improve the configuration anyway. In the TS, we do not need long or medium term memory because the diversification and the intensification are achieved by the remainder of the COSEARCH method.

The DA finds new configurations in the unexplored area by examining the frequency matrix of the AM and minimising the function g. This new optimisation problem is $\mathcal{NP}$-hard. We do not try to solve it exactly because we do not need the farthest configuration, we just need a far enough configuration to contribute to the search. To solve this problem, we use a GA because they are well suited to dynamic environments [2]. Indeed, the frequency memory is continuously updated. The DA is steady state GA, using a SUS selection, the both evolutionary operators (FX and SX) involved concurrently in the GA, and the mutation operator which consists in changing a frequency of a antenna.

According to the fact that good configurations are concentrated in few regions, the IA refers to the set of elites and provides new configurations in good regions for the SA to start with. The IA is implemented by a short random walk, to make the SA exploit the good region represented by an elite.

5   Evaluation of COSEARCH

For solving the problem, a large spectrum span is allowed at the beginning. Then a feasible (no interference) configuration of this spectrum is searched. If such a configuration is found, the search is restarted reducing the spectrum by one. This process is iterated until no improvement is found or a time limit is exceeded. Each time the span is reduced, the new search is launched from the best configurations previously found in which the greatest values of frequencies are replaced by an uniform random smaller one.

The different parameters of the different agents are based on the diameter[4] of the instances. For the used distance, the diameter of an instance is the number of

---

[3] N is the number of stations and F the current number of frequencies.

[4] the greatest distance between any two configurations.

transceivers N'. The parameters of the agents of COSEARCH are set as follows:
the searching agent: SA

length of the tabu list $= \alpha_1 \times N' + rand(\alpha_1 \times N')$
number max of iterations $= \alpha_2 \times N'$

the intensifying agent: IA

length of the random walk $= \alpha_3 \times N'$

the diversifying agent: DA

number of generations $= \alpha_4 \times N'$
size of the population $= \frac{Log(1 - {}^{F \times N}\sqrt{\mu})}{Log(1 - \frac{1}{F})}$
where F is the first number of frequencies used, N is the number of
cells, μ is a coefficient of confidence. So we can guarantee with the
confidence μ that the whole diversity of the search space is represented
in the initial random population.
number of selections $= k$
mutation probability $= p$

The parameters are experimentally tuned:
$\alpha_1 = 1/8; \quad \alpha_2 = 2; \quad \alpha_3 = 1/8; \quad \alpha_4 = 1; \quad k = 20; \quad p = 0.1.$

COSEARCH was tested on several benchmarks provided by the France
Télécom R&D Department. In this instances, each BTS is equipped with 2
transceivers, the separation needed between transceivers of the same BTS is
always 3 and the separation needed between transceivers of different BTS is 1
for distant sites, 2 on the same site. The name of the instances are composed
by 4 elements $ms.nc..d.p.$
ms : information about the lower bounds.
nc : the number of station of the network.
d : the density of the network.
p : the average degree of a node.

The results on large instances are grouped in the table 1. We made, for each
benchmark, runs which number of iterations are bounded by 50 000, in ordred
to notice, the significance of diversification cycle and intensification cycle in the
FAP. For each instances, this table provides the minimum spectrum span, the
minimum excess found by COSEARCH, and the number of movements effected
by COSEARCH. the minimum spectrum span is a theorical lower bound. This
bound is know at the construction of the instance. So we reported in the table 1
the results with only the SA, then with SA and DA, finally with SA and IA. When
COSEARCH found a zero excess, it guarantees, that the found configuration is
the optimal one.

We found for each instance, a zero excess configuration as Dorne in [5]. We can
notice that for several instances, we found the best configuration without using

Table 1. Results on instances of STCP from France télécom R&D.

problem	ms	excess	nb moves	SA	SA+DA	SA+IA
15.150..05.30	30	0	128	0	0	0
15.150..05.60	30	0	116	0	0	0
15.150..15.30	30	0	2260	0	0	0
15.150..15.60	30	0	255	0	0	0
15.150..25.30	30	0	34152	3	2	1
15.150..25.60	30	0	10776	0	0	0
15.300..05.60	30	0	316	0	0	0
15.300..05.90	30	0	325	0	0	0
15.300..15.60	30	0	22359	-	-	0
15.300..15.90	30	0	8755	0	0	0
30.300..05.60	60	0	177	0	0	0
30.300..05.90	60	0	198	0	0	0
30.300..15.60	60	0	520	0	0	0
30.300..15.90	60	0	463	0	0	0
30.300..25.60	60	0	2516	0	0	0
30.300..25.90	60	0	1696	0	0	0

ms: minoring spectrum span;
excess: minimum_spectrum_span found by COSEARCH − ms;
nb moves: number of movements in the search space;
SA: the excess using only the searching agent several times with random initialization;
SA+DA: the excess using the searching agent several times with initialization provided by the diversifying agent;
SA+IA: the excess using the searching agent several times with initialization provided the intensifying agent.

any divrsification or intensification agent. In fact, for most of this instances only one run of the searching agent found the best result. They are relatively easy to solve. The "-", in the table, maks, that the procedure do not find a configuration with out any interferences using a spectrum larger of ten frequencies of the lower bound (ms). This studies shows that the intensifying cycle is more interesting for COSEARCH (with this parameters). We can notice that diversification is helpful for 15.150..25.30; Moreover in some other cases, it reduces the total number of iterations of the searching agent.

6   Conclusion

In this paper, we have presented the co-evolutionist meta-heuristic COSEARCH, which is based on the co-evolution of three complementary agents. We also see how it can be implemented for the Frequency Assignment Problem, using a genetic algorithm and a short tabu search. Advanced mechanism to reduce the neighbourhood in the tabu search and two complementary X-overs in the genetic algorithm are used. The reported experiments show the efficiency of our approach for several instances of the FAP in the mobile telecommunication networks. We are investigating now others adaptation of the COSEARCH's model, around the elite. Moreover we will parallelise COSEARCH for the Frequency Assignment Problem, and will implement it on the 64 processors of the IBM SP2 parallel machine.

References

1. V. Bachelet and E.G. Tabli. Co-search: A parallel co-evolutionary metaheuristic. In *Congress on Evolutionary Computation CEC'2000*, pages 1550–1557, San Diego, USA, 2000.
2. P. Bessiere, J.M. Ahuactzin, E-G. Talbi, and E. Mazer. The ariadne's clew algorithm: global planning with local methods. *IEEE International Conference on Intelligent Robots Systems IROS, Yokohama, Japan*, pages 1373–1380, july 1993.
3. A. Bouju, J.F. Boyce, C.H.D. Dimitropoulos, G. Vom Scheid, and J.G. Taylor. Tabu search for the radio links frequency assignment problem. In *Applied Decision Technologies*, pages 233–250, London ADT'95, april 1995.
4. A. Bouju, J.F. Boyce, C.H.D. Dimitropoulos, G. Vom Scheidt, and J.G. Taylor. Intelligent search for the radio links frequency assignment problem. In *Digital Signal Processing DSP'95*, pages –, University of Cyprus, june 1995.
5. R. Dorne. *Etude des méthodes heuristiques pour la coloration, la T-coloration et l'affectation de fréquences.* PhD thesis, Université de Montpellier II, 1998.
6. R. Dorne and J.K. Hao. An evolutionary approch for frequency assigment in cellular radio networks. In *IEEE Intl. Conf. on Evolutionary Computation (IEEE ICEC'95)*, pages 539–544, Perth, Australia, 1995.
7. R. Dorne and J.K. Hao. Constraint handling in evolutionary search: A case study of the frequency assignment. In *Parallel Problem Solving from Nature PPSN'96*, pages 801–810, Berlin Germany, sept 1996. Lecture Notes in Computer Science.
8. J.K. Hao and R. Dorne. Study of genetic search for the frequency assignment problem. In *Artifical Evolution AE'95*, pages 333–344, Brest France, 1996. Lecture Notes in Computer Science.
9. J.K. Hao, R. Dorne, and P. Galinier. Tabu search for frequency assignment problem in mobile radio networks. *Journal of Heuristics*, pages 47–62, june 1998.
10. S. Hurley, D. Smith, and C. Valenzuela. A permutation based genetic algorithm for minimum span frequency assignment. In *Parallel Problem Solving from Nature PPSN'5*, pages 907–916, Amsterdam, sept 1998. Springer-Verlag.
11. T.L. Lau and E.P.K. Tsang. Solving the radiolink frequency assignment problem with the guided genetic algorithm. In *NATO Symposium on Radio Length Frequency Assignment, Sharing and Conservation Systems (Aerospace)*, Aalborg, Denmark, Oct 1998.
12. V. Maniezzo and A. Carbonaro. An ants heuristic for the frequency assignment problem. In *Ants'98*, pages 927–935, North-Holland/Elsevier, Amsterdam, oct 1998. M.Dorigo.
13. R. Steuer. *Multiple criteria optimization: Theory, computation and application.* Wiley, New York, 1986.

A Simulated Annealing Algorithm for Extended Cell Assignment Problem in a Wireless ATM Network

Der-Rong Din[1] and Shian–Shyong Tseng[2]

Department of Computer and Information Science
National Chiao-Tung University
Hsinchu 300, Taiwan R.O.C.
deron@aho.cis.nctu.edu.tw, sstseng@cis.nctu.edu.tw

Abstract. In this paper, we investigate the *extended cell assignment problem* which optimally assigns new adding and splitting cells in PCS (Personal Communication Service) to switches in a wireless ATM (Asynchronous Transfer Mode) network. Given cells in a PCS and switches on an ATM network (whose locations are fixed and known), we would like to do the assignment in an attempt to minimize a cost criterion. The cost has two components: one is the cost of handoffs that involve two switches, and the other is the cost of *cabling*. This problem is modeled as a complex integer programming problem, and finding an optimal solution to this problem is *NP-hard*. A simulated annealing algorithm are proposed to solve this problem. The simulated annealing algorithm, ESA (enhanced simulated annealing), generates constraint-satisfy configurations, and uses three configuration perturbation schemes to change current configuration to a new one. Experimental results indicate that ESA algorithm has good performances.

Keywords: wireless ATM, PCS, optimization, simulated annealing, cell assignment problem.

1 Introduction

The rapid worldwide growth of digital wireless communication services motivates a new generation of mobile switching networks to serve as infrastructure for such services. Mobile networks being deployed in the next few years should be capable of smooth migration to future broadband services based on high-speed wireless access technologies, such as *wireless asynchronous transfer* mode (*wireless ATM*)[1]. The architecture shown in Fig. 1 was presented in [1]. In this architecture, the base station controllers (BSCs) in traditional PCS network are omitted, and the base stations (BSs or cells) are directly connected to the ATM switches. The mobility functions supported by the BSCs will be moved to the BSs and/or the ATM switches. In this paper, we address the problem that is currently faced by designers of mobile communication service and in the future, it is likely to be faced by designers of personal communication service (PCS).

E.J.W. Boers et al. (Eds.) EvoWorkshop 2001, LNCS 2037, pp. 150–159, 2001.

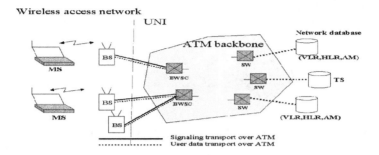

Fig. 1. Architecture of wireless ATM PCS.

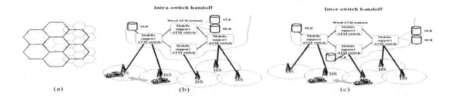

Fig. 2. (a)Cell splitting, (b)intra-switch handoff, (c)inter-switch handoff.

In the designing process of wireless ATM system, first the telephone company estimates the demand of mobile users and devided the global service area into local coverage areas defined by a mesh of hexagonal cells. Second, the cellular system, base stations, are constructed and connected to the switches on the ATM network. This topology may be out of date, since more and more users may use the PCS communication system. Some areas, which have not been covered in the originally designing plan may now have mobile users to traverse. The services requirement of some areas, which covered by original base stations may be increased and the capacities of the original base stations may be exceeded. Though, the wireless ATM system must be extended so that the system can provides higher quantity of services to the mobile users. Two methods can be used to extend the capacities of system and provide higher quantity of services. The first one is: adding new cells or base-stations (BSs) to the system so that the non-covered area can be covered by one of the new cells. The other is: reducing the size of the cells so that the total number of channels available per unit area and the capacity of a system can be increased. In practice, this can be achieved by performing *cell splitting*[2] process. The cell splitting process establishes new base stations at specific points in the cellular pattern and reduces the cell size by a factor of 2 (or more) as shown in Fig. 2(a).

In this paper, we are given a two-level wireless ATM network as shown in Fig. 3. In the PCS network, cells are divided into two sets. One is the set of cells, which are built originally, and each cell in this set has been assigned to a switch

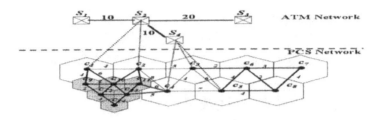

Fig. 3. Two-level wireless ATM network.

on the ATM network. The other is the set of cells which are newly added or established by performing the cell splitting process. Moreover, the locations of cells and switches are fixed and known. To simplify the discussion, we assumed that the number of cells and switches are fixed. The problem is to assign new adding and splitting cells in the PCS to switches on the ATM network in an optimum manner. We would like to do the assignment in as attempt to minimize a cost criterion. The cost has two components: one is the cost of (inter-switch) handoffs that involve two switches (as shown in Fig. 2(c)), and the other is the cost of *cabling* (or *trucking*)[3][4][5]. In the wireless environment, two types of handoffs should be considered in the design of PCS. They are inter-switch handoff and inter-switch handoff as shown in Fig. 2(b) and (c), respectively. We assume that the cost of (intra-switch) handoff involving only one switch is negligible[3][4][5]. Through this paper, we assume each cell to be connected to only one switch.

Merchant and Sengupta[3] considered the *cell assignment problem* which assigns cells to switches in PCS network. They formulated the problem and proposed a heuristic algorithm to solve it so that the total cost can be minimized. The total cost consists of cabling and location update. The location update cost considered in [3], which depend only on the frequency of handoff between two switches, is not practical. Since the switch of the ATM backbone is wide spread, the communication cost between two switches should be considered in calculating the location update cost. In [4][5], this model was extended to solve the problem that grouped cells into clusters and assigned these clusters to switches on the ATM network in an optimum manner by considering the communication cost between two switches. A three-phase heuristic algorithm and two genetic algorithms are proposed to solve the *cell assignment problem* in wireless ATM network, respectively. In this paper, we follow the objective function, which was formulated in [4] and [5], and solve the problem of assigning the new adding and the splitting cells to the switches on the ATM network so that the total cost can be minimized. This problem is defined as *extended cell assignment problem*[6].

2   Problem Formulation

Let $CG(C,L)$ be the PCS network, where C is a finite set of cells with $|C| = n$ and L is the set of edges such that $L \subseteq C \times C$. We assume that $C^{new} \cup C^{old} =$

C, $C^{new} \cap C^{old} = \emptyset$, C^{new} be the set of new adding and splitting cells where $|C^{new}| = n'$, cells in C^{new} have not yet been assigned to switches on the ATM, and C^{old} be the set of original cells where $|C^{old}| = n - n'$. Without loss of generality, we assume that cells in C^{old} and C^{new} are indexed from 1 to $n - n'$ and $n - n' + 1$ to n, respectively. If cells c_i and c_j in C are assigned to different switches, then a handoff cost is incurred. Let f_{ij} be the frequency of handoff per unit time that occurs between cells c_i and c_j, $(i, j = 1, ..., n)$ and is fixed and known. We assume that all edges in CG are undirected and weighted; and assume cells c_i and c_j in C are connected by an edge $(c_i, c_j) \in L$ with weight w_{ij}, where $w_{ij} = f_{ij} + f_{ji}$, $w_{ij} = w_{ji}$, and $w_{ii} = 0$[4][5][6]. Let $G(S, E)$ be the ATM network, where S is the set of switches with $|S| = m$, $E \subseteq S \times S$ is the set of edges, s_k, s_l in S and (s_k, s_l) in E and G is connected. We assume that the locations of cells and switches are fixed and known. The topology of the ATM network $G(S, E)$ is also fixed and known. Let d_{kl} be the minimal communication cost between the switches s_k and s_l. Let l_{ik} be the cost of cabling per unit time and between cell c_i switch s_k, $(i = 1, ..., n; k = 1, ..., m)$ and assume l_{ik} is the function of Euclidean distance between cell c_i and switch s_k.

Assume the number of calls that can be handled by each cell per unit time is equal to 1. Let Cap_k be the number of remaining cells that can be used to assigned cells to switch s_k. Our objective is to assign cells in C^{new} to switches so that the total cost (sum of cabling cost and handoffs cost) per unit time of whole system can be minimized.

To formulate this problem, let us define the following variables. Let $x_{ik} = 1$ if cell $c_i \in C$ is assigned to switch s_k; $x_{ik} = 0$, otherwise. Since each cell should be assigned to only one switch, we have the constraint $\sum_{k=1}^{m} x_{ik} = 1$, for $i = 1, .., n$. Further, the constraint on the capacity is $\sum_{i=n-n'+1}^{n} x_{ik} \leq Cap_k, k = 1, ..., m$. Also, the sum of cabling costs is $\sum_{i=1}^{n} \sum_{k=1}^{m} l_{ik} x_{ik}$.

To formulate handoff cost, variables $z_{ijk} = x_{ik} x_{jk}$, for $i, j, = 1, ..., n$ and $k = 1, ..., m$ are defined in [3]. Thus, z_{ijk} equals 1 if both cells c_i and c_j are connected to a common switch k; otherwise it is zero. Further, let $y_{ij} = \sum_{k=1}^{m} z_{ijk}, i, j = 1, ..., n$ Thus, y_{ij} takes a value of 1 if both cells c_i and c_j are connected to a common switch and 0 otherwise. With this definition, it is easy to see that the cost of handoffs per unit time is given by [4] [5] $\sum_{i=1}^{n} \sum_{j=1}^{n} \sum_{k=1}^{m} \sum_{l=1}^{m} w_{ij}(1 - y_{ij}) x_{ik} x_{jl} d_{kl}$

This, together with our earlier statement about the sum of cabling costs, gives us the objective function[4][5][6] is to $minimize \sum_{i=1}^{n} \sum_{k=1}^{m} l_{ik} x_{ik} + \alpha \sum_{i=1}^{n} \sum_{j=1}^{n} \sum_{k=1}^{m} \sum_{l=1}^{m} w_{ij}(1 - y_{ij}) x_{ik} x_{jl} d_{kl}$, where α is the ratio of the cost between cabling and handoff costs.

The following assumptions will be satisfied:

(1) We assume that the number of cells in C^{new} is less or equal to $\sum_{k=1}^{m} Cap_k$. That is, there is no need for adding new switches into the ATM network.

(2) The structures and locations of the ATM network and the PCS network are fixed and known.

(3) Each cell in the PCS network will be directly assigned and connected to only one switch in ATM network.

(4) To simplify the discussion, we assumed that $Cap_k > 0$, for $k = 1, ..., m$.
Example 1. Consider the two graphs shown in Fig. 3. There are 14 cells in CG which should be assigned to four switches in S. In CG, cells are divided into two sets, one is the set C^{old} of cells which are built originally, and cells in C^{old} have been assigned to switches in the ATM network (*e.g.*, $\{c_1, c_2, c_3, c_4, c_5\}$ in Fig. 3). The other is the set C^{new} of cells which are new adding cells (*e.g.*, $\{c_6, c_7, c_8\}$) or splitting cells (*e.g.*, $\{c_9, c_{10}, c_{11}, c_{12}, c_{13}, c_{14}\}$). The edge weight between two cells is the frequency of handoffs per unit time that occurs between them. Four switches are positioned at the center of cells c_1, c_2, c_4, and c_6.

3   Simulated Annealing Algorithm

Due to the complexity of the extended cell assignment problem in two-level wireless ATM network, the provision of an optimal solution in reasonable time is not guaranteed. In this respect, the usual step is to devise an approximate algorithm for solving this problem. The SA technique is applied to solve the extended cell assignment problem in this section.

In the design of simulated annealing algorithm[7], if the traditional-SA approach is used to solve the extended cell assignment problem, it may generate a significant number of configurations, but only a small fraction of these are indeed constraint-satisfied (10% or 20%). Thus, the performance of the traditional-SA algorithm is not promising. In this paper, we attempt to develop an *enhanced-SA* approach to solve the extended cell assignment problem by generating configurations which are feasible and satisfy all the constraints. The key elements in simulated annealing are a cost function, a configuration space, a perturbation mechanism, and a cooling schedule. In our case, the solution method is shown in follows.

3.1   Configuration Space and Perturbation Mechanism

The objective of the extended cell assignment problem in two-level wireless ATM network is to find an optimal assignment of new adding and splitting cells to switches so that the object function value is minimized. To solve the extended cell assignment problem, the configuration space is the set of feasible solutions which can be defined as a binary $\sum_{k=1}^{m} Cap_k \times m$ matrix X. Define x_{ik} in X to be 1 if the cell c_i is assigned to switch s_k; $x_{ik} = 0$, otherwise. Consider a feasible assignment of the example shown in Fig. 4(a), the configuration matrix of the assignment is shown in Fig. 4(a). In this example, two dummy cells c_{15} and c_{16} introduced to the configuration.

It is worth noting that the configuration can be divided into two parts, fixed and variable parts, the first part of matrix, which represents the assigning status of cells in C^{old}, is fixed in running of SA. Also, the first part of matrix can be ignored since it is never changed during experiments. For the reason of easily understanding, the fixed part of matrix is still kept in configuration in the rest of the paper. If the initial configuration is randomly generated or new configuration is formed by changing the assigning status of cell by way of randomly choosing cells and switches, then, there is a large chance that the configuration generated is not a feasible one. To avoid generating infeasible configurations, constraint-satisfying configurations and perturbation schemes must be constructed. To generate the constraint-satisfying initial configuration, we propose the following algorithm:

Algorithm: Initial Configuration Generating Algorithm(ICGA).

Step   1. Let $A = S$, $B = C^{new} \bigcup DM$.
Step   2. Repeat Steps 2.1, 2.2 and 2.3 until B is empty.
Step   2.1 Randomly select a switch s_a from A.
Step   2.2 Randomly assign $|Cap_k|$ cells in B to switch s_a and remove these cells form B.

Step   2.3 Remove s_a form A.

In the traditional-SA algorithm, the search process may stuck at a local minimum due to the small change moves, particularly as the barrier is high and the temperature is low. Hence, we introduce an idea of *large perturbation schema*[9] whose function is the same as the *mutation operation* in *genetic algorithm*. The main idea of large perturbation schema is to leap over the barrier during a search and to explore another region of the search space. This can be achieved by applying a certain number of moves consecutively, special problem domain perturbations, or local search heuristic perturbation. Three types of perturbations are introduced to the enhanced-SA algorithm which are shown in follows.

- *cells exchanging schema:* First, the cells exchanging schema randomly selects two cells c_i and c_j in $C^{new} \bigcup DM$, which have been assigned to different switches s_k and s_l, respectively. Then, reassign the cell c_i to the switch s_l and the cell c_j to the switch s_k. For example, assume cells c_6 and c_{10} are randomly selected; after performing cells exchanging perturbation, the configuration matrix is shown in Fig. 4(b).
- *multiple cells exchanging schema:* First, the multiple cells exchanging schema randomly selects two switches s_k and s_l from S. Then reassigns cells in the switch s_k to the switch s_l and *vice versa*. Since the original configuration is constraint-satisfied, after perturbation, the resulting configuration must be a feasible one. If we reassign cells in two switches directly, the resulting configuration may violate the constraints. Let COS_k be the set of cells that contains cells which are currently assigned to switch s_k. If $n_k > Cap_l$ and $n_l < Cap_k$ then randomly select Cap_l cells from the set COS_k and reassigned these cells to switch s_l; at the same time, all cells in COS_l are reassigned to the switch s_k. If $n_l > Cap_k$ and $n_k < Cap_l$, randomly select Cap_k cells from the set COS_l and reassigned these cells to the switch s_k; at the same time, all cells in COS_k are reassigned to the switch s_l. Otherwise, all cells being assigned to the switch s_k are directly reassigned to the switch s_l and *vice versa*. For example, assume switches s_1 and s_2 are randomly selected; after performing multiple cells exchanging, the configuration matrix is shown in Fig. 4(c) (Assume cells c_6 and c_7 are randomly selected from the switch s_1.)

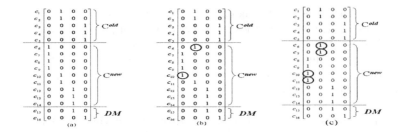

Fig. 4. (a)Possible feasible configuration for Example 1, (b)cell exchanging for Example 1c (c)multiple cells exchanging for Example 1.

It is worth noting that if the probabilistic decision mechanism of SA is disable, and two perturbation schemes are used to perturb the initial configuration to generate a new one, then, it is easily to prove that all possible feasible configurations can be reached by applying a sequence of perturbations. In the experiment, let p_1 and p_2 be the probabilities of transforming current configuration to a new one by applying cell exchanging schema and multiple cells exchanging schema, respectively. We assume $p_1 + p_2 = 1$ and the values of p_1 and p_2 will be empirically determined and described in later section.

3.2   Cooling Schedule

One of the most important problems involved in the simulated annealing algorithm implementation is the definition of a proper cooling schedule, which is based on the choice of the following parameters: starting temperature, final temperature, length of Markov chains, the way of decreasing temperature. A correct choice of these parameters is crucial because the performances of the algorithm strongly depend on it. In the following we describe these parameters.

- Initial value of the control parameter: The rule used in our enhanced-SA is determine starting temperature c_0 by calculating the average increasing in cost, $\overline{\Delta C}^+$, for 50 random transitions and solve c_0 from $c_0 = \overline{\Delta C}^+ / ln(\chi_0^{-1})$, where *accepted ratio* χ_0 defined as the number of accepted transitions divided by the number of proposed transitions. In this paper, the accepted ratio χ_0 is empirically set to value of 0.99.
- Decrement of the control parameter: The decreasing rate of the temperature needs to be small enough to reach thermal equilibrium for each temperature value. The decrement rule in enhanced-SA is defined as follows: $T_{k+1} = \gamma T_k$, where γ is empirically set to value of 0.99.
- The final value of the control parameter: The iterative procedure is terminated when there is no significant improvement in the solution after a prespecified number of iterations. It can also be terminated when the maximum number of iterations is reached.

– The length of Markov Chains: In general, a chain length of more than 100 transitions is reasonable. In this paper, the chain length is empirically set to value of n.

3.3 Enhanced-SA Algorithm of Extended Cell Assignment Problem

The details of the simulated annealing is described as follows:

Algorithm: Enhanced-SA Algorithm.

Step	1.	Perform **Initial Configuration Generating Algorithm** to generate initial configuration IC and an initial temperature T. The currently best configuration (CBC) is IC, i.e., $CBC = IC$, and the current temperature value (CT) is T, i.e., $CT = T$.
Step	2.	If $CT = 0$ or the stop criterion is satisfied then go to **Step 7**.
Step	3.	Generate a random number p in $[0, 1)$, if $p \leq p_1$ then new configuration (NC) is generated by applying cells exchanging schema; if $p_1 < p \leq p_1 + p_2$ then NC is generated by applying multiple cells exchanging schema; otherwise NC is generated by applying by applying local search heuristic schema.
Step	4.	The difference of the costs of the two configurations, CBC and NC is computed, i.e., $\Delta C = E(CBC) - E(NC)$.
Step	5.	If $\Delta C \geq 0$ then the new configuration NC becomes the currently best configuration, i.e., $CBC = NC$. Otherwise, if $e^{-(\Delta C/CT)} > random[0, 1)$, the new configuration NC becomes the currently best configuration, i.e., $CBC = NC$. Otherwise, go to **Step 2**.
Step	6.	The cooling schedule is applied, in order to calculate the new current temperature value CT and go to Step 1.
Step	7.	End.

4   Experimental Results

In this paper, a simulated annealing algorithm is proposed to solve the extended cell assignment problem. In order to evaluate its performances, we have implemented the algorithm and applied it to a number of examples with randomly positioned cells and switches, The results of these experiments are reported below. For all experiments, the implementation language is C, and some experiments have been made on Windows NT with a Pentium II 450MHz CPU and 256MB RAM. We simulated a hexagonal system in which cells are configured as an H-mesh. The handoff frequency f_{ij} of two cells were generated by a normal random number generator with mean 100 and variance 20. To examine effects of different number of cells, Cell Graph CG with $the number of cells n = 50$, 100, 150, and 200 cells were tested. $|C^{new}| = 3n/4$, $|C^{old}| = n/4$, the number of switches $m = 10$, $\alpha = 1$ and the Cap/n, which is the ratio of capacity to the number of cell, of each problem is 0.2.

To know the efficiency of the enhanced-SA algorithm to the traditional-SA algorithm, we also implement a traditional-SA algorithm that does not guarantee to generate constraint-satisfying configurations. The experimental results shown in Fig. 5(a) explain that the enhanced-SA algorithm have better performance than the traditional-SA algorithm. In other words, constraint-satisfying configurations and perturbation schemes of enhanced-SA algorithm are indeed more efficient than the traditional-SA algorithm and enhanced-SA algorithm has better convergent behaviors. To evaluate the effect of the probabilities of different perturbation schemes of enhanced-SA algorithm described in Section 3, we test the enhanced-SA algorithm with different values of probabilities p_1 and p_2.

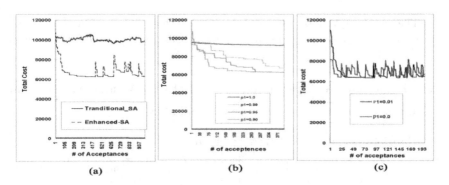

Fig. 5. (a) Comparison of the result of traditional-SA and enhanced-SA algorithms. (b) Comparison of the result of different probabilities of p_2. (c) Comparison of the result of different probabilities of p_2.

In the experiments, assume $p_2 = 1 - p_1$, and the value of p_1 is in $\{1.00,$ $0.99, 0.95, 0.90, 0.01, 0.00\}$. Figures 5(b) and 5(c) show the results of the experiment, where x-axis represents the number of acceptances when the configuration is perturbed in enhanced-SA algorithm and y-axis represents the total cost of the problem instance. When $p_1 = 1.00$ ($p_2 = 0.00$), i.e., the multiple cells exchanging perturbation does not activate in the experiment. We found that the enhanced-SA algorithm converges very slow and traps into local minima, since the cell exchanging perturbation only exchange two cells at one time. When $p_1 = 0.99$, $p_1 = 0.95$, or $p_1 = 0.90$, that is, only a small chance that the multiple cells exchanging perturbation may activate in the experiment, the enhanced-SA algorithm converges quicker than the case with $p_1 = 0$ and has very good performance as shown in Fig. 5(b). When $p_1 = 0.00$ or $p_1 = 0.01$, the single cell exchanging perturbation dose not activate or has very small chance to activate. As seen in Fig. 5(c), the current configuration up and down rapidly and hardly converges to the global minima. Thus, we can conclude that if the probability of cells exchanging perturbation is higher ($p_1 = 0.90$–0.99), the enhanced-SA algorithm has very good performance.

The results of the executions of the ESA for the above networks are given in Table 1. The ESA was run with fifty random seeds on each problem in order to get some statistical information about the quality of their solutions. The ESA column of Table 1 shows the minimum, mode, average, maximum, and standard deviation of the cost of solutions for fifty runs. The most repeated solution or the mode is close to the minimum cost solution for all example problems.

5   Conclusions

A simulated annealing algorithm ESA (enhanced-SA) is proposed to solve the extended cell assignment problem. Owing to the inability of *simulated annealing (SA)* to generate solutions that always satisfy all the constraints, the performance of a traditional-SA approach is not so promising. The SA technique is, however, easy to implement, requires little expert knowledge and is not memory intensive. Hence, in this paper, we have developed an *enhanced-SA* algorithm to solve the *extended cell assignment problem*. The enhanced-SA algorithm constructs constraint-satisfying configurations and perturbation mechanism to ensure that the candidate configurations produced are feasible and satisfy all the constraints. The performance of the enhanced-SA algorithm is demonstrated through simulation. The experimental results indicate that the proposed algorithm run efficiently.

Acknowledgment. This work was supported in part by MOE program of Excellence Research under Grant 89-E-FA04-1-4.

References

1. M. Cheng, S. Rajagopalan, L. F. Chang, G. P. Pollini, and M. Barton, "PCS Mobility Support over Fixed ATM Networks," *IEEE Communication Magazine*, Nov. 1997, pp. 82–91.
2. R. C. V. Macario, Cellular Radio. McGraw-Hill, New York (1993).
3. A. Merchant and B. Sengupta, "Assignment of Cells to Switches in PCS Networks", *IEEE/ACM Trans. on Networking,* Vol. 3, no. 5, 1995, pp. 521-526.
4. Der-Rong Din and S. S. Tseng, "Genetic Algorithms for Optimal design of two-level wireless ATM network," Technical Report, Department of Computer Science, NCTU, TR-WATM-9902 Taiwan, R.O.C, 1999; to appear in *Proceeding of NSC*.
5. Der-Rong Din and S. S. Tseng, "Heuristic Algorithm for Optimal design of two-level wireless ATM network, "Technical Report, Department of Computer Science, NCTU, TR-WATM-9901 Taiwan, R.O.C, 1999; to appear in *JISE*.
6. Der-Rong Din and S. S. Tseng, "Genetic Algorithm for Extended Cell Assignment Problem in Wireless ATM Network," accepted by ASIAN'00, Asian Computing Science Conference, Penang, Malaysia, November 25-27, 2000, to appear in *Lecture Note on Computer Science*.
7. Kirkpatrick S., Gelatt C. D. and Vecchi M. P. "Optimization by simulated annealing", *Science* 220, 671–680, 1983.
8. D. Raychaudhuri and N. Wilson, "ATM Based Transport Architecture for Multi-services Wireless Personal Communication Network," *IEEE JSAC*, Oct. 1994.
9. Ravinda K Ahuja, James B. Orlin and Dushyant Sharma, "Very large-scale neighbourhood search", *Int'l. Trans. in Op. Res.* 7 (2000) 301-317.

Table 1. Experiments of ESA algorithm.

				ESA			
$\mid n\mid$	$\mid m\mid$	Cap	Minimum	Mode	Average	Maximum	Std. Dev.
50	10	10	2476.7	2478.2	2480.4	2485.7	352.26
100	10	20	6011.3	6026.9	6022.9	6035.7	1141.1
150	10	30	14219.9	14236.8	14237.0	14257.0	4266.5
200	10	40	22052.6	22075.8	22075.9	22099.8	7289.7

7. Kirkpatrick S., Gelatt C. D. and Vecchi M. P. "Optimization by simulated annealing", *Science* 220, 671–680, 1983.
8. D. Raychaudhuri and N. Wilson, "ATM Based Transport Architecture for Multiservices Wireless Personal Communication Network," *IEEE JSAC*, Oct. 1994.
9. Ravinda K Ahuja, James B. Orlin and Dushyant Sharma,"Very large-scale neighbourhood search", *Int'l. Trans. in Op. Res.* 7 (2000) 301-317.

On Performance Estimates for Two Evolutionary Algorithms

Pavel A. Borisovsky[1] and Anton V. Eremeev[2]

[1] Omsk State University, Mathematical Department,
55 Mira str. 644077, Omsk, Russia
borisovski@mail.ru
[2] Omsk Branch of Sobolev Institute of Mathematics,
13 Pevtsov str. 644099, Omsk, Russia.
eremeev@iitam.omsk.net.ru
http://iitam.omsk.net.ru/~eremeev

Abstract. In this paper we consider the upper and lower bounds on probability to generate the solutions of sufficient quality using evolutionary strategies of two kinds: the (1+1)-ES and the (1,λ)-ES (see e.g. [1,2]). The results are obtained in terms of monotone bounds [3] on transition probabilities of the mutation operator. Particular attention is given to the computational complexity of mutation procedure for the NP-hard combinatorial optimization problems.

1 Introduction

The estimates for the behaviour of evolutionary strategies (ES) obtained in this paper are based on the a priori known parameters of the mutation operator, called monotone transition bounds. Using a model analogous to the one proposed in [3] we obtain in Sect. 2 the upper and lower bounds on the probability to generate a solution with fitness above given threshold on any iteration of the (1,λ)-ES and the (1+1)-ES. In this section some conditions are formulated when the latter heuristic has definite advantages over the first one. A connection with the complexity theory in Sect. 3 demonstrates some possible limitations on the a priori transition bounds of mutation operators implementable as polynomial-time randomized algorithms [4].

Let the optimization problem consist in finding a feasible solution $y \in Sol \subseteq D$, which maximizes the objective function $f : D \to R$, where D is the *space of solutions*, and $Sol \subseteq D$ is a *set of feasible solutions*. In general an evolutionary algorithm is searching for the optimal or near-optimal solutions using a population of individuals, which is driven by the principles observed in biological evolution. In both simplified versions of ES considered here the population consists of a single individual: in each iteration a new individual is constructed on the basis of the current one, and then the current individual is called a *parent*.

We assume that an individual is represented by a *genotype*, which is a fixed length string g of genes $g^1, g^2, \ldots, g^n$, and all genes are the symbols of some finite alphabet A. For example, the alphabet $\Sigma = \{0, 1\}$ is used in many applications.

E.J.W. Boers et al. (Eds.) EvoWorkshop 2001, LNCS 2037, pp. 161–171, 2001.

Each genotype g represents an element $y = y(g)$ of space D, which may not necessarily be a feasible solution.

The search process is guided by evaluations of nonnegative fitness function $\Phi(g)$, which defines the fitness of an individual with genotype g. Usually in case $y(g) \in Sol$, Φ is a monotone function of $f(y(g))$, and for simplicity in this paper we shall assume that $\Phi(g) = f(y(g))$. In case $y(g) \notin Sol$, the fitness function may incorporate a penalty for violation of constraints defining the set Sol. In what follows we will assume that for any $g \in Sol$ and $g' \in D\backslash Sol$, $\Phi(g) > \Phi(g')$ if $D\backslash Sol \neq \emptyset$.

The genotype of the current individual on iteration t of the $(1,\lambda)$-ES will be denoted by $b^{(t)}$, and in the $(1+1)$-ES it will be denoted by $g^{(t)}$. The initial genotypes $b^{(0)}$ and $g^{(0)}$ are generated with some a priori chosen probability distribution. The stopping criterion of the heuristics is the limit of the maximum number of iterations.

The only difference between the $(1,\lambda)$-ES and the $(1+1)$-ES consists in the method of construction of an individual for iteration $t + 1$ using the current individual of iteration t as a parent. In both algorithms the new individual is built with the help of a random *mutation operator* $Mut : A^n \to A^n$, which adds some random changes to the parent genotype. In case of the $(1,\lambda)$-ES the mutation operator is independently applied λ times to the parent genotype $b^{(t)}$ and out of λ offspring a single genotype with the highest fitness value is chosen as $b^{(t+1)}$. In case there are several offspring with the highest fitness, the new individual $b^{(t+1)}$ is chosen with uniform distribution among them. In the $(1+1)$-ES heuristic the mutation operator is applied to $g^{(t)}$ once. If $g = Mut(g^{(t)})$ is such that $\Phi(g) > \Phi(g^{(t)})$, then we set $g^{(t+1)} := g$; otherwise $g^{(t+1)} := g^{(t)}$.

In this paper the mutation operator will be viewed as a randomized algorithm which has an input g and a random output $Mut(g) \in A^n$ with probability distribution depending on g. We assume that all data of the given problem instance is available in mutation as well. One of the frequently used mutation operators consists in randomly changing each gene of g with a fixed mutation probability p_m. Another example is the operator which chooses a random position i and replaces the gene g^i by a new random symbol (see e.g. [5]).

The analysis of ES in principle could be carried out by the means of Markov chains theory (see e.g. [5]). However, the size of the transition matrix of a Markov chain grows exponentially as the genotype length increases, and the applicability of this approach appears to be limited when studying the optimization problems with large cardinality of solutions space. In order to overcome this difficulty we use a grouping of the states into larger classes on the basis of fitness.

1.1   Notations and Assumptions of the Model

The model of mutation operator used in this paper has been introduced in [3]. The information about fitness of distinct genotypes is not used explicitly. Instead, model makes use of certain a priori known parameters of probability distribution of the mutation operator described below.

Assume that there are d level lines of the fitness function fixed so that $\Phi_0 = 0 < \Phi_1 < \Phi_2 \ldots < \Phi_d$. The number of the level lines and the fitness values corresponding to them may be chosen arbitrarily, but they should be relevant to the given problem and the mutation operator to yield a meaningful model. Let us introduce the following sequence of subsets of the set A^n:

$$H_i = \{g : \Phi(g) \geq \Phi_i\}, \quad i = 0, \ldots, d.$$

Due to nonnegativity of the fitness function, H_0 equals to the set of all genotypes. Besides that, for the sake of convenience, let us define the set $H_{d+1} = \emptyset$.

Now suppose that for all $i = 0, \ldots, d$ and $j = 1, \ldots, d$ the a priori lower bounds α_{ij} and upper bounds β_{ij} on mutation transition probability from subset $H_i \backslash H_{i+1}$ to H_j are known, i.e. for every $g \in H_i \backslash H_{i+1}$ holds $\alpha_{ij} \leq P\{Mut(g) \in H_j\} \leq \beta_{ij}$, where $P\{Mut(g) \in H_j\} = \sum_{g' \in H_j} P\{Mut(g) = g'\}$.

Let $\mathbf{A}$ denote the matrix with elements α_{ij} where $i = 0, \ldots, d$, and $j = 1, \ldots, d$. The matrix of upper bounds β_{ij} is denoted by $\mathbf{B}$.

If for all $i = 0, \ldots, d$, and $j = 1, \ldots, d$ the probability $P\{Mut(g) \in H_j\}$ does not depend on choice of $g \in H_i \backslash H_{i+1}$ then there exist such matrices of probability bounds that $\mathbf{A} = \mathbf{B}$. In this case the mutation operator is called a *step mutation operator with respect to the sequence of subsets* $H_0, H_1, \ldots, H_d$ (or step mutation operator for short), and the matrix $\mathbf{A} = \mathbf{B}$ is called *the threshold transition matrix*. Here we should note that the step mutation operators do not seem to be natural for the real-life optimization problems, however they will be useful in further analysis as a sort of ideal operators.

A matrix $\mathbf{M}$ with elements $m_{ij}, i = 0, \ldots, d$, and $j = 1, \ldots, d$ will be called *monotone* if $m_{i-1,j} \leq m_{ij}$ for all i, j from 1 to d. In other words, the matrix of bounds on transition probabilities is monotone if for any $j = 1, \ldots, d$, the genotypes from any subset H_i have the bounds on transition probabilities to H_j not less than the bounds of the genotypes from any $H_{i'}$, $i' < i$.

In what follows we will focus the attention on the monotone matrices. Obviously, for any mutation operator the monotone bounds exist. (For example $\mathbf{A} = \mathbf{0}$ where $\mathbf{0}$ is a zero matrix and $\mathbf{B} = \mathbf{U}$ where $\mathbf{U}$ is the matrix with all elements equal 1.) The problem may be only with the absence of bounds which are sharp enough to evaluate the mutation operator properly. In fact if we conjecture that the probability distribution of a mutation operator is closely approximated by some monotone matrix of bounds on transition probabilities, this will imply that all solutions with high fitness values are relatively "easy" to reach from other solutions of high fitness. So the assumption that such "good" approximation exists, appears to be resembling the well-known "big valley" conjecture.

The distribution of the current individual in the $(1,\lambda)$-ES will be characterized (though not completely) by the *vector of probabilities*

$$P^{(t)} = (p_1^{(t)}, \ldots, p_d^{(t)}) = (P\{b^{(t)} \in H_1\}, \ldots, P\{b^{(t)} \in H_d\}),$$

which reflects the chances to have "good enough" genotypes on iteration t. By "good" genotypes we mean the genotypes with fitness above a certain threshold.

In [3] for a genetic algorithm (GA) based on the tournament selection and mutation it was shown that in the case of step mutation with monotone threshold transition matrix and a sufficiently large population, a GA with a smaller tournament size does not "outperform" the same GA with larger tournament size. Namely, for the first algorithm the probability that the offspring generated on iteration t belongs to $H_j, j = 1, ..., d$ is not greater than that for the latter one. This conclusion motivated our present study of the $(1,\lambda)$-ES, since its behavior is identical to the behavior of the GA with infinite tournament size in the framework of our model.

2   Bounds for the Probability of Obtaining "Good" Solutions

Our goal now is to estimate $P\{b^{(t)} \in H_j\}$ for all j. Note that there always exists such $0 \le i \le d$ that $b^{(t-1)} \in H_i \setminus H_{i+1}$, so by the total probability formula

$$p_j^{(t)} = \sum_{i=0}^{d} P\{b^{(t)} \in H_j | b^{(t-1)} \in H_i \setminus H_{i+1}\} P\{b^{(t-1)} \in H_i \setminus H_{i+1}\}.$$

Let us assume for convenience that $p_0^{(t-1)} = 1$ and $p_{d+1}^{(t-1)} = 0$. Then for all $i = 0, ..., d$ we have $P\{b^{(t-1)} \in H_i \setminus H_{i+1}\} = p_i^{(t-1)} - p_{i+1}^{(t-1)}$. Thus

$$P\{b^{(t)} \notin H_j | b^{(t-1)} \in H_i \setminus H_{i+1}\} = \sum_{b \in H_i \setminus H_{i+1}} \frac{P\{b^{(t)} \notin H_j | b^{(t-1)} = b\} P\{b^{(t-1)} = b\}}{P\{b^{(t-1)} \in H_i \setminus H_{i+1}\}} \le$$

$$\sum_{b \in H_i \setminus H_{i+1}} \frac{(1 - \alpha_{ij})^\lambda P\{b^{(t-1)} = b\}}{P\{b^{(t-1)} \in H_i \setminus H_{i+1}\}} = (1 - \alpha_{ij})^\lambda.$$

Proposition 1. *The components of $P^{(t)}$ are bounded as follows:*

$$p_j^{(t)} \ge 1 - (1 - \alpha_{0j})^\lambda + \sum_{i=1}^{d} ((1 - \alpha_{i-1,j})^\lambda - (1 - \alpha_{ij})^\lambda) p_i^{(t-1)} \qquad (1)$$

for $j = 1, \dots, d$, and (1) is an equality in case of a step mutation operator.

Let us define for $(d \times d)$-matrix $\mathbf{W}$ a matrix norm $||\mathbf{W}|| = \max_j \sum_{i=1}^{d} |w_{ij}|$. To convert recurrent bound (1) into a bound for a general term of sequence $\{P^{(t)}\}$ we will use the following result from [3].

Lemma 1. *Let $\mathbf{W}$ be a $(d \times d)$-matrix with nonnegative elements such that $||\mathbf{W}|| < 1$. Suppose that $\alpha \in R^d$, $\beta \in R^d$ and $\{\zeta^{(t)}\}, t = 0, 1, \dots$ is a sequence of vectors from R^d.*
(a) If $\zeta^{(t+1)} \ge \zeta^{(t)} \mathbf{W} + \alpha$ for all $t \ge 0$, then

$$\zeta^{(t)} \ge \zeta^{(0)} \mathbf{W}^t + \alpha(\mathbf{I} - \mathbf{W})^{-1}(\mathbf{I} - \mathbf{W}^t). \qquad (2)$$

(b) If $\zeta^{(t+1)} \le \zeta^{(t)} \mathbf{W} + \beta$ for all $t \ge 0$, then

$$\zeta^{(t)} \le \zeta^{(0)} \mathbf{W}^t + \beta(\mathbf{I} - \mathbf{W})^{-1}(\mathbf{I} - \mathbf{W}^t), \tag{3}$$

were $\mathbf{I}$ denotes the identity matrix.

Note that the right-hand side of (2) approaches $\alpha(\mathbf{I} - \mathbf{W})^{-1}$ when t tends to infinity, regardless of the initial vector $\zeta^{(0)}$.

Now we apply Lemma 1 to matrix $\mathbf{W}'$ with elements $w'_{ij} = (1 - \alpha_{i-1,j})^\lambda - (1 - \alpha_{i,j})^\lambda$ (which are nonnegative if $\mathbf{A}$ is monotone), and $\alpha' = (1 - (1 - \alpha_{01})^\lambda, \ldots, 1 - (1 - \alpha_{0d})^\lambda)$. Also let us require that $\alpha_{dj} - \alpha_{0j} < 1, j = 1, \ldots, d$ which is sufficient for $\|\mathbf{W}\| < 1$.

Theorem 1. *If $\mathbf{A}$ is monotone and $\alpha_{dj} - \alpha_{0j} < 1$ for $j = 1, \ldots, d$, then $P^{(t)} \ge \zeta^{(t)}$ where $\zeta^{(t)}$ is bounded according to (2) with $\zeta^{(0)} = P^{(0)}, \mathbf{W} = \mathbf{W}'$, and $\alpha = \alpha'$.*

Note that in many practical algorithms an arbitrary given genotype may be produced with non-zero probability during mutation. In such cases the condition $\alpha_{dj} - \alpha_{0j} < 1$ is obviously satisfied for all j.

The upper bounds for the components of $P^{(t)}$ can be obtained from matrix $\mathbf{B}$ in the same way. Let us denote $\hat{w}'_{ij} = (1 - \beta_{i-1,j})^\lambda - (1 - \beta_{i,j})^\lambda, i = 1, \ldots, d, j = 1, \ldots, d$, and $\beta' = (1 - (1 - \beta_{01})^\lambda, \ldots, 1 - (1 - \beta_{0d})^\lambda)$.

Theorem 2. *If $\mathbf{B}$ is monotone and $\beta_{dj} - \beta_{0j} < 1$ for $j = 1, \ldots, d$, then $P^{(t)} \le \zeta^{(t)}$ where $\zeta^{(t)}$ is bounded according to (3) with $\zeta^{(0)} = P^{(0)}, \mathbf{W} = \widehat{\mathbf{W}}'$, and $\beta = \beta'$.*

In case of step mutation operator we have $\mathbf{A} = \mathbf{B}$, and the lower and upper bounds of theorems 1 and 2 coincide. In this case we can consider $P^{(t)}$ as a function on $P^{(0)}$, and denote $P^{(t)} = P^{(t)}(P^{(0)})$.

One can note that in the case of step monotone mutation on any iteration t there is no use to continue generating the offspring genotypes if we have already found a genotype with higher fitness than $\Phi(b^{(t)})$ (then it is reasonable to use the new genotype as a parent). Besides that if all offspring have lower fitness than $\Phi(b^{(t)})$, there is no sense to substitute $b^{(t)}$ by the best offspring.

This observation leads us to investigation of the (1+1)-ES and to its comparison to the (1,λ)-ES in a formal setting. For the (1+1)-ES we define the following vector of probabilities:

$$Q^{(t)} = (q_1^{(t)}, \ldots, q_d^{(t)}) = (P\{g^{(t)} \in H_1\}, \ldots, P\{g^{(t)} \in H_d\}).$$

By means of the total probability formula we conclude that

$$q_j^{(t)} \ge \sum_{i=0}^{j-1} \alpha_{ij} P\{g^{(t)} \in H_i \setminus H_{i+1}\} + q_j^{(t-1)} = \sum_{i=0}^{j-1} \alpha_{ij}(q_i^{(t-1)} - q_{i+1}^{(t-1)}) + q_j^{(t-1)}.$$

Proposition 2. *The components of $Q^{(t)}$ are bounded as follows:*

$$q_j^{(t)} \geq \alpha_{0j} + \sum_{i=1}^{j-1}(\alpha_{ij} - \alpha_{i-1,j})q_i^{(t-1)} + (1 - \alpha_{j-1,j})q_j^{(t-1)} \tag{4}$$

for $j = 1, \ldots, d$, and (4) is an equality in case of step mutation operator.

We can also apply Lemma 1 defining matrix $\mathbf{W}''$ and vector α'' as follows:

$$w_{ij}'' = \begin{cases} \alpha_{i,j} - \alpha_{i-1,j}, & i < j, \\ 1 - \alpha_{j-1,j}, & i = j, \\ 0, & i > j, \end{cases} \qquad \alpha'' = (\alpha_{01}, \ldots, \alpha_{0d}).$$

Theorem 3. *If $\mathbf{A}$ is monotone and $\alpha_{0j} > 0$ for $j = 1, \ldots, d$, then $Q^{(t)} \geq \zeta^{(t)}$ where $\zeta^{(t)}$ is bounded according to (2) with $\zeta^{(0)} = Q^{(0)}, \mathbf{W} = \mathbf{W}''$, and $\alpha = \alpha''$.*

It is easy to verify that $\mathbf{1} = \alpha'' + \mathbf{1}\mathbf{W}''$, where $\mathbf{1} = (1, \ldots, 1)$. Thus, naturally, $P\{g^{(t)} \in H_d\} \to 1$ when $t \to \infty$.

Analogously let us denote

$$\hat{w}_{ij}'' = \begin{cases} \beta_{i,j} - \beta_{i-1,j}, & i < j, \\ 1 - \beta_{j-1,j}, & i = j, \\ 0, & i > j, \end{cases} \qquad \beta'' = (\beta_{01}, \ldots, \beta_{0d}).$$

Theorem 4. *If $\mathbf{B}$ is monotone and $\beta_{0j} > 0$ for $j = 1, \ldots, d$, then $Q^{(t)} \leq \zeta^{(t)}$ where $\zeta^{(t)}$ is bounded according to (3) with $\zeta^{(0)} = Q^{(0)}, \mathbf{W} = \widehat{\mathbf{W}}''$, and $\beta = \beta''$.*

2.1   Comparison of the (1+1)-ES with the (1,λ)-ES

The superiority of the (1+1)-ES over the (1,λ)-ES in case of monotone step mutation has been mentioned above and now it will be formally supported. To compare the performance of both algorithms here we will require to have the same number of evaluations of the fitness function in each of them.

Theorem 5. *Suppose that the same step mutation operator with a monotone threshold transition matrix is used in the (1+1)-ES and in the (1,λ)-ES. If $Q^{(0)} \geq P^{(0)}$ then $Q^{(t\lambda)} \geq P^{(t)}$ for any t.*

Proof. It is sufficient to give a proof for $t = 1$, since the statement for the general case will follow by induction on t.

a) Consider the moment when k offspring of $b^{(0)}$ have been generated and let $b^{(1,k)}$ denote an offspring with the highest fitness among them. Let us first assume that $b^{(0)}$ and $g^{(0)}$ are fixed so that $b^{(0)} \in H_i \setminus H_{i+1}$ and $g^{(0)} \in H_i \setminus H_{i+1}$ for some i. Then for the (1,λ)-ES for arbitrary $j = 1 \ldots d$ we have:

$$P\{b^{(1,k)} \notin H_j\} = P\{b^{(1,k-1)} \notin H_j\}P\{Mut(b^{(0)}) \notin H_j\}. \tag{5}$$

Now for the (1+1)-ES if $j \leq i$ then $P\{g^{(k)} \notin H_j\} = 0$, and if $j > i$ then

$$P\{g^{(k)} \notin H_j\} = P\{g^{(k-1)} \in H_i \setminus H_j\} \times$$

$$P\{Mut(g^{(k-1)}) \notin H_j | g^{(k-1)} \in H_i \setminus H_j \}. \tag{6}$$

Assume that $P\{b^{(1,k-1)} \notin H_j\} \geq P\{g^{(k-1)} \notin H_j\}$, and thus, $P\{b^{(1,k-1)} \notin H_j\} \geq P\{g^{(k-1)} \in H_i \setminus H_j\}$. From the monotonicity of threshold transition matrix we obtain $P\{Mut(b^{(0)}) \notin H_j\} \geq P\{Mut(g^{(k-1)}) \notin H_j | g^{(k-1)} \in H_i \setminus H_j\}$. Hence, by induction on k, using (5) and (6), we conclude that $P\{b^{(1,k)} \notin H_j\} \geq P\{g^{(k)} \notin H_j\}$ for all $k = 1, ..., \lambda$.

b) To prove that $Q^{(\lambda)} \geq P^{(1)}$ for arbitrary initial distribution with $P^{(0)} = Q^{(0)}$ we use the total probability formula, the step mutation definition and the conclusion of the case a):

$$P\{b^{(1)} \notin H_j\} = \sum_{i=0}^{d} P\{b^{(1)} \notin H_j | b^{(0)} \in H_i \setminus H_{i+1}\}(p_i^{(0)} - p_{i+1}^{(0)}) \geq$$

$$\sum_{i=0}^{d} P\{g^{(\lambda)} \notin H_j | g^{(0)} \in H_i \setminus H_{i+1}\}(q_i^{(0)} - q_{i+1}^{(0)}) = P\{g^{(\lambda)} \notin H_j\}. \tag{7}$$

c) In general, when $Q^{(0)} \geq P^{(0)}$ let us recall that in case of step mutation we can consider $P^{(1)}$ as a function on vector $P^{(0)}$. By (7) we have $Q^{(\lambda)} \geq P^{(1)}(Q^{(0)})$, and due to nonnegativity of the multipliers of probabilities $p_1^{(t-1)}, ..., p_d^{(t-1)}$ in (1), which is an equality in our case, we conclude that $P^{(1)}(Q^{(0)}) \geq P^{(1)}(P^{(0)})$. $\square$

In [3] we showed that the bit-flip mutation operator for the ONEMAX fitness function is an example of a monotone step mutation operator. Analogous result has been obtained there for a family of vertex cover problems of special structure. Of course the situation when we have a monotone step mutation operator is not likely to occur while solving difficult optimization problems, unless we use some "artificial" mutation, especially adjusted for the particular problem instance. For example, if the the set of fitness levels $\Phi_0, ..., \Phi_d$ includes all possible values of the fitness function, then existence of a local optimum (which is not global) in terms of Hamming distance, implies that the bit-flip mutation is not a monotone step mutation operator (with respect to the corresponding subsets $H_1, ..., H_d$).

However the step mutation operators may be considered as convenient theoretical constructions since for any given problem and mutation operator Mut, a monotone step mutation operator Mut' can be constructed to have the threshold transition probabilities equal to the monotone lower bounds of Mut. Then Mut' may be viewed as a realization of the worst-case situation for a given set of lower bounds. In this respect Theorem 5 shows that the best possible lower bound on $P^{(t)}$ for the (1,λ)-ES within our framework is not better than the best lower bound on $Q^{(t)}$ for the (1+1)-ES.

A similar result to Theorem 5 can be also obtained to show that the GA based on mutation and selection does not "outperform" the (1, λ)-ES in terms of probability vectors if a monotone step mutation operator is used.

3   Some Connections with Complexity Theory

The computational complexity theory has been a source of numerous pessimistic results related to optimization algorithms (see e.g. [6,7]), in particular, it is shown that there are some classes of combinatorial optimization problems for which even computing good approximate solutions is NP-hard. In this section we are going to consider the a priori lower bounds for transition probabilities with respect to hardness of finding the approximate solutions. While doing this we shall use the notation analogous to the one in [8]. Recall that Σ denotes the alphabet $\{0, 1\}$. By Σ^* we denote the set of all strings with symbols from Σ.

Definition 1. *An NP maximization problem P_{max} is a triple $P_{max} = (I, Sol, f_x)$, where $I \subseteq \Sigma^*$ is the set of instances of P_{max} and:*

1. I is recognizable in polynomial time.

2. Given an instance $x \in I$, $Sol(x) \subseteq \Sigma^*$ denotes the set of feasible solutions of x. Given x and y the decision whether $y \in Sol(x)$ may be done in polynomial time, and there exists a polynomial h such that given any $x \in I$ and $y \in Sol(x)$, $|y| \leq h(|x|)$. (Without loss of generality in what follows we assume that the equality $|y| = h(|x|)$ holds for any $x \in I$ and $y \in Sol(x)$.)

3. Given an instance $x \in I$ and $y \in Sol(x)$, $f_x(y)$ is a positive integer objective function (to be maximized), which is computable in polynomial time.

A similar definition may be given for NP minimization problems, and all the following statements could be properly adjusted for the minimization case as well. Here we will consider the maximization problems for convenience. The optimal objective function value for x is denoted by $f_x^* = \max_{y \in Sol(x)} f_x(y)$.

We have made no specific assumptions concerning the method of encoding the solutions in genotype strings. Given the encoding scheme for the feasible solutions of an NP maximization problem, one can use the string y as a genotype with the fitness function identical to $f_x(y)$ on set $Sol(x)$ and equal to 0 on all other genotypes from $\Sigma^{h(|x|)}$. For simplicity the representation with mapping $y(g) \equiv g$ will be assumed below (the same way we could consider a wider class of representations with the mappings computable in polynomial time).

Definition 2. *Given an instance x of an NP maximization problem and $y \in Sol(x)$, the performance ratio of y with respect to x is $R(x, y) = f_x^*/f_x(y)$.*

Definition 3. *Let P_{max} be an NP maximization problem and let T be an algorithm that for any instance x of P_{max} such that $Sol(x) \neq \emptyset$, returns a feasible solution $T(x)$ in polynomial time. Given a function $r : \mathbf{Z}^+ \to [1, \infty)$, T is called an r-approximation algorithm for P_{max} if the performance ratio of $T(x)$ with respect to x is such that $R(x, T(x)) \leq r(|x|)$. The solution $T(x)$ is called an $r(|x|)$-approximate solution then.*

Producing for any instance $x \in P_{max}$ such solution y that $f_x(y) \geq f_x^*/r(|x|)$ (provided that $Sol(x) \neq \emptyset$) is called *approximation of P_{max} within a factor r*. For example it is proven that the maximum cut problem is NP-hard to approximate within a factor 1.012, approximation of the maximum clique problem within $|V|^{1-\varepsilon}$ is most likely to be impossible (where $|V|$ is the number of vertices in graph and $\varepsilon > 0$ is an arbitrary constant), etc. For more details see e.g. [6].

We will also need the formal definitions of a randomized algorithm and of class BPP of languages recognizable with bounded probability in polynomial time (see e.g. [4]). By a randomized algorithm we mean an algorithm which may be executed by a *probabilistic Turing machine*, i.e. the Turing machine which has a special state for "tossing a coin". When the machine enters this state it receives a bit which is 0 with probability 1/2 and 1 with probability 1/2. A *polynomial-time probabilistic Turing machine* is a probabilistic Turing machine which always halts after a polynomial number of steps.

Definition 4. *BPP is the class of languages $L \subseteq \Sigma^*$ for which there exists a polynomial-time probabilistic Turing machine M, such that:*
1) *For all $x \in L$ holds $P\{M \, gives \, an \, output \, 1\} \geq 3/4$.*
2) *For all $x \notin L$ holds $P\{M \, gives \, an \, output \, 0\} \geq 3/4$.*

Studying the complexity of mutation operators with certain a priori lower bounds for transition probabilities we will consider the possibility of implementing the mutation by a polynomial-time randomized algorithm. The following supplementary result will be of use.

Lemma 2. *Let P_{max} be an NP maximization problem and the approximation of P_{max} within a factor r is NP-hard. Then unless $NP \subseteq BPP$, no polynomial-time randomized algorithm obtains the r-approximate solutions to all instances x of P_{max} with probability more than $1/\mathrm{poly}(|x|)$, where $\mathrm{poly}(|x|)$ is a polynomial in the length of input x. (The proof is omitted for brevity.)*

The conjecture $NP \not\subseteq BPP$ is equivalent to $NP \neq RP$ and widely believed to be true [9] (here RP denotes the class of languages L recognizable by polynomial-time probabilistic Turing machine with the probability of error equal to zero for inputs $x \notin L$ and not more than $\frac{1}{2}$ if $x \in L$).

Assuming $r = 1$ in Lemma 2 we see that for any NP-hard NP maximization problem most likely there are no randomized polynomial-time algorithms finding the optimum with a probability bounded below by a positive constant.

Using Lemma 2 we can give a pessimistic estimate of the probability of obtaining "good" solutions by a mutation operator computable in polynomial time, when the genotype under mutation represents a relatively "poor" solution.

Theorem 6. *Suppose that P_{max} is an NP maximization problem, the approximation of P_{max} within a factor r is NP-hard, and there is an r'-approximation algorithm for it. Then unless $NP \subseteq BPP$, no polynomial-time mutation operator exists such that for some polynomial $\mathrm{poly}(|x|)$ for all instances x there would be monotone transition bounds with $\Phi_i \leq f_x^*/r'$, $\Phi_j \geq f_x^*/r$ and $\alpha_{ij} \geq 1/\mathrm{poly}(|x|)$.*

Proof. Consider a polynomial-time randomized algorithm which obtains an approximate solution y by the r'-approximation algorithm, and applies the mutation operator to genotype $g = y$. Now since $g \in H_i$ there must exist such k, $i \le k \le d$ that $g \in H_k \backslash H_{k+1}$. Due to monotonicity of the transition bounds, $\alpha_{kj} \ge \alpha_{ij} \ge 1/\text{poly}(|x|)$. Thus Lemma 2 yields the required statement. $\square$

For example setting $r = 1$ here we conclude that for any NP-hard NP maximization problem most likely no polynomial-time mutation operator exists to guarantee for all instances x the monotone transition bounds with $\Phi_i \le f_x^*/r'$, $\Phi_d = f_x^*$ unless α_{id} is a quickly vanishing function of $|x|$.

Theorem 6 suggests that solving the NP-hard problems one should not expect to be able to find polynomial-time mutation operators with monotone lower bounds much better than the guaranteed performance bounds of deterministic polynomial-time algorithms (unless $NP \subseteq BPP$).

4   Conclusions

In this paper we presented the lower and upper bounds for probability of obtaining solutions of sufficient quality for the $(1,\lambda)$-ES and the $(1+1)$-ES evolutionary strategies. In terms of these probabilities the algorithm $(1+1)$-ES is proved to be always at least as good as the $(1,\lambda)$-ES when a step mutation operator with monotone threshold transition matrix is used. Finally, provided that the conjecture $NP \nsubseteq BPP$ holds, the best possible lower bounds on transition probability for polynomial-time mutation operator are shown to approach zero in many cases, when dealing with NP-hard combinatorial optimization problems.

Further research is expected to address the estimation of the average number of iterations until the $(1+1)$-ES obtains the optimum. Besides that it will be interesting to study some benchmark optimization problems in the framework used in this paper.

References

1. Bäck, T., Schwefel H.-P.: An overview of evolutionary algorithms for parameter optimization, Evolutionary Computation, **1** (1) (1993) 1–23
2. Rechenberg I.: Evolutionsstrategie'94. Formann-Holzboog Verlag, Stuttgart (1994)
3. Eremeev A.V.: Modeling and Analysis of Genetic Algorithm with Tournament Selection. Proc. of Artificial Evolution Conference (AE'99). Lecture Notes in Computer Science, Vol. 1829. Springer Verlag (2000) 84–95
4. Motwani, R. and Raghavan, P.: Randomized Algorithms. Cambridge University Press (1995)
5. Rudolph, G.: Finite Markov Chain Results in Evolutionary Computation: A Tour d'Horizon, Fundamenta Informaticae **35** (1-4) (1998) 67–89
6. Arora, S., Lund, C.: Hardness of approximations. Approximation Algorithms for NP-Hard Problems. Ed. by S.D.Hochbaum. PWS Publishing Company, (1995) 399–446
7. Garey, M.R. and Johnson, D.S.: Computers and intractability. A guide to the theory of NP-completeness. W.H. Freeman and Company, San Francisco (1979)

8. Ausiello, G., Protasi, M.: Local search, reducibility and approximability of NP-optimization problems. Information Processing Letters, **54** (1995) 73–79
9. Ko, K.: Some observations on the probabilistic algorithms and NP-hard problems, Information Processing Letters, **14** (1982) 39–43

A Contribution to the Study of the Fitness Landscape for a Graph Drawing Problem

Rémi Lehn and Pascale Kuntz

IRIN – Université de Nantes – France

Abstract. These past few years genetic algorithms and stochastic hill-climbing have received a growing interest for different graph drawing problems. This paper deals with the layered drawing of directed graphs which is known to be an NP-complete problem for the arc-crossing minimization criterium. Before setting out a (n+1)th comparison between meta-heuristics, we here prefer to study the characteristics of the arc-crossings landscape for three local transformations (greedy switching, barycenter, median) adapted from the Sugiyama heuristic and we propose a descriptive analysis of the landscape for two graph families. First, all the possible layouts of 2021 small graphs are generated and the optima (number, type, height, attracting sets) are precisely defined. Then, a second family of 305 larger graphs (up to 90 vertices) is examined with one thousand hill-climbers. This study highlights the diversity of the encountered configurations and gives leads for the choice of efficient heuristics.

1 Introduction

The problem of drawing a graph G is generally set as a combinatorial optimization problem: producing a layout of G on a given support (grid, ...) according to a drawing convention (layered drawing, ...) that optimizes some measurable aesthetics (e.g. arc-crossing or arc-length minimization). Unfortunately, numerous criteria lead to NP-complete problems for general graphs, and aesthetic requirements often conflict which each other. The importance of the problem in numerous application fields, in particular in Computer Aided Design and more recently for Web visualization, has stimulated the development of various heuristics from classical divide-and-conquer approaches to meta-heuristics (see the recent survey in [1]).

Among the meta-heuristics, Genetic Algorithms (GA) and Evolutionary Algorithms (EA) seem to have attracted a particular attention these past few years. Indeed, several developments can be found in the literature for different graph characteristics and drawing constraints: directed graphs ([2], [3], [4], [5],[6]), undirected graphs ([7],[8], [9],[10], [11], [12]), interactive layouts [3], dynamic layouts [13]. The comparisons either with problem-specific heuristics or other meta-heuristics (simulated-annealing, ...) show that GAs are particularly promising for this class of problems.

E.J.W. Boers et al. (Eds.) EvoWorkshop 2001, LNCS 2037, pp. 172–181, 2001.

Yet, a recent empirical comparison [12] has shown that Stochastic Hill Climbing (SHC) may sometimes outperform EA for the undirected graph layout on a grid. This work is along the same line as other comparative studies of these two approaches for combinatorial optimization problems [14][15] [16] which have followed the seminal work of Mitchell et al [17] with hill climbing. They show that some revised versions of the simple basic SHC may lead to significantly better results than GA or EA. However, most authors recognize that these comparisons closely depend on the implementation of these techniques.

Instead of setting out an $(n + 1)$th comparison, we here prefer to tackle the problem at the root by studying the features of the arc-crossing landscape. We focus on a type of layout: the layered drawing of directed graphs where vertices are arranged in vertical layers and arcs are represented by line segments. In this case, the arc-crossing number does not depend on the precise position of the vertices but only on the ordering within each layer. Although this problem seems *a priori* easier than the general problem of minimizing arc crossings on a plane, it remains NP-complete even if there are only two layers [18]. The most widely used optimization techniques have been developed in the so-called Sugiyama-heuristic for drawing directed acyclic graphs [19]. As they allow to define local transformations on the graph, they have recently been adapted to GA for building mutation operators [6], [13].

In this paper, we analyze the arc-crossing landscape for these transformations. It is well-known that the structures of the landscapes influence the performances of the search for the optimal solutions, and several techniques have been proposed in the literature to define their features (e.g. [20] for a recent overview). Numerous authors introduce a measure of correlation between fitness values on the landscape. It allows to assess the smoothness or the ruggedness of a landscape and often aims at characterizing the ability of a GA to solve the problem. But, instead of comparing such measures whose interpretation for real-life problems may be questionable, we thought it better to start our landscape investigation by a descriptive analysis. First, we study all the possible layouts for a given transformation of 2021 "small" graphs and precisely define the optima (number, type, height, attracting sets). Then, we extend our analysis to a set of 300 larger graphs (with a maximum of 93 vertices). As, in this case, an exhaustive examination is beyond reach in a reasonable computation time, we resort to one thousand hill-climbers to mine the landscape structures.

2   Definition of a Local Transformation Landscape

In the definition of a fitness landscape proposed by Jones [21][22] which is now taken up by many authors, three components are essential: a solution coding, a mapping that associates each solution with a numerical value (fitness function) and an operator that defines a neighborhood relationship within the set of solutions.

We here restrict ourselves to the study of arc-crossing landscapes associated with some local transformations. Let $G = (V, E)$ be an acyclic digraph, $l_1, ..., l_K$

be a set of K layers for the layout and Ω^G be the set of all layered drawing of G for a given distribution of the vertex set on these layers. The "fitness" value $f(L_i^G)$ of any drawing $L_i^G \in \Omega^G$ (figure 1) is the number of arc-crossings on L_i^G. A neighbor of L_i^G is a layout $\widetilde{L}_i^G$ of G deduced from L_i^G by a local operator (greedy switching, barycenter, median) described below. The interest of these operators is that they can be used for local optimization in an SHC or in multi-start hill-climbing, as well as in GA combined with a bitflip operator to introduce a local improvement in the "mutation" phase -in the sense of a hybridization strategy with a local search.

$$L_j^G = \qquad f(L_j^G) = 1.$$

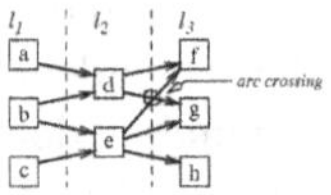

Fig. 1. A layered drawing of a graph.

2.1   Local Transformations

Each local operator is here defined for any layer l_k of a layout L_i^G of Ω^G . Let us denote by σ_k the vertex ordering on l_k.

Greedy switching (S)

The basic idea of this heuristic is to switch adjacent pairs of vertices on l_k when this exchange makes the arc-crossing number decrease. More precisely, all the vertices of l_k are sequentially scanned and two adjacent vertices are permuted whenever the fitness of the layout associated with this new ordering is improved.

Barycenter (B)

Roughly speaking, a vertex is repositioned at the average position of its neighbors. Let u be a vertex on l_k and $N_{k+1}(u)$ (resp. $N_{k-1}(u)$) be the set of its neighbors on the adjacent layer l_{k+1} (resp. l_{k-1}). The average position $avg(u)$ of its neighbors is defined by

$$avg(u) = \frac{1}{|N_{k+1}(u)| + |N_{k-1}(u)|} \left(\sum_{v \in N_{k-1}(u)} \frac{\sigma_{k-1}(v)}{n_{k-1}} + \sum_{v \in N_{k+1}(v)} \frac{\sigma_{k+1}(v)}{n_{k+1}} \right)$$

where n_{k-1}(resp. n_{k+1}) is the number of vertices on l_{k-1} (resp. l_{k+1}). These ratio are introduced to normalize the vertex position on each layer according to their cardinality. For instance, in figure 2, $N_{k-1}(u) = \{a, b, c\}$ and $N_{k+1}(u) = \{d, e, f, g\}$ and $avg(u) = 0.62$.

The average position is computed for each vertex of l_k and the local transformation **B** is the new vertex ordering σ_k' obtained after sorting the average values $\{avg(u); u \in l_k\}$ ($\sigma_k'(u) > \sigma_k'(v)$ if $avg(u) > avg(v)$). This transformation is applied on L_i^G only if it improves the fitness.

Median (M)

This operator is very close to the barycenter. Let us recall that a median of a set of sorted numbers -by decreasing or increasing order- is the number m so that

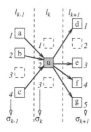

$$n_{k-1} = 4, \qquad n_k = 3, n_{k+1} = 5,$$
$$|N_{k-1}(u)| = 3, \qquad\qquad |N_{k+1}(u)| = 4.$$

$$avg(u) = \tfrac{1}{3+4}\left(\tfrac{1+2+4}{4} + \tfrac{1+3+4+5}{5}\right) = 0.62.$$

sorted adjacent vertices positions : $\left(\tfrac{1}{5}, \tfrac{1}{4}, \tfrac{2}{4}, \tfrac{3}{5}, \tfrac{4}{5}, \tfrac{4}{4}, \tfrac{5}{5}\right)$.
$med(u) = \tfrac{3}{5} = 0.6.$

Fig. 2. Barycenter and median : example.

half of the numbers are smaller than m and the other half are greater than m. If the neighbors of $u \in l_k$ on l_{k-1} (resp. l_{k+1}) are $v_1, v_2, ..., v_p$ (resp. $w_1, w_2, ..., w_q$) then the median position $med(u)$ of u is the median of the following sorted values $\left\{\frac{\sigma_{k-1}(v_1)}{n_{k-1}}, \frac{\sigma_{k-1}(v_2)}{n_{k-1}}, ..., \frac{\sigma_{k-1}(v_p)}{n_{k-1}}, \frac{\sigma_{k+1}(w_1)}{n_{k+1}}, \frac{\sigma_{k+1}(w_2)}{n_{k+1}}, ..., \frac{\sigma_{k+1}(w_q)}{n_{k+1}}\right\}$. The median position is computed for each vertex of l_k. And as previously, the local transformation $\mathbf{M}$ is the new vertex ordering σ'_k obtained after sorting the median values $\{med(u); u \in l_k\}$:$\sigma'_k(u) > \sigma'_k(v)$ if $med(u) > med(v)$.

2.2 First Characteristics

A local operator $\mathbf{O}$ acts on a layout L_i^G of Ω^G to produce a new layout $\widetilde{L}_i^G$. The set $N(L_i^G)$ of the neighbors of L_i^G is composed of the layouts of Ω^G which are obtained from L_i^G by applying $\mathbf{O}$ once, and we denote by $\Omega_{\mathbf{O}}^G$ the set of all the layouts deduced from one another with $\mathbf{O}$. This set is similar to the vertex set of the graph with characterizes a landscape in the definition proposed by Jones [21], but we here restrict ourselves to local transformations which improve the fitness. A *local optimum* (here a minimum) of the landscape is a layout $L_i^G \in \Omega_{\mathbf{O}}^G$ s.t. $f(L_i^G) \le f(L_j^G)$ for $L_j^G \in N(L_i^G)$ and a *global optimum* is a layout $\widehat{L}^G \in \Omega_{\mathbf{O}}^G$ s.t. $f(\widehat{L}^G) \le f(L_i^G)$ for $L_i^G \in \Omega_{\mathbf{O}}^G$.

Before getting into details, we can even deduce three characteristics of the landscapes associated with the operators $\mathbf{S}$, $\mathbf{B}$ and $\mathbf{M}$: *1)* there can be some local optima (figures 3a. and 3b.); *2)* due to symmetries in the layouts, the global optimum is not necessarily unique; *3)* there are layouts for which different transformations improve the fitness in the same way (figure 3c.).

It is clear that, for this drawing problem, the fitness landscape is multimodal in the general case. But now different questions arise: are the local optima (resp. global optima) numerous ? what is the quality of the local optima ? what is the size of the basins of attraction ?

3 Exhaustive Exploration for Small Graphs

Let D be the set of acyclic digraphs which are not trees, and so that each arc is incident to vertices on adjacent layers and $\prod_{k=1}^K n_k! \le 2000$ where n_k is the

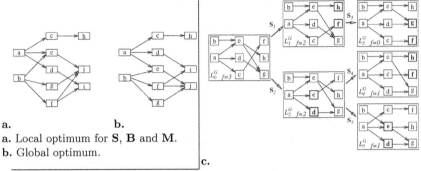

a.

b.

a. Local optimum for **S**, **B** and **M**.

b. Global optimum.

c.

c. Two transformations (greedy switching) S_1 and S_2 can be applied on L_0^G leading to two layouts L_1^G and L_2^G with the same fitness. However a global optimum L_3^G can be reached from L_1^G whereas any **S** transformation on L_2^G leads to a local optimum (L_4^G or L_5^G) only.

Fig. 3. A local optimum for the three local transformations and equivalent local transformations on a layout.

vertex number on the layer k and K the maximum number of layers, we have analyzed all the possible layouts for 2021 different graphs from D, and for the study of local and global optima we restrict ourselves to the greedy switching operator **S**. The average number of explored layouts for each graph is equal to $(\sum_{i=1}^{2021} \prod_{k_i=1}^{K_i} n_{k_i}!)/2021 \simeq 917$.

3.1 Global and Local Optima

Even for these small graphs, there is a local optimum in 70% of the cases. The mean number of local optima is equal to 25.2 and that of the global optima is equal to 92.9. This value partly results from the numerous symmetries of the graphs. It is very important to note that the distribution of the optima greatly varies from graph to graph: the standard deviation of the number of local optima (resp. global optima) is equal to 31.9 (resp. 264.5). However, there is no correlation between the number of local optima and the number of global optima (the correlation coefficient $\rho \simeq -0.16$).

In an optimization perspective, it is important to know whether the local optimum values are close or not to the global optimum values. Let $f(L_i^G)$ (resp. $f(\widehat{L}^G)$) be the arc-crossing number on a layout L_i^G associated with a local optimum (resp. any global optimum) and $f(L_w^G)$ the arc-crossing number of the worst layout, then the relative height $h(L_i^G)$ of a local optimum L_i^G is $f(L_i^G) = 1 - \frac{f(L_i^G)-f(\widehat{L}^G)}{f(L_w^G)-f(\widehat{L}^G)}$. Intuitively, if $h(L_i^G)$ is very close to 1, then the local optimum L_i^G can be considered as a "good" solution.

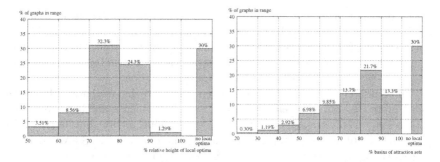

a. Relative height of local optima.   **b.** Size of the global optima attraction sets.

Fig. 4. Relative height of local optima and size of the global optima attraction sets.

3.2 Global Optimum Attracting Sets

The basin of attraction $b(\widehat{L}_j^G)$ of a global optimum $\widehat{L}_j^G$ is the set of layouts of $\Omega_\mathbf{S}^G$ which can be reached from $\widehat{L}_j^G$ by a descent method: $b(\widehat{L}_j^G) = \{L_0^G; \exists L_1^G, ..., L_{n-1}^G \in \Omega_\mathbf{S}^G \text{ with } L_i^G \in N(L_{i+1}^G) \text{ and } f(L_i^G) < f(L_{i+1}^G) \ \forall i = 0, n\}$ where $L_n^G = \widehat{L}_j^G$ (e.g. [20]). In practice, starting from a global optima, we look for its successive neighbors while the fitness strictly decreases.

The basins of attraction $b(\widehat{L}^G)$ of the set L^G of all global optima $\widehat{L}^G = \{\widehat{L}_1^G, \widehat{L}_2^G, ..., \widehat{L}_p^G\}$ is the set $b(\widehat{L}_1^G) \cup b(\widehat{L}_2^G) \cup ... \cup b(\widehat{L}_p^G)$ of layouts which can be reached from any global optimum $\widehat{L}_1^G, \widehat{L}_2^G, ..., \widehat{L}_p^G$. The relative size $s(\widehat{L}^G)$ of $b(\widehat{L}^G)$ is defined by $s(\widehat{L}^G) = | b(\widehat{L}^G) | / | \Omega_\mathbf{S}^G |$. Note that for the operator $\mathbf{S}$, $\Omega_\mathbf{S}^G = \Omega^G$ and then $| \Omega_\mathbf{S}^G |= \Pi_{k=1}^K n_k!$.

On average the basins of attraction sets represent 83.1% of the lanscape and this average drops to 75.9% if we disregard cases where there are global optima only. However, there are large disparties among the graphs (figure 4**b.**): 30% of the graphs with one local optimum at least have $s(\widehat{L}^G) < 70\%$, and 70% have $s(\widehat{L}^G) \geq 70\%$.

4 Exploration with a Set of Hill-Climbers

As the computation time quickly becomes prohibitive for an exhaustive exploration of larger graphs, we resort to an extended multi-start hill-climbing to discover the main characteristics of the fitness landscape. A set of one thousand layouts[1] is randomly generated, and each layout is improved by an iterative application of an operator **O**: **O** is applied on each layer $l_1, ...l_K$ of a random layout $L_i^G \in \Omega_0^G$ taken one after the other and this loop starts again until the

---

[1] This number has been empirically fixed after some experiments; it has to be confirmed with statistical tests.

fitness does not move any more. In this case, the term "best solution" refers to the best layouts found by the set of hill-climbers and we improperly call "local optima" the other layouts on which the hill-climbers converge. Moreover, as we cannot now compute the exact basins of attraction, we only retain the proportion $n_h(\widehat{L}^G)/1000$ of hill-climbers which reach a best solution.

Unfortunately, as far as we know, there is no benchmark base of large size well-adapted to our problem, therefore, our experiments are here concerned with a set of 305 graphs built like in the previous section but larger (with $\prod_{k=1}^{K} n_k! \leq 10^{14}$). We present results where the operator $\mathbf{O}=\mathbf{S}+\mathbf{M}+\mathbf{B}$ is the successive application on each layer of $\mathbf{S}$ then of $\mathbf{M}$ and then $\mathbf{B}^2$.

There are some differences with the previous results due to the graph sizes but the main tendencies are confirmed. Here, 72.14% of the graphs have at least a local optimum with an average equal to 551.19. This means that, on average, among the 1000 climbers a little more than half reach a local optimum and a little less than half find a best solution. However, there are very important differences between graphs: the standard deviation of the global optimum number is equal to 354.27 and the histogram (figure 5) shows that the distribution of the proportion of hill-climbers reaching a global optimum is spread out.

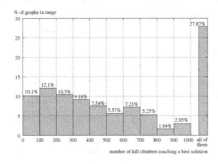

Fig. 5. Number of hill climbers reaching a best solution.

5   Discussion

It is now well-known for some combinatorial optimization problems that the instance characteristics may vary a lot from one to another (e.g. [23] for the quadratic assignment problem),   and that consequently the heuristic performances closely depend on the processed instance. Our study leads to similar results for a graph drawing problem. It is obvious that a descent method or a

---

[2] The mean computation time is equal to 33m. 14s. (CPU time, i686 433MHz 128MB, Linux 2.2.5, *perl* version 5.005_03)

simple multi-start hill-climbing is sufficient for landscapes with no, or few, local optima. But, for the other cases analyzed above, they are numerous local optima with a high relative height and the basins of attraction of the global optima are rather extensive. Such characteristics could justify the choice of hill-climbing approaches for these landscapes. Nevertheless, some problem-specific remarks can inflect this conclusion. Due to the high relative height of the majority of local optima, SHCs are *a priori* expected to reach solutions of "good" quality quickly. However, recent psychological experiments have confirmed the importance of the arc-crossing criterium for drawing intelligibility [24], and the addition of just a few crossings can notably disturb the readability of the layouts in some cases (e.g. when they are in the middle of the visual support on which the user focuses). Consequently, for this problem, local optima may easily be associated with debased solutions in terms of usefulness.

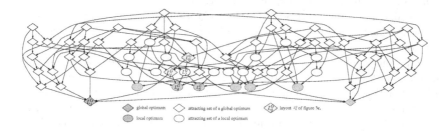

Fig. 6. Fitness graph for the 72 possible layouts of the graph on figure 3c. for the operator **S.**

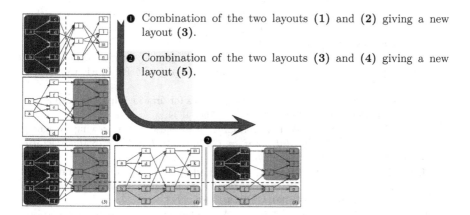

❶ Combination of the two layouts (1) and (2) giving a new layout (3).

❷ Combination of the two layouts (3) and (4) giving a new layout (5).

Fig. 7. Reconstruction by recombination.

To study this multimodality more precisely, the definition of the fitness landscape based on graphs [21] is very convenient. Let us recall that the vertex set is the set Ω_O^G of all the layouts deduced from one another with an operator $\mathbf{O}$ and the arcs represent these deductions. For very small graphs this "fitness graph" can be entirely drawn. The example given figure 6 shows that, for the majority of the layouts, different local transformations can lead to a better drawing : for this very simple instance the average vertex degree is 2.05 and the standard deviation is 1.15. In particular, due to the highly combinatorial structure of the landscape, the basins of attraction of global and local optima often have a non-empty intersection. Consequently, starting from a layout in the basin of attraction of a global optima (e.g. L_0^G on figure 6) the probability of reaching a local optimum (e.g. L_4^G on figure 6) is not nil. The strongly combinatorial structure of the landscape associated with local optima close in value to the global ones lead to think that the probability for a hill-climbing approach -in particular an SHC- to be stuck on a local optimum is no more negligible.

In this paper we have been focusing on landscapes associated with operators of local transformations only. It is clear that a similar study should be carried out for the crossover landscapes. Different operators adapted from the "ordering genetic algorithms", can be envisaged (e.g. [13]). Roughly speaking, a nice property of the layered drawing problem for the GAs seems to be the condition of reconstruction by recombination (e.g. [25]): the combination of two solutions, or sub-solutions, can lead to a better solution and not necessarily to a worst one. This condition is illustrated on an example in figure 7 but more general cases remain to be studied.

References

1. G. Di-Battista, P. Eades, R. Tamassia, and I.-G. Tollis. *Graph drawing – Algorithms for the visualization of graphs.* Prentice-Hall, 1999.
2. L.-J. Groves, Z. Michalewicz, P.-V. Elia, and C.-Z. Janikow. Genetic algorithms for drawing directed graphs. In *Proc. of the 5th Int. Symp. on Methodologies for Intelligent Systems*, pages 268–276. Elsevier, 1990.
3. T. Masui. Graphic object layout with interactive genetic algorithms. In *Proc. of the 1992 IEEE Workshop on Visual Languages*, pages 74–80. IEEE Comp. Soc. Press, 1992.
4. E. Makinen and M. Sieranta. Genetic algorithms for drawing bipartite graphs. *Int. J. of Computer Mathematics*, 53(3-4):157–166, 1994.
5. A. Ochoa-Rodrìguez and A. Rosete-Suàrez. Automatic graph drawing by genetic search. In *Proc. of the 11th Int. Conf. on CAD, CAM, Robotics and Manufactories of the Future*, pages 982–987, 1995.
6. J. Utech, J. Branke, H. Schmeck, and P. Eades. An evolutionary algorithm for drawing directed graphs. In *Proc. of the Int. Conf. on Imaging Science, Systems and Technology*, pages 154–160. CSREA Press, 1998.
7. C. Kosak, J. Marks, and S. Shieber. A parallel genetic algorithm for network-diagram layout. In *Proc. of the 4th Int. Conf. on Genetic Algorithms*, pages 458–465. Morgan-Kaufmann, 1991.

8. C. Kosak, J. Marks, and S. Shieber. Automating the layout of network diagrams with specified visual organization. In *IEEE Transactions on System, Man and Cybernetics*, volume 24, pages 440–454, 1994.

9. A. Rosete and A. Ochoa. Genetic graph drawing. In *Proc. of the 13th Int. Conf. on Appl. of Artificial Intelligence in Engineering*, 1998.

10. D. Kobler and A. Tettamanzi. Recombination operators for evolutionary graph drawing. In *Proc. of Parallel Problem Solving from Nature PPSN-V*, volume 1498 of *Lect. Notes in Comp. Sc.*, pages 988–997. Springer-Verlag, 1998.

11. A. Rosete-Suarez, M. Sebag, and A. Ochoa-Rodriguez. A study of evolutionary graph drawing. Rapport de Recherche 1228, LRI, UMR 8623, Bat. 490, Université Paris-Sud XI, 91405 Orsay-CEDEX, France, Septembre 1999. http://www.lri.fr/~rosette-type.html.

12. A. Rosete-Suárez, A. Ochoa-Rodríguez, and M. Sebag. Automatic graph drawing and stochastic hill climbing. In *Proc. of the Genetic and Evolutionary Conf.*, *GECCO'99*, volume 2, pages 1699–1706. Morgan Kaufmann, 1999.

13. P. Kuntz, R. Lehn, and H. Briand. Dynamic rule graph drawing by genetic search. In *Proc. of the IEEE Int. Conf. on System Man and Cybernetics*, 2000.

14. S. Baluja. An empirical comparison of seven iterative and evolutionary heuristics for static functions of optimization. In *Proc. of the 11th Int. Conf. on System Engineering*, pages 692–697. Univ. of Nevada, 1996.

15. A. Juels and M. Wattenberg. Hillclimbing as a baseline method for the evaluation of stochastic optimization algorithms. In D.S. Touretsky and al, editors, *Advances in Neural Information Processing Systems*, pages 430–436. MIT Press, 1995.

16. O.J. Sharpe. *Towards a rational methodology for using evolutionary search algorithms*. PhD thesis, Univ. of Sussex, Brighton, UK, 2000.

17. M. Mitchell, J. Holland, and S. Forrest. When will a genetic algorithm outperform hill climbing ? In J. Cowan, G. Tesauro, and J. Alspector, editors, *Advances in Neural Information Processing Systems*. Morgan Kauffman, 1994.

18. P. Eades and N. Wormald. Edge crossings in drawings of bipartite graphs. *Algorithmica*, 11:379–403, 1994.

19. K. Sugiyama, S. Tagawa, and M. Toda. Methods for visual understanding of hierarchical systems. *IEEE Trans. Sys. Man and Cybernetics*, 11(2):109–125, 1981.

20. V.K. Vassilev. *Fitness Landscapes and Search in the Evolutionary Design of Digital Circuits*. PhD thesis, Napier Univ., UK, 2000.

21. T.-C. Jones. *Evolutionary Algorithms, Fitness Landscapes and Search*. PhD thesis, University of New Mexico, Alburquerque, 1995.

22. T. Jones and S. Forrest. Genetic algorithms and heuristic search. In *Santa Fe Institute Tech. Report 95-02-021*. Santa Fe Institute, 1995.

23. E. Taillard. Comparison of iterative searches for the quadratic assignment problem. *Location Science*, 3:87–105, 1995.

24. Purchase H. Which aesthetic has the greatest effect on human understanding ? In *Proc. Graph Drawing'97*, volume 1353 of *Lect. Notes in Comp. Sc.*, pages 248–261. Springer Verlag, 1997.

25. D.-E. Goldberg. *Genetic algorithms in search, optimization and machine learning*. Addison-Wesley, 1989.

Evolutionary Game Dynamics in Combinatorial Optimization: An Overview

Marcello Pelillo

Dipartimento di Informatica
Università Ca' Foscari di Venezia
Via Torino 155, 30172 Venezia Mestre, Italy

Abstract. Replicator equations are a class of dynamical systems developed and studied in the context of evolutionary game theory, a discipline pioneered by J. Maynard Smith which aims to model the evolution of animal behavior using the principles and tools of noncooperative game theory. Because of their dynamical properties, they have been recently applied with significant success to a number of combinatorial optimization problems. It is the purpose of this article to provide a summary and an up-to-date bibliography of these applications.

1 Replicator Equations and Their Properties

Consider a large, ideally infinite population of individuals belonging to the same species which compete for a particular limited resource, such as food, territory, etc. In evolutionary game theory [1,2], this kind of conflict is modeled as a game, the players being pairs of randomly selected population members. In contrast to traditional application fields of game theory, such as economics or sociology, players here do not behave "rationally," but act instead according to a pre-programmed behavior pattern, or *pure strategy*. Reproduction is assumed to be asexual, which means that, apart from mutation, offspring will inherit the same genetic material, and hence behavioral phenotype, as its parent. Let $J = \{1, \cdots, n\}$ be the set of pure strategies and, for all $i \in J$, let $x_i(t)$ be the relative frequency of population members playing strategy i, at time t. The *state* of the system at time t is simply the vector $\mathbf{x}(t) = (x_1(t), \cdots, x_n(t))^T$, which is clearly constrained to lie in the standard simplex:

$$S_n = \left\{ \mathbf{x} \in \mathbb{R}^n \; : \; x_i \geq 0 \; \forall i \in J, \; \mathbf{e}^T\mathbf{x} = 1 \right\} \; .$$

Here and in the sequel, the letter $\mathbf{e}$ is reserved for a vector of appropriate length, consisting of unit entries (hence $\mathbf{e}^T\mathbf{x} = \sum_i x_i$).

One advantage of applying game theory to biology is that the notion of "utility" is much simpler and clearer than in human contexts. Here, a player's utility can simply be measured in terms of Darwinian fitness or reproductive success, i.e., the player's expected number of offspring. Let $W = (w_{ij})$ be the $n \times n$ fitness (or payoff) matrix. Specifically, for each pair of strategies $i, j \in J$, w_{ij} represents the payoff of an individual playing strategy i against an opponent

E.J.W. Boers et al. (Eds.) EvoWorkshop 2001, LNCS 2037, pp. 182–192, 2001.

playing strategy j. Without loss of generality, we shall assume that the fitness matrix is nonnegative, i.e., $w_{ij} \geq 0$ for all $i, j \in J$. At time t, the average payoff of strategy i is given by:

$$\pi_i(t) = \sum_{j=1}^{n} w_{ij} x_j(t) \tag{1}$$

while the mean payoff over the entire population is $\sum_{i=1}^{n} x_i(t)\pi_i(t)$.

In evolutionary game theory the assumption is made that the game is played over and over, generation after generation, and that the action of natural selection will result in the evolution of the fittest strategies. If successive generations blend into each other, the evolution of behavioral phenotypes can be described by the following continuous-time *replicator equations* [1,2]:

$$\dot{x}_i(t) = x_i(t) \left(\pi_i(t) - \sum_{j=1}^{n} x_j(t)\pi_j(t) \right) \tag{2}$$

where a dot signifies derivative with respect to time. The basic idea behind this model is that the average rate of increase $\dot{x}_i(t)/x_i(t)$ equals the difference between the average fitness of strategy i and the mean fitness over the entire population. It is straightforward to show that the simplex S_n is invariant under equation (2), i.e. any trajectory starting in S_n will remain in S_n.

Similar arguments provide a rationale for the following discrete-time dynamics, assuming non-overlapping generations, which can be obtained from (2) by setting $1/\Delta t = \sum_{j=1}^{n} x_j(t)\pi_j(t)$:

$$x_i(t + \Delta t) = \frac{x_i(t)\pi_i(t)}{\sum_{j=1}^{n} x_j(t)\pi_j(t)} \tag{3}$$

Because of the non-negativity of the fitness matrix W and the normalization factor, this system too makes the simplex S_n invariant.

Equations (2) and (3) arise independently in different branches of theoretical biology [1]. In population ecology, for example, the famous Lotka-Volterra equations for predator-prey systems turn out to be equivalent to the continuous-time dynamics (2), under a simple barycentric transformation and a change in velocity. In population genetics they are known as *selection equations* [3]. In this case, each x_i represents the frequency of the i-th allele A_i and the payoff w_{ij} is the fitness of genotype $A_i A_j$. Here the fitness matrix W is always symmetric.

The following result is known in mathematical biology as the fundamental theorem of natural selection [3,1,2] and, in its original form, traces back to R. A. Fisher [4].

Theorem 1. *If W is symmetric then the population's average fitness $\mathbf{x}^T W \mathbf{x}$ is strictly increasing along any non-constant trajectory of replicator equations (2) and (3), and any such trajectory converges to a stationary point.*

Because of their dynamical properties, replicator dynamics have been recently applied with significant success to a number of combinatorial optimization problems. It is the purpose of this article to provide a summary and an up-to-date bibliography of these applications.

2   Maximum Clique Problems

Let $G = (V, E)$ be an undirected graph, where $V = \{1, \cdots, n\}$ is the set of vertices and $E \subseteq V \times V$ is the set of edges. The *order* of G is the number of its vertices, and its *size* is the number of edges. Two vertices $i, j \in V$ are said to be *adjacent* if $(i, j) \in E$. The *adjacency matrix* of G is the $n \times n$ symmetric matrix $A_G = (a_{ij})$ defined as $a_{ij} = 1$ if $(i, j) \in E$, and $a_{ij} = 0$ otherwise.

A subset C of vertices in G is called a *clique* if all its vertices are mutually adjacent. i.e., for all $i, j \in C$ A clique is said to be *maximal* if it is not contained in any larger clique, and *maximum* if it is the largest clique in the graph. The *clique number*, denoted by $\omega(G)$, is defined as the cardinality of the maximum clique. The maximum clique problem is to find a clique whose cardinality equals the clique number. It is known to be NP-hard for arbitrary graphs and so is the problem of approximating it within a constant factor. We refer to [5] for a recent survey of results concerning algorithms, complexity and applications of this problem.

In 1965, Motzkin and Straus [6] established a remarkable connection between the maximum clique problem and a certain quadratic programming problem. Consider the following quadratic function, called the *Lagrangian* of G:

$$f_G(\mathbf{x}) = \mathbf{x}^T A_G \mathbf{x} \tag{4}$$

and let $\mathbf{x}^*$ be a global maximizer of f_G on S_n, n being the order of G. In [6] it is proved that the clique number of G is related to $f_G(\mathbf{x}^*)$ by the formula:

$$\omega(G) = \frac{1}{1 - f_G(\mathbf{x}^*)} . \tag{5}$$

Additionally, it is shown that a subset of vertices C is a maximum clique of G if and only if its *characteristic vector* $\mathbf{x}^C$, which is the vector of S_n defined as $x_i^C = 1/|C|$ if $i \in C$ and $x_i^C = 0$ otherwise, is a global maximizer of f_G on S_n. In [7,8], the Motzkin-Straus theorem has been extended by providing a characterization of *maximal* cliques in terms of *local* maximizers of f_G on S_n.

Once that the maximum clique problem is formulated in terms of maximizing a quadratic polynomial over the standard simplex, the use of replicator dynamics naturally suggests itself [9,10]. In fact, consider a replicator system with fitness matrix defined as:

$$W = A_G .$$

From the fundamental theorem of natural selection, we know that the replicator dynamical systems, starting from an arbitrary initial state, will iteratively

maximize the Lagrangian f_G in S_n, and will eventually converge to a local maximizer which, by virtue of the Motzkin-Straus formula provides an estimate of
the clique number of G. Additionally, if the converged solution happens to be
a characteristic vector of some subset of vertices of G, then we are also able to
extract the vertices comprising the clique from its nonzero components. Clearly,
in theory there is no formal guarantee that the converged solution will be a *global*
maximizer of f_G.

In [9,10], experiments with the previous approach over thousands of randomly
generated graphs are presented. The results obtained suggests that the basins
of attraction of global maximizers are quite large, and frequently the algorithm
converges to one of them. The approach was found to be competitive with more
sophisticated neural network heuristics, both in terms of quality of solutions and
speed.

One drawback associated with the original Motzkin-Straus formulation, however, relates to the existence of spurious solutions, i.e., maximizers of f_G which
are not in the form of characteristic vectors [8]. In principle, spurious solutions
represent a problem since, while providing information about the cardinality of
the maximum clique, they do not allow us to easily extract its vertices. This
problem has recently been solved by Bomze [11]. Consider the following regularized version of function f_G:

$$\hat{f}_G(\mathbf{x}) = \mathbf{x}^T A_G \mathbf{x} + \frac{1}{2} \mathbf{x}^T \mathbf{x} \qquad (6)$$

which is obtained from (4) by substituting the adjacency matrix A_G of G with

$$\hat{A}_G = A_G + \frac{1}{2} I_n$$

where I_n is the $n \times n$ identity matrix. Unlike the Motzkin-Straus formulation, it
can be proved that *all* maximizers of $\hat{f}_G$ on S_n are strict, and are characteristic
vectors of maximal/maximum cliques in the graph [11].

Theorem 2. *Let C be a subset of vertices of a graph G, and let $\mathbf{x}^C$ be its
characteristic vector. Then, C is a maximum (maximal) clique of G if and only
if $\mathbf{x}^C$ is a global (local) maximizer of $\hat{f}_G$ in S_n. Moreover, all local (and hence
global) maximizers of $\hat{f}_G$ over S_n are strict.*

Preliminary experiments with this regularized formulation (6) on random
graphs are reported in [11], and a more extensive empirical study on DIMACS
benchmark graphs is presented in [12]. The emerging picture is the following.
The solutions produced by the replicator models are typically very close to the
ones obtained using more sophisticated continuous-based heuristics. Moreover,
the original version of the Motzkin-Straus problem performs slightly better than
its regularized counterpart, but the former often returns spurious solutions. This
may be intuitively explained by observing that, since all local maxima of $\hat{f}_G$ are
strict, its landscape is certainly less smoothed than the one associated to the non-
regularized version. This therefore enhances the tendency of local optimization
procedures to get stuck into local maxima.

In order to study the effects of varying the starting point of clique finding replicator dynamics, Bomze and Rendl [13] implemented various sophisticated heuristics and compared them with the usual (less expensive) strategy of starting from the simplex barycenter. Surprisingly, they concluded that the amount of sophistication seems to have no significant impact on the quality of the solutions obtained. Additionally, they showed that using (Runge-Kutta discretizations of) the continuous-time dynamics (2) instead of (3) does not improve efficiency. This analysis indicates that to improve the performance of replicator dynamics on the maximum clique problem one has necessarily to resort to some escape strategies. Various attempts along this direction can be found in [11,14,15,16].

Recently, the Motzkin-Straus theorem has been generalized to the weighted case [7,17]. Let $G = (V, E, \mathbf{w})$ be a weighted graph, where $V = \{1, \cdots, n\}$ is the vertex set, $E \subseteq V \times V$ is the edge set and $\mathbf{w} \in \mathbb{R}^n$ is the *weight* vector, the i-th component of which corresponds to the weight assigned to vertex i. It is assumed that $w_i > 0$ for all $i \in V$. Given a subset of vertices C, the weight assigned to C is defined as

$$W(C) = \sum_{i \in C} w_i .$$

A *maximal weight* clique C is one that is not contained in any other clique having weight larger than $W(C)$. Since we are assuming that all weights are positive, it is clear that the concepts of maximal clique and maximal weight clique coincide. A *maximum weight* clique is one having largest total weight, and the maximum weight clique problem is to find one such clique (see [5] for a recent review). The classical (unweighted) version of the maximum clique problem arises as a special case when all vertices have the same weight. Hence the maximum weight clique problem has at least the same computational complexity as its unweighted counterpart.

Given a weighted graph $G = (V, E, \mathbf{w})$, let $\mathcal{N}(G, \mathbf{w})$ be the class of $n \times n$ symmetric matrices $M = (m_{ij})_{i,j \in V}$ defined as $m_{ij} \geq m_{ii} + m_{jj}$ if $(i, j) \notin E$ and $m_{ij} = 0$ otherwise, and $m_{ii} = \frac{1}{2w_i}$ for all $i \in V$, and consider the program:

$$\begin{aligned} \text{minimize} \quad & g(\mathbf{x}) = \mathbf{x}^T M \mathbf{x} \\ \text{subject to} \quad & \mathbf{x} \in S_n . \end{aligned} \tag{7}$$

Also, for a given subset of vertices C, denote by $\mathbf{x}^C(\mathbf{w})$ its *weighted characteristic vector*, which is the vector in S_n with coordinates $x_i^C(\mathbf{w}) = w_i/W(C)$ if $i \in C$ and $x_i^C(\mathbf{w}) = 0$ otherwise. The following theorem is the weighted counterpart of Theorem 2 [17].

Theorem 3. *Let C be a subset of vertices of a weighted graph $G = (V, E, \mathbf{w})$, and let $\mathbf{x}^C(\mathbf{w})$ be its characteristic vector. Then, for any matrix $M \in \mathcal{N}(G, \mathbf{w})$, C is a maximum (maximal) weight clique of G if and only if $\mathbf{x}^C(\mathbf{w})$ is a global (local) solution of program (7). Moreover, all local (and hence global) solutions of (7) are strict.*

The previous result suggests using replicator equations to approximately solve the maximum weight clique problem. Experiments with this approach on both random graphs and DIMACS benchmark graphs are reported in [17]. Weights were generated randomly in both cases. The results obtained with replicator dynamics (3) were compared with those produced by a very efficient maximum weight clique algorithm of the branch-and-bound variety. The algorithm performed remarkably well especially on large and dense graphs, and it was typically an order of magnitude more efficient than its competitor.

3   Graph Isomorphism

Given two graphs $G' = (V', E')$ and $G'' = (V'', E'')$, an *isomorphism* between them is any bijection $\phi : V' \to V''$ such that $(i, j) \in E' \Leftrightarrow (\phi(i), \phi(j)) \in E''$, for all $i, j \in V'$. Two graphs are said to be *isomorphic* if there exists an isomorphism between them. The graph isomorphism problem is therefore to decide whether two graphs are isomorphic and, in the affirmative, to find an isomorphism. This is one of those few combinatorial optimization problems which still resist any computational complexity characterization [18,19]. The current belief is that it lies strictly between the P and NP-complete classes.

The *association graph* of $G' = (V', E')$ and $G'' = (V'', E'')$ is the undirected graph $G = (V, E)$ where

$$V = V' \times V''$$

and

$$E = \{((i, h), (j, k)) \in V \times V \ : \ i \neq j, \ h \neq k \text{ and } (i, j) \in E' \Leftrightarrow (h, k) \in E''\} \ .$$

The following easily proven result establishes an equivalence between the graph isomorphism problem and the maximum clique problem [20].

Theorem 4. *Let $G' = (V', E')$ and $G'' = (V'', E'')$ be two graphs of order n, and let G be the corresponding association graph. Then, G' and G'' are isomorphic if and only if $\omega(G) = n$. In this case, any maximum clique of G induces an isomorphism between G' and G'', and vice versa.*

This, together with Theorems 1 and 2, allows one to use replicator dynamics as a heuristic for finding isomorphisms between G' and G''. This approach has been tested over hundreds of random 100-vertex graphs with expected densities ranging from 1% to 99%. Except for very sparse and very dense instances, the algorithm was always able to obtain a correct isomorphism very efficiently. In terms of quality of solutions, the result compare favorably with those obtained using more sophisticated state-of-the-art deterministic annealing heuristics which, in contrast to replicator dynamics, are explicitly designed to escape from poor local solutions. As far as computational time is concerned, replicator dynamics turned out to be significantly faster.

In [20] experiments were also done using an exponential version of replicator equations, which arises as a model of evolution guided by imitation [2]. The extensive experiments conducted show that this dynamics may be considerably faster and even more accurate than the standard, first-order model.

The approach just described is general and can easily be extended to deal with subgraph isomorphism or relational structure matching problems. Preliminary experiments, however, seem to indicate that local optima may represent a problem here, especially in matching sparse and dense graphs. In these cases escape procedures like those presented in [11,14,15,16] would be helpful.

4   Subtree Isomorphism

Let $T_1 = (V_1, E_1)$ and $T_2 = (V_2, E_2)$ be two rooted trees. Any bijection ϕ : $H_1 \rightarrow H_2$, with $H_1 \subseteq V_1$ and $H_2 \subseteq V_2$, is called a *subtree isomorphism* if it preserves the adjacency and hierarchical relationships between the vertices and, in addition, the subgraphs obtained when we restrict ourselves to H_1 and H_2 are trees. The former condition amounts to stating that, given $i, j \in H_1$, we have that i is the parent of j if and only if $\phi(i)$ is the parent of $\phi(j)$. A subtree isomorphism is *maximal* if there is no other subtree isomorphism $\phi' : H_1' \rightarrow H_2'$ with H_1 a strict subset of H_1', and *maximum* if H_1 has largest cardinality. The maximal (maximum) subtree isomorphism problem is to find a maximal (maximum) subtree isomorphism between two rooted trees. This is a problem solvable in polynomial time [18].

Let i and j be two distinct vertices of a rooted tree T, and let $i = x_0 x_1 \ldots x_n = j$ be the (unique) path joining them. The *path-string* of i and j, denoted by $\text{str}(i, j)$, is the string $s_1 s_2 \ldots s_n$ on the alphabet $\{-1, +1\}$ where, for all $k = 1 \ldots n$, $s_i = \text{lev}(x_k) - \text{lev}(x_{k-1})$. By convention, when $i = j$ we define $\text{str}(i, j) = \varepsilon$, where ε is the null string (i.e., the string having zero length).

The *tree association graph* (TAG) of two rooted trees $T_1 = (V_1, E_1)$ and $T_2 = (V_2, E_2)$ is the graph $G = (V, E)$ where

$$V = V_1 \times V_2$$

and, for any two vertices (i, h) and (j, k) in V, we have

$$((i, h), (j, k)) \in E \Leftrightarrow \text{str}(i, j) = \text{str}(h, k) .$$

The following theorem establishes a one-to-one correspondence between the maximum subtree isomorphism problem and the maximum clique problem [21].

Theorem 5. *Any maximal (maximum) subtree isomorphism between two rooted trees induces a maximal (maximum) clique in the corresponding TAG, and vice versa.*

Theorem 5 provides a formal justification for applying replicator dynamics to find maximal subtree isomorphisms. In [21] this approach has been applied in computer vision to the problem of matching articulated and deformed visual shapes described by "shock" trees, an abstract representation of shape based on the singularities arising during a curve evolution process. The experiments, conducted on a number of shapes representing various object classes, yielded very good results. The system typically converged towards the globally optimal solutions in only a few seconds, and compared favorably with another powerful tree matching algorithm.

In many practical applications the trees being matched have vertices with an associated vector of symbolic and/or numeric attributes. The framework just described has also been extended in a natural way for solving attributed tree matching problems [21].

5   Multi-population Models

The single-population replicator equations discussed so far can easily be generalized to the case where interactions take place among $n \geq 2$ individuals randomly drawn from n distinct populations [1,2]. In this case, the continuous-time dynamics (2) becomes

$$\dot{x}_i^\mu(t) = x_i^\mu(t) \left(\pi_i^\mu(t) - \sum_\nu x_i^\nu(t)\pi_i^\nu(t) \right), \tag{8}$$

and its discrete-time counterpart is

$$x_i^\mu(t + \Delta t) = \frac{x_i^\mu(t)\pi_i^\mu(t)}{\sum_\nu x_i^\nu(t)\pi_i^\nu(t)} . \tag{9}$$

The function π can either be linear, as in (1), or can take a more general form. If there exists a polynomial F such that

$$\pi_i^\mu = \frac{\partial F}{\partial x_i^\mu} ,$$

then it can be proven that F strictly increases along any trajectory of both dynamics. Note that these dynamics work in a product of standard simplices.

Mühlenbein et al. [22] used multi-population replicator equations to approximately solve the graph partitioning problem, which is NP-hard [18]. Given a graph $G = (V, E)$ with edge weights w_{ij}, their goal was to partition the vertices of G into a predefined number of clusters in such a way as to maximize the overall intra-partition traffic

$$F = \prod_\mu K^\mu ,$$

where

$$K^\mu = \sum_i \sum_j w_{ij} x_i^\mu x_j^\mu$$

is the intra-partition traffic for cluster μ. Here, x_i^μ can be interpreted as the probability that vertex i belongs to cluster μ.

By putting

$$\pi_i^\mu = \frac{2F \sum_j w_{ij} x_j^\mu}{K^\mu}$$

the replicator equations seen above will indeed converge toward a maximizer of F. However, in so doing the system typically converges towards an interior attractor, thereby giving an infeasible solution. To avoid this problem, Mühlenbein et al. [22] put a "selection pressure" parameter S on the main diagonal of the weight matrix, and altered it during the evolution of the process. Intuitively, $S = 0$ has no influence on the system. Negative values of S prevent the vertices to decide for a partition, whereas positive values force the vertices to take a decision. The proposed algorithm starts with a negative value of S, and makes the discrete-time dynamics (9) evolve. After convergence, if an infeasible solution has been found, S is increased and the algorithm is started again. The entire procedure is iterated until convergence to a feasible solution. A similar, but more principled, strategy for the maximum clique problem can be found in [15]. The results presented in [22] on a particular problem instance are fairly encouraging. However, more experiments on larger and diverse graphs are needed to fully assess the potential of the approach.

Multi-population replicator models have also been used in [22,23] to solve the traveling salesman problem, which asks for the shortest closed tour connecting a given set of cities, subject to the constraint that each city be visited only once. The results presented on small problem instances, i.e., up to 30 cities, are encouraging but it seems that they do not scale well with the size of the problem.

6 Conclusions

Despite their simplicity and inherent inability to escape from local solutions, replicator dynamics have proved to be a useful heuristic for attacking a number of combinatorial optimization problems. They are completely devoid of operational parameters, which typically require a lengthy, problem-dependent tuning phase, and are especially suited for parallel (analog) hardware implementation.

Due to lack of space we cannot discuss recent developments aimed at improving their performance, and simply refer the reader to the original papers [14,15, 16,24,25]. We also refer to [26] for intriguing connections among replicator equations, majorization theory and genetic algorithms.

References

1. J. Hofbauer and K. Sigmund. *Evolutionary Games and Population Dynamics.* Cambridge University Press, Cambridge, UK, 1998.
2. J. W. Weibull. *Evolutionary Game Theory.* MIT Press, Cambridge, MA, 1995.

3. J. F. Crow and M. Kimura. *An Introduction to Population Genetics Theory.* Harper & Row, New York, 1970.

4. R. A. Fisher. *The Genetical Theory of Natural Selection.* Oxford University Press, London, UK, 1930.

5. I. M. Bomze, M. Budinich, P. M. Pardalos, and M. Pelillo. The maximum clique problem. In D.-Z. Du and P. M. Pardalos, editors, *Handbook of Combinatorial Optimization (Suppl. Vol. A)*, pages 1–74. Kluwer, Boston, MA, 1999.

6. T. S. Motzkin and E. G. Straus. Maxima for graphs and a new proof of a theorem of Turán. *Canad. J. Math.*, 17:533–540, 1965.

7. L. E. Gibbons, D. W. Hearn, P. M. Pardalos, and M. V. Ramana. Continuous characterizations of the maximum clique problem. *Math. Oper. Res.*, 22:754–768, 1997.

8. M. Pelillo and A. Jagota. Feasible and infeasible maxima in a quadratic program for maximum clique. *J. Artif. Neural Networks*, 2:411–420, 1995.

9. M. Pelillo. Relaxation labeling networks for the maximum clique problem. *J. Artif. Neural Networks*, 2:313–328, 1995.

10. M. Pelillo and I. M. Bomze. Parallelizable evolutionary dynamics principles for solving the maximum clique problem. In H.-M. Voigt, W. Ebeling, I. Rechenberg, and H.-P. Schwefel, editors, *Parallel Problem Solving from Nature—PPSN IV*, pages 676–685. Springer-Verlag, Berlin, 1996.

11. I. M. Bomze. Evolution towards the maximum clique. *J. Glob. Optim.*, 10:143–164, 1997.

12. I. M. Bomze, M. Pelillo, and R. Giacomini. Evolutionary approach to the maximum clique problem: Empirical evidence on a larger scale. In I. M. Bomze, T. Csendes, R. Horst, and P. M. Pardalos, editors, *Developments in Global Optimization*, pages 95–108. Kluwer, Dordrecht, The Netherlands, 1997.

13. I. M. Bomze and F. Rendl. Replicator dynamics for evolution towards the maximum clique: Variations and experiments. In R. De Leone, A. Murlí, P. M. Pardalos, and G. Toraldo, editors, *High Performance Algorithms and Software in Nonlinear Optimization*, pages 53–67. Kluwer, Dordrecht, The Netherlands, 1998.

14. I. M. Bomze. Global escape strategies for maximizing quadratic forms over a simplex. *J. Glob. Optim.*, 11:325–338, 1997.

15. I. M. Bomze, M. Budinich, M. Pelillo, and C. Rossi. A new "annealed" heuristic for the maximum clique problem. In P. M. Pardalos, editor, *Approximation and Complexity in Numerical Optimization: Continuous and Discrete Problems*, pages 78–95. Kluwer, Dordrecht, The Netherlands, 2000.

16. I. M. Bomze and V. Stix. Genetic engineering via negative fitness: Evolutionary dynamics for global optimization. *Ann. Oper. Res.*, 89:279–318, 1999.

17. I. M. Bomze, M. Pelillo, and V. Stix. Approximating the maximum weight clique using replicator dynamics. *IEEE Trans. Neural Networks*, 11(6):1228–1241, 2000.

18. M. R. Garey and D. S. Johnson. *Computers and Intractability: A Guide to the Theory of NP-Completeness.* W. H. Freeman, San Francisco, CA, 1979.

19. D. S. Johnson. The NP-completeness column: An ongoing guide. *J. Algorithms*, 9:426–444, 1988.

20. M. Pelillo. Replicator equations, maximal cliques, and graph isomorphism. *Neural Computation*, 11(8):2023–2045, 1999.

21. M. Pelillo, K. Siddiqi, and S. W. Zucker. Matching hierarchical structures using association graphs. *IEEE Trans. Pattern Anal. Machince Intell.*, 21(11):1105–1120, 1999.

22. H. Mühlenbein, M. Gorges-Schleuter, and O. Krämer. Evolution algorithms in combinatorial optimization. *Parallel Computing*, 7:65–85, 1988.

23. M. Pelillo. Relaxation labeling processes for the traveling salesman problem. In *Proc. Int. J. Conf. Neural Networks*, pages 2429–2432, Nagoya, Japan, 1993.
24. A. Menon, K. Mehrotra, C. K. Mohan, and S. Ranka. Optimization using replicators. In *Proc. 6th. Int. Conf. Genetic Algorithms*, pages 209–216. Morgan Kaufmann, 1995.
25. C. Rossi. A replicator equations based evolutionary algorithm for the maximum clique problem. In *Congress on Evolutionary Computation*, pages 1565–1570, 2000.
26. A. Menon, K. Mehrotra, C. K. Mohan, and S. Ranka. Replicators, majorization and genetic algorithms: New models and analytical tools. In *Proc. FOGA '96—Found. of Genetic Algorithms*, Aug. 1996.

A Parallel Hybrid Heuristic for the TSP

Ranieri Baraglia[1], José Ignacio Hidalgo[2], and Raffaele Perego[1]

[1] CNUCE - Institute of the Italian National Research Council
Via Alfieri 1, 56010 S. Giuliano Terme, Pisa (Italy)
{Ranieri.Baraglia, Raffaele.Perego}@cnuce.cnr.it
[2] Dpto. Arquitectura Computadores y Automatica
Universidad Complutense, Madrid 28040
hidalgo@dacya.ucm.es

Abstract. In this paper we investigate the design of a coarse-grained parallel implementation of *Cga-LK*, a hybrid heuristic for the Traveling Salesman Problem (TSP). Cga-LK exploits a compact genetic algorithm in order to generate high-quality tours which are then refined by means of an efficient implementation of the Lin-Kernighan local search heuristic. The results of several experiments conducted on a cluster of workstations with different TSP instances show the efficacy of the parallelism exploitation.

Keywords: Parallel algorithms, TSP, compact genetic algorithm, Lin-Kernighan algorithm, hybrid GA.

1 Introduction

The Traveling Salesman Problem (TSP) is the problem of finding the shortest closed tour through a given set of cities visiting each city exactly once. Thus, given a set of cities $C = \{c_1, c_2, ..., c_k\}$, for each pair (c_i, c_j), $i \neq j$, let $d(c_i, c_j)$ be the distance between city c_i and c_j. Solving the TSP entails finding a permutation $\pi\prime$ of the cities $(c_{\pi'(1)}, ..., c_{\pi'(k)})$, such that

$$\sum_{i=1}^{k} d(c_{\pi'(i)}, c_{\pi'(i+1)}) \leq \sum_{i=1}^{k} d(c_{\pi(i)}, c_{\pi(i+1)}) \qquad \forall \pi \neq \pi', (k+1) \equiv 1 \quad (1)$$

In the *symmetric* TSP $d(c_i, c_j) = d(c_j, c_i), \forall i, j$, while in the *asymmetric* TSP this condition is not satisfied. In this work we consider the symmetric TSP.

Since the TSP is probably the most well-known NP-hard combinatorial optimization problem, researchers have proposed many heuristics for searching the space of all permutations π. Problem-independent heuristics such as simulated annealing (SA) [1] and genetic algorithms (GA) [2,3] perform quite poorly with this particular problem. They require high execution times for solutions whose quality is not comparable with those achieved in much less time by their domain-specific local search counterparts. Domain-specific heuristics such as *2-Opt* [4],

E.J.W. Boers et al. (Eds.) EvoWorkshop 2001, LNCS 2037, pp. 193–202, 2001.
© Springer-Verlag Berlin Heidelberg 2001

3-Opt [5], and Lin-Kernighan (LK) [6] are effective. In particular LK is considered to be the heuristic that leads to the best solutions. Efficient implementations have been devised for LK which take just a few seconds to compute a high-quality solution for problems with hundreds of cities [7,8].

Several published results demonstrate that combining a problem-independent heuristic with a local search method is a viable and effective approach for finding high-quality solutions of large TSPs. The problem-independent part of the hybrid algorithm drives the exploration of the search space focusing on the global optimization task, while the local search algorithm allows to search in depth the subregions of the solution space.

In [9] the *Chained local optimization* algorithm is proposed. It exploits a special type of *4-opt* moves under the control of a SA mechanism to escape from the local optima found with LK. In [10] genetic operators to search the space of the local optima determined with LK are proposed.

Martina Georges-Schleuter experimented with the exploitation of simple *k*-Opt moves within her *Asparagos96* parallel genetic algorithm [11]. She concluded that, for large problem instances, the strategy of producing many fast solutions might be almost as effective as using powerful local search methods with fewer solutions.

In this paper we propose a coarse-grained parallel implementation of Cga-LK, a previously proposed hybrid heuristic for the TSP which exploits a compact genetic algorithm (Cga) in order to generate high-quality tours which are then refined by means of the Lin-Kernighan local search heuristic. The parallel implementation allowed us to enhance the encouraging results obtained with the sequential version of Cga-LK [12,13].

The rest of the paper is organized as follows: Section 2 gives a brief introduction to the Cga. Section 3 describes the parallel implementation of Cga-LK, and discusses the experimental results. Finally, Section 4 outlines some conclusions.

2   The Compact Genetic Algorithm

The Cga does not manage a population of solutions but only mimics its existence [14]. The Cga represents the population by means of a vector of values $p_i \in [0,1], \forall i = 1,\ldots,l$, where l is the number of alleles needed to represent the solutions. Each value p_i indicates the proportion of individuals in the simulated population which has a 0 or 1 in the ith locus of their representation. By treating these values as probabilities, new individuals can be generated and, based on their fitness, the probability vector can be updated in order to favor the generation of better individuals.

The values for probabilities p_i are initially set to 0.5 to represent a randomly generated population in which the value for each allele has equal probability. At each iteration the Cga generates two individuals on the basis of the current probability vector and compares their fitness. Let W be the representation of the individual with a better fitness and L the individual whose fitness is worse.

The two representations are used to update the probability vector at step $k + 1$ in the following way:

$$
p_i^{k+1} = \begin{cases} p_i^k + 1/n & \text{if } w_i = 1 \wedge l_i = 0 \\ p_i^k - 1/n & \text{if } w_i = 0 \wedge l_i = 1 \\ p_i^k & \text{if } w_i = l_i \end{cases} \tag{2}
$$

where n is the dimension of the population simulated, and w_i (l_i) is the value of the ith allele of W (L). The Cga ends when the values of the probability vector are all equal to 0 or 1. At this point vector p represents the solution obtained.

In order to represent a population of n individuals, the Cga updates the probability vector by a constant value equal to $1/n$. Only $\log_2 n$ bits are thus needed to store the finite set of values for each p_i. The Cga therefore requires $l \cdot \log_2 n$ bits with respect to the $n \cdot l$ bits needed by a classic GA. Larger populations can be thus exploited without significantly increasing memory requirements, but only slowing Cga convergence. This peculiarity makes the use of Cga attractive to address problems for which the huge memory requirements of GAs is a constraint.

On order-one problems the Cga and the simple GA with uniform crossover are approximately equivalent in terms of solution quality and in the number of function evaluations needed. To solve higher than order-one problems GAs with both higher selection rates and larger population sizes have to be used [15]. The Cga selection rate can be increased by adopting the following mechanism: (I) generate at each iteration s individuals from the probability vector; (II) choose among the s individuals the one with best fitness; (III) compare the individual with best fitness with the other $s - 1$ individuals and update the probability vector according to (2). Such an increase on the selection rate helps the Cga to converge to better solutions since it increases the survival probability of higher-order building blocks.

3   A Parallel Hybrid Heuristic for the TSP

Cga-LK combines a Cga with *Chained LK* [1], an efficient implementation of LK proposed by Applegate Bixby, Chvatal, and Cook [8]. In Cga-LK the Cga is used to explore the more promising part of the TSP solution space in order to generate "good" initial solutions which are refined with Chained LK. The refined solutions are also exploited to improve the quality of the simulated population as the execution progresses. A detailed description of the sequential implementation of Cga-LK can be found in [12]. Here we will concentrate the attention on the parallelism exploitation.

The coarse-grained parallelization model [16,17] was used to design the parallel version of Cga-LK. According to this model the whole population is divided into a few demes, which evolve in parallel. Each deme is assigned to a different

---

[1] This routine is available in the CONCORDE library at
url http://www.keck.caam.rice.edu/concorde.

processor and the evolution process takes place only among individuals belonging to the same deme. This feature means that a greater genetic diversity can
be maintained with respect to the exploitation of a panmitic population, thus
improving the solution space exploration. Since the size of the demes is smaller
than the population used by the correspondent serial GA, in general, a parallel
GA converges faster. Moreover, it is also true that the quality of the solution
might be poorer than that of the sequential case. Therefore, in order to improve
the deme genotypes, a migration operator that periodically exchanges the best
solutions among different demes is used. Depending on the migration operator chosen we can distinguish between the *island* model and the *stepping stone*
model. In the island model migration occurs among every deme, while in stepping stone model the migration occurs only between neighboring demes. Studies
have shown that there are two critical factors [18]: the number of solutions migrated each time and the interval time between two consecutive migrations. A
large number of migrants leads to the behavior of the island model similar to
the behavior of a panmitic model. A few migrants prevent the GA from mixing
the genotypes, and thus reduce the possibility to bypass the local optima inside
the islands. Implementations of coarse grained parallel GAs can be found in [18,
19,20,21,22,23].

To implement Cga-LK the island model was adopted. Moreover, Cga-LK
exploits the MPI message-passing library [24], and the SPMD programming
model [25]. According to this programming model all the processing nodes run
the same code which simulates a different population. We can consider each simulated population elaborated in parallel an island of the same larger population.
To improve the sub-population genotypes, a migration operator that periodically
exchanges the best solution among different islands was adopted.

To extend the Cga to the TSP, we considered the frequencies of the edges
occurring in the simulated population. A $k \times k$ triangular matrix of probabilities
P was used to store these frequencies. Each element $p_{i,j}, i > j$, of P represents the
proportion of individuals whose tour contains edge (c_i, c_j). If n is the population
dimension, our Cga thus requires $(k^2/2) \cdot \log_2 n$ bits to represent the population,
compared with the $k \cdot n \log_2 k$ bits required by a classical GA. Figure 1 shows
the pseudo-code of our parallel Cga-LK. Its main functions are discussed in the
following.

After setting the MPI environment, the matrix P is initialized. To this end,
first we randomly generate a tour to which the Chained LK routine is applied
to carry out a local optimum. Then the probability values associated to all the
edges belonging to the local optimum are increased by $1/n$. This procedure is
iteratively applied n times to represent in P the whole simulated population.

To differentiate its behavior each parallel process uses a different seed (loc-
al_seed) for the pseudo-random generation. It is obtained by adding to the same
seed (global_seed) the process identifier.

At each generation k of Cga-LK, a single individual L (current_tour) is
generated from the probability matrix. To this end a *greedy* algorithm is used. A
starting city c_a is randomly selected and inserted in the tour V. Then, another

```
Program Par-Cga-LK
  begin
    /* Setting of the MPI environment, me is the process identifier */
    MPI_Init(...);
    MPI_Comm_rank(MPI_COMM_WORLD,&me);
    /* Initialization of the probability matrix P */
    local_seed := global_seed + me;
    Initialize(P, local_seed);
    best_tour_length = MAX_INT;
    generations := 0;
    repeat
        generations := generations + 1;
        current_tour := Generate(P);
        /* Apply Chained LK algorithm */
        optimized_tour := Chained_LK(current_tour);
        optimized_tour_length := Tour_length(optimized_tour);
        Update(P,optimized_tour,current_tour);

        /* Store the best tour found so far */
        if (optimized_tour_length < best_tour_length) then
            count := 0;
            best_tour_length := optimized_tour_length;
            best_tour := optimized_tour;
        end if
        count := count + 1;

        if (mod(generations, F_mig) = 0) then
            /* Perform migration */
            MPI_Allreduce(best_tour,best_global_tour,.......)
            if (best_global_tour = termination_signal) then
                Output_Results();
                exit;
            else
                Update(P,best_global_tour, best_tour);
            end if
        end if
    until (Local_Termination());

    MPI_Allreduce(termination_signal,.....);
    Output_Results();
end
```

Fig. 1. Pseudo-code of the parallel Cga-LK.

city $c_b \notin V$ is randomly chosen. City c_b is inserted in V as successor of c_a with probability $p_{a,b}$ (i.e. the probability associated to edge (c_a, c_b)). Otherwise c_b is discarded and the process is repeated by choosing another city not belonging to V. Clearly, this process may fail to find the successor of some city $c_{\overline{a}}$. This takes place when all the cities not already inserted in the current tour have been analyzed, but the probabilistic selection criterion failed to choose one of them. In this case the city $c_{\overline{b}}$ successor of $c_{\overline{a}}$ is selected according to the following formula:

$$\overline{b} = \text{argmax}\{p_{\overline{a},j} : c_j \in \{c_1, c_2, \dots, c_k\} \setminus V\} \qquad (3)$$

When (3) is satisfied by several cities, i.e. edges $(c_{\overline{a}}, c_j)$ have the same probability for different cities $c_j \notin V$, the city which minimizes the distance $d(c_{\overline{a}}, c_j)$ is selected. The generation process ends when all the cities have been inserted in V, and a feasible tour has been thus generated.

The current_tour is then used as the starting solution for the Chained LK routine which produces an individual W (optimized_tour). Then the probability matrix is updated comparing current_tour with optimized_tour as follows:

$$p_{i,j}^{k+1} = \begin{cases} p_{i,j}^k + \frac{1}{n} & \text{if } (c_i, c_j) \vee (c_j, c_i) \in W \text{ and } (c_i, c_j) \vee (c_j, c_i) \notin L \\ p_{i,j}^k - \frac{1}{n} & \text{if } (c_i, c_j) \vee (c_j, c_i) \in L \text{ and } (c_i, c_j) \vee (c_j, c_i) \notin W \\ p_{i,j}^k & \text{otherwise} \end{cases} \qquad (4)$$

Every F_mig generations a migration of individuals among different islands takes place to improve sub-population genotypes. The function MPI_Allreduce provides an efficient way to perform migration. It implements an all-to-all reduction operation where the tour achieving minimal length (best_global_tour) is broadcast to all the processes. The global optimum found so far is then used to update the probability matrix local to all the processes. The migration mechanism is used also to manage distributed termination. To this purpose, when a process reaches the termination condition, it broadcasts a termination_signal which is intercepted by the other processes that accordingly terminate their execution. Local termination is decided when a threshold is reached on the number of generations performed without an improvement of the best solution achieved, or on the elapsed execution time.

3.1   Experimental Results

The parallel Cga-LK algorithm was tested on some TSP instances defined in TSPLIB [26]. We used instances: att532, gr666, rat783, pr1002 which have optimal solutions equal to $27686, 294358, 8806, 259045$, respectively. The experiments were conducted on a cluster of three Linux Pentium II 200 MHz PCs with 128 Mbytes of memory, and each test was repeated ten times to obtain an average behavior. Each Cga-LK process simulates a population of 128 individuals. In all the tests performed the parallel Cga-LK algorithm carried out solutions with optimal length, independently from the parallelism degree and the migration frequency exploited. To make the algorithm performance evaluation independent from the computer architecture used, the comparison of the results obtained on each test was based on the number of generations performed. Figure 2 shows the average number of generations required to get the optimal tour as a function of the number of parallel processes used, for different values of the migration parameter. As it can be seen from the plots, the average number of generations required by the parallel algorithm to get the optimal tour is always lower than in the sequential case. Moreover, such number, in general decreases when the number of processes increases. With regard to the migration parameter, on smaller TSP instances a low value seems to work better than a large one, while on larger instances it affects slightly the results achieved. For the same tests, Figure 3 shows the minimal number of generations needed to achieve optimal solutions. Also in this case the benefits of parallelism exploitation are clear. The parallel algorithm always needs a lower number of iterations to obtain the optimal solution than those needed by the sequential version of the algorithm.

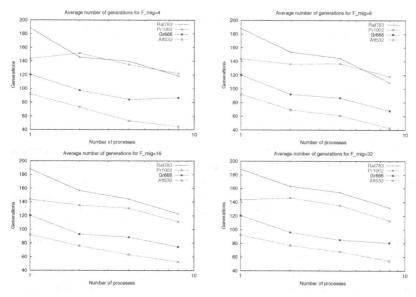

Fig. 2. Average number of generations required to carry out the optimal tour of various TSP instances as a function of the migration parameter and the number of parallel processes used.

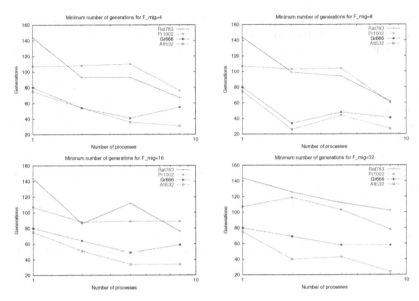

Fig. 3. Minimum number of generations required to carry out the optimal tour of various TSP instances as a function of the migration parameter and the number of parallel processes used.

4   Conclusions

In this paper we proposed a coarse-grained parallel hybrid heuristic to solve TSP. It combines a compact genetic algorithm to generate high-quality tours which are then refined by means of the Lin-Kernighan local search heuristic. The refined solutions are also exploited to improve the quality of the simulated population as the execution progresses. The parallel algorithm was implemented according to the SPMD programming paradigm by using the MPI message-passing library. Our parallel algorithm was evaluated on medium TSP instances. The results achieved were satisfactory if compared to those obtained by the sequential version of the algorithm on the same instances. The average number of generations required by the parallel algorithm to get the optimal tour was always lower than in the sequential case. As future work we plan to investigate either the behavior of the parallel algorithm on large TSP instances and extensions of the hybrid approach to solve other optimization problems.

Acknowledgments. This work was supported by the Italian National Research Council and the Spanish Government, Grant TYC 1999-0474.

References

1. S. Kirkpatrick, C.D. Gelatt, and M.P. Vecchi. Optimization by simulated annealing. *Science*, 220:671–680, 1983.
2. J. Grefenstette, R. Gopal, B. Rosimaita, and D. van Gucht. Genetic algorithms for the traveling salesman problem. In *Proceedings of the International Conference on Genetics Algorithms and their Applications*, pages 160–168, 1985.
3. H. C. Braun. On solving traveling salesman problems by genetic algorithm. In H. P. Schwefel and R. Männer, editors, *Parallel Problem Solving from Nature*, volume 496 of *Lecture Notes in Computer Science*, pages 129–133, Berlin, 1991. Springer-Verlag.
4. G. A. Croes. A method for solving traveling salesman problems. *Operations Research*, 6:791–812, 1958.
5. S. Lin. Computer solution of the traveling salesman problem. *Bell System Technical Journal*, 44:2245–2269, 1965.
6. S. Lin and B. W. Kernighan. An effective heuristic algorithm for the traveling salesman problem. *Operations Research*, 21:498–516, 1973.
7. D. S. Johnson and L. A. McGeoch. *Local Search in Combinatorial Optimization*, chapter The Traveling Salesman Problem: A Case Study in Local Optimization. John Wiley and Sons, New York, 1996.
8. D. Applegate, R. Bixby, V. Chvátal, and W. Cook. Finding tours in the tsp. Preliminary chapter of a planned monograph on the TSP, available at URL: `http://www.caam.rice.edu/~keck/reports/lk_report.ps`, 1999.
9. O. Martin and S.W.Otto. Combining simulated annealing with local search heuristic. *To appear on Annals of Operation Research*.
10. P. Merz and B. Freisleben. Genetic local search for the TSP: New results. In *Proceedings of the 1997 IEEE International Conference on Evolutionary Computation*, pages 159–163, Indianapolis, USA, 1997. IEEE press.
11. M. Gorges-Schleuter. Asparagos96 and the travelling salesman problem. In T. Bäck, editor, *Proceedings of the Fourth International Conference on Evolutionary Computation*, pages 171–174, New York, IEEE Press, 1997.
12. R. Perego R. Baraglia, J. I. Hidalgo. A hybrid approach for the TSP combining genetics and the Lin-Kerninghan local search. Technical Report CNUCE-B4-2000-007, CNUCE - Institute of the Italian National Research Council, 2000.
13. R. Perego R. Baraglia, J. I. Hidalgo. A hybrid approach for the Traveling Salesman Problem. *submitted paper*.
14. G. Harik, F. Lobo, and D. Goldberg. The compact genetic algorithm. *IEEE Transactions on Evolutionary Computation*, 3(4):287–297, 1999.
15. D. Thierens and D. Goldberg. Mixing in genetic algorithms. In S. Forrest, editor, *Proceedings of the Fifth International Conference on Genetic Algorithms*, pages 38–45, San Mateo, CA, 1993. Morgan Kaufmann.
16. E. Cantu-Paz. A survey of parallel genetic algoritms. Technical Report 97003, University of Illinois at Urbana-Champaign, Genetic Algoritms Lab. (IlliGAL), http://gal4.ge.uiuc.edu/illigal.home.html, July 1997.
17. M. Tomassini. A survey of genetic algorithms. Technical Report 95/137, Department of Computer Science, Swiss Federal Institute of Technology, Lausanne, Switzerland, July 1995.
18. P. Grosso. *Computer Simulations of Genetic Adaptation: Parallel Subcomponent Interaction in a Multilocus Model*. PhD thesis, University of Michigan, 1985.

19. R. Tanese. Parallel genetic algorithms for a hypercube. In *Proceedings of the Second International Conference on Genetic Algorithms*, pages 177–183. L. Erlbaum Associates, 1987.

20. R. Tanese. Distribuited genetic algorithms. In *Proceedings of the Third International Conference on Genetic Algorithms*, pages 434–440. M. Kaufmann, 1989.

21. S. Cohoon, J. Hedge, S. Martin, and D. Richards. Punctuated equilibria: a parallel genetic algorithm. *IEEE Transaction on CAD*, 10(4):483–491, April 1991.

22. C. Pettey, M. Lenze, and J. Grefenstette. A parallel genetic algorithm. In *Proceedings of the Second International Conference on Genetic Algorithms*, pages 155–161. L. Erlbaum Associates, 1987.

23. H. Muhlenbein, M. Schomisch, and J. Born. The parallel genetic algorithm as function optimizer. *Parallel Computing*, 17:619–632, 1991.

24. W. Gropp, E. Lusk, and A. Skjellum. *Using MPI*. Massachusetts Institute of Technology, 1999.

25. D. E. Culler and J. P. Singh. *Parallel Computer Architecture a Harware/Sotware Approach*. Morgan Kaufmann Publishers, Inc., 1999.

26. G. Reinelt. TSPLIB—A traveling salesman problem library. *ORSA Journal on Computing*, 3:376–384, 1991.

Effective Local and Guided Variable Neighbourhood Search Methods for the Asymmetric Travelling Salesman Problem

Edmund K. Burke, Peter I. Cowling, and Ralf Keuthen

Automated Scheduling, Optimization, and Planning Group (ASAP),
School of Computer Science & IT,
University of Nottingham,
Jubilee Campus,
Nottingham, NG8 1BB, UK
{ekb, pic, rxk}@cs.nott.ac.uk
http://www.asap.cs.nott.ac.uk

Abstract. In this paper we present effective new local and variable neighbourhood search heuristics for the asymmetric Travelling Salesman Problem. Our local search approach, HyperOpt, is inspired by a heuristic developed for a sequencing problem arising in the manufacture of printed circuit boards. In our approach we embed an exact algorithm into a local search heuristic in order to exhaustively search promising regions of the solution space. We propose a hybrid of HyperOpt and 3-opt which allows us to benefit from the advantages of both approaches and gain better tours overall. Using this hybrid within the Variable Neighbourhood Search (VNS) metaheuristic framework, as suggested by Hansen and Mladenović, allows us to overcome local optima and create tours of very high quality. We introduce the notion of a "guided shake" within VNS and show that this yields a heuristic which is more effective than the random shakes proposed by Hansen and Mladenović. The heuristics presented form a continuum from very fast ones which produce reasonable results to much slower ones which produce excellent results. All of the heuristics have proven capable of handling the sort of constraints which arise for real life problems, such as those in electronics assembly.

1 Introduction

The Travelling Salesman Problem (TSP) is one of the best studied combinatorial optimization problems of our time. The TSP consists of a set of n cities $\{1, 2, \ldots, n\}$, associated with a cost matrix (c_{ij}), with $i, j \in \{1, 2, \ldots, n\}$, defining the travelling costs between cities i and j. The aim of the TSP is to determine a tour of minimum cost visiting each city exactly once and returning to the starting city. In the case that the travelling cost from city i to city j is not necessarily equal to the travelling cost from j to i we speak of an *asymmetric* TSP (ATSP).

Symmetric TSPs (STSPs), especially Euclidean problems where cities lie in a two dimensional plane and use the Euclidean distance metric, have been very well researched in the literature [1][2] and fast and powerful heuristic approaches

E.J.W. Boers et al. (Eds.) EvoWorkshop 2001, LNCS 2037, pp. 203–212, 2001.

have been proposed [3][4]. The ATSP has been less well studied and the heuristic approaches proposed so far do not match their symmetric counterparts in effectiveness. Many heuristics that are successful for symmetric instances cannot be applied efficiently to asymmetric problems.

The heuristic approaches for the ATSP can be loosely divided into two groups. The first group contains the constructive approaches which create a feasible tour from scratch. These include the well known *Nearest Neighbour*, *Nearest Insertion* and *Multiple Fragment* (often referred to as *Greedy*) algorithms [2][3][5] as well as some approaches based on the *Assignment Problem* (AP)[6]. The Assignment Problem is equivalent to finding a minimum weight matching of the bipartite graph formed by repeating each vertex on both sides of the bipartition and orienting all directed edges in the original city graph in the same direction. This represents a generalization of the ATSP with either a tour or a minimum cycle cover as solution. The *Repeated Assignment Algorithm* is one of these algorithms [7]. It is based on the solutions of the assignment problem where an AP algorithm is repeatedly applied until a valid tour is found. *Patching Algorithms*, are also based on minimum cycle covers where the cycles are repeatedly *patched* (combined) to create a tour. Various patching strategies, like the Karp-Steele Patching heuristic [8], Greedy Karp-Steele Patching or Contract-or-Patch [9] have been suggested in the literature. A truncated branch-and-bound subtour elimination procedure has been introduced in [10].

The second group includes the local search heuristics 3-opt [5], the Kanellakis and Papadimitriou algorithm [11], which is based on the successful Lin-Kernighan heuristic for the STSP [4], and their variants. Further approaches to extend local search either by restarting (iterative or chained approaches) or evolutionary algorithms [12] have been suggested in the literature as well. It has been shown in [13] that an ATSP of n cities can be transformed to a symmetric problem of $3n$ cities. However, since many industrial applications of the TSP are asymmetric further research is necessary and good heuristic approaches are desirable, especially those that can handle the additional constraints that arise in industrial ATSP applications. These applications range from problems in flow shop scheduling [11][14], steel hot strip mill scheduling [15], to sequencing problems for numerically controlled punch press [16] or drilling machines [17] where the asymmetry of the problem occurs due to asymmetry in machine set ups.

The heuristics we describe were inspired by such an industrial application arising in the manufacture of printed circuit boards [18][19]. This application concerns the movement of a placement head for a numerically controlled component placement machine. The placement head moves between a feeder magazine, where the electronic components are stored, and placement locations on the circuit board. When considering placement locations as cities and the movement times between two placement locations as the distance, this sequencing problem can be modelled as a TSP. However, since different component types are usually assigned to different feeder slots in the magazine, the head movement time between two placement locations is dependent on the movement direction, making the problem asymmetric. The machine we were considering in [18][19] is equipped with multiple placement heads. This allows the machine to pick up as many components as heads are available from the feeder magazine, move back to the circuit board and place them without having to return to the feeder magazine. This feature inspired our heuristic approach where an exact algorithm is used to ensure optimality of these small pick-and-place subsequences. The optimal subsequences are then re-embedded into the original placement sequence. The resulting heuristic represents a novel approach where an exact algorithm is used within the framework of a local search heuristic.

This paper is structured as follows. In the next section we introduce our new heuristic approaches. First we are going to describe the HyperOpt approach for the asymmetric Travelling Salesman Problem. We then propose an approach which is a hybrid of HyperOpt and 3-opt, to benefit from the advantages of both methods. Then we introduce Variable Neighbourhood Search (VNS), as described by Hansen and Mladenović in [20][21], and show how the hybrid method proposed earlier can be used efficiently for local search in a VNS framework. We further introduce guided shakes to accelerate convergence for our VNS approach gaining tours of near optimal quality for the test instances considered. The third section presents the computational results for the ATSP. In the last section we present the conclusions we have drawn from our experiments and give a brief insight into planned future research.

2   Heuristic Methods for the ATSP

In this section we propose new heuristic approaches for the ATSP based on local search. The local search routine we introduce is based on splitting the original problem into small subproblems which are then solved to optimality using an exact algorithm. We discuss advantages and disadvantages of this approach in comparison to local search 3-opt for the ATSP, and introduce a hybrid between 3-opt and our new approach. In order to enable local search to create solutions of very high quality we demonstrate how this hybrid approach can be embedded efficiently into the Variable Neighbourhood Search metaheuristic framework proposed by Hansen and Mladenović in [20][21]. We further investigate how the "shaking" process of the Variable Neighbourhood Search method can be guided using information gained in a phase of pre-processing to improve the metaheuristics performance and find solutions of near optimal quality.

2.1   Local Search for the ATSP

HyperOpt belongs to the class of local improvement algorithms which, starting from an initial solution, repeatedly perform tour modifications for as long as improvements can be found. A variant of the algorithm described here has already been applied successfully to very large instances of the Euclidean Travelling Salesman Problem. This variant compares well to established local search approaches such as 2- and 3-opt [1][3].

206   E.K. Burke, P.I. Cowling, and R. Keuthen

The basic idea of HyperOpt local search for the ATSP is as follows. Starting from an initial tour, first two hyperedges are defined by the cities p_0 and q_0 and their h successors in the tour, as illustrated in Fig. 1(a). Next all inner edges of the hyperedges are removed from the tour leaving only the starting and finishing cities of each hyperedge attached to the tour, Fig. 1(b). The unconnected cities are then reconnected to the rest of the tour optimally using a dynamic programming algorithm. This feature of our local search routine, which we refer to as a HyperOpt move, was introduced to ensure optimality of the small pick-and-place sequences arising for multi-headed placement machines. Since the number of placement heads available is limited in practice (it typically lies between 2 and 4) enforcing optimality on these small pick-and-place sequences is computationally still tractable. Two possible outcomes of an asymmetric HyperOpt move with $h = 4$ are shown in Fig. 1(c) and (d). The orientation of the fixed tour segments is not changed, since evaluating the impact of such a change has time complexity $O(n)$, and is very expensive in comparison to the evaluation of a HyperOpt move which requires $(2h - 2)^2 \cdot 2^{(2h-2)}$ iterations, and is thus independent of n. Computational results of the HyperOpt approach with parameter h taking values of 3 and 4 will be discussed later in Section 3.

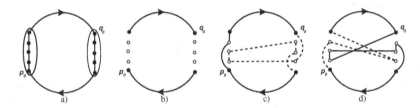

Fig. 1. HyperOpt move

An obvious drawback of the HyperOpt approach for the ATSP is its inability to reorder tour segments involving more than $h-1$ successive cities. However, HyperOpt can perform complex reorderings of the unconnected cities which would be difficult to find using other approaches. In order to overcome this problem we suggest a hybrid of HyperOpt with the simplest tour segment reordering algorithm for the TSP, 3-opt [5]. Having deleted 3 edges from an ATSP tour, there is only one way to reconnect which does not reverse any tour segments, illustrated in Fig. 2 below.

In our hybrid approach we first choose a hyperedge with starting city p_0 and evaluate the possible HyperOpt moves involving this hyperedge. If no improvement is found the tour is searched for an improving 3-opt move involving the edge connecting p_0 and its successor. In section 3 we discuss the computational results achieved by this hybrid approach for 3- and 4-HyperOpt.

To keep computational time low we made use of implementation techniques which proved to work well for local search approaches for the TSP [1][3]. Instead of a *steepest descent* approach, where the best move for a given tour is performed,

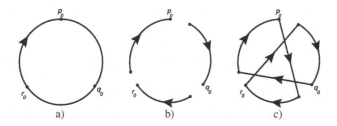

Fig. 2. 3-opt move for ATSPs

we successively select one hyperedge and evaluate for it all possible HyperOpt moves. When one or more improving moves are found, the best move is chosen and performed. The same low cost approach is taken when the tour is searched for a 3-opt move. To further speed up the heuristic approaches we made use of a stack to store hyperedges which are to be searched for improving moves [3][22]. Starting with a stack consisting of all available hyperedges in the tour, only hyperedges which have been affected by a HyperOpt or 3-opt move are stored on this stack and will be considered in later iterations. The aim of this approach, similar to the *no-look-bits* of [3], is to avoid the re-evaluation of moves which have been unsuccessful in previous iterations.

2.2   Variable Neighbourhood Search

In this section we propose an efficient variant of the Variable Neighbourhood Search heuristic proposed by Hansen and Mladenović [20][21]. This extends the work in [22] for symmetric TSPs and in [19] for electronic assembly. We introduce the notion of *guided shakes* for Variable Neighbourhood Search methods as a method to restart local search in order to overcome local optima, improving on the performance of random shaking strategies suggested in the original work by Hansen and Mladenović [20][21].

The aim of Variable Neighbourhood Search approaches is to avoid poor local optima by systematically changing neighbourhood in order to explore an increasingly larger region of the solution space. Hansen and Mladenović suggest various VNS strategies for a wide range of combinatorial optimization problems [20][21]. The VNS strategy we consider here is given in Fig. 3. The major difference between our approach and the VNS approach proposed by Hansen and Mladenović in [20][21] lies in the change of neighbourhoods used to restart the local search. Once a heuristic has been chosen for local search and an initial tour has been determined, the pivotal component of VNS is the *shaking* process. In contrast to the random shakes proposed by Hansen and Mladenović for the STSP we suggest a different strategy which guides the shakes based on information gained about the tour prior to the first shake.

Immediately after the first local search has been performed a list of suspect hyperedges is determined which we will use in our shakes. The quality of a hyperedge $H = (p_0, p_1, \ldots, p_h)$ is estimated by

$$\sum_{i=0}^{h-1} c_{p_i,p_{i+1}} - \min_{q\in\{1,2,...,n\}}(c_{p_i,q}),$$

where a high value indicates a hyperedge of poor quality. We store the $\frac{n}{3}$ hyperedges having the highest value in the list of low quality hyperedges.

For shaking we have chosen a variant of the HyperOpt neighbourhood. Instead of two hyperedges, three hyperedges are removed from the tour and the tour is restored in a random fashion. Reversal of a tour segment is avoided as otherwise too much information from the tour may be destroyed. An illustration of this shaking strategy is shown in Fig. 4.

```
Choose Neighbourhood Structures {N₁, ..., Nₘ};
Select Local Search procedure;
Create an initial tour T;
Set: T* = T, k = 1;
Do{
    Apply Local Search procedure to T;
    IF (length(T) < length(T*))
        Set: T* = T, k = 1;
    ELSE{ Set T = T*;
            IF (k < m)
                Set k = k + 1;
    }
    Shake Tour T using neighbourhood Nₖ;
} Until stopping criterion is met;
OUTPUT best found tour: T*;
```

Fig. 3. Variable Neighbourhood Search

To guide the shaking process and destroy segments of the tour of low quality, one of the hyperedges we shake is chosen at random from the list of low quality hyperedges. The remaining two hyperedges for shaking are then selected randomly out of the set of 20 nearest neighbours of the starting city of the low quality hyperedge.

To prevent the local search from falling back to a local optimum visited earlier, the length of the hyperedges used for shaking starts at $k = (h + 1)$, where h denotes the neighbourhood size used for the local search HyperOpt. When local search does not lead to an improvement in tour quality, the length of the shaking hyperedges k is increased by one and the tour is abandoned. In order not to destroy too much of the current tour and considering the size of the test problems we bounded the maximal length of shaking hyperedges by 5 for instances of less than 50 cities and by 6 otherwise.

For local search we have chosen the HyperOpt/3-opt hybrid with HyperOpt parameter $h = 3$ which provided the best compromise between high quality tours

hyperedges to shake on   deleting hyperedges   random reconnection

Fig. 4. Schema of a hyper-shake

and low computing time. The implementation of this approach is as described earlier except that we are also using nearest neighbour lists in order to keep computing times low. This means that for HyperOpt as well as 3-opt, we are not considering all possible moves but only a subset of nearest neighbours. This subset consists of the 20 nearest neighbours to and from each city. As a stopping criterion for VNS we limit the number of shakes performed by VNS to the number of cities of the problem.

3   Computational Results

In this section we apply our local and Variable Neighbourhood Search methods to all 27 ATSP instances in Reinelts Travelling Salesman Problem library TSPlib [23]. The solution quality of the heuristic approaches is expressed in percent excess over optimum. Seven of the instances ftv90, ftv100 up to ftv160 represent submatrices of problem ftv170. The largest instances provided by TSPlib, rbg323 up to rbg458, represent stacker crane problems [5] . However, the performance of the heuristics for the 27 instances should provide us with a good indication of their ability to find good solutions for other ATSPs. To construct initial solutions from which local search is started, we have chosen the *Nearest neighbour* heuristic [2]. This lead on the average to solutions of better quality for all the heuristic approaches considered here than the *Multiple Fragment (Greedy)* algorithm [3].

All the heuristic calculations we refer to below have been carried out on a Pentium II 400 MHz PC with 128 Mb memory running a Linux 2.2.13 kernel. The algorithms were coded in C++ using the GNU g++ compiler with -O2 compiler option. The computational results given in Table 1 show the average excess over optimum and average CPU-time over 10 runs for each problem, using different Nearest Neighbour starting tours and different seeds for the randomized VNS heuristics.

The first column of Table 1 lists the names of the TSPlib instances considered, where the number indicates the problem size except for problem class ftv for which this number is equal to #*cities* − 1. Because of space limitations we used the following abbreviations. HO is used to abbreviate HyperOpt where the number preceding HO indicates the parameter h used. The 3-opt/HyperOpt hybrid approaches are christened HY where the preceding number again indicates parameter h used in HyperOpt. As above VNS abbreviates Variable Neighbourhood Search using the random shaking strategy proposed by Hansen and Mladenović [20][21], while GVNS indicates the use of our VNS strategy which uses guided shakes. Both VNS strategies considered here use HY3 for local search.

Out of the five local search approaches 3opt, 3HO, 4HO, HY3 and HY4 we can see that the hybrid approaches HY yield significantly better results for nearly all problems, significantly outperforming its constituent heuristics 3-opt and HyperOpt (HO). The fastest hybrid approach HY3 outperforms 3-opt on the average by about 0.7% while the much slower hybrid HY4 can improve on this by another percent gaining an average excess rate over all instances of about 5%.

Table 1. Computational Results

TSP	Excess in % over optimum							CPU-times in sec.						
	3opt	3HO	4HO	HY3	HY4	VNS	GVNS	3opt	3HO	4HO	HY3	HY4	VNS	GVNS
br17	0.00	0.00	0.00	0.00	0.00	0.00	0.00	0.01	0.01	0.05	0.01	0.16	1.31	1.25
ft53	7.13	12.31	7.16	5.57	4.37	0.33	0.12	0.09	0.14	1.05	0.28	1.16	10.6	10.6
ft70	3.06	3.74	2.51	2.32	1.81	0.23	0.13	0.22	0.23	1.62	0.50	2.00	20.3	20.8
ftv33	5.81	9.24	5.33	4.76	4.90	0.12	0.29	0.02	0.05	0.29	0.07	0.33	3.21	3.23
ftv35	2.89	4.96	3.68	3.99	3.04	0.33	0.09	0.02	0.05	0.33	0.08	0.36	3.20	3.36
ftv38	3.20	5.73	4.95	5.04	4.36	0.66	0.37	0.03	0.06	0.47	0.08	0.50	3.57	3.45
ftv44	6.29	6.83	5.40	4.86	3.67	1.02	0.60	0.05	0.09	0.51	0.14	0.58	4.96	4.92
ftv47	5.02	10.48	6.70	4.87	3.23	0.20	0.10	0.06	0.11	0.72	0.16	0.91	5.52	5.58
ftv55	7.08	10.14	7.13	6.35	6.16	0.24	0.08	0.10	0.13	0.95	0.23	0.94	7.05	6.58
ftv64	7.10	10.99	7.48	5.09	4.18	0.54	0.44	0.16	0.18	1.14	0.32	1.33	10.5	10.6
ftv70	8.15	11.33	8.28	8.17	6.47	0.81	0.82	0.19	0.23	1.50	0.48	1.75	12.5	12.7
ftv90	14.02	21.33	16.85	15.34	10.84	0.30	0.32	0.37	0.34	2.39	0.66	2.93	19.8	18.9
ftv100	10.91	10.50	11.42	8.78	9.09	0.46	0.41	0.50	0.49	2.96	1.02	3.45	23.9	22.3
ftv110	12.72	16.76	13.19	10.09	9.67	1.36	1.48	0.65	0.58	3.12	1.18	3.75	28.0	28.4
ftv120	11.78	16.52	11.02	10.93	8.01	1.13	0.63	0.84	0.53	3.96	1.44	4.77	32.5	31.6
ftv130	10.19	19.95	13.00	12.11	8.91	1.66	1.09	1.13	0.85	5.03	1.74	6.42	36.1	34.1
ftv140	12.20	14.80	12.97	11.31	9.84	1.45	0.98	1.41	0.91	6.26	2.31	7.71	40.7	40.1
ftv150	9.52	15.44	13.60	10.75	9.49	2.03	1.23	1.73	1.01	7.10	2.86	9.05	45.5	45.9
ftv160	14.65	17.77	12.50	11.71	9.68	3.02	2.41	2.25	1.38	8.48	3.88	10.1	50.7	49.9
ftv170	11.64	19.73	17.01	11.85	9.45	2.07	1.83	2.71	1.48	8.82	4.50	11.8	55.5	54.4
kro100	7.07	8.61	7.72	5.16	3.71	0.81	0.78	0.61	0.56	3.37	1.15	4.32	24.6	24.0
pr43	0.29	0.34	0.47	0.19	0.19	0.01	0.01	0.04	0.08	0.50	0.12	0.56	4.22	4.25
rbg323	0.33	1.75	0.82	0.42	0.39	0.09	0.08	23.0	5.87	32.3	27.0	49.6	280.1	282.2
rbg358	1.00	2.14	1.76	0.82	0.98	0.04	0.08	32.3	5.54	37.7	36.0	62.0	306.5	300.1
rbg403	0.08	0.20	0.17	0.07	0.09	0.00	0.00	47.3	7.34	50.4	49.0	82.7	288.6	284.6
rbg443	0.09	0.11	0.04	0.04	0.05	0.00	0.00	63.1	11.1	53.5	63.5	90.6	347.0	338.6
ry48p	4.27	3.01	2.99	2.65	2.77	0.28	0.28	0.06	0.10	0.69	0.16	0.78	5.29	4.97
av.	6.78	9.43	7.19	6.04	5.01	0.71	0.54	–	–	–	–	–	–	–

We can see that both VNS approaches, the random shaking strategy (VNS) suggested by Hansen and Mladenović [20][21] and our guided VNS approach, are capable of improving significantly on the results gained by pure local search, gaining average excess rates of well below 1%. For nearly all instances our guided VNS (GVNS) produced tours of same or better quality than the original, random VNS strategy (VNS). The guided restart mechanism of GVNS enabled the heuristic to create tours with an average excess rate of about 0.5% which represents a major improvement over the excess rates gained by all of the other heuristic approaches. In particular for the "hard" instances (ftv170 and its subproblems) which proved difficult for the local search guided VNS was able to produce tours of good quality.

4   Conclusions

In this paper we presented heuristic approaches which embed an exact algorithm within the framework of a local search heuristic for the asymmetric Travelling Salesman Problem. This approach, HyperOpt, was inspired by a problem arising in electronics manufacture and is a development of heuristics that have previously been applied successfully to electronics manufacture as well as symmetric TSPs [19][22]. We further proposed a hybrid of HyperOpt with the well known 3-opt local search technique in order to benefit from the advantages of both approaches, yielding a fast, effective approach which is significantly greater than the sum of its parts.

The computational results are encouraging. The HyperOpt/3-opt hybrid approaches proved able to find tours of better or same quality than 3-opt for nearly all of the 27 test problems considered in this paper. We further used the fastest HyperOpt/3-opt hybrid as a local search heuristic in a Variable Neighbourhood Search framework, as suggested by Hansen and Mladenović [20][21], and show how we may use a "guided shakes" strategy to improve on the random shakes in [20][21]. This metaheuristic approach yields very good results with an average excess rate over optimum of about 0.5% for the test problems considered here. We think that this approach represents a powerful method to determine near optimal solutions for asymmetric TSPs. All our approaches have shown themselves robust enough to handle a wide range of constraints such as occur in industrial problems.

References

1. Johnson, D.S., McGeoch, L.A.: The travelling salesman problem: A case study. In: Aarts, E.H.L. and Lenstra, J.K. (eds.): Local Search in Combinatorial Optimization. John Wiley & Sons, New York (1997) 215–310
2. Reinelt, G.: The travelling salesman: Computational solutions for TSP applications. Lecture Notes in Computer Science, Springer-Verlag, Berlin Heidelberg New York (1994)
3. Bentley, J.L.: Fast algorithms for geometric traveling salesman problems. ORSA Journal on Computing, 4 (1992) 387–411
4. Lin, S., Kernighan, B.W.: An effective heuristic algorithm for the travelling salesman problem. Operat. Res. 21 (1973) 498–516

5. Lawler, E.J., Lenstra, J.K., Rinnoy Kan, A.H.G., Shmoys, D.B.: The travelling salesman problem: A guided tour of combinatorial optimization. Wiley, New York (1985)
6. Martello, S., Toth, P.: Linear assignment problems. Annals of Discrete Mathematics **31** (1987) 259–282
7. A. Frieze A., Galbiati, G., Maffioli, F: On the worst-case performance of some algorithms for the asymmetric traveling salesman problem. Networks **12** (1982) 23–39
8. Karp, R.M.: A patching algorithm for the nonsymmetric traveling salesman problem. SIAM Journal on Computing **8** (1979) 561–573
9. Glover, F., Gutin, G., Yeo, A., Zverovich, A.: Construction heuristics for the asymmetric TSP. European Journal of Operational Research **129** (2000) 555-568
10. Zhang, W.: Depth-first branch-and-bound versus local search: A case study. In: Proceedings of the 17th National Conference on Artificial Intelligence (AAAI 2000), Austin, TX, USA (2000) 260–266
11. Kanellakis, P.C., Papadimitriou, C.H.: Local search for the asymmetric traveling salesman problem. Oper. Res. **28** (1980) 1086–1099
12. Merz, P., Freisleben, B.: Genetic local search for the TSP: New results. In: Proceedings of the 1997 IEEE International Conference on Evolutionary Computation (1997) 159–164
13. Papadimitriou, C.H., Steiglitz, K.: The complexity of local search for the traveling salesman problem. SIAM Journal on Computing **6** (1977) 76–83
14. Pekney, J.F., Miller, D.L.: Exact solution of the no-wait flowshop scheduling problem with a comparison to heuristic methods. Computers & Chemical Engineering **15** (1991) 741–748
15. Cowling, P.I.: Optimization in steel hot rolling. In: Optimization in Industry. John Wiley & Sons, Chichester, England, (1995) 55–66
16. Kolohan, F., Liang, M.: A tabu search approach to optimization of drilling operations. Comp. in Eng. **31** (1996) 371–374
17. Walas, R.A., Askin, R.G.: An algorithm for NC turret punch press tool location and hit sequencing. IIE Transactions **16** (1984) 280–287
18. Burke, E.K., Cowling, P.I., Keuthen, R.: New models and heuristics for component placement in printed circuit board assembly. In: Proceedings of the 1999 International Conference on Information, Intelligence and Systems (ICIIS99), Bethesda, MD, USA. IEEE Computer Society Press, (1999) 133–140
19. Burke, E.K., Cowling, P.I., Keuthen, R.: Effective heuristic and metaheuristic approaches to optimize component placement in printed circuit board assembly. In: Proceedings of the Congress on Evolutionary Computation CEC2000, San Diego, CA, USA. IEEE Computer Society Press, (2000) 301–308
20. Hansen, P., Mladenović, N.: An introduction to variable neighborhood search. In: S. Voss, S. Martello, I.H. Osman and C. Roucairol (eds.): Meta-Heuristics: Advances and Trends in Local Search Paradigms for Optimization, pages 433-458, Kluwer Academic Publishers, Boston, MA (1999)
21. Hansen, P., Mladenović, N.: Variable neighborhood search: Principles and applications. In: Invited papers at Euro XVI. Brussels, Belgium, (1998)
22. Burke, E.K., Cowling, P.I., Keuthen, R.: Embedded local search and variable neighborhood search heuristics applied to the travelling salesman problem. University of Nottingham, Technical Report (2000)
23. Reinelt, G.: TSPLIB - A travelling salesman problem library. ORSA-Journal of the Computing **3** (1991) 376–384

Pheromone Modification Strategies for Ant Algorithms Applied to Dynamic TSP

Michael Guntsch and Martin Middendorf

Institute for Applied Computer Science and Formal Description Methods,
University of Karlsruhe, D-76135 Karlsruhe, Germany
{guntsch,middendorf}@aifb.uni-karlsruhe.de

Abstract. We investigate strategies for pheromone modification of ant algorithms in reaction to the insertion/deletion of a city of Traveling Salesperson Problem (TSP) instances. Three strategies for pheromone diversification through equalization of the pheromone values on the edges are proposed and compared. One strategy acts globally without consideration of the position of the inserted/deleted city. The other strategies perform pheromone modification only in the neighborhood of the inserted/deleted city, where neighborhood is defined differently for the two strategies. We furthermore evaluate different parameter settings for each of the strategies.

1   Introduction

Dynamic optimization problems play an important role in practical applications and are a challenging field for optimization methods. Problem instances that may change over time require that optimization methods adapt the solution to the changing optimum. One key aspect is whether solutions found for older stages of a problem instance can be re-used to quickly find a new good solution after the problem has changed.

In this paper we investigate several strategies for ant algorithms applied to optimization problems with changing instances (see [1,2] for an overview of ant algorithms). So far, the only dynamic problems that have been studied with ant algorithms are routing problems in communication networks where the traffic in the network continually changes (e.g. [3,4]), but no explicit reaction occurring to a single change.

In this paper we study a dynamic problem were a change occurs at a certain time point and the ant algorithm reacts explicitly to this change. In particular, we study an ant algorithm for a Traveling Salesperson Problem (TSP) where an instance may change through the deletion or insertion of a city. Ant algorithms have been applied successfully for the static TSP problem by several authors [5,6,7,8,9,10]. In Bonabeau et al. [1], it is suggested that ant algorithms should in general exhibit particularly good performance for dynamic versions of TSP, however they did not confirm this.

The simplest way to handle the change of a problem instance would be to restart the ant algorithm after the change has occurred. However, if one assumes

E.J.W. Boers et al. (Eds.) EvoWorkshop 2001, LNCS 2037, pp. 213–222, 2001.

that the change of the problem is relatively small, it is likely that the new optimum will be in some sense related to the old one, and it would probably be beneficial to transfer knowledge from the old optimization run to the new run. On the other hand, if too much information is transferred, the run basically starts near a local optimum, and will be stuck there. Thus, a reasonable compromise between these two opposing approaches has to be found. Based on this general idea, we propose and test three different strategies to make ant algorithms more suitable for the optimization in dynamic environments.

The paper is structured as follows: in Section 2, we introduce the ant algorithm used. In Section 3, the three strategies for modifying the values in the pheromone matrix are described. These approaches are examined empirically in Section 4. The paper concludes with a summary and ideas for future work in Section 5.

2   Ant Algorithms

The general approach of our algorithm for the TSP follows the ant algorithm of Dorigo et al. [9]. In every generation each of m ants constructs one tour through all the given n cities. Starting at a random city an ant selects the next city using heuristic information as well as pheromone information, which serves as a form of memory by indicating which choices were good in the past. The heuristic information, denoted by η_{ij}, and the pheromone information, denoted by τ_{ij}, are indicators of how good it seems to move from city i to city j. The heuristic value is $\eta_{ij} = 1/d_{ij}$ where d_{ij} is the distance between city i and city j.

With probability q_0, where $0 \leq q_0 < 1$ is a parameter of the algorithm, an ant at city i chooses next city j from the set S of cities that have not been visited so far which maximizes $[\tau_{ij}]^{\alpha} [\eta_{ij}]^{\beta}$, where α and β are constants that determine the relative influence of the heuristic values and the pheromone values on the decision of the ant. With probability $1 - q_0$ the next city is chosen according to the probability distribution over S determined by

$$p_{ij} = \frac{[\tau_{ij}]^{\alpha} [\eta_{ij}]^{\beta}}{\sum_{h \in S} [\tau_{ih}]^{\alpha} [\eta_{ih}]^{\beta}}$$

Before doing global pheromone update some of the old pheromone is evaporated on all edges according to

$$\tau_{ij} \mapsto (1 - \rho) \cdot \tau_{ij}$$

where parameter ρ determines the evaporation rate. For pheromone update an elitist strategy is used where one elitist ant updates pheromone along the best solution found so far, i.e. for every city i some amount of pheromone is added to element (i, j) of the pheromone matrix when j is the successor of i in the so far best found tour. Observe that pheromone is added to exactly two edges incident to a node i. The amount of pheromone added is $\rho/4$, that is

$$\tau_{ij} \mapsto \tau_{ij} + \frac{1}{4}\rho$$

The same is done also for the best solution found in the current generation.

For initialization we set $\tau_{ij} = 1/(n-1)$ for every edge (i,j). Observe, that for every city i the sum of the pheromone values on all incident edges is one, which is not changed by the pheromone update.

3   Pheromone Modification Strategies

In this section we introduce three strategies for modifying the pheromone information in reaction to a change of the problem instance, i.e. the insertion or deletion of a city. The difficulty with modifying the pheromone information is to find the right balance between resetting enough information to give the search process the flexibility to find a new, good solution for the changed problem instance, and keeping enough of the old information to speed up the process of finding this solution. Resetting information is achieved by equalizing the pheromone values to some degree, which effectively reduces the influence of experience on the decisions an ant makes to build a solution.

Strategies for the modification of pheromone information have been proposed before to counteract stagnation of ant algorithms. The approach used by Gambardella et al. [11] was to reset all elements of the pheromone matrix to their initial values. Stützle and Hoos [10] suggested to increase the pheromone values proportionately to their difference to the maximum pheromone value.

Similar to these approaches, we propose a global pheromone modification strategy which reinitializes all the pheromone values by the same degree. This method will be called the "Restart-Strategy". The Restart-Strategy is limited, however, because it does not take into account where the change of the problem instance actually occurred. Often, good solutions to the changed instance will differ only locally from good solutions to the old one. Therefore, the most extensive resetting of pheromone values would need to occur in the close vicinity of the inserted/deleted city. In this spirit, we define two more locally oriented update strategies, each based on one of the factors contributing to an ant's local decisions. The "η-Strategy" uses heuristic based information, distances between cities in this case, to decide to what degree equalization is done on the pheromone values on all edges incident to a city j. The "τ-Strategy" uses pheromone based information, i.e. the pheromone values on the edges, to define another concept of "distance" between cities. Equalization of pheromone values is then again performed to a higher degree on the edges of "closer" cities.

All our strategies work by distributing a reset-value $\gamma_i \in [0:1]$ to each city i. These reset-values are then used to reinitialize the pheromone values on edges incident to i according to the equation

$$\tau_{ij} \mapsto (1 - \gamma_i)\tau_{ij} + \gamma_i \frac{1}{n-1} \tag{1}$$

In case of a problem with symmetric η-values like Euclidean TSP, the average of the reset-values $(\gamma_i + \gamma_j)/2$ is used instead of γ_i in equation 1 for modifying the pheromone value on the edge connecting cities i and j. An inserted city i always receives an unmodifiable reset-value of $\gamma_i = 1$, resulting in all incident edges to i having the initial pheromone value of $1/(n-1)$. We will now describe in more detail how the different strategies assign the values γ_i.

3.1   Restart

The Restart-Strategy assigns each city i the strategy-specific parameter $\lambda_R \in [0, 1]$ as its reset-value:

$$\gamma_i = \lambda_R. \tag{2}$$

3.2 η-Strategy

In the η-Strategy, each city i is given a value γ_i proportionate to its distance from the inserted/deleted city j. This distance d_{ij}^{η} is derived from η_{ij} in such a way that a high η_{ij} implies a high d_{ij}^{η} and that scaling the η-values has no effect:

$$d_{ij}^{\eta} = 1 - \frac{\eta_{avg}}{\lambda_E \cdot \eta_{ij}}$$

with $\eta_{avg} = \frac{1}{n(n-1)} \sum_{i=1}^{n} \sum_{k \neq i} \eta_{ik}$ and the strategy-specific parameter $\lambda_E \in [0, \infty)$ scaling the width of the distance-cone. A city i then receives

$$\gamma_i = \max(0, d_{ij}^{\eta}). \tag{3}$$

3.3 τ-Strategy

The τ-Strategy uses a distance measure based on pheromone information to calculate the reset-values. The pheromone-distance d_{ik}^{τ} between two cities i and k is basically defined as the maximum over all paths P_{ik} from i to k of the product of pheromone-values on the edges in P_{ik}. To enable an equal treatment of symmetric as well as asymmetric problems and to have all distances be able to attain any value of $[0, 1]$, the pheromone-values on the edges are scaled by the maximum possible pheromone value on an edge τ_{max} [1]. Formally,

$$d_{ik}^{\tau} = \max_{P_{ik}} \prod_{(x,y) \in P_{ik}} \frac{\tau_{xy}}{\tau_{max}}.$$

For the case of insertion, we set the pheromone value of the edges to the two closest cities, i.e. those with the highest value for η_{ij}, to τ_{max} during the application of this strategy, since the new city does not yet have any pheromone information. The pheromone-distance multiplied with a strategy-specific parameter $\lambda_T \in [0, \infty)$, with the result limited to 1 for application of equation 1, is the reset-value for a city i if j is inserted/deleted:

$$\gamma_i = min(1, \lambda_T \cdot d_{ij}^{\tau}) \tag{4}$$

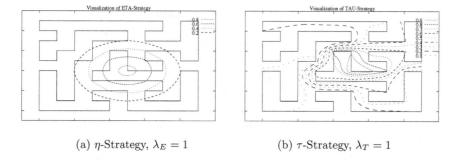

(a) η-Strategy, $\lambda_E = 1$ (b) τ-Strategy, $\lambda_T = 1$

Fig. 1. Visualization of reset-value distribution for (a) the η-Strategy and (b) the τ-Strategy after the deletion of city (5,4), with the best path found by the ant algorithm on the 10x10 grid before the change depicted in the background.

To give an illustration of how the η- and τ-Strategies work, we used a 10x10 grid of cities ((0,0) to (9,9)) with Euclidean distances to visualize how the distribution of reset-values takes place. The best path found before the deletion of city (5,4) at iteration 300 is shown in the background. One can see clearly in Figure 1 (a) that the distribution for the η-Strategy is proportionate to Euclidean distance, while 1 (b) shows that the τ-Strategy tends to distribute along the path.

4   Empirical Evaluation

We evaluated the three strategies proposed in Section 3 on the Euclidean TSP eil101 from the TSP-LIB [13]. Insertion and deletion of a city were evaluated either after 250 iterations (del250, ins250) or after 500 iterations (del500, ins500). Insertion was done by reinserting a city that was removed before the algorithm started. The total runtime of all test runs was 1500 iterations. To eliminate a possible bias by the position of the inserted/deleted city, all test results are averaged over the insertion respectively deletion of all 101 cities.

The parameter values of the ant algorithm were $m = 10$ ants, $\alpha = 1$, $\beta = 5$, $q_0 = 0.5$, and $\rho = 0.01$. The heuristic weight of $\beta = 5$ has been used by several authors (e.g. [5,12]) for TSP. We also performed all tests with $\beta = 1$ and $q_0 \in \{0.0, 0.9\}$, with equivalent or worse performance. The elitist ant was dropped when the insertion/deletion occurred, and redetermined in the first iteration thereafter.

---

[1] τ_{max} is 0.5 for symmetric and 1.0 for asymmetric TSP.

To obtain an understanding for the total amount of equalization done in the pheromone matrix we measure the average row-/column-entropy E:

$$E = \frac{1}{n \log n} \sum_{i=1}^{n} \sum_{j=1}^{n} -\tau_{ij} \log(\tau_{ij})$$

of the pheromone matrix for each iteration of every run. E is normalized to be in $[0,1]$ independently of n. The entropy of the pheromone matrix was also used by Merkle et al. in [14] as a measure for the behavior of ant algorithms.

In the following we concentrate on the results for the deletion of a city since the results for insertion were quite similar. For all three strategies we tested the influence of the λ parameters (i.e. λ_R, λ_E, λ_T) which influence the height of the reset-values. We compare the best solutions of the ant algorithm when using a specific strategy for different λ-parameter values at 5, 50, and 250 iterations after the change has occurred, and the final results after 1500 iterations.

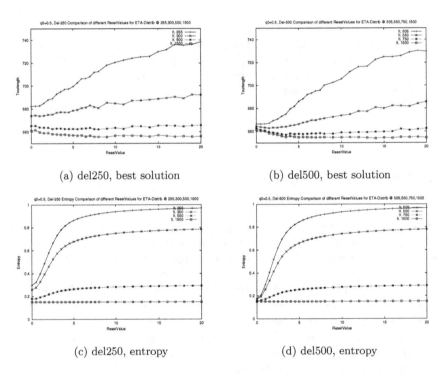

(a) del250, best solution

(b) del500, best solution

(c) del250, entropy

(d) del500, entropy

Fig. 2. Best results and entropy of η-Strategy with $\lambda_E \in [0:16]$ on del250, del500.

Figure 2 shows the results obtained by the η-Strategy for parameter-values of $\lambda_E \in [0:20]$ in 0.5 to 1.0 increments. As can be seen, a higher value of λ_E

entails a worse starting solution and a better final solution, the latter however holding only for $\lambda \leq 8$, after which no significant difference between the final tourlengths exists. This suggests that "good" values for λ_E are not too large. The respective entropy-curves asymptotically approach their maximum value of 1 after the turning point around $\lambda_E = 2$. In conformity with the development of the best solution, the difference in entropy for $\lambda_E > 8$ is only marginal.

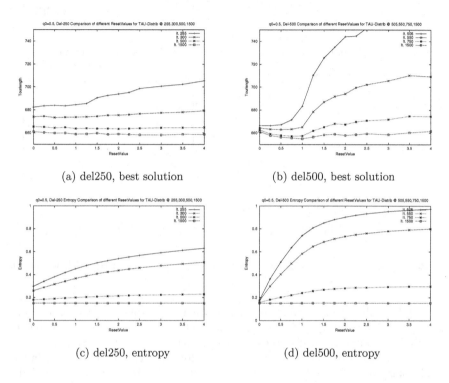

(a) del250, best solution

(b) del500, best solution

(c) del250, entropy

(d) del500, entropy

Fig. 3. Best results and entropy of τ-Strategy with $\lambda_T \in [0 : 4]$ on del250, del500.

The results for the τ-Strategy with $\lambda_T \in [0 : 4]$, in increments of 0.25 to 0.5, are shown in Figure 3. The results for a deletion after 250 iterations are somewhat similar to those for the η-Strategy, but a deletion after 500 iterations leads to different results. This is due to the mechanism of pheromone-based distances that is used by the τ-Strategy, which leads to $\frac{\tau_{ij}}{\tau_{max}} \to 1$ and therefore $d_{ij}^{\tau} \to 1$ for advanced iterations. It seems that here the τ-Strategy is quite efficient when $\lambda_T \approx 1$. In this case not too much pheromone is reset so that the results are good even at iteration 505, but enough to be comparable to a total reinitialization (see Figure 4) after 1500 iterations.

The results for the Restart-Strategy are shown in Figure 4, with $\lambda_R \in [0 : 1]$ in 0.025 to 0.1 increments. Similarly to the η-Strategy, the two cases del250 and

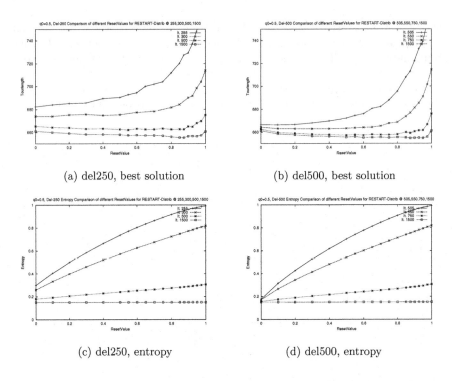

(a) del250, best solution   (b) del500, best solution

(c) del250, entropy   (d) del500, entropy

Fig. 4. Best results and entropy of Restart-Strategy with $\lambda_R \in [0:1]$ on del250, del500.

del500 look almost identical , with a slightly higher entropy- and worse solution-level for del250. For $\lambda_R \leq 0.5$ the solution quality is roughly the same as before the change 50 iterations after the deletion has occurred. For higher λ_R the trade-off of worse solution quality in the beginning for a better solution quality in the end holds up to $\lambda_r \approx 0.9$, after which virtually all information is reset and the ant algorithm needs more time to "rediscover" a good solution.

Now, the goal is to find an optimal parameter for each of the strategies and compare their performance. However, determining which parameter is best is not possible without knowledge of how many iterations after the change the best found solution is needed. Figure 5 (a)-(c) shows which parameter for each individual strategy resulted in the best average solution over the indicated number of iterations after the deletion. If we assume that the probability for needing the new best solution at a specific iteration is equal for all iterations between the change and iteration 1500, then we only need the rightmost value in each of these subfigures to determine the best λ-setting (i.e. for del250 $\lambda_E = 7.5$, $\lambda_T = 1.0$, and $\lambda_R = 0.85$).

Subfigure 5 (d) shows the curve for each strategy with their respective optimal parameter for del250. As can be seen, the τ-Strategy performs best immediately after the change while the η- and Restart-strategy sacrifice good immediate performance for a better solution quality towards the end. This fact also results in the latter two strategies having a slightly better average solution quality over the 1250 iterations considered (η-Strategy : 660.396 and Restart-Strategy : 660.316 vs. τ-Strategy : 661.674).

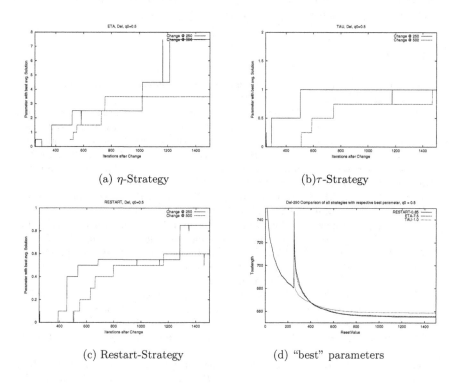

(a) η-Strategy

(b)τ-Strategy

(c) Restart-Strategy

(d) "best" parameters

Fig. 5. Figures (a)-(c) show which parameter had the best average solution how many iterations after the deletion, (d)-(f) compares the three strategies for their respective best parameter λ for del250.

5   Conclusion and Future Work

In this paper, we have proposed three parameterized strategies for Ant Colony Optimization algorithms to deal with dynamic TSP instances. In particular, we suggested two approaches to locally reset the pheromone matrix after a change. Different parameters for the individual strategies were evaluated, and suggestions made as to which parameters seem better suited than others.

In summary, the η- and Restart-Strategies perform best, closely followed by the τ-Strategy. We expect the very good performance of the Restart-Strategy to be due to the singular insertion/deletion performed and plan to verify this in future work by examining environments where multiple insertions and deletions, exclusively as well as in combination with one another, occur. Furthermore, we will implement and analyze combinations of the proposed strategies. Since at least some of the results depended upon the state of convergence of the ant algorithm, using the entropy of the pheromone matrix or parts thereof as a guide for choosing a parameter also seems justified.

References

1. E. Bonabeau, M. Dorigo, G. Theraulaz: *Swarm Intelligence: From Natural to Artificial Systems*, Oxford University Press, New York, 1999.
2. M. Dorigo, G. Di Caro, "The ant colony optimization meta-heuristic", in D. Corne, M. Dorigo, F. Glover (Eds.), *New Ideas in Optimization*, McGraw-Hill, 11-32, 1999.
3. G. Di Caro, M. Dorigo, "AntNet: Distributed Stigmergetic Control for Communications Networks," *Journal of Artificial Intelligence Research*, 9: 317-365, 1998.
4. R. Schoonderwoerd, O. Holland, J. Bruten, L. Rothkrantz, "Ant-based Load Balancing in Telecommunications Networks," *Adaptive Behavior*, 1996.
5. B. Bullnheimer, R.F. Hartl, C. Strauss, "A New Rank Based Version of the Ant System - A Computational Study," *CEJOR*, 7: 25-38, 1999.
6. M. Dorigo, "Optimization, Learning and Natural Algorithms (in Italian)," PhD Thesis, Dipartimento di Elettronica, Politecnico di Milano, Italy, pp.140, 1992.
7. M. Dorigo, L. M. Gambardella, "Ant-Q: A Reinforcement Learning approach to the traveling salesman problem," *Proceedings of ML-95, Twelfth Intern. Conf. on Machine Learning*, Morgan Kaufmann, 252-260, 1995.
8. M. Dorigo, and L.M. Gambardella, "Ant colony system: A cooperative learning approach to the travelling salesman problem," *IEEE TEC*, 1: 53-66, 1997.
9. M. Dorigo, V. Maniezzo, A. Colorni, "The Ant System: Optimization by a Colony of Cooperating Agents," *IEEE Trans. Systems, Man, and Cybernetics – Part B*, 26: 29-41, 1996.
10. T. Stützle, H. Hoos, "Improvements on the ant system: Introducing MAX(MIN) ant system," in G. D. Smith et al. (Eds.), *Proc. of the International Conf. on Artificial Neutral Networks and Genetic Algorithms*, Springer-Verlag, 245-249, 1997.
11. L.-M. Gambardella, E. D. Taillard, M. Dorigo, "Ant Colonies for the Quadratic Assignment Problem," *Journal of the Operational Research Society*, 50: 167-76, 1999.
12. T. Stützle, H. Hoos, "*MAX-MIN* Ant System," *Future Generation Computer Systems*, 16: 889-914, 1999.
13. http://www.iwr.uni-heidelberg.de/iwr/comopt/software/TSPLIB95/
14. D. Merkle, M. Middendorf, H. Schmeck, "Ant Colony Optimization for Resource-Constrained Project Scheduling," *Proc. GECCO-2000*, 893-900, 2000.

Conventional and Multirecombinative Evolutionary Algorithms for the Parallel Task Scheduling Problem

Susana Esquivel, Claudia Gatica, and Raúl Gallard

Laboratorio de Investigación y Desarrollo en Inteligencia Computacional,
Universidad Nacional de San Luis, Argentina
{esquivel, crgatica, rgallard}@unsl.edu.ar

Abstract. The present work deals with the problem of allocating a number of non identical tasks in a parallel system. The model assumes that the system consists of a number of identical processors and that only one task may be executed on a processor at a time. All schedules and tasks are nonpreemptive. Graham's [1] well-known list scheduling algorithm (LSA) is contrasted with different evolutionary algorithms (EAs), which differ on the representations and the recombinative approach used. Regarding representation, direct and indirect representation of schedules are used. Concerning recombination, the conventional single crossover per couple (SCPC) and a multiple crossover per couple (MCPC) are used [2]. Outstanding behaviour of evolutionary algorithms when contrasted against LSA was detected. Results are shown and discussed.

1 Introduction

The problem of how to find a schedule on $m > 2$ processors of equal capacity that minimises the whole processing time of independent tasks has been shown as belonging to the NP-complete class [3]. Parallel task scheduling is important from both the theoretical and practical points of view [4], [3], [5], [6], [7], [8]. From the theoretical viewpoint, it is a generalisation of the single machine scheduling problem. From the practical point of view the occurrence of resources in parallel is common in real world. Evolutionary algorithms have also been used to solve scheduling problems, [3], [9], [10], [11], [12], [13], [14], [15], [16], [17], [18].

In Uniform Memory Access (UMA) multiprocessors, a dynamic scheduling of parallel processes is provided. Nevertheless, there are many reasons for applying static scheduling. First, static scheduling sometimes results in lower execution times than dynamic scheduling. Second static scheduling allows only one process per processor, reducing process creation, synchronisation and termination overhead. Third, static scheduling can be used to predict speedup that can be achieved by a particular parallel algorithm on a target machine, assuming that no preemption of processes occur.

A parallel program is a collection of tasks, some of which must be completed before others begin. In a deterministic model, the execution time for each task and the precedence relations between them are known in advance. This information is depicted in a directed graph, usually known as the task graph. In Fig. 1 different tasks graphs are shown. Even if the *task graph* is a simplified representation of a parallel program

E.J.W. Boers et al. (Eds.) EvoWorkshop 2001, LNCS 2037, pp. 223-232, 2001.
© Springer-Verlag Berlin Heidelberg 2001

execution, (e.g., overheads due to interrupts for accessing resources are ignored) it provides a basis for static allocation of processors. A *schedule* is an allocation of tasks to processors, which can be depicted by a *Gantt chart*. In a Gantt chart, the initiation and ending times for each task in the available processors are indicated and the makespan (total execution time of the parallel program) of the schedule can be easily derived. Other performance variables, such as individual processor utilisation or evenness of load distribution [19] can be considered. Some simple scheduling problems can be solved to optimality in polynomial time while others can be computationally intractable. As we are interested in the scheduling of arbitrary tasks graphs onto a reasonable number of processors we would be content with polynomial time scheduling algorithms that provide good solutions even though optimal ones can not be guarantee.

The paper is organised as follows. Section 2 describes the List Scheduling Algorithm and some of its anomalies are indicated. Section 3 deals with evolutionary approaches to the problem including a brief description of the MCPC multirecombinative approach. Section 4 shows experiments and results and section 5 present the conclusions and future work.

2 The List Scheduling Algorithm (LSA)

For a given list of tasks ordered by priority, it is possible to assign tasks to processors by always assigning each available processor to the first unassigned task on the list whose predecessor tasks have already finished execution.
Let us denote,
   $T=\{T_1,....,T_n\}$ a set of tasks,
   $\mu: T \rightarrow (0, \infty)$ a function which associates an execution time to each task,
   $\leq$ a partial order in T and
   L a priority list of tasks in T.
Each time a processors is idle, it immediately removes from L the first ready task; that is, an unscheduled task whose ancestors under $\leq$ have all completed execution. In the case that two or more processors attempt to execute the same task, the one with lowest identifier succeeds and the remaining processors look for another adequate task.
Using this heuristic, contrary to the intuition, some anomalies can happen. For example, increasing the number of processors, decreasing the execution times of one or more tasks, or eliminating some of the precedence constraints can actually increase the makespan

3 Evolutionary Approaches for the Task Scheduling Problem

We devised different evolutionary approaches to task scheduling. First, we addressed two representation schemes: *direct* and *indirect* [20], [21]. Then we addressed two recombinative approaches: SCPC and MCPC.

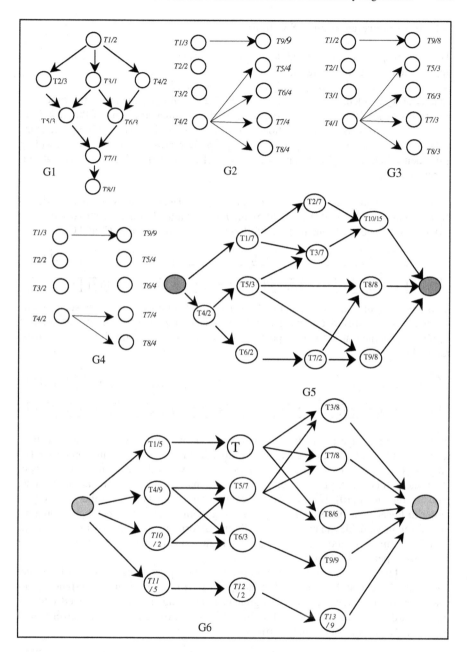

Fig. 1. Some of the task graphs used for testing

3.1 Direct Representation of Solutions

Here we propose to use a schedule as a chromosome. A gene in the chromosome is represented by the following four-tuple:

$$<task\_id, proc\_id, init\_time, end\_time>$$

where, *task_id*, identifies the task to be allocated; *proc_id*, identifies the processor where the task will be allocated; *init_time*, is the commencing time of the *task_id* in *proc_id*, *end_time*, is the termination time of the *task_id* in *proc_id*. The precedence relation described in the task graph can be properly represented in the corresponding precedence matrix A, where element a_{ij} is set to 1 if task i precedes task j, otherwise it is set to 0.

With this structure the list of the corresponding predecessor tasks is easily retrieved by entering the column of A indexed by the *task_id* value. For instance, two schedules (a) and (b) can be represented by the following two chromosomes C_a and C_b:

C_a :

1,1,0,2	2,1,2,5	3,2,2,3	4,2,3,5	5,1,5,8	6,2,5,8	7,1,8,9	8,1,9,10

C_b:

1,1,0,2	2,1,2,5	3,2,2,3	4,1,5,7	5,2,5,8	6,1,7,10	7,1,10,11	8,2,11,12

This representation has a problem. If we use a conventional crossover such as one point crossover invalid offspring could be created. For example, if we decided to apply this operator after the fifth position we would obtain two invalid chromosomes: $C_{a'}$ and $C_{b'}$.

$C_{a'}$:

1,1,0,2	2,1,2,5	3,2,2,3	4,2,3,5	**5,1,5,8**	**6,1,7,10**	7,1,10,11	8,2,11,12

$C_{b'}$:

1,1,0,2	2,1,2,5	3,2,2,3	4,1,5,7	**5,2,5,8**	**6,2,5,8**	7,1,8,9	8,1,9,10

Both of them violate the restriction that a processor must process a task at a time. Genes 5 and 6 in $C_{a'}$ and $C_{b'}$ describe invalid schedules where the same processor (P_1 in case of $C_{a'}$ and P_2 in case of $C_{b'}$) processes two tasks at some time interval. To remedy this situation, it can either be used penalty functions or repair algorithms. Penalty functions [10], [22], of varied severity can be applied to invalid offspring in order to lower their fitness values but allowing them to remain in the population aiming to retain valuable genetic material. Repair algorithms attempt to build up a valid solution from an invalid one. This approach is embedded in the *knowledge-augmented crossover* operator [21] proposed by Bruns.

Here a *collision* occurs if an operation (task processing) inherited from one of the parents cannot be scheduled in the specified time interval on the assigned processor. In this case the processor assignment is unchanged and it is delayed into the future until the processor is available. In our example, this advanced crossover would generate the chromosomes that follow:

$C_{a''}$:

1,1,0,2	2,1,2,5	3,2,2,3	4,2,3,5	5,1,5,8	6,1,8,11	7,1,11,12	8,2,12,13

$C_{b''}$:

1,1,0,2	2,1,2,5	3,2,2,3	4,1,5,7	5,2,5,8	6,2,8,11	7,1,11,12	8,1,12,13

As expected, both children have a larger makespan but are still feasible. In Bruns's proposed *knowledge-augmented* crossover, only a child is generated where the part taken from the first parent builds a consistent schedule. Then the assignment of the missing tasks is chosen from the second parent maintaining the assignment order and the processor allocations to tasks. Timing adjustments are included if necessary. The latter decision can imply, as we have showed, larger makespans for the children.

In our case we adopted an *as-soon-as-possible* (ASAP) approach similar to Brun's proposal but modified to avoid delays, moving the assignment to the earliest possible time, by random selection of one available processor at the ready time of the unassigned task. In this way no processor will remain idle if a task is available for execution and the precedence constraints are satisfied. The available processor is selected as to minimise assignment changes in the second parent part of the offspring. In our example this decision provides only one alternative and would give us the following:

$C_{a'''}$:

| 1,1,0,2 | 2,1,2,5 | 3,2,2,3 | 4,2,3,5 | 5,1,5,8 | 6,2,5,8 | 7,1,8,9 | 8,2,9,10 |

$C_{b'''}$:

| 1,1,0,2 | 2,1,2,5 | 3,2,2,3 | 4,1,5,7 | 5,2,5,8 | 6,1,7,10 | 7,1,10,11 | 8,1,11,12 |

These chromosomes differ from their parents only in the assignments of tasks T_7 and T_8. A similar operator, *switch processors,* was conceived for mutation. If the chromosome undergoes mutation then a search is done, from left to right, until one gene is modified either by choosing an alternative free processor or by moving the assignment to the earliest possible time. This would imply modifying subsequent genes of the chromosome to create a valid offspring.

3.2 Indirect Representation of Solutions

Under this approach a schedule is encoded in the chromosome in a way such that the task indicated by the gene position is assigned to the processor indicated by the corresponding allele, as shown in Fig. 2:

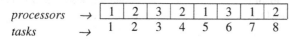

processors →

| 1 | 2 | 3 | 2 | 1 | 3 | 1 | 2 |

tasks   →   1   2   3   4   5   6   7   8

Fig. 2. Chromosome structure for the task allocation problem

The idea is to use a decoder. A *decoder* is a mapping from the space of representations that are evolved to the space of feasible solutions that are evaluated. Here the chromosome gives instructions to a decoder on how to build a feasible schedule.

Regarding the task allocation problem, to build a feasible schedule, a decoder is instructed in the following way: By following the priority list, traverse the chromosome and assign the corresponding task to the indicated processor as soon as the precedence relation is fulfilled. Under this approach the restriction on avoiding processor idleness while a task is ready is relaxed. We believe that this less restrictive approach will contribute to population diversity. One advantage of decoders resides in their ability to create valid offspring even by means of simple conventional operators.

One disadvantage is a slower evaluation of solutions. Also, simplest mutation operators can be implemented by a simple swapping of values at randomly selected chromosome positions or by a random change in the allele. The new allele value identifies any of the involved processors.

3.3 Multiple Crossover Per Couple (MCPC)

Conventional approaches to crossover, independently of the method being used, involve applying the operator only once on the selected parents. Such a procedure is known as the Single Crossover Per Couple (SCPC) approach. In earlier works we devised a different approach to crossover which allows multiple offspring per couple, as often happens in nature. This allowed to deeply explore the recombination possibilities of previously found solutions, and several experiments in which more than one crossover operation for each mating pair was permitted were implemented. The number of children per couple was fixed or granted as a maximum number. Different variants selected the best, a random chosen or all created offspring to be inserted in the next generation. The idea of MCPC was tested on a set of well-known testing functions [2] as well as in different single and multiple objective scheduling problems [23].

4 Experiments and Results

The experiments implemented evolutionary algorithms under SCPC and MCPC with randomised initial population of size fixed at 50 individuals. Twenty series of ten runs each were performed on ten testing cases, using elitism. For *direct representation ASAP* SCPC and *switch-processor* mutation were used. For *indirect representation* one point crossover and big creep mutation were used. In the case of MCPC, tests with 2, 3 and 4 crossovers were run and after the multiple crossover operation the best created child was selected for insertion in the next generation. The maximum number of generations was fixed at 100, but a stop criterion was used to accept convergence when after 20 consecutive generations, mean population fitness values differing in $\varepsilon \leq 0.001$ were obtained. Probabilities for crossover and mutation were fixed at conventional values of 0.65 and 0.001, respectively. The ten testing cases were:

1. Task graph G1 (8 tasks, 3 processors).
2. Task graph G2 (9 tasks, 3 processors).
3. Task graph G2 (9 tasks, 4 processors).
4. Task graph G3 (9 tasks, 3 processors, but decreasing task's duration).
5. Task graph G4 (9 tasks, 3 processors, eliminating precedence constrains).
6. Task graph G5 (10 tasks and 2 processors).
7. Randomly generated task graph G6 (13 tasks and 3 processors).
8. Randomly generated task graph (25 tasks and 5 processors).
9. Randomly generated task graph (50 tasks and 5 processors).
10. Randomly generated task graph (25 tasks and 5 processors).

The first 6 task graphs can be seen in fig. 1. Cases 3 to 5 test the anomalies presented by Graham's algorithm. Opposite to other scheduling problems, after an intensive search in the literature we could find few benchmarks. The first 6 cases were extracted from the literature [1], [24] and they have known optima. Cases 7 to 10 were generated by random assignment of relations and task durations. Their optimum values are unknown. Nevertheless, for these cases initial trials were run to determine the best quasi-optimal solution under the whole set of contrasted heuristics. This value will be referred in what follows as an estimated optimal value (the best known value is assumed as optimal).

The following evolutionary algorithms were developed: EA1 (SCPC) and EA2 (MCPC), both with direct representation, ASAP crossover and switch processor mutation and EA3 (SCPC) and EA4 (MCPC) both with indirect representation, one point crossover and big creep mutation.

To measure the *quality of solutions* provided by the EAs we used:
Ebest: (Abs(opt_val – best value)/opt_val)/100. It is the percentile error of the best found individual in one run when compared with the known, or estimated, optimum value opt_val. It gives us a measure of how far are we from that opt_val.
MEbest: is the mean value, over the total number of runs, of the error.
Best: is the best makespan value found throughout all the runs. This value has been reached by different solutions (schedules).
MBest: is the mean value, over the total number of runs, of the best makespan found in each run (average best individual).

To measure the *versatility* of evolutionary algorithms, we used:
Alt: Number of alternative solutions. It is the mean number the *distinct* alternative solutions found by the algorithm in one run including optimum and non-optimal solutions.
Opt: Number de optimal solutions. It is the mean number of *distinct* optimum or quasi-optimum solutions found by the algorithm in one run.
Totb: Total number of best solutions. It is the mean total number of *distinct* best solutions found by the algorithm throughout all runs.

The *versatility* property shows the EAs inherent ability to provide multiple distinct solutions and nothing in particular to promote this, such as crowding, have been done. As with direct representation of solutions the evolutionary algorithms behave similarly but better than with indirect-decode representation, we show here results of EA1 and EA2 only.

Regarding *quality* of results table 1 shows that MCPC (EA2) outperforms SCPC (EA1) in each considered performance variable. Also, the average best makespan throughout all instances of the problem (136.1) is the nearest to the average optimal value (132.6) provided by any of the contrasted heuristics. Moreover, all the anomalies observed with LSA do not hold when the EA is applied, because: a) when the number of processors is increased, the minimum (optimum) makespan is also found, b) when the duration of tasks is reduced, this is reflected in a reduced optimum makespan and, finally, c) when the number of precedence restrictions is reduced the optimum makespan is preserved. Also EA1 outperforms LSA (except case 9). This fact indicates that the evolutionary algorithm and, specially the multirecombinative approach, are good alternatives to find good quality results.

Table 1. *Quality* of solutions (EA1:SCPC – EA2: MCPC, both with direct representation)

C	Otim. values	Best			MBest		Ebest			MEbest	
		EA1	EA2	LSA	EA1	EA2	EA1	EA2	LSA	EA1	EA2
1	9	9	9	9	9	9	0.0	0.0	0.0	0.0	0.0
2	12	12	12	12	12.3	12.1	0.0	0.0	0.0	2.5	0.83
3	12	12	12	15	12	12	0.0	0.0	25.0	0.0	0.0
4	10	10	10	13	10	10	0.0	0.0	30.0	0.0	0.0
5	12	12	12	16	12.1	12.1	0.0	0.0	33.3	0.8	0.83
6	31	31	31	38	33.5	32.4	0.0	0.0	22.58	8.06	4.51
7	30	32	30	33	32.9	32.8	6.66	0.0	10.0	9.66	9.33
8	309	329	331	375	341.6	344.9	6.47	11.7	21.3	10.5	11.61
9	591	615	621	591	639.9	631.3	4.06	5.07	0.0	8.27	6.81
10	310	310	315	412	325.5	337.3	0.0	1.61	32.9	13.7	8.80
Av.	132.6	137.2	**136.1**	151.4	145.6	**143.4**	1.71	**0.66**	17.51	5.35	**4.27**

Regarding *versatility*, average results in table 2 show that as expected SCPC provides a greater number of alternative solutions (Alt) due to its intrinsic genetic diversity. Regarding the mean number of optimal solutions found in a run (Opt) MCPC provides a slight superior average behaviour than SCPC, and also found a greater number of distinct optimal solutions in the whole experiment (Topb).

Table 2. *Versatility* of solutions (EA1:SCPC – EA2: MCPC, both with direct representation)

Case	Alt			Opt			Totb		
	LSA	EA1	EA2	LSA	EA1	EA2	LSA	EA1	EA2
1	1	20.9	20.5	1	20.9	20.5	1	194	181
2	1	5.5	3.3	1	1.1	2.1	1	11	21
3	1	5.3	5.9	-	5.3	5.9	-	53	59
4	1	3.1	4.4	-	2.9	4.2	-	29	76
5	1	3.2	2.8	-	1.2	1.4	-	12	14
6	1	8.3	5.8	-	0.1	0.4	-	1	4
7	1	9.9	7.2	-	0	0.1	-	0	1
8	1	19.6	20.4	-	0	0.1	-	0	1
9	1	30.1	20.0	1	0	0	1	0	0
10	1	30.7	13.6	-	0.1	0	-	1	0
Aver.	1	**13.66**	10.39	0.3	3.16	**3.47**	0.3	30.1	**35.7**

A more detailed analysis on each run detected that in most cases alternative solutions do not include, or include a low percentage, of non optimal alternative solutions. That means that the final population is composed of many replicas of the optimal solutions due to a loss of diversity. This fact stagnates the search and further improvements are difficult to obtain. To avoid this behaviour we currently conduct new experiments with other multirecombinative approaches.

5 Conclusions

In this work we approached allocation of a number of parallel tasks in parallel supporting environments attempting to minimise the makespan. As we are interesting in scheduling of arbitrary task graphs onto a reasonable number of processors, in many cases we would be content with polynomial time scheduling algorithms that provide good solutions even though optimal ones can not be guaranteed. The list scheduling algorithm (LSA) satisfies this requirement.

Two variants of representations and two approaches of recombination were undertaken to contrast their behaviour with the LSA. Preliminary results on the selected test suite showed three important facts. Firstly, EAs provide not a single, but a set of optimal solutions, providing for fault tolerance when system dynamics must be considered. Secondly, EAs are free of the LSA anomalies. Finally, two variants of recombination were undertaken SCPC and MCPC for each representation. The behaviours of the EAs were similar and all of them showed better results that LSA.

When we compare their performance it is clear that the approaches including multirecombination behave better than the conventional ones (in both representations) but yet it would be necessary to continue experimentation with different parameter settings, self-adaptation of parameters, and contrasting with newer non-evolutionary heuristics. Current research includes diverse multirecombination schemes and hybridisation of EAs by means of local search applied on different stages of the evolutionary process.

6 References

1. Graham R. L.: Bounds on Multiprocessing Anomalies and Packing Algorithms. Proceedings of the AFIPS 1972 Spring Joint Computer Conference, pp 205-217, (1972).
2. Esquivel S., Leiva A., Gallard R: Multiplicity in Genetic Algorithms to Face Multicriteria Optimization. Proceedings of the Congress on Evolutionary Algorithms (IEEE), Washington DC, pp 85 – 90, (1999).
3. Horowitz E. and Sahni S.: Exact and Approximate Algorithms for Scheduling non Identical Processors. Journal of the ACM, vol. 23, No. 2, pp 317-327, (1976).
4. Ercal F.: Heuristic Approaches to Task Allocation for Parallel Computing. Doctoral Dissertation, Ohio State University, (1988).
5. Reeves C.R., Karatza H.: Dynamic Sequencing of a Multiprocessor System; a Genetic Algorithnm, Proc. of 1st International Conference on Artificial Neural Nets and Genetic Algorithms. Springer Verlag (1993).
6. Seredynski F.: Task Scheduling with use of Classifier Systems. AISB. International Workshop 1997: Selected Papers, Lecture Notes in Computers Sciences 1305, pp 287 – 306, Springer, (1997).
7. Tsang E.P.K., Voudouris C.: Fast Local Search and Guided Local Search and their Applications to British Telecom's Workforce Scheduling Problem. Operations Research Letters 20 (3), pp 129 –137 (1997)
8. Yue K.K., Lilja D.J.: Designing Multiprocessor Scheduling Algorithms using a Distributed Genetic Algorithm. Evolutionary Algorithms in Engineering Applications, pp 207 – 222. Springer (1997).
9. Kidwell M. :Using Genetic Algorithms to Schedule Tasks on a Bus-based System. Proceedings of the 5th International Conference on Genetic Algorithms, pp 368-374, (1993).

10. Krause M., Nissen V.,: On Using Penalty Functions and Multicriteria Optimization Techniques in Facility Layout. Evolutionary Algorithms for Management Applications, ed J. Biethahn and V. Nissen (Berling: Springer), pp 153-166, (1995).
11. Lin S-C, Goodman E.D, Punch W.F,: Investigating Parallel Genetic Algorithms on Job Shop Scheduling Problems. Evolutionary Programming VI, Lecture Notes in Computer Sciences 1213, pp 383 –393, Springer (1997).
12. Murata T., Ishibuchi H.: Positive and Negative Combinations and Effects of Crossover and Mutation Operators in Sequencing Problems. Proc. of 1996 IEEE International Conference on Evolutionary Computation, pp 170- 175. IEEE (1996).
13. Sannomiya N., Iima H.: Applications of Genetic Algorithms to Scheduling Problems in Manufacturing Processing. Proc. of 1996 IEEE International Conference on Evolutionary Computation, pp 523 – 528, IEEE (1996).
14. Syswerda G.,: Scheduling Optimisation using Genetic Algorithms. Davis, L., Editor, Handbook of Genetic Algorithms, chapter 21, pp 332 – 349, Van Nostrand Reinhold, New York, (1991).
15. Yamada T., Nakano R.: Scheduling by Genetic Local Search with Multi-step crossover. Parallel Problem Solving from Nature, PPSN IV, Lecture Notes in Computer Sciences 1141, pp 960 –969, Springer (1996).
16. Yamada T., Reeves C.R.: Solving the C_{sum} Permutations Flow Shop Scheduling Problem by Genetic Local Search. ICEC (1998).
17. Yamada, T., Reeves C.R.: Permutation Flow Scheduling by Genetic Local Search. Proc. of the 2nd International Conference on Genetic Algorithms in Engineering Systems: Innovations and Applications, pp 232 – 238 (1997).
18. Withley D., Starkweather T., Fuquay D'A: Scheduling Problems and Travelling Salesman: The Genetic Edge Recombination Operator. Proceedings of the 3th International Conference on Genetic Algorithms, pp 133-140. Morgan Kaufmann Publishers, Los Altos CA, (1989).
19. Fox G. C.: A Review of Automatic Load Balancing and Decomposition Methods for the Hipercube. In M Shultz, ed., Numerical Algorithms for Modern Parallel Computer Architectures, Springer Verlag, pp 63-76, (1988).
20. Bagchi S., Uckum S., Miyabe Y., Kawamura K.: Exploring Problem Specific Recombination Operators for Job Shop Scheduling. Proceedings of the 4th International Conference on Genetic Algorithms, pp 10 – 17 (1991)
21. Bruns R.: Direct Chromosome Representation and Advanced Genetic Operators for Production Scheduling. Proceedings of the 5th International Conference on Genetic Algorithms, pp 352-359, (1993).
22. Michalewicz Z., Genetic Algorithms + Data Structures = Evolution Programs. Springer Verlag, Third, Extended Edition, (1996).
23. Esquivel S., Ferrero S., Gallard R., Alfonso H. Salto C., Schütz M.:Enhanced evolutionary algorithms for single and multiobjective optimization in the job shop scheduling problem. To appear in the Knowledge Based System Journal, Elsevier 2001.
24. Pinedo M.,: Scheduling: Theory, Algorithms and Systems. Prentice Hall International Series in Industrial and Systems Engineering, (1995).

Two-Sided, Genetics-Based Learning to Discover Novel Fighter Combat Maneuvers

Robert E. Smith[1], Bruce A. Dike[2], B. Ravichandran[3], Adel El-Fallah[3], and
Raman K. Mehra[3]

[1] The Intelligent Computing Systems Centre, Bristol, UK, robert.smith@uwe.ac.uk
[2] The Boeing Company, St. Louis, MO, USA, bruce.a.dike@boeing.com
[3] Scientific Systems, Woburn, MA, USA, {ravi, adel, rkm}@ssci.com

Abstract. This paper reports the authors' ongoing experience with a system for discovering novel fighter combat maneuvers, using a genetics-based machine learning process, and combat simulation. In effect, the genetic learning system in this application is taking the place of a test pilot, in discovering complex maneuvers from experience. The goal of this work is distinct from that of many other studies, in that *innovation, and discovery of novelty (as opposed to optimality), is in itself valuable.* This makes the details of aims and techniques somewhat distinct from other genetics-based machine learning research.

This paper presents previously unpublished results that show two co-adapting players in similar aircraft. The complexities of analyzing these results, given the *red queen effect* are discussed. Finally, general implications of this work are discussed.

1 Introduction

New technologies for fighter aircraft are being developed continuously. Often, aircraft engineers can know a great deal about the aerodynamic performance of new fighter aircraft that exploit new technologies, even before a physical prototype is constructed or flown. Such aerodynamic knowledge is available from design principles, from computer simulation, and from wind tunnel experiments.

Evaluating the impact of new technologies on actual combat can provide vital feedback to designers, to customers, and to future pilots of the aircraft in question. However, this feedback typically comes at a high price. While designers can use fundamental design principles (i.e., tight turning capacity is good) to shape their designs, often times good maneuvers lie in odd parts of the aircraft performance space, and in the creativity and innovation of the pilot.

Therefore, the typical process would be to develop a new aircraft, construct a one-off prototype, and allow test pilots to experiment with the prototype, developing maneuvers in simulated combat. Clearly, the expense of such a prototype is substantial. Moreover, simulated combat with highly trained test pilots has a substantial price tag. Therefore, it would be desirable to discover the maneuver utility of new technologies, without a physical prototype.

E.J.W. Boers et al. (Eds.) EvoWorkshop 2001, LNCS 2037, pp. 233–242, 2001.

The approach that is pursued in the authors' past work [1,2], and in new work presented here, an adaptive, machine learning system takes the place of the test pilot in simulated combat. This approach has several advantages.

As in a purely analytical approach, this approach requires a model. However, in this case the model need only be accurate for purposes of combat simulation. It need not present a mathematically tractable form.

Moreover, the approach is similar to that of man-in-the-loop simulation, except in this case the machine learning "pilot" has no bias dictated by past experiences with real aircraft, no prejudices against simulated combat, and no tendency to tire of the simulated combat process after hundreds or thousands of engagements. Also, one overcomes the constraints of real-time simulation.

This paper considers ongoing work in this area in terms of its unique character as a machine learning and adaptive systems problem. To recognize the difference between this problem and more typical machine learning problems, one must consider its ultimate goal. This work is directed at filling the role of test pilot in the generation of innovative, novel maneuvers. The work is not directed at online control. That is to say, the machine learning system is not intended to generate part of an algorithm for controlling a real fighter aircraft. Like the test pilot in simulated combat, the machine learning system can periodically fail, without worry that the associated combat failure will result in possible loss of hardware and personnel. In many ways, the machine learning system is even less constrained than the test pilot, in that it is more willing to experiment with maneuvers that would be dangerous in fighter combat with real aircraft. The goal of this work is the process of innovation and novelty, rather than discovering optimality.

2   The LCS Used Here

Details of the LCS used here are briefly provided below. For a more detailed discussion, see previous papers [1][3].

The LCS interacts in simulated, 1-versus-1 combat, through AASPEM, the Air-to-Air System Performance Evaluation Model. AASPEM is a U.S. Government computer simulation of air-to-air combat, and is one of the standard models for this topic. The classifier actions directly fire effectors (there are no internal messages).

In our system if no classifiers are matched by the current message, a default action for straight, level flight is used. There is no "cover" operator [4].

At the end of an engagement, the "measure of effectiveness" score for the complete engagement is calculated. This score is *assigned* as the fitness for *every* classifier that acted during the engagement (and to any duplicates of these classifiers). Note that this score *replaces* the score given by averaging the parent scores when the GA generated the rule. Thus, rules that do not fire simply "inherit" the averaged fitness of their GA parents [2].

Our efforts have included an evaluation of different measures of effectiveness within the genetics-based machine learning system, to determine the relative sensitivity of the process. Initial candidate measures included exchange ratio, time on advantage, time to first kill, and other relevant variables.

The measure of effectiveness ultimately selected to feedback into the GA fitness function was based on the following steps. The base score was based on a linear function of average angular advantage (opponent target aspect angle minus ownship target aspect angle). To encourage maneuvers that might enable gun firing opportunities, an additional score was added when the target was within 5 degrees of the aircraft's nose. A tax was applied to non-firing classifiers to discourage the proliferation of parasite classifiers that contain elements of high-performance classifiers but have insufficient material for activation. All non-firing classifiers that were identical to a firing classifier were reassigned the firing classifier's fitness.

The GA acts at the end of each 30-second engagement. The GA is panmictic (it acts over the entire population). In some of our experiments, the *entire* classifier list is replaced each time the GA is applied. This has been surprisingly successful, despite the expected disruption of the classifier list. In recent experiments, we have used a generation gap of 0.5 (replacing half of the classifier population with the GA). A new, GA-created classifier is assigned a fitness that is the average of the fitness values of its "parent" classifiers. The GA used employed tournament selection, with a tournament size ranging from 2 to 8. Typical GA parameters are a crossover probability of 0.95, and a mutation rate of 0.02 per bit position. When a condition bit is selected for mutation, it is set to one of the three possible character values (1, 0, or $\#$), with equal probability. Note that this actually yields an effective mutation probability of $(0.02)(2/3)=0.0133$. Children rules replaced randomly selected rules in the population.

The matching rule with the highest fitness/strength is selected to act *deterministically*.

We have used a number of starting conditions for the 1-v.-1 combat simulations considered here. The primary source document for these conditions was the X-31 Project Pinball II Tactical Utility Summary, which contained results from manned simulation engagements conducted in 1993 at Ottobrunn, Germany [5]. The starting condition we will consider in this paper is the Slow-Speed Line Abreast (SSLA, where the aircraft begin combat side-by-side, pointing in the same direction.

3   Two-Sided Learning Results

In recent work with the fighter combat LCS, we have allowed both opponents to adapt under the action of a GA [3]. This ongoing effort complicates the fighter combat problem, and the interpretation of simulation results. Because of the red queen effect [6], the dynamic system created by two players has several possible attractors. These include fixed points, periodic behaviors, chaotic behaviors, and arms races. The latter is clearly the behavior we want our simulations to encourage. Our current results have (qualitatively) shown promise in this area (i.e., we have seen an escalation of strategies between the two aircraft).

A number of approaches to two-sided learning have been considered. In each approach, a "run" consists of 300 simulated combat engagements. Results in this paper consider the following approach:

Alternate freeze learning with memory (MEM): This learning scheme can be viewed as an extended version of the ALT learning. At the end of each run, the results of the 300 engagements are scanned to obtain the highest measure of effectiveness. The rules from the highest scoring engagement are used for the frozen strategy in the next run. Furthermore, these rules are memorized and are added to the population in the upcoming learning sequence runs. Thus, the system has memory of its previously learned behavior.

3.1   Similar Aircraft (X-31 v. X-31)

This section presents results where two-players in similar aircraft (both X-31s) co-adapt to one another.

Before examining the results graphically, it is useful to consider the progression of raw scores observed. These results are shown in Table 1. We will distinguish the two X-31s by their initial configurations. Relative to their SSLA starting conditions, we will call the player initially on the right player R and the player initially on the left player L.

Table 1. Progression of Scores for one player (Player R) in a simulation with two X-31 aircraft co-adapting with LCSs.

Learning Run	Best Score of Player R
1	49.53379
2	-38.88130
3	48.49355
4	1.854810
5	72.52103
6	-21.01414
7	87.11726
8	-7.360970
9	79.42159
10	30.43967

Note the nature of this progression. Player R's relative superiority alternates as a result of the system's learning. In other words, the player that is adapting has a continual advantage. Note that the player's interactions do not seem to evolve towards a fix-point compromise, but seem to continue to adapt. This leaves the possibility of period, chaotic, or (the desirable) arms race behavior. We can gain some insight by examining plots of the "best" (learning) player's dominance in each run. Note that these are typical results, and that each figure is shown from a slightly different angle, for clarity.

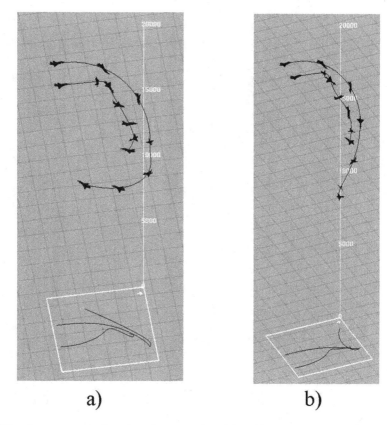

a)   b)

Fig. 1. "Best" maneuver in learning a) run 1 where Player R is learning and b) run 2 where Player L is learning.

Figure 1 a) shows the "best" maneuver discovered in learning run 1, where the player starting on the right (player R) has been learning under the action of the GA, and player L has followed standard combat logic. This maneuver is best in the sense of player R's raw score. Player R has learned to dominate player L, by staying inside player L's turning radius, and employing a helicopter gun maneuver. This is one of the post-stall tactic (PST) maneuvers often discovered by the LCS in out one-sided learning experiments. Figure 1 b) shows the results from the next learning run, where player R follows the strategy dictated by the rules employed in maneuver shown in Figure 1 a). Note the shadow trace of player L shown at the bottom of this figure. Player L has learned to respond to player's R's helicopter gun maneuver with a J-turn (a turn utilizing Herbst-like maneuvering) to escape. This is our first evidence of one player trumping the advanced, PST maneuver learned by its opponent, by learning a PST maneuver of its own.

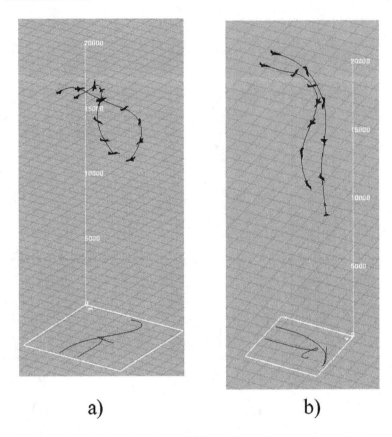

Fig. 2. "Best" maneuver in learning a) run 3 where Player R is learning and b) run 4 where Player L is learning.

Figure 2 a) shows the response when player R returns to the learning role in run 3. Player R learns to abandon the helicopter gun maneuver given player L's J-turn escape. In this run, both players are exhibiting Herbst or J-turn type maneuvers. Note that player L, while not learning, remains responsive to changes in player R's maneuver, due to activation of different rules at different times in the run. At this stage, both players have reached similar strategy levels, by exploiting so-called "out-of-plane" behavior (three-dimensional maneuvering, with drastic movement out of the common plane the players occupy in space) . Figure 2 b) shows player L learning to alter the end of its J-turn, such that it turns to target player R near the end of the maneuver. Note that player R has clearly remained responsive, despite not learning, and altered part of its maneuver.

Figure 3 a) shows a much more advanced strategy emerging on the part of player R, once again in the learning role. This maneuver combines features of a

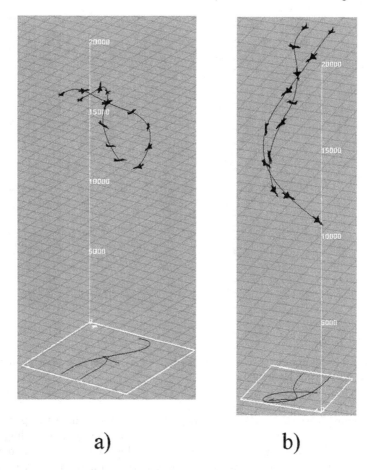

a)   b)

Fig. 3. "Best" maneuver in learning a) run 5 where Player R is learning and b) run 6 where Player L is learning.

Herbst maneuver (high angles of attack and rolling to rapidly change directions) and features of a helicopter gun attack (thrust-vectored nose pointing inside the opponent's turn). Given this advanced maneuver, player L learns in run 6 to extend its J-turn, and escape the fight (Figure 3 b)).

In run 7, player R refines its Herbst turn, putting the two players in parallel PST turns, resulting in a steeply diving chase (Figure 4 a)). In run 8 (Figure 4 b)), player L learns to gain a few critical moments of advantage early in the maneuver, through a brief helicopter gun attack, before it extending a dive out of the fight. Note that, as before, player R remains reactive, despite its lack of learning in this run. In reaction to player L's early attack, it maintains altitude to escape, rather than following the parallel diving pursuit shown in Figure 4 a).

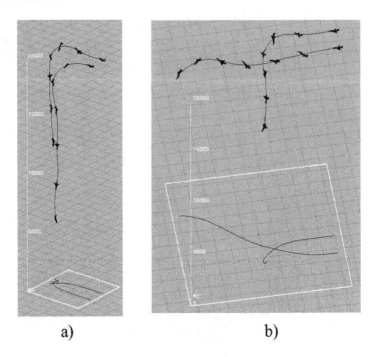

a)   b)

Fig. 4. "Best" maneuver in learning a) run 7 where Player R is learning and b) run 8 where Player L is learning.

Figure 5 a) shows the emergence of a maneuver where the players swing and cross one another's paths in the air, in a complex sort of "rolling scissors" maneuver [7]. Note the shadow traces in this plot, and compare the maneuver's complexity to that of the diving pursuit shown in Figure 4 a). In Figure 5 b), player L once again learns to escape player R's advanced strategy, through a full inversion in a rolling turn. However, note that player R has remained reactive, and, despite its lack of learning in this run, executes an effective helicopter gun attack early in the run.

Throughout these runs, player R (which had the advantage of being "first to learn") assumes a somewhat more aggressive posture. However, note that there is a definite progression in the complexity of both players' strategies, in reaction to each other's learning. This is the desired "arms race" behavior that we are attempting to encourage, such that the system discovers increasingly interesting and novel maneuver sets.

4 Final Comments

Many conclusions and areas for future investigation can be drawn from the work presented here. However, as a concluding focus of this paper, one should consider

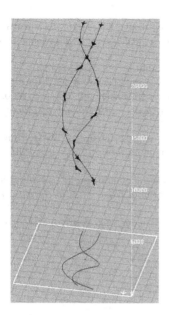 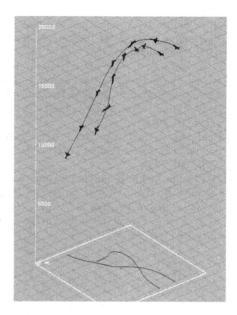

Fig. 5. "Best" maneuver in learning a) run 9 where Player R is learning and b) run 10 where Player L is learning.

the goal of the LCS approach in the fighter aircraft LCS, as a guideline for future applications of the LCS, and other adaptive systems technologies. Since there is a real, quantifiable value to the discovery of innovative, high utility fighter combat maneuvers, one can concentrate on the exploration and synthesis aspects of the LCS, without particular consider for the long term stability of any given rule set. One should not overlook the utility of the LCS approach for generating novel, innovated approaches to problems. In many domains (like the fighter aircraft task), such open-ended machine innovation can have a real world, hard-cash value. The applicability of the adaptive systems approach to such tasks deserves further consideration.

Acknowledgements. The authors gratefully acknowledge that this work is sponsored by The United States Air Force (Air Force F33657-97-C-2035 and Air Force F33657-98-C-2045). The authors also gratefully acknowledge the support provided by NASA for the early phases of this project, under grant NAS2-13994.

References

1. R. E. Smith and B. A. Dike. Learning novel fighter combat maneuver rules via genetic algorithms. *International Journal of Expert Systems*, 8(3):247–276, 1995.

2. R. E. Smith, B. A. Dike, and Stegmann. Inheritance in genetic algorithms. In *Proceedings of the ACM 1995 Symposium on Applied Computing*, pages 345–350. ACM Press, 1994.
3. R. E. Smith, B. A. Dike, R. K. Mehra, B. Ravichandran, and A. El-Fallah. Classifier systems in combat: Two-sided learning of maneuvers for advanced fighter aircraft. *Computer Methods in Applied Mechanics and Engineering*, 186:421–437, 2000.
4. S. W. Wilson. Zcs: A zeroth-level classifier system. *Evolutionary Computation*, 2(1):1–18, 1994.
5. P. M. Doane, C. H. Gay, and J. A. Fligg. Multi-system integrated control (music) program. Technical Report Final Report, Wright Laboratories, Wright-Patterson AFB, OH, 1989.
6. D. Floriano and S. Nolfi. God save the red queen: Competition in co-evolutionary robotics. In *Proceedings of the Second International Conference on Genetic Programming*, pages 398–406. MIT Press, 1997.
7. R. L. Shaw. *Fighter Combat: Tactics and Maneuvering*. United States Naval Institute Press, 1998.

Generation of Time-Delay Algorithms for Anti-air Missiles Using Genetic Programming

Henry O. Nyongesa

School of Computing and Management Sciences,
Sheffield Hallam University,
Sheffield S1 1WB,
United Kingdom
HNyongesa@aol.com

Abstract. This paper describes the application of genetic programming to generate algorithms for control of time-delays in anti-air missiles equipped with proximity fuzes. Conventional algorithms for determining these delay-times rely on human effort and experience, and are generally deficient. It is demonstrated that by applying genetic programming determination of the timing can be automated and made near-optimal.

1 Introduction

In the final stages of an anti-air missile's engagement with an airborne target, it is often the case that the missile will pass near the target without physically hitting it. For this reason missile warheads are equipped with a proximity fuze, a device which detects the target, and explodes the warhead at a suitable moment to maximise the probability of destroying the target or disabling vital systems on the the target. The vulnerability of the target varies along its frame due to positioning and shielding of components, hence the optimum moment to burst the warhead depends on many parameters including [1], the velocity of the missile, the approach angle of the missile relative to the target, miss distance and direction, orientation of the missile, and kill capacity of the missile warhead (fragment density and velocity). From the available information, an "optimum" time can be determined at which to detonate the warhead to maximise the probability of kill (P_k). As a missile travels past its target, P_k can vary greatly, from a minimum ($\approx$ 0.0) to a maximum value ($\approx$ 1.0) over relatively very short distances (typically, 20cm to 30cm), which at supersonic speeds represents a time period of less than 1 millisecond. Furthermore, the distribution function of P_k is also greatly multi-modal, such that, over the fly-by distance several P_k peaks (not necessarily all maximal) are traversed. Therefore, an algorithm for optimal delay-time algorithm should allow the missile to travel over one or more sub-optimal peaks before eventually detonating the warhead. Thus, it is required to model the P_k from the information that is provided.

E.J.W. Boers et al. (Eds.) EvoWorkshop 2001, LNCS 2037, pp. 243–247, 2001.
© Springer-Verlag Berlin Heidelberg 2001

2   Application of GP for a Time-Delay Algorithm

Genetic programming (GP) is a machine learning technique based on the principles of natural evolution including *survival of the fittest* and *natural selection*. GP was developed by Koza [2] and has its roots in the better known field of Genetic Algorithms [3] with which it shares many characteristics. GP has been chosen for the time-delay problem because it does not rely on human understanding of the complexities of the problem. Secondly, GP is not known to have been previously applied to this type of problem, although a previous study [4] aplied it to improve aircraft survivability in an environment with surface-air missile. GP is also particularly suitable because it is able to evolve explicit mathematical expressions which can be examined for understandability after the learning process, unlike alternative learning algorithms such as neural networks.

GP evolution relies on an iterative process of generation and testing of solutions. An initial population of trial algorithms is randomly generated using a set of functions and a set of terminals, which has been chosen for the problem at hand. In the present case, the function set F, is a selection of mathematical functions comprises common arithmetic and trigonometric functions, as shown in Table 1. The terminal set T, also shown in Table 1, comprises the independent variables of the problem, namely, the velocity of the missile v, the approach angle of the missile relative to the target θ_1, miss distance d and miss direction θ_2, and orientation of the missile relative to its velocity vector θ_3, and $\mathcal{R}$, a random constant between 0.0 and +5.0. Each feasible solution is a complex expression involving parameters of the terminal and function sets, which models the distribution of P_k.

Table 1. GP Tableau for delay-time problem

Objective:	Evolve an algorithm which outputs a delay-time t which maximises the P_k for given d, θ_1, θ_2, θ_3
Function Set:	$F = \{+, -, *, /, \texttt{PWR}, \texttt{SIN}, \texttt{COS}, \texttt{EXP}\}$
Terminal Set:	$T = \{v, \theta_1, \theta_2, \theta_3, d, \mathcal{R}\}$
GP Parameters:	population = $500 \times 10 = 5000$, max. generations =250, crossover = 0.8, reproduction = 0.2, mutation = 0.00, migration frequency = 5 generations, migration rate = 5%
Success predicate:	Fitness=0

The evolutionary learning process was carried out to a maximum of 250 generations, with 10 separate populations. Every fifth generation, the top 5% of the population from each processor migrated to a neighbouring processor, displacing the bottom 5%. The evolution process in this study took approximately 2 hours on a university-network ten processor system, which at the best of times is quite slow. However, this in fact is an indication that with superior technology the learning process is feasible in real-time applications.

3 Simulation Results

The GP system was trained using a dataset of 12,500 combinations of $(v, \theta_1, \theta_2, \theta_3, d)$, against the delay time t that corresponded to the highest value of P_k for the given target. Another data set of 12,500 entries was used to test the evolved algorithms. The data was made available through a Defence establishment, representing realistic missile end-game scenarios. A demonstration of the performance of the evolved algorithms are shown in Figures 1 and 1. The plots show a cross-section of P_k in 3D space for a specific set of $v, \theta_1, \theta_2, \theta_3$. The value of the miss distance d was varied, and the required delay-time t determined by the learned algorithm. The "color intensity" indicates the values of P_k in the cross-sectional plane. The curve imposed on each cross-section corresponds to the time delays determined by the algorithm as d was varied. For example in Figure 1, with a miss distance of 25cm there are two possible values of time delay that would result in a high P_k, namely, 10ms ansd 45ms. The value determined by the algorithm, however, is 45ms.

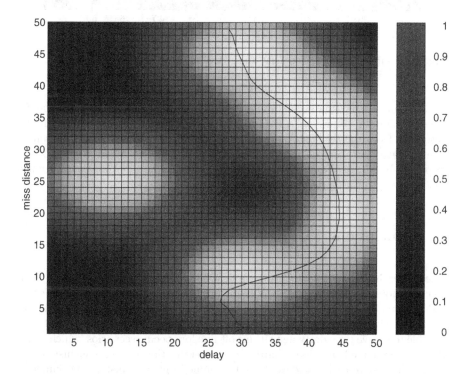

Fig. 1. Example 1 of an evolved delay-time curve superimposed on a simulated P_k plot

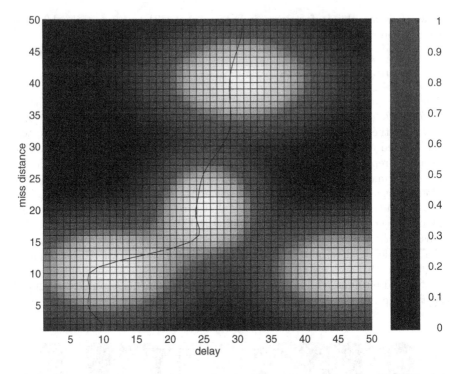

Fig. 2. Example 2 of an evolved delay-time curve superimposed on a simulated P_k plot

It is shown that the evolved delay time algorithms accurately track the regions of high P_k. However, a root mean square difference (rmsd) between the actual P_k from the test data and that suggested by the algorithm, is chosen as a quantitative measure of performance. This was found to be an average of approximately 0.01%. We can glean that this implies a near-optimal performance with respect to this type of application.

4   Conclusions

The paper has demonstrated the applicability of GP to an application in a complex nonlinear environment. While it would appear desirable to incorporate GP-learning within a missile in order to adapt to real-time changes, this is still infeasible with the current technologies. However, the study suggests that it is feasible to evolve time delay algorithms off-line for specific scenarios, which could then be embedded in missile hardware. Generalisation of this conclusion will require further research studies of the technique in more realistic environments, and with practical constraints.

References

1. G. Payne, "Anti-air lethality modelling," *Military Technology*, vol. 19, no. 11, pp. 46–49, 1995.
2. J. R. Koza, *Genetic Programming: on the Programming of Computers by Means of Natural Selection*, MIT Press, Cambridge, MA, USA, 1992.
3. J. H. Holland, *Adaption in Natural and Artificial Systems*, University of Michigan Press, 1975.
4. F. W. Moore, "Genetic programming solves the three-dimensional missile counter-measures optimization problem under uncertainty," in *Genetic Programming 1998: Proceedings of the Third Annual Conference*, John R. Koza, et al Eds., University of Wisconsin, 22-25 July 1998, pp. 242–245, Morgan Kaufmann.

Surface Movement Radar Image Correlation Using Genetic Algorithm

Enrico Piazza (member, IEEE)

Navia Aviation AS, division NOVA
P.O.Box 1080, 3194 Horten - Norway
enrico.piazza@naviaav.no

Abstract. The goal of this work is to describe an application of Genetic Algorithms to to a real aeronautical problem involving radar images. The paper presents the aeronautical problem, the specific implementation of the Genetic Algorithm and the result of the variation of some of the parameters of the Genetic Algorithm in term of time employed by the process, and ability to reach a useful solution of the aeronautical problem in a given time. The aeronautical problem is to find the position, orientation and dimension of a radar observed target. All the methods used here involve the correlation between an actual radar image and a template image. The Genetic Algorithm itself is not standard since it involve a dynamic computation of the best value for the probability of mutation. The probability of mutation (Pm) is dynamically adjusted according to the fitness of the best individual so that a worse fitness gives a greater probability of mutation and a better individual gives a lower probability of mutation.

Keywords: Radar, Image Correlation, Genetic Algorithms, A-SMGCS, Air Traffic Control

1 Introduction

Nowadays, because of the ever-growing number of aircraft in controlled areas; the rising demand for safety of passengers and cargo; and the greater growing environmental concern, the problem of improving the performance of traditional Airport Control systems must be faced [1].

Research trend in future years is oriented towards the development of integrated systems, which collect information coming from various sensors and distribute it both to ground Air Traffic Controllers and to moving vehicles. The realisation of a system for the automatic control of airport surface traffic can play an important role in this context. The systems and procedures that ensure the above could be included in the "Advanced-Surface Movement Guidance and Control System" (A-SMGCS).

Despite the new technologies developed to monitor the Airport Surface, including Global Positioning System (GPS) or Multilateration and co-operative sensors, the radar will keep its role of main surveillance sensor for many years.

E.J.W. Boers et al. (Eds.) EvoWorkshop 2001, LNCS 2037, pp. 248-256, 2001.

The automatic extraction in real time of the target features can be carried out through innovative approaches to the processing of radar data, based on image processing techniques.

While the aeronautical problem is to find the position, orientation and dimension of the observed target, the scope of this work is to show how this can be done with the use of a Genetic Algorithm. The goal of this work is to present a Genetic Algorithm and a study of the parameters of the Algorithm in order to use it in the best possible way.

Previous works have addressed the problem of image matching both with the use of Genetic Algorithms, for example, in [2] and [3] and without the use of Genetic Algorithms, for example in [4] and [5]. In both cases neither the direct application to the aeronautical problem, nor the time requirement was addressed.

2 Test Images

All the methods used here to find the position, orientation and dimension of the observed target, involve the correlation between an actual radar image and a template image. In this chapter the radar images and template images are described.

2.1 Radar Images

The images used in this work are radar images obtained by courtesy of Oerlikon Contraves Italiana SpA. The radar is a prototype operating at 95 GHz, developed by Oerlikon Contraves Italiana in the framework of the research project on transportation (PFT2) with a grant awarded by the Italian National Research Council (CNR) [6] [7].

It is important to note that the target echoes, as they appear in a bidimensional form, do not resemble the targets real shape exactly (See Fig 1a). The appearance of the target in a 2D radar image, in fact, will strongly depend on its 3D shape, on the radar resolution and on the position of the target with respect to the sensor.

The images were recorded at Rome-Fiumicino Airport, Italy in 1997 with Pulse Repetition Frequency PRF = 2500 Hz, antenna rotation time Tra = 1.81 and range binary sampling frequency f_{RBC} = 50 MHz [7], leading to dr = 3 m/sample and da = 0.0796 deg/sweep. They have been resampled on a square grid with a resolution of 3 m/pixel. As an example, Fig. 1a shows a 65 m long B747 radar image, made by 64 x 64 pixels, sampled at 3.0 m/pixel.

2.2 Template Images

The template T (i, j) is computed from a generic aircraft shape, given a resolution dr, a scale s, an angle a and a position x , y. Since the Genetic Algorithm is being applied to airplanes, the scale parameter represent the wingspan or the length of the aircraft in metres. The heading angle is a value in degrees from north, clockwise. The resolution is the sampling step in m/pixel.

In order to have a template useful for the Genetic Algorithm, a blur function is implemented with a simple convolution filter. As an example, Fig. 1b shows a template made of by 64 x 64 pixels, rotated by 45 degrees, 70 m long, and sampled at 3.0 m/pixel.

(a)   (b)

Fig. 1 - (a) Example of B747 radar image 65 m long, made by 64 x 64 pixels, sampled at 3.0 m/pixel. (b) Example of a template 64 x 64 pixel. Template rotated by 45 degrees and 70 m long, sampled at 3.0 m/pixel.

3 Genetic Algorithms

These kind of Algorithms are said to be generational because the solution is found by letting a population evolve generation by generation until the best individual is finally able to perform the given task, or a maximum time is reached.

In this work termination is set after 300 generations or if the best individual reaches 90% of the best-expected fitness value.

3.1 Creation of Initial Population

The initial population consists of entirely random individuals to avoid convergence to local optima. The total number of individuals is kept constant over all generations.

3.2 Selection

Selection is the process of choosing individuals for the next generation from the individuals in the current generation. The selection procedure is a stochastic procedure that guarantees that the number of offspring of any individual is bound to the expected number of offspring. The idea is to allocate to each individual a portion of a spinning roulette wheel, proportional to the individual's relative fitness [8]. The selection used here is the classic Roulette wheel, where best-fit individuals have a higher probability of being chosen without any elitist behaviour, that is, the best individuals are not copied from one generation to the next.

3.3 Crossover

Because of the number of parameter involved, the single point crossover operation may not be useful [9]. So crossover is applied to each sub string corresponding to each of the parameters. This crossover is similar to the standard single point crossover operator, but it operates on each parameter. Therefore, there are NP single point crossover operation taking place between two parents.

3.4 Mutation

The mutation used here is a classic mutation operator in which each bit of each individual is complemented with a small mutation probability. The classic Genetic Algorithm implementation uses a fixed mutation-rate. To avoid Genetic Algorithm stagnation, defined as the number of generations in which the best performer remains the same [2], the probability of mutation (Pm) is dynamically adjusted according to the fitness of the best individual so that, a worse fitness gives a greater probability of mutation and a better individual gives a lower probability of mutation. The used formula is

$$Pm(t) = 0.6 \ (1\text{-}\max\{fitness[population(t)]\}).$$

3.5 Genes Mapping

Genes are 16 bit integer numbers coded with the reflected Grey code. It has the property that adjacent integer values differ by exactly one bit [8]. As expected, the use of the reflected Gray code mapping had a positive effect on the generation-to-generation performance of the Genetic Algorithm, resulting in a better convergence after a given number of generations.

3.6 Fitness Evaluation

The individual's fitness is computed at the end of each generation. If an individual was copied from one generation to another by the selection, crossover and mutation and remains unchanged, then its fitness is preserved and there is no need to waste processor time to evaluate it again.

The fitness value of an individual is chosen in the range 0.0 to 1.0 where the best individuals have higher values.

In order to constrain the range of some of the parameters, the fitness is chosen to be zero for those individual whose parameter values are out of the allowed range.

4 Parameter Optimisation

According to the description of the Genetic Functions given above, the Genetic Algorithm has a number of parameters. Different problems have different sets of parameters, which lead to the "best" solution. The definition of "best" is often not so clear. Sometimes it means faster, sometimes it means having a better chance to get to a valid solution in a given time.

The parameters which are free in the Genetic Functions used in this work, are:
- Probability of Crossover (Pc);
- Number of individuals in the population (Ni).

Here follow few tests run on the Genetic Algorithm software. The complete search algorithm was run 100 times for each couple of values of Probability of Crossover (P_c) in (0.1, 0.2, 0.3, 0.4, 0.5, 0.6, 0.7, 0.8, 1.0) and Number of individuals in the population (N_i) in (50, 100, 200, 300, 400, 500). The whole run for each test, took several days on an SGI Indy 180 MHz processor. For each search, the termination is set after 300 generations or if the best individual reaches 90% of the best-expected fitness value.

4.1 Four Coefficients Polynomial
The first test was to find a specific four coefficients polynomial given a set of eight points.

$$y = a_0 + a_1 x + a_2 x^2 + a_3 x^3;$$

given a set of point (x_k, y_k), k=0, ..., 7.
where a_0=-2.5, a_1=2.0, a_2=1.5 and a_3=-1.0.

The coefficients are coded as 16 bit fixed-point real values into the chromosomes. One coefficient for each gene limited to the range -327.2, 327.2. The genes are then coded with the reflected Grey code.

The fitness of each individual was evaluated by the quadratic error between the expected polynomial and the polynomial encoded into the individual's genes. The fitness value of an individual is chosen to range between 0.0 and 1.0 where the best individuals have higher values.

$$Ro = \frac{1}{1 - \sum_k \left| y_k - \sum_j a_j x_k^j \right|^2}$$

This test revealed that the Reflected Gray Code is necessary, and was useful to set the dynamically adjusted probability of mutation (P_m) to the present formula.

The results shown below in Fig. 2 are the average performance of the Genetic Algorithm over 100 runs, as number of generation needed to accomplish the task and the time used by the search process, versus the two free parameters.

It is possible to see that the number of generation needed decreases when N_i increases and when P_c decreases. The time increases when N_i increases and when P_c decreases. Both the functions show a kind of flex around N_i = 400 and P_c = 0.4 which are then considered the local optima. This study reveals that the Algorithm performs best for low values of Pc and a given range of N_i. Higher numbers of individuals take

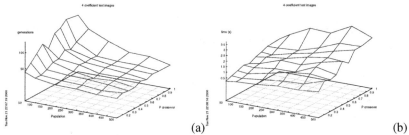

(a)   (b)

Fig. 2 - (a) Number of generations required to get to a solution. (b) Time required getting to a stop: either a solution or 300 generations. Case of four coefficients polynomial.

too long time to evaluate the fitness and lower numbers do not reach a solution in the given number of generations, that is, in the maximum available time.

4.2 Correlation Between a Template and a Real Image

The second test was to find the best correlation between a template and a real image. The image is the one shown in Fig. 1a, that is a 65 m long B747 radar image, made by 64 x 64 pixels, sampled at 3.0 m/pixel extracted from an actual radar image [10]. The template T(i, j) is made by 32 x 32 pixels, computed given the set of four coefficients, which are coded as 16 bit integers into the chromosomes. The four genes contain one coefficient for each gene and their value is limited to the following ranges: scale s between 1 and 120 m, angle a between 0 and 360 deg and position x, y inside the template dimension that is between -32 and 32. The resolution is not a coefficient but is assumed to be the same for template and image, dr=3.0 m/pixel.

The genes are then coded with reflected Grey code. The fitness of each individual was evaluated by the NCC correlation [3] between a fixed template and the one encoded in the genes. The fitness value of an individual is chosen to range between 0.0 and 1.0 where the best individuals have higher values.

$$NCC(u,v) = \frac{\sum_{i,j} I(i,j)T(i+u,j+v)}{\sqrt{\sum_{i,j} I^2(i,j) \sum_{i,j} T^2(i+u,j+v)}}$$

This test has not completely confirmed the basic evaluations about the performance of the algorithm discovered during the four coefficient polynomial test.

Fig. 3, Fig. 4 and Fig. 5, below, show the performance of the Genetic Algorithm as the number of generations needed to accomplish the task versus the two free parameters, averaged over 100 runs.

These results show that there is no clear correlation between P_c and the number of searches that reach 300 generation without finding a solution, that is, without developing a good individual. As a general behaviour the function has a minimum between $P_c = 0.4$ and $P_c = 0.6$. On the (b) side of Fig. 4 it is shown that

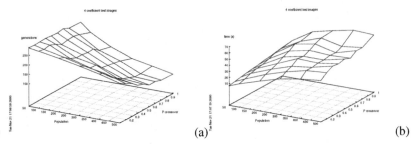

Fig. 3 - (a) Number of generations required to get to a solution. (b) Time required to get to a stop: either a solution or 300 generations. Case of correlation between a template and a real image.

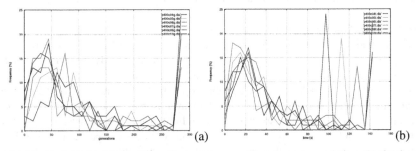

Fig. 4 - Distribution of (a) the number of generations required to get to a solution, (b) the time required to get to a stop: either a solution or 300 generations over a number of different searches, varying the probability of crossover between 04 and 1.0.

Increasing P_c, affects the speed of the process and, actually, there is a minimum between $P_c = 0.4$ and $P_c = 0.6$. The full time required now varies and is no more due mainly to Ni = 400 as it was in the four coefficients polynomials case.

On the contrary, Fig. 5 shows the distribution of the same quantities against the Number of individuals in the population (Ni) for the generic "best" Probability of Crossover (Pc) = 0.4. These results show that by increasing Ni, in fewer searches the line of 300 generation is reached without finding a solution. On the (b) side of Fig. 5 it is shown that the minimum time to reach a solution is now much bigger than in the case of the four coefficients polynomials. With Ni = 50, for example, no solutions are ever found and the search time is about 20 s. With Ni = 300 the search time for an unsuccessful search is 160 s while the average time is about 50 s. At the same time, the figure shows the peaks due to the unsuccessful searches which, despite the fact that they contain less searches, take more and more time.

This test was also used to check the different time performance of the available machines at Navia. All the machines were tested with the same unsuccessful run (a search that, after 300 generations, didn't find a good individual) with a population of 300 individuals and a Pc = 0.6. The program and its data are small

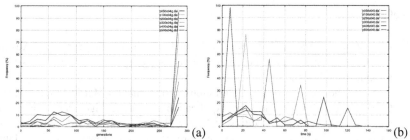

Fig. 5 - Distribution of (a) the number of generations required to get to a solution, (b) the time required to get to a stop: either a solution or 300 generations over a number of different searches, varying the number of individual between 50 and 500.

enough to fit into the smallest processor cache memory, which is the 512 kb on the R5000IP22 of the SGI Indy.

Table 1 - Time performance on an unsuccessful run with 300 individuals and a Pc = 0.6.

SGI	Indy	R5000IP22	150 MHz	IRIX 6.5	166 s
SGI	O2	R5000IP32	300 MHz	IRIX 6.5	101 s
SGI	Origin2000	R10000IP27	225 MHz	IRIX 6.5	61 s
Toshiba	Satellite	PentiumMMX	200 MHz	Win 98	215 s

5 Conclusions

This paper has presented a possible solution to the problem of characterising a target on a radar scene. The proposed solution involves the use of a Genetic Algorithm. The Genetic Algorithm introduced here has a new feature: it automatically chooses a suitable probability of mutation according to the actual fitness of the present generation of individuals.

A study devoted to the tuning of the two free parameters of the Genetic Algorithm set has been presented with a choice of "best" parameters.

When properly tuned, that is, when the best set of parameters is used, the Genetic Algorithm has a time performance better than the exhaustive search for the maximum of image correlations, but still too long for the heavy use in Air Traffic Control environment. According to Fig 4b, applied to the correlation of a template with a real image, no search is shorter than 20 s and all the successful searches take less than 80 s with an average of 50 s. This time, compared with the 1.81 s radar update time is still too long. However, optimizing the algorithm and running it on more powerful processors, it could still be used, from time to time, whenever safety requires a deeper investigation of some of the aircraft and traffic conditions, under looser time constraints.

References

1. EUROCAE, "Minimum Aviation System Performance Standards for Advanced Surface Movement Guidance and Control Systems", EUROCAE WG41 Final Report, Bruxelles 1997.
2. Dickens Thomas P., "Image-Calibration Transformation Matrix Solution Using a Genetic Algorithm", in Industrial Application of Genetic Algorithms, Karr Charles and Freeman Michael Editors, Chapter 2, CRC press, ISBN 0849398010, http://corpitk.earthweb.com/ reference/ pro/ 0849398010/ ewtoc.html, 12 January 1998
3. Banks Jasmine, Bennamoun Mohammed, Corke Peter, "Fast and Robust Stereo Matching Algorithms for Mining Automation", Digital Signal Processing, Academic Press, Vol. 9, No. 3, p. 137-148, http://www.idealibrary.com/ links/ doi/ 10.1006/ dspr.1999.0337 , July 1999
4. Pellegrini P.F., Piazza E., "Airport Surface Radar Signal Analysis for Target Characterization. A Model Validation", IEEE IECON-95 Conference, Orlando, Florida, November 1995
5. Sezgin M., Birecik S., Demir D., Bucak I.O., Cetin S., Kurugollu F., "A Comparison of Visual Target Tracking Methods in Noisy Environments", IEEE IECON-95 proceedings, p. 1360, Orlando, Florida, November 1995
6. Ferri M., Galati G., Marti F., Pellegrini P.F., Piazza E., "Design and Field Evaluation of Millimetre-wave Surface Movement Radar", IEE Radar 97 Conference, Edinburgh, Scotland, Oct 1997
7. Galati G., Naldi M., Ferri M., "Airport Surface Surveillance with a Network of Miniradars", IEEE Transactions on Aerospace and Electronic Systems, Vol. 35, No. 1, p. 331-338, January 1999
8. Schraudolph Nicol N., Grefenstette John J., "A User's Guide to GAucsd 1.4", ftp://cs.ucsd.edu/ pub/ GAucsd , 7 July 1992
9. Chakraborty Samarjit, De Sudipta, Deb Kalyanmoy, "Model-Based Object Recognition from a Complex Binary Imagery Using Genetic Algorithm", First European Workshop, EvoIASP'99 and EuroEcTel'99, Goteborg, Sweden, May 1999
10. Piazza Enrico, "Adaptive Algorithms for Real Time Target Extraction from a Surface Movement Radar", Proceedings of SPIE 4118-07, Parallel and Distributed Methods for Image Processing IV, SPIE 2000 Annual Meeting, San Diego, CA, July 2000

A Conceptual Approach for Simultaneous Flight Schedule Construction with Genetic Algorithms

Tobias Grosche, Armin Heinzl, and Franz Rothlauf

University of Bayreuth
Department of Information Systems
D-95440 Bayreuth, Germany
{Grosche, Heinzl, Rothlauf}@uni-bayreuth.de
http://wi.oec.uni-bayreuth.de

Abstract. In this paper, a new conceptual approach for flight schedule construction will be developed. Until now, the construction of a flight schedule is performed by decomposing the overall problem into subproblems and by solving these subproblems with various optimization techniques. The new approach constructs flight schedules with the help of genetic algorithms. Each individual of a population represents a complete flight schedule. This encoding avoids the artificial decomposition of the planning problem. With the help of genetic operators, flight schedule construction may be conducted simultaneously, efficiently searching for better solutions.

1 Introduction

Although positive market trends accompany the airline industry in the new millennium, ongoing deregulation continuously increases competition and rivalry within the industry. Recent developments in the commodity markets, especially the rise in oil prices, volatile demand, and extension of existing respectively new strategic alliances facilitates more effective ways of flight schedule construction.

Since every instance of a flight schedule affects the cost and revenue structure of an airline and its competitors [1], its construction is of paramount importance for every airline. The construction of a schedule is a complex planning activity. Until now, it is performed by decomposing the overall planning task into various subproblems of less complexity. Well known examples are fleet assignment or rotation building etc. [1]. Many of these subproblems are well structured and, therefore, solvable with linear optimization techniques. The subproblems will be solved sequentially where a preceding subproblem delivers the input data for the subsequent subproblems. Since the sum of partial optimal solutions does not imply an overall optimal solution, a planning methodology which facilitates a simultaneous (not sequential) optimization of flight schedules subproblems bears a high potential. This paper outlines a conceptual approach for simultaneous flight schedule construction with genetic algorithms (GA). We propose that a GA-based scheduling approach will generate better flight schedules than the sequential planning approach.

E.J.W. Boers et al. (Eds.) EvoWorkshop 2001, LNCS 2037, pp. 257–267, 2001.

Our paper is organized as follows. The next section focuses on the traditional planning approach, e.g. the sequential optimization of flight schedules. We will briefly elaborate the subproblems and finally discuss the effectiveness of this traditional approach. In the third section, we will develop a new approach for simultaneous flight schedule construction. A coding scheme will be explicated which possesses the ability to represent real-life flight schedules. Then, genetic operators for schedule construction will be described and illustrated with examples. Finally, a concept for fitness evaluation will be stepwise modeled and exemplified. In section four, we will briefly discuss the properties of our conceptual approach and activities which help to improve the concept in the near future.

2 Traditional Schedule Construction

2.1 Sequential Planning Activities

In order to effectively develop flight schedules, it is important to represent airline operations in substantial detail. This, of course, creates complexity and leads to a large number of variables in the overall planning model. Since it is almost impossible to formulate the schedule construction problem as a mathematical optimization model in a closed form, schedule construction has been undertaken through a structured planning process which involves many activities and parts of the airline. The planning process will be decomposed into subproblems with less complexity which, in return, will be optimized sequentially.

In figure 1 an overview of the subproblems and possible aggregations to higher problem levels is illustrated [2], [3].

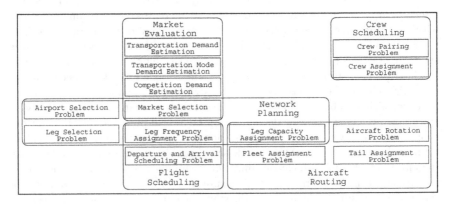

Fig. 1. Planning activities

2.2 Major Planning Activities

Market Evaluation. The quality of flight schedules is significantly influenced by the demand for flights. The estimation of the demand can be performed by first estimating the transportation demand between an origin-destination pair (market), then splitting up this demand on different transportation modes and finally partition the demand for flight service on each airline in the market [2]. Different models exist to estimate the demand for each step. Most of these models require statistical data or facts about the socio-economic structure of the specific market. One of the most important input for demand estimation is historical data about bookings from computerized reservation systems.

Network Planning. The result of the market evaluation will allow the airline to decide which markets to serve. This may already indicate a flight network, but furthermore, a flight network is defined by its exact routes and airports. Thus, the airline has to determine which airport in a region to serve. Moreover, the airline has to select flight legs to operate in the chosen markets. This part of the network planning process is affected mostly by political strategies, geographic and socio-economic structures in the region, and ground resources of the airline.

Leg frequency assignment and leg capacity assignment are closely related. The airline needs to decide about the total number of flights on one leg for the planning period. This number influences the leg's capacity and vice versa. For some solution approaches see for example [4] and [5].

Flight Scheduling. Besides the frequency assignment in flight scheduling, the airline has to specify exact departure times for its flights. The departure times and the block time estimation (block time is the sum of flight time and the time for taxiing on the ground) determine the arrival times. However, during flight scheduling the fleet has not yet been assigned to the flights. The block time is affected by the cruising speed of an aircraft type, so as a consequence, it is unknown which block time should be taken into consideration. Thus, most approaches which determine departure times are combined with fleet assignment.

Aircraft Routing. The main tasks in aircraft routing are fleet assignment and aircraft rotation building. In fleet assignment, the airline faces the problem to assign specific aircraft types to every flight leg. After fleet assignment, flights of the same fleet are grouped into rotations of logical aircraft. A rotation is a sequence of flights which can be flown in succession by the same aircraft. Required maintenance restrictions usually are considered in this step. At the end of the aircraft routing, the airline assigns real aircraft to the rotations (tail assignment). Some approaches for this problem are proposed in [6], [7], [8], [9].

Crew Scheduling. Crew scheduling has minor influences on the schedule construction problem. In this planning step, it will be determined which crew has to operate which flight. This is done by first creating crew pairings and then assign crews to the pairings. A crew pairing is a sequence of flights originating and ending at the same crew base. See [10] for a solution approach.

2.3   Overall Assessment of Existing Flight Schedule Construction Process and Methodologies

Since researches have devoted much effort in developing optimization techniques for the subproblems illustrated, it is possible to find good subsolutions. These solutions might be (near) optimal for the subproblems, but this does not imply a good overall solution. Local optimal solutions for subproblems can turn out to be unfavorable for later planning steps. Interdependencies between the scheduling steps are cut-off, e.g. the airline may realize during crew scheduling or aircraft routing that the scheduled flights may not be carried out with the available resources. Fleet assignment models are not able to consider the number of aircraft per fleet. This problem is solved later in rotation building. Thus, it may happen that the fleet assignment has to be modified after schedule construction.

The literature provides few approaches which integrate two or more subproblems [11]. Unfortunately, there is no model that takes all subproblems for flight schedule construction into consideration. Thus, the state-of-the-art of flight schedule construction will not support the development of acceptable schedules. This might explain why many airlines - if at all - construct their schedules by trial-and-error.

3   Simultaneous Flight Schedule Construction with Genetic Algorithms

In this section, we will develop an integrated, GA-based flight schedule construction approach which simultaneously permits multiple planning activities like airport selection, leg selection, leg frequency assignment, leg capacity assignment, departure as well as arrival scheduling, aircraft rotation, and fleet assignment.

3.1   Basic Representation Concept

A flight schedule may simply be coded as a list. For example, the list [FRA0.20 AMS40 ZRH30 OSL40 ... MIL30] represents the weekly flight schedule of one single aircraft. This aircraft departs in Frankfurt (FRA) at 0.20 am to Amsterdam (AMS). Block times and minimal ground times will be stored in matrices or tables. The suffix of the city code indicates that there is an additional waiting time in AMS of 40 minutes. Thus, a flight program may be calculated as $DepTimeB = DepTimeA + FlightTimeAB + MinGroundTimeB + AdditionalWaitingTimeB$.

If the aircraft has started its service in FRA on Monday morning, it arrives in Milan (MIL) on Sunday. From MIL the aircraft will fly back to FRA in order to begin its weekly service in FRA again.

With this basic representation concept we can generate flight schedules that consist of weekly rotations only. But in reality, rotations may be longer than one week. We solve this issue by simply appending multiple week plans while introducing two markings ∘ and ∗. The symbol ∘ indicates that the rotation will be continued. A * denotes the end of a rotation. Using these two markings, we are able to represent rotations of varying length. Consider the flight schedule in figure 2.

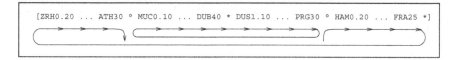

Fig. 2. Rotations

It indicates four week plans which are made up of two rotations, lasting two weeks each. The first rotation starts Monday morning with an aircraft departing from Zurich (ZRH) at 0.20 am and arriving at Athens (ATH) at the end of the week. Instead of returning to ZRH, the marking ∘ indicates that it will fly to Munich (MUC) where it continues its service on Monday morning in the second week. At the end of the second week, the aircraft will arrive in Dublin (DUB) from where it will return to ZRH to end its two week rotation. The end of the rotation is symbolized by the marking *. The second rotation starts Monday morning in Dusseldorf (DUS). At the end of the first week, this aircraft arrives at Prague (PRG) and continues the rotation to Hamburg (HAM) on Sunday night. In the second week, this aircraft departs HAM on Monday morning on 0.20 am and completes its program in Frankfurt (FRA) where the second rotation ends.

Note that the two markings introduced represent multiple aircraft. In this example, there are four aircraft deployed in the flight schedule. The first aircraft departs ZRH Monday morning at 0.20 am, the second MUC at 0.10 am, the third DUS at 1.10 am, and the fourth HAM at 0.20 am.

The last issue to resolve is the representation of multiple aircraft types for fleet assignment. In this context the weekly plans of one aircraft type will simply be joined until the next aircraft type follows. Consider the example in figure 3 which consists of two aircraft types and two individuals.

```
Type A                              Type B
 [ZRH0.20 ... ATH30 ° MUC0.10 ... DUB40 * DUS1.10 ... PRG ° HAM0.20 ... FRA25 *]


Type A                                        Type B
 [ZRH0.20 ... ATH30 ° MUC0.10 ... DUB40 * DUS1.10 ... PRG ° HAM0.20 ... FRA25 *]
```

Fig. 3. Fleet assignment

Each individual represents a different flight schedule. In the first schedule, every single aircraft type has to perform a two week rotation. In the second schedule, aircraft type A will be rotated twice whereas type B will be rotated once. The first rotation for type A has a length of one week and the second rotation one of two weeks. Note, that the various aircraft types will not be explicitly coded in the list notation. Since the number of aircraft of each type is known in advance (for example two aircraft of type A and two aircraft of type B) it is determined that the week plans for type B start after two week plans of type A.

3.2   Context Specific Genetic Operators

As we use a GA for our approach, we need recombination and mutation operators. Recombination should create individuals that inherit the properties from their parents. As the information of the schedule is mainly the order of the cities in the schedule, crossover should work well, if it preserves the order of the cities when creating an offspring. Mutation operators should in general create offspring with very similar properties as its parents. To ensure that our offspring has the desired properties, we define the genetic operators directly on the structure of the problem.

We believe that the one-point crossover is a good choice for preserving the structure of the problem. We simply need to determine the crossover site. We suggest to use the position where an exact point of time (e.g. 11.50 am) will determine the crossover site in the recombination process. As a result, we will obtain new flights at the crossover site.

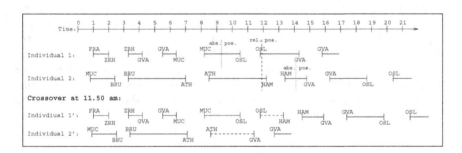

Fig. 4. One-point crossover

In our example the flights Oslo (OSL) - Geneva (GVA) and Athens (ATH) - Hamburg (HAM) will be replaced by OSL - HAM and ATH - GVA. By changing parts of individuals that contain markings of rotations this operator could produce different rotations.

Another option is the introduction of an operator which we call string crossover (see figure 5). With the help of this operator, segments of identical length may be exchanged among different week plans and will be inserted at an identical position in a week plan. The string-crossover permits the assignment of (multiple) flights to other aircraft types. Since this operator permits a better

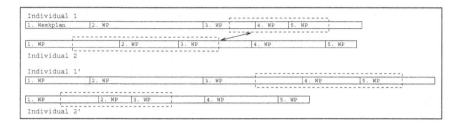

Fig. 5. String crossover

mixing of existing solution structures while not overemphasizing the effect of a pure random search, we advocate its use for future implementation efforts.

Developing mutation operators is much more straightforward. We randomly change a city, the length of a waiting time interval, or the rotation marking. The replacement of a ∗ by a ∘ will change one flight. The alteration of a city will change two flights. The change of the additional waiting time will not change any destinations but rather modify the length of a flight schedule.

Since the application of the single-point crossover, the string crossover, and the mutation operator will change arrival respectively departure times of flights between child and parent individuals, a specific repair mechanism needs to be developed. This mechanism should be able to adjust the departure times in way that the differences in comparison to the parent individuals are minimized. For this reason, the additional waiting time will serve as a buffer which will gradually be shortened or prolonged through the process of time adjustment.

In some cases, this time adjustment mechanism will not permit a proper re-alignment of the flight schedule. Thus, the deletion or insertion of distinct flights is inevitable. We introduce the sum of waiting time intervals as a criterion for flight insertion/deletion. If the sum of waiting time intervals in a given planning period is low then the deletion of a flight is likely to occur. In contrast, if this sum is high then the insertion of new flights will be likely. We will model the insertion (deletion) probabilities as a (inverse) sigmoid function.

3.3   Fitness Evaluation

As the problem is quite large and difficult, approaches like GAs need a reasonable population size and run over some 50 or hundred generations. Because current, state-of-the-art flight schedule evaluation tools are very time-consuming heuristic search could only be used, if we develop a simplified model for evaluating the individuals.

Since the large time consumption of the flight schedule evaluation tools is caused by the application of an open model which considers the effects on the organization's flight schedule on its competitors and vice versa, we propose the development of a closed profit estimation model for fitness evaluation.

Since profit is defined by revenues minus cost, we need to elaborate the de-termination of both variables. The basis for revenue estimation is the number of passengers or the number of seats sold. We approximate the number of passengers as following:

$$s_{ij}(t) = h_{ij}(t)^b \cdot f^{e_{ij}} \cdot p_{ij}(t) \cdot (m_{ij} + m_{j_{ik}}). \qquad (1)$$

The main factor in this expression is the time-dependent attractiveness $p_{ij}(t)$ of a flight from city i to city j. It will be calculated with the help of a time-of-week curve. Since the time-of-week curve can be derived from historical data, to every flight pair in a schedule a specific attractiveness could be assigned according to the departure time. For example, if a flight is scheduled for $t =$ Monday, 4.00 pm, the time-of-week curve will deliver an attractiveness of $p_{ij}(t) = 0.1$. This indicates that this particular flight captures 10% of an airline's entire market volume m_{ij} between these two cities.

In addition, we assume that the number of connecting flights at the destination j will increase the attractiveness of a flight. Thus, we model the additional demand with the help of the variable $m_{j_{ik}}$. For $i =$ HAM, $j =$ ATH, and $k =$ OSL, $m_{j_{ik}}$ expresses the additional market size of a connecting flight from ATH to OSL if the passenger has started it's trip in HAM.

Another factor to consider is the frequency $f^{e_{ij}}$. The higher the number of flights e between two cities i and j, the more passengers per flight will be attracted.

The last factor to include is the homogeneity $h_{ij}(t)^b$. It reflects the issue whether a flight takes place daily or on special weekdays only. If it takes place every day, b would be 7. If this flight is offered only on Monday, Wednesday, and Friday, b would be 3. So each flight schedule will be inspected for recurring flights for a given t and the result will bounded to the value for variable b.

Applying this fitness function, the GA would generate flight schedules containing a high proportion of flights at peak hours. Thus, the fitness evaluation needs to be extended with regard to the following issues:

  – How can we distribute the demand across conflicting flights?
  – How can we model limitations in capacity?

As stated, there are flights between two cities i and j which depart close together and compete for identical passengers. Therefore, we need to develop a mechanism that distributes the overall demand in a time interval among the departing flights. We assume that the demand for one flight is described by a normal distribution. For our purpose, we model a normal distribution curve for each flight in the same market. The area below the curve corresponds to $s_{ij}(t)$. The maxima of the curves represent the departure times, the standard deviation depends on $s_{ij}(t)$. A curve has the following function:

$$f_{t_x}(t) = \frac{s_{ij}(t_x)}{\sigma_{s_{ij}(t_x)}\sqrt{2\pi}} e^{-\frac{(t-t_x)^2}{2\sigma^2_{s_{ij}(t_x)}}}, \qquad (2)$$

where
  – $s_{ij}(t_x)$ is the expected number of seats sold for a flight at the exact time t_x between two cities i and j,

− $\sigma_{s_{ij}(t_x)}$ is the standard deviation of the flight x with the departure time t_x and its demand $s_{ij}(t_x)$.

Depending on the departure times and the demand curve there are areas which comprise two or more flights. These areas may be separated, indicating the specific demand of a distinct flight. For example, four aircraft with identical destinations depart from the same city at $t_1 = 6.00$ am, $t_2 = 10.50$ am, $t_3 = 12.00$ pm, and at $t_4 = 12.05$ (see figure 6). Without consideration of this conflict, there would be a demand for each flight of $s(t_1) = 80$, $s(t_2) = 50$, $s(t_3) = 70$, and $s(t_4) = 69$ passengers. Unfortunately, the estimated total number of seats of 269 seems to be unrealistic since the flights at t_2, t_3, and t_4 compete for the same passengers. If we apply the function as listed above, the demand will be distributed as follows: $s(t_1) = 80$, $s(t_2) = 47.1$, $s(t_3) = 31.1$ and $s(t_4) = 29.9$.

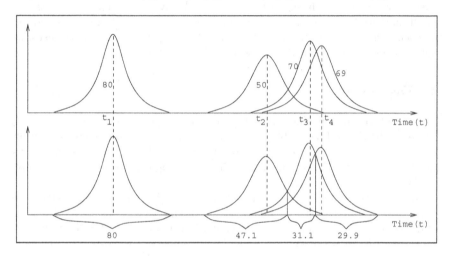

Fig. 6. Conflicting flights

This formula has the advantage that it may also be applied for modeling limited transportation capacity of an aircraft. Capacity restrictions may be represented as left and right boundaries of the area below the normal distribution curve. The demand external this area represent capacity shortages. We need to consider these shortages in order to model customer retention or compensation effects which may take place due to more than one flight. In this situation, the neighboring flights might be able to satisfy this demand peaks if they offer enough additional seats (see figure 7).

In figure 7, flight f_1 will not be able to serve all customers. This limitation is modeled with the help of the boundaries t_{1_l} and t_{1_r}. The capacity barriers of flight f_2 are modeled with the help of t_{2_l} and t_{2_r}. Since f_1 is already able to serve the demand of f_2 within the area of t_{2_l} and t_{1_r}, it will add additional capacity for f_2. This case is illustrated with the help of an area which is adjacent to the right of t_{2_r}. With out this spill-over effect, the area to the right of t_{2_r} could not be utilized.

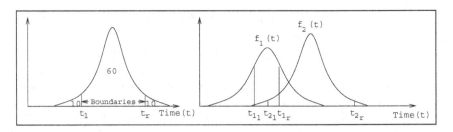

Fig. 7. Limited transportation capacity

Using these evaluation schemes, we are now able to estimate the numbers of passengers on each flight leg. Since we know the average revenue per passenger in all markets from the past, we will be able to compute the revenues induced by each flight schedule. Since the costs involved is also available from historical data, the estimated profit of each schedule can be calculated.

4   Discussion

The complexity of constructing flight schedules has led to a sequential planning process. The scheduling problem has been decomposed into numerous subproblems of less complexity. These subproblems may be solved with exact and heuristic approaches. But the optimal local solutions of the subproblems will not imply a good overall solution.

In order to achieve good results, flight schedule construction needs to be performed simultaneously. We believe that our concept sketches a promising way of doing this. In our concept, an individual in a population of a GA represents a complete flight schedule. We have defined genetic operators and derived a fast fitness evaluation method. We have not yet found similar approaches in the literature.

Since we are currently preparing the implementation of our concept, we need to define an appropriate test scenario. Within this scenario, we need to independently test the validity of our profit proxy as well as its underlying assumptions. If this proves to be applicable, we need to test our prototype against the existing planning approach. The genetic operators may be refined. If our GA-based concept will meet or exceed current standards, additional planning activities like tail assignment, crew assignment, and maintenance planning might be integrated.

References

1. Etschmaier, M. M., Mathaisel, D. F. X., *Airline Scheduling: An Overview*, in: Transportation Science, May 1985, pp. 127 - 138.
2. Suhl, L., *Computer-Aided Scheduling, An Airline Perspective*, Wiesbaden, 1995.
3. Antes, Jürgen, *Structuring the Process of Airline Scheduling*, in: Operations Research Proceedings, 1997, pp. 515 - 520.

4. Teodorovic, D., Krcmar-Nozic, E., *Multicriteria Model to Determine Flight Frequencies on an Airline Network under Competitive Conditions*, in: Transportation Science, February 1989, pp. 14 - 25.

5. Dobson, G., Lederer, P. J., *Airline Scheduling and Routing in an Hub-and-Spoke System*, in: Transportation Science, August 1993, pp. 281 - 297.

6. Desaulniers, G. et al., *Daily Aircraft Routing and Scheduling*, in: Management Science, June 1997, pp. 841 - 855.

7. Kontogiorgis, S., Acharya, S., *US Airways Automates Its Weekend Fleet Assignment*, in: Interfaces, May/June 1999, pp. 52 - 62.

8. Clarke, L. et al., *The Aircraft Rotation Problem*, in: Annals of Operations Research, No. 69, 1997, pp. 33 - 46.

9. Talluri, K. T., *The Four-Day Aircraft Maintenance Routing Problem*, in: Transportation Science, February 1998, pp. 43 - 53.

10. Mellouli, T., *Standardisierte Decision-Support-Komponenten für Ressourceneinsatzplanung und Störungsmanagement bei Bahn-, Regionalbus- und Fluggesellschaften*, Proceedings of the 3. Meistersingertreffen, November 25./26. 1999, Schloss Thurnau, Germany.

11. Barnhardt, C. et al., *Flight String Models for Aircraft Fleeting and Routing*, in: Transportation Science, August 1998, pp. 208 - 220.

Genetic Snakes for Color Images Segmentation

Lucia Ballerini

Centre for Image Analysis
Swedish University for Agricultural Sciences
Lägerhyddvägen 17, 752 37 Uppsala, Sweden
lucia@cb.uu.se

Abstract. The world of meat faces a permanent need for new methods of meat quality evaluation. Recent advances in the area of computer and video processing have created new ways to monitor quality in the food industry. In this paper we propose a segmentation method to separate connective tissue from meat. We propose the use of Genetic Snakes, that are active contour models, also known as snakes, with an energy minimization procedure based on Genetic Algorithms (GA). Genetic Snakes have been proposed to overcome some limits of the classical snakes, as initialization, existence of multiple minima, and the selection of elasticity parameters, and have both successfully applied to medical and radar images. We extend the formulation of Genetic Snakes in two ways, by exploring additional internal and external energy terms and by applying them to color images. We employ a modified version of the image energy which considers the gradient of the three color RGB (red, green and blue) components. Experimental results on synthetic images as well as on meat images are reported. Images used in this work are color camera photographs of beef meat.

1 Introduction

There is a permanent need for new methods of meat quality evaluation. The demand of researchers for improved techniques to deepen the understanding of meat features, as well as the demand of consumers for high meat quality products, induces the necessity of new techniques [1].

Fat content in meat influences some important meat quality parameters. The quantitative fat content has been shown to influence the palatability characteristics of meat. There are several methods to analyze quantitative fat content and its visual appearance in meat. However, few of them are satisfactory enough in terms of fat quantification in a cross section of a consumer size meat slice, without using large amounts of organic solvents or being too time consuming.

Recent advances in the area of computer and video processing have created new ways to monitor quality in the food industry.

In a previous paper we have developed image analysis methods for the specific problem of measuring the percentage of fat [2]. In particular, a segmentation algorithm (i.e. classification of different substances) has been optimized for color camera photographs of meat. There is still an open problem with these images:

E.J.W. Boers et al. (Eds.) EvoWorkshop 2001, LNCS 2037, pp. 268–277, 2001.

fat and connective tissue present almost the same color and therefore they are almost indistinguishable by any color segmentation technique.

In this paper we propose an automatic segmentation procedure to identify the separation of connective tissue from meat. We propose the use of Genetic Snakes [3], that are active contour models, also known as snakes [4], with an energy minimization procedure based on Genetic Algorithms (GA) [5].

Snakes are a robust global segmentation method combining constraints derived from the image data with *a priori* knowledge about the position, size, and shape of the structure to be segmented [6]. However, the application of snakes to extract region of interest suffers from some limitations. In fact, there may be a number of problems associated with this approach such as algorithm initialization, existence of local minima, and the selection of model parameters.

We proposed the use of GAs to overcome some of these limits. GAs offer a global search procedure that has shown its robustness in many tasks, and they are not limited by restrictive assumptions on the objective function, such as the existence of derivatives. The usefulness of GAs in pattern recognition and image processing has been demonstrated [7,8,9,10,11,12]. GAs operate on a coding of the free variables (the positions of the snake) and their fitness function is the total snake energy. Snakes optimization through Genetic Algorithms proved particularly useful in order to overcome problems related to initialization, parameter selection and local minima, and they have been successfully applied to medical and radar images [3,13].

The purpose of this paper is to extend the Genetic Snakes model, and to present additional energy functionals found to be useful in specific applications.

The organization of the paper is as follow: in Section 2 we discuss active contours, the basic notions, their limitations and some improvements proposed in literature. In Section 3 we review the Genetic Snakes model. In Section 4 we extend our previous formulation by experimenting new energy functionals. Experimental results on synthetic and meat images are reported in Section 5, in particular for meat images we proposed an internal area energy functional, which will allow snakes to give accurate separation of muscle and fat from connective tissue.

2   Active Contours (Snakes)

Snakes are planar deformable contours that are useful in several image analysis tasks. They are often used to approximate the locations and shapes of object boundaries on the basis of the reasonable assumption that boundaries are piecewise continuous or smooth.

Representing the position of a *snake* parametrically by $\mathbf{v}(s) = (x(s), y(s))$ with $s \in [0, 1]$, its energy functionals can be written as:

$$E_{snake} = \int_0^1 E_{int}\left[\mathbf{v}(s)\right] ds + \int_0^1 E_{ext}\left[\mathbf{v}(s)\right] ds \qquad (1)$$

where

- E_{int} represents the internal energy of the snake due to bending and it is associated with *a priori* constraints
- E_{ext} is an external potential energy which depends on the image and accounts for *a posteriori* information.

The final shape of the contour corresponds to the minimum of this energy.

In the original technique of Kass et al. [4] the internal spline energy is defined as:

$$E_{int}\left[\mathbf{v}(s)\right] = \frac{1}{2}\left[\alpha(s)\left|\frac{\partial \mathbf{v}(s)}{\partial s}\right|^2 + \beta(s)\left|\frac{\partial^2 \mathbf{v}(s)}{\partial s^2}\right|^2\right]. \tag{2}$$

The spline energy is composed of a first order term controlled by $\alpha(s)$ and a second order term controlled by $\beta(s)$. The two parameters $\alpha(s)$ and $\beta(s)$ dictate the simulated physical characteristics of the contour: $\alpha(s)$ controls the *tension* of the contour while $\beta(s)$ controls its *rigidity*.

The external energy couples the snake to the image. It is defined as a scalar potential function whose local minima coincide with intensity extrema, edges, and other image features of interest.

The external energy, which is commonly used, is defined as:

$$E_{ext}\left[\mathbf{v}(s)\right] = -\gamma|\nabla G_\sigma * I(x,y)|^2 \tag{3}$$

where $G_\sigma * I(x,y)$ denotes the image convolved by a Gaussian filter with a standard deviation σ, ∇ is the gradient operator and γ a weight associated with image energies. This edge functional is used by many researchers.

The application of snakes and other similar deformable contour models to extract regions of interest is, however, not without limitations. For example, snakes were designed as interactive models. In non-interactive applications, they must be initialized close to the structure of interest to guarantee good performance. The internal energy constraints of snakes can limit their geometric flexibility and prevent a snake from representing long tube-like shapes or shapes with significant protrusions or bifurcations. Furthermore, the topology of the structure of interest must be known in advance since classical deformable contour models are parametric and are incapable of topological transformations without additional machinery.

Various methods have been proposed to improve and further automate the deformable contour segmentation process. A review of some of them can be found in [14].

3   Genetic Snakes

In this section we review the Genetic Snakes model, i.e. our previous model of active contours, where the energy minimization procedure is based on Genetic Algorithms [3].

The parameters that undergo genetic optimization are the positions of the snake in the image plane $\mathbf{v}(s) = (x(s), y(s))$. The coordinates x and y are codified in the chromosomes using a Gray-code [15,16]. To simplify the implementation we used polar coordinates.

The fitness function is the total snake energy as previously defined in Equation (1), where E_{int} and E_{ext} are defined in Eqs. (2) and (3). The sigma scaling option is used [5].

The genetic optimization requires the definition of a region of interest (see Fig. 1), given by r and R (the minimum and the maximum magnitude allowed for each $\mathbf{v}(s)$). The initial population is randomly chosen in such region, and each solution lies in this region (r and R are user defined). This replaces the original initialization with a region-based version, enabling a robust solution to be found by searching the region for a global solution. This overcomes the problems associated with sensitivity to initialization which was a crucial problem for "hill climbing" techniques. As a result, the new optimization criterion is better at extracting non-convex shapes compared to conventional snakes. This helps to overcome the difficulties related to initialization and local minima. In addition, we have observed a noticeable improvement of the segmentation with respect to standard snake algorithm.

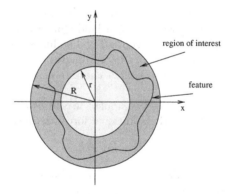

Fig. 1. Genetic Snakes Initialization.

An accurate description of implementation details along with a discussion on the choice of the model coefficients can be found in [3].

4   Evolution of Genetic Snakes

In this section we extend our previous formulation of Genetic Snakes by experimenting new internal and external energy functionals.

The classical optimization techniques impose different restrictions on the type of image functional that can be employed (for example the existence of derivatives); the use of Genetic Algorithms gives us more freedom on the choice of such functional. The internal energy term, E_{int}, defined in Equation (2), controls the properties of the snake. This internal energy provides an efficient interpolation mechanism for recovering missing data.

In the present work, we have investigated an energy proportional to the area enclosed by the snake which has the effect of causing the contour to expand or contract.

The area energy we propose is:

$$E_{area}\left[\mathbf{v}(s)\right] = \delta A \tag{4}$$

where A is the area enclosed by the snake. The sign of δ determines whether the snake will tend to expand or contract. For the case of a positive δ, this energy will be a positive term in the fitness and will cause the snake to choose regions enclosing small areas. On the other hand, for a negative δ, the snake will tend to prefer regions having larger area.

The area is calculated as:

$$A = \frac{1}{2}\left|\sum_{i=1}^{N}(x_i y_{i+1} - x_{i+1} y_i)\right| \tag{5}$$

where (x_i, y_i) denotes the coordinates of a vertex and by convention $(x_{N+1}, y_{N+1}) = (x_1, y_1)$. This expression holds for any polygon provided that the vertices are ordered around the contour and no line segments joining the vertices intersect. Our representation of the coordinates, along with the use of polar coordinates, does not require any explicit check to ensure that this relationship is applicable.

The effect of this area energy is similar to the *balloon* force employed by Cohen [17,18]. In their model the additional force has been formulated as:

$$F = k\mathbf{n}(s) \tag{6}$$

where $\mathbf{n}(s)$ is the unit vector normal to the curve at point $v(s)$ and k is the amplitude of this force. Note that F depends on not only the position $v(s)$, but also on the normal at this position. In our formulation the area energy term depends on the position, but not on derivatives.

We experimented also an area energy which force the snake to have a determinate area enclosed:

$$E_{area}\left[\mathbf{v}(s)\right] = \delta(A - A_{ref})^2. \tag{7}$$

This energy has the form of an harmonic potential with a minimum when the area enclosed by the snake is equal to the reference area A_{ref}.

In order to apply our model to color images, we also employed a modified version of the image energy. When the goal is to fit a snake to a boundary within an image, it is useful to preprocess the image with an edge detector so that the points of maximum gradient are emphasized. The edge detector most commonly employed uses the gradient of the image convolved with a Gaussian smoothing function.

We considered the gradient of the three color RGB components. Thus, the proposed image energy functional is composed of three terms and can be expressed as:

$$E_{ext}\left[\mathbf{v}(s)\right] = -\gamma_R|\nabla G_\sigma * I_R(x,y)|^2 - \gamma_G|\nabla G_\sigma * I_G(x,y)|^2 - \gamma_B|\nabla G_\sigma * I_B(x,y)|^2 \tag{8}$$

where $I_R(x,y), I_G(x,y), I_B(x,y)$ are the three components of the image intensities. The weights are negatives so that local minima of E_{ext} correspond to maxima of the gradient, i.e. strong edges.

5   Results

5.1   Experiments on Synthetic Images

In this section, the behavior of the new area energy term is examined using artificial images.

The experiment uses synthetic images containing circular shapes having an inner and an outer boundary as shown in Figure 2. The intensity images are generated by setting the pixel value to 100 if it is internal to the shape, and 0 otherwise.

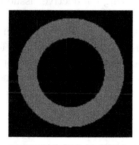

Fig. 2. Example of synthetic test image.

This kind of images is constructed to study the snake ability to be attracted by the inner or the outer boundary according to the area energy term.

On these images we performed experiments using snakes having 50 points, varying the energy weighting coefficients ($\alpha = 0.1, 0.15, 0.2$, $\beta = 0.1, 0.15, 0.2$, $\gamma = 0.1, 0.2, 0.3$ and $\delta = -0.3, -0.2, -0.1, 0.1, 0.2, 0.3$), running the GA for 2000000 iterations each time on a population of 10000 individuals. The population size was computed according to the length of genome, as suggested by Goldberg [5]. We have used the standard two-point crossover. The crossover rate and mutation rate are set respectively to 0.6 and 0.000006 based on previous experimental observations.

Figure 3 reports one of the results obtained, using the following weight values: $\alpha = 0.1$, $\beta = 0.2$, $\gamma = 0.2$ and $\delta = \pm 0.3$. When the sign of δ is positive the snake chooses the inner boundary, while when it is negative the snake is attracted by the outer boundary.

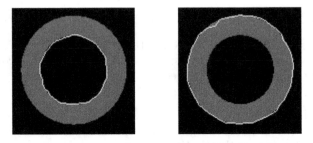

Fig. 3. Simulation results on synthetic test images with different area energy weights.

5.2 Applications to Meat Images

Genetic Snakes are then applied to meat images in order to segment them, with the special purpose of separating connective tissue from the remaining parts of the meat.

Color images of many samples of M. longissimus dorsi were captured by a Sony DCS-D700 camera. The same exposure and focal distance were used for all images. Digital color photography was carried out on both sides of each slice on a green background. Images are 1024 x 1024 pixel matrices with a resolution of 0.13 mm x 0.13 mm. They are represented in an RGB (red, green, and blue) format (see Figure 4).

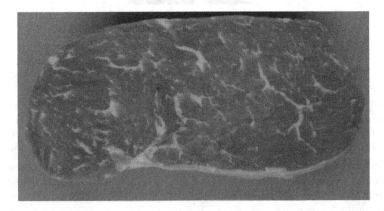

Fig. 4. Digital camera photograph of the *longissimus dorsi* muscle from a representative beef meat, (original is in color).

Images were preprocessed to suppress the background; the choice of green color as background was very helpful for this stage. Three values of threshold (RGB) were used for this phase. An example of background suppression is shown in Figure 5.

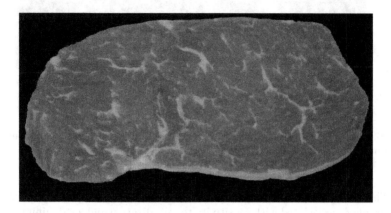

Fig. 5. Background suppression from image shown in Figure 4.

On the background suppressed images we computed the reference area A_{ref}. Then we applied the Genetic Snakes. For simplicity, the origin of the coordinates was located at the center of the image.

The energy functionals are chosen according to meat properties. We observed that the connective tissue is usually located close to the border of the meat and that there is a strong edge between it and the muscle. Moreover the percentage of connective tissue is usually around $4 \div 5\%$.

Thus, the proposed area energy functional is expressed as:

$$E_{area}\left[\mathbf{v}(s)\right] = \delta(A - 0.95 A_{ref})^2. \tag{9}$$

On these images we used the internal energy defined in Equation (2) and the image energy functional defined in Equation (8) with $\sigma = 0.2$. We used snakes having 50 points, running the GA for 40000000 iteration each time on a population of 100000 individuals. The crossover rate and mutation rate are set respectively to 0.43 and 0.000001 based on some experimental observations.

In Figure 6 we can see an example of original image and the corresponding meat outlines segmented by our snake model. Note the connective tissue in the bottom part of the image is separated from the rest of the meat.

6   Conclusions

In this paper we have discussed Genetic Snakes and presented an extension of our previous model. The snake employed in this work has several advantages with respect to the standard snake algorithm [4].

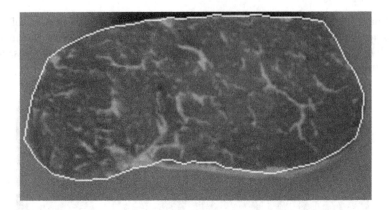

Fig. 6. Digital camera photograph of the *longissimus dorsi* muscle from a representative beef meat. The final position of our Genetic Snake model is superimposed.

The energy minimization procedure based on Genetic Algorithms overcomes the problems associated with sensitivity to initialization and local minima, which was a crucial problem of classical techniques.

Other energy terms may easily added using the genetic optimization procedure. This was not possible in the classical snake formulation. In our model there is no restriction on the form of the energy functionals.

The new area energy we proposed exhibits interesting properties in the localization of meat boundary.

Compared to current methods to separate meat from connective tissue (manual selection or threshold methods), this snake-based approach is expected to provide significant improvements.

As each individual image needs to be processed from the scratch, the main problem with this method is the time required and the impossibility to exploit the ability of GAs to generalize.

In this work we applied GAs to the positions of the snake. The management of the weight controls of the energy function is an open important problem. Further work on this technique could be the evolution of the parameters and the functional governing the snake behavior.

Acknowledgments. This work is performed within the National FOOD21 research programme, financed by MISTRA.

I would like to thank the meat-group at the Dept. of Food Science, Swedish University of Agricultural Sciences, Uppsala for their friendly collaboration, especially Anders Högberg for providing the camera pictures and for general discussions.

I would also like to thank Professor Gunilla Borgefors, Centre for Image Analysis, SLU, Uppsala, for her valuable comments on the manuscript.

References

1. G. Monin, "Recent methods for predicting quality of whole meat", *Meat Science*, vol. 49, no. Suppl. 1, pp. S231–S243, 1998.
2. L. Ballerini, A. Högberg, K. Lundström, and G. Borgefors, "Colour image analysis technique for measuring of fat in meat: An application for the meat industry", To appear in: Proc. Electronic Imaging, 2001.
3. L. Ballerini, "Genetic snakes for medical images segmentation", *Lectures Notes in Computer Science*, vol. 1596, pp. 59–73, 1999.
4. M. Kass, A. Witkin, and D. Terzopoulos, "Snakes: Active contour models", *International Journal of Computer Vision*, vol. 1, no. 4, pp. 321–331, 1988.
5. D. E. Goldberg, *Genetic Algorithms in Search, Optimization, and Machine Learning*, Addison-Wesley, Reading, MA, 1989.
6. T. McInerney and D. Terzopoulos, "Deformable models in medical image analysis: A survey", *Medical Image Analysis*, vol. 1, no. 2, pp. 91–108, 1996.
7. Poli, Voigt, Cagnoni, Corne, Smith, and Fogarty, Eds., *Evolutionary Image Analysis, Signal Processing and Telecommunications*, vol. 1596 of *Lectures Notes in Computer Science*, Goteborg, Sweden, 1999. Springer–Verlag.
8. "Special issue on genetic algorithms", *Pattern Recognition Letters*, vol. 16, no. 8, 1995.
9. C. Bounsaythip and J. Alander, "Genetic algorithms in image processing - a review", in *Proc. 3nwga (3rd Nordic Workshop on Genetic Algorithms)*, Helsinki, Finland, 1997.
10. D. N. Chun and H. S. Yang, "Robust image segmentation using genetic algorithm with a fuzzy measure", *Pattern Recognition*, vol. 29, no. 7, pp. 1195–1211, 1996.
11. B. Bhanu, S. Lee, and J. Ming, "Adaptive image segmentation using a genetic algorithm", *IEEE Transactions on Systems, Man and Cybernetics*, vol. 25, no. 12, pp. 1543–1567, December 1995.
12. P. Andrey, "Selectionist relaxation: Genetic algorithms applied to image segmentation", *Image and Vision Computing*, vol. 17, pp. 175–187, 1999.
13. L. Ballerini and E. Piazza, "Genetic snakes for radar images segmentation", in *proc. IEEE International Symposium on Intelligent Signal Processing and Communication Systems*, Phucket, Thailand, December 1999, pp. 621–624.
14. L. Ballerini, *Computer Aided Diagnosis in Ocular Fundus Images*, PhD thesis, Università di Firenze, Italy, 1998.
15. D. Beasley, D. R. Bull, and R. R. Martin, "An overview of genetic algorithms: Part 1, fundamentals", *University Computing*, vol. 15, no. 2, pp. 58–69, 1993.
16. D. Beasley, D. R. Bull, and R. R. Martin, "An overview of genetic algorithms: Part 2, research topics", *University Computing*, vol. 15, no. 4, pp. 170–181, 1993.
17. L. D. Cohen and I. Cohen, "Finite element methods for active contour models and balloons for 2D and 3D images", *IEEE Transactions on Pattern Analysis and Machine Intelligence*, vol. 15, no. 11, pp. 1131–1147, November 1993.
18. L. D. Cohen, "On active contour models and balloons", *Computer Vision, Graphics, and Image Processing: Image Understanding*, vol. 53, no. 2, pp. 211–218, March 1991.

A Distributed Genetic Algorithm for Parameters Optimization to Detect Microcalcifications in Digital Mammograms

Alessandro Bevilacqua[1,2], Renato Campanini[2,3], and Nico Lanconelli[2,3]

[1] Department of Electronics, Computer Science and Systems,
University of Bologna, viale Risorgimento, 2,
40136 Bologna, Italy
abevilacqua@deis.unibo.it
[2] INFN (National Institute for Nuclear Physics), viale Berti Pichat, 6/2,
40127 Bologna, Italy
{bevila, lanconel, campanini}@bo.infn.it
[3] Department of Physics, University of Bologna, viale Berti Pichat, 6/2,
40127 Bologna, Italy

Abstract. In this paper, we investigate the improvement obtained by applying a distributed genetic algorithm to a problem of parameter optimization in medical images analysis. We setup a method for the detection of clustered microcalcifications in digital mammograms, based on statistical techniques and multiresolution analysis by means of wavelet transform. The optimization of this scheme requires multiple runs on a set of 40 images, in order to obtain relevant statistics. We aim to evaluate how fluctuations of some parameters values of the detection method influence the performance of our system. A distributed genetic algorithm supervising this process allowed to improve of some percents previous results obtained after having "hand tuned" these parameters for a long time. At last, we have been able to find out parameters not influencing performance at all.

1 Introduction

The presence of microcalcifications in breast tissue is one of the most important signs considered by radiologist for an early diagnosis of breast cancer, which is one of the most common form of cancer among women. Statistical analyses show how errors in microcalcifications detection are very high in population screening programmes. A feasible solution, in order to reduce these kind of errors, consists in providing doctors with a computer aided system, which could act as a "second radiologist". Experiments showed that these systems can significantly improve the accuracy in the detection task. Our Computer Aided Diagnosis (CAD) scheme described in [1] is quite complex and its effectiveness depends on the values of different parameters. Therefore it is necessary to optimize the choice of these parameters, in order to achieve good performance. Unfortunately, their number is very high (about thirty) and they are correlated with each other.

E.J.W. Boers et al. (Eds.) EvoWorkshop 2001, LNCS 2037, pp. 278–287, 2001.
© Springer-Verlag Berlin Heidelberg 2001

Consequently, it is difficult to get an optimal choice of them. In the earlier study the selection were performed manually; we refer to this procedure as the "hand tuned" one. In this paper, we present an automated method for the selection of the parameters values by means of a genetic algorithm. Genetic Algorithms (GAs) search the solution space to maximize (minimize) a fitness (or cost) function by using simulated evolutionary operators such as mutation and sexual recombination. In this study the fitness function to be maximized reflects the goal of maximizing the number of true-positive detections while minimizing the number of false-positive detections.

GAs are currently applied to many diverse and difficult optimization problems (see [2] and [3]). In a number of applications where the search space was too large for other heuristic methods or too complex for analytic treatment GAs produced favorable results. Other researchers in [4] and [5] have shown that GAs could improve the performance of a CAD scheme.

In the present study, we will evaluate how the parameter values fluctuations influence the performance of the CAD scheme and which parameters more affect the cost function; our goal is as well to select, by using a GA, the most significant parameters. The GA needs to evaluate several generations, in order to obtain a good optimization. Due to the very long time required for one run, it would be almost impracticable to execute the GA on a sequential architecture. We therefore implement a distributed GA on a small Network Of Workstations (NOWs), by realizing a global parallelized GA. In this type of parallel GAs, there is only one population, as in the serial GA, and even if the evaluation of individuals is parallelized explicitly, the algorithm remains unchanged. In this way, we could easily apply the existing principles for sequential GAs.

We accomplish the optimization of our CAD scheme by using the 40 digitized mammograms of the Nijmegen database. Performances of the detection scheme are shown by means of Free Response Operating Characteristic (FROC) curves: they display the number of true positive clusters of microcalcifications detected versus the average number of false positives per image.

2   The Detection Method

Microcalcifications are very small spots which appear brighter than the surrounding normal tissue. Typically they are between 0.1 mm and 1 mm in size and are of particular clinical significance when found in clusters of five or more in a 1 cm^2 area. Most of the clusters consist of at least one evident microcalcification and other more hidden signals.

Our approach to the detection task includes two different methods: the first one (coarse) is able to detect the most obvious signals, while the second one (fine), based on multiresolution analyses, discovers more subtle microcalcifications (see [1]). Signals coming out from these two methods are combined through a logical OR operation and then clusterized to give the final result. There are some steps, common to both coarse and fine methods:

- *pre-processing*, which isolates breast tissue;

- *filtering*, in which structured background is removed;
- *signal extraction*, to find out microcalcifications candidates signals;
- *false positive reduction*, where microcalcifications are separated, by calculating a set of features, as described in [1] and [6] from false signals extracted.

In all these tasks there are several parameters to be tuned; we used twenty-nine of them (listed in Table 1) for the optimization process with the GA. A very critical phase of every CAD system is the false positive reduction step: here a detected signal is considered either microcalcification or false signal, according to the value of a set of its features.

Table 1. Parameters used in the optimization process

Parameters of the coarse method	Parameters of the fine method
Size of the local thresholding window	
Threshold for Gaussianity test h_t	
Values for local thresholding k	
Minimum edge gradient (EG)	
Maximum EG	
Minimum average local gradient (ALG)	
Maximum ALG	
Minimum area of signal	Minimum area of signal
Maximum area of signal	Maximum area of signal
Minimum gray level (GL)	Minimum gray level (GL)
Maximum GL	Maximum GL
Minimum degree of linearity (DL)	Minimum degree of linearity (DL)
Maximum DL	Maximum DL
ct_{11}, in: GL $> ct_{11}$ * EG $+ ct_{12}$	p_1, in: EG $< p_1$ * tanh(p_2 * GL)
ct_{12}, in: GL $> ct_{11}$ * EG $+ ct_{12}$	p_2, in: EG $< p_1$ * tanh(p_2 * GL)
ct_{21}, in: DL $< ct_{21}$ * ALG $+ ct_{22}$	p_3, in: EG $> p_3$ * GL $+ p_4$
ct_{22}, in: DL $< ct_{21}$ * ALG $+ ct_{22}$	p_4, in: EG $> p_3$ * GL $+ p_4$
ct_{31}, in: DL $> ct_{31}$ * ALG $+ ct_{32}$	
ct_{32}, in: DL $> ct_{31}$ * ALG $+ ct_{32}$	

Most of the parameters are thresholds to choose the range of features values in this false positive reduction phase; others are used for selecting regions of interest or extracting signals (see [1]). Any individuals of the population considered in GA optimization is therefore described by a string (gene) of twenty-nine values. Each one of them represents a parameter value and can be a real or an integer number, according to the domain of the parameter itself. The purpose of the optimization of a CAD scheme is to find out the set of parameters which gets the highest number of true positive clusters of microcalcifications with the lowest rate of false positive clusters, i.e. the best tradeoff between sensitivity and specificity. This tradeoff is described by the design of the fitness function.

3   The Genetic Algorithm

3.1   Design

The advantages in using GAs are that they require no knowledge or gradient information about the response surface, they are resistant to becoming trapped in local optima and they can be employed for a wide variety of optimization problems. On the other hand GAs could have trouble in finding the exact global optimum and they require a large number of fitness functions evaluations. It is very difficult to achieve analytic relationship between the sensitivity of the CAD and the parameters values to be optimized. Since a GA does not need this kind of information, it should be suitable in our optimization task.

If there is an explicit knowledge about the system being optimized, that information can be included in the initial population. In this study we initialize the population to the best "hand tuned" results.

An evolutionary strategy needs to be adopted in order to generate individuals for the next generation. We chose an *elitist generation* as replacement operator. Namely the individuals are ranked by their fitness and only the best of them (10% of the population) are taken unchanged into the next generation. In this way, we guarantee that good individuals are not lost during a run. Other children come from crossover and mutation, (their associated probabilities are p_{CO} and p_{MUT} respectively).

The aim of the fitness function is to numerically represent the performance of an individual. In our case a couple (true positives, false positives) is mapped by this function to a real number normalized between 0 and 1. That number encodes the excellence of a couple obtained by a particular individual (i.e. a set of parameters in the CAD scheme). We designed the cost function as a 2D gaussian, with the maximum in the most desired point (100% of true positives and 0 false positives) and variance equal to 15% true positives and 2.2 false positives.

In order to end the evolution of the population we choose the following termination criterion: we stop the evolution when the average of the fitness of the best six individuals has reached a plateau. The final result of the GA optimization is the best individual of the last iteration.

3.2   Implementation

The GA supervises the executions of an existent "basic program", which solves the domain problem and provide fitness evaluation. In the sequential program, the fitness evaluation step is computed independently for each individual of the population, by means of a *for* loop. We have one more loop that is innate in the "basic program" (the detection algorithm), which performs the detection scheme described in Sect. 2 over a whole database of mammographic images: the evaluation of one individual so requires the independent execution of the detection program *for each* image in the database. These loops are exploited in the parallel development of the algorithm.

Global parallelization is one of the most common way to realize a parallel GA. In global parallelization (see [7]), any individual can mate with any other because the operators and the evaluation of the individuals are explicitly parallelized, often by a "master processor" that sends individuals to other processors for the evaluation and applies genetic operator itself. The master program stores the entire population and performs an iterative decomposition: on each generation it sends a fraction of the population (one or more individuals) to each slave processors and waits results from the slaves to come back.

Slaves are self-scheduled, they ask the master for more work as their task ends. This algorithm behaves in a synchronous manner, since the master waits to receive the fitness values for *all* the individuals, before proceeding into the next generation. Once all individuals have been computed, the master performs replacement, crossover and mutation operations to create a new generation. In this way, the GA operations keep global and the existing design guidelines for simple (sequential) GAs are directly applicable.

In our GA, individuals are short strings of bytes and they are not time consuming, from a communication point of view. For this reason, the data parallelism of the detection algorithm can be exploited too. We recall that in the sequential algorithm an individual processes a whole images database. The easiest distributed implementation is to let an individual to sequentially process the entire set of images too. In this way, for each generation the program waits for the slowest slave to end its computation: in the worst case this time is the one necessary to process a database. Thus we create a new item, named *chunk*, which is constituted by an image identifier and by an individual. In this way, it is not necessary that one slave compute a whole database, but this task may be assigned to different slaves. Therefore, the master program sends a chunk, instead of one individual, every generation to each slave process, which sends back results and the individual identifier. In this way, the maximum idle time of the program shrinks from the time needed for computing a whole database to the time to compute one image. Here we use a modified version of the manager-workers paradigm, the *"working-manager"* model defined in [8], in which the manager uses its idle time to process data itself, by so increasing the overall performance.

4   Results

4.1   The Analysis of Performance

NOWs are cluster of workstations with quite slow communication links, that anyway can be suitable for a large number of applications. In addition, small Symmetric MultiProcessor (SMP) systems can be found within many of the modern computers which may lead to a powerful distributed computer if connected to each other.

The cluster we used is a heterogeneous computers network consisting of 6 workstations (10 computing nodes), connected to a LAN by a 100 Mbit

Ethernet. Workstations are listed below, according to their performance on the sequential algorithm:

W_1) 1 PentiumIII 450Mhz, 512MB RAM;
W_2) 1 SMP: 2 PentiumIII 450Mhz, 256MB RAM;
W_3) 1 SMP: 2 PentiumII 450Mhz, 512MB RAM;
$W_{4,5}$) 2 SMPs: 2 PentiumII 400Mhz, 512MB RAM;
W_6) 1 Mobile PentiumII 366Mhz, 128MB RAM;

We adopt static mapping and do not introduce any virtual parallelism degree, by assigning one task to each processor, for a total amount of 10 processes.

All the code is written in C, *Linux* is the operating system, *PVM* libraries supply the communication routines and *gcc* is the C compiler.

Each population is constituted by 30 individuals, and each generation required 27 individuals to be computed. Each generation takes about 1 h of elapsed time to be calculated on this cluster.

We get both *CPU time* and *wall clock time* measures necessary to obtain final results. This data parallel application has a very coarse grain structure and in addition small amount of data are transferred when communication takes place. For this reason, time due to communication between master and slaves is irrelevant and it has not been considered.

We then focus our attention on the "weighted" efficiency (Fig. 1) of each workstation:

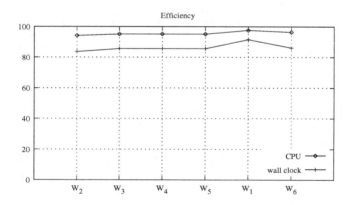

Fig. 1. The "Weighted" efficiency of each workstation

$$Eff_i = \frac{S_i}{I_i} \qquad (1)$$

where S_i and I_i are the relative speedup and the "ideal" (theoretical) speedup, respectively. S_i is defined as:

$$S_i = \frac{T_s(W_i)}{T_p} \qquad (2)$$

where $T_s(W_i)$ is the time it takes to the sequential algorithm on the workstation W_i, and T_p is the time of the parallel algorithm. Finally, I_i:

$$I_i = \frac{P_T}{P_i}. \tag{3}$$

Here, $P_i = T_s(W_1)/T_s(W_i)$ is the *power weight* of W_i compared with that of the fastest workstation W_1. P_T is the power weight of the *whole* cluster and its value is obtained by summing over all P_i.

We observe how the intrinsically parallel nature of this problem has been successfully exploited, by dividing original data in chunks. We obtain excellent results in term of efficiency, if we consider CPU time. Experiments showed a mean idle time value of about 10 sec.

Regarding with the time needed to read images, it would be quite significant, since images are only local to W_3. Each image (8 MB) takes about 1.5 sec to be read, and for each generation 27×40 images should be read. This sequential step causes a loss in terms of efficiency, even if we tried to keep in memory a few images, and it becomes more visible if we consider wall clock time.

4.2   Experimental Results

The main goal of the present study is to show that the performance of our CAD detection scheme improves due to the optimization based on the GA. To this

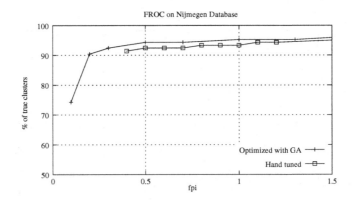

Fig. 2. FROC of the detection methods on the 40 images (Nijmegen) database

end we depict in Fig. 2 two FROC curves: one related to the "hand tuned" method and one for the optimized one. The optimized curve is derived from the best individual, by varying the value of the parameter p_1. Both the curves are obtained on the entire database of 40 images: here we do not divide the data into training and test groups, rather we use all the images for training and testing. We can see that the sensitivity of the optimized scheme is always greater than in the "hand tuned" case. The GA allows to outperform the previous method of some percents. Even if at first look it could seem a slight improvement, nevertheless this is extremely important because here it is necessary to minimize the losses of clusters of microcalcifications, maintaining at the same time a low rate of false alarms. Indeed, to avoid any losses of suspect cases is a vital point in issues regarding the detection of lesions for early breast cancer diagnosis; therefore, any little step towards a sensitivity of 100% is crucial. The best solution obtained by the GA is an individual with a fitness value of 0.827, which corresponds to 94.3% of true clusters with 0.47 false positives per image. To obtain these results we utilize uniform crossover p_{CO}=0.8 and p_{MUT}=0.1. The convergence of the GA evolution has required the computation of 1974 individuals, corresponding to 73 generations: it took roughly 3 days (76 h) on our NOW. Let's take a look at the peculiarity of the best individual: focusing our attention on its genes values, we can find out the differences between them and the parameters values of the "hand tuned" results. We can notice that these changes reduce the range of the values which identify the true signals. In particular, regarding the coarse method, the ranges of area, gray level, edge gradient and degree of linearity are narrower than those achieved in the "hand tuned" study (e.g. the minimum area of signal increases from 3 to 5 pixels, whereas the maximum area decreases from 30 to 22). In Fig. 3 it is possible to observe a similar effect about the fine method. There are plotted the curves, described by parameters p_1 and p_2, which separate true signals from false detections. Signals above the curve are kept,

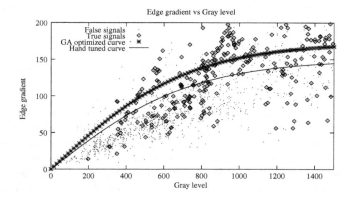

Fig. 3. Scatter plot of edge gradient versus gray level. The two curves, described by p_1 and p_2, separate false signals from true ones

while signals below it are eliminated. Also in this case the GA is more selective in maintaining true signals: if a signal has a given gray level, the GA keep it only if the edge gradient is higher, with respect to the "hand tuned" case. We can therefore summarize that the GA optimization tends to restrict the range of the features, which characterize the true signals. That allows to cut the number of false positive signals, without losing too many true ones.

Another issue investigated is how the parameters values fluctuations influence the performance of the CAD system. Starting from the best individual we vary the value of the first gene around its best value (the one discovered by the GA), keeping the other genes fixed. The variation of the parameter ranges around ±50% its best value. We repeat this process for all the genes, each time maintaining the other values fixed at their best solution. In this way, we can see which parameter more affect the fitness value.

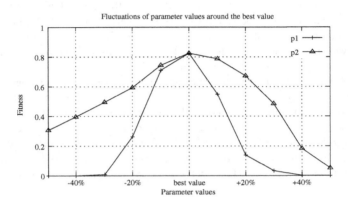

Fig. 4. Variation of fitness due to the change of parameters p_1 and p_2, with the other parameters fixed at their best values

In Fig. 4 we can see an example of how the fitness changes, due to the variation of p_1 and p_2 separately. We notice that p_1 is a very significant parameter, because the fitness goes to zero, with a modification of p_1 of only ±30% around its best value. It turns out that the k value for local thresholding (coarse method), p_1, p_2 and the maximum of the degree of linearity (fine method) are the most significant parameters. A little fluctuation of their values indeed implies a rapid fall of the fitness. On the other hand, there are some genes which do not affect the value of the fitness. They are, in particular: the maximum of the edge gradient, the minimum and the maximum of the average local gradient, ct_{31} and ct_{32} for the coarse method and p_4 for the fine method. A variation up to 50% of them around their best value does not imply any change in the fitness value. This fact has allowed us to perform an optimization without these parameters (keeping them fixed at their best value). With this reduced set of parameters (23 genes instead of 29) we obtain the same results of those cited above (fitness value of 0.827), by analyzing only 1623 individuals instead of 1974 (60 generations instead of 73).

5   Conclusion

We have been developing a Computer Aided Diagnosis system for the detection of clustered microcalcifications in digital mammograms. The performance of the system strongly depends on the settings of a set of several parameters. The goal of the present research is to optimize those parameters, in order to restrict the range of the features, which characterize the true signals, allowing the cut of the number of false alarms, without losing too many true signals detected. In an early version, this task was performed manually, and it took to us a very long time. In this paper, we presented an heuristic approach to this optimization problem by means of a distributed genetic algorithm. We have obtained better results in few generations, which correspond to a really short interval time. Finally, we have been able to discover the most significant parameters, being those that more affect the cost function.

Notes and Comments. Images were provided by courtesy of the National Expert and Training Center for Breast Cancer Screening and the Department of Radiology at the University of Nijmegen, the Netherlands.

This work is supported by the Italian National Institute for Nuclear Physics (CALMA project).

References

1. Bazzani, A., Bevilacqua, A., Bollini, D., Brancaccio, R., Campanini, R., Lanconelli, N., Romani, D.: System for automatic detection of clustered microcalcifications in digital mammograms. Int. J. Mod. Phys. C **11** (2000) 901–912
2. Dhawan, A. P., Chitre, Y., Kaiser-Bonasso, C., Moskowitz, M.: Analysis of mammographic microcalcifications using gray-level image structure features. IEEE Trans. Med. Imag. **15** (1996) 246–259
3. Dokur, Z., Olmez, T., Yazgan, E.: Classification of MR and CT images using genetic algorithms. Proceedings of the 20th Annual International Conference of the IEEE Engineering in Medicine and Biology Society **20** (1998)
4. Anastasio, M., Yoshida, H., Nagel, R., Nishikawa, R. M., Doi, K.: A Genetic Algorithm-Based Method for Optimizing the Performance of a Computer-Aided Diagnosis Scheme for Detection of Clustered Microcalcifications in Mammograms. Med. Phys. **25** (1998) 1559–1566
5. Yoshida, H., Anastasio, M., Nagel, R., Nishikawa, R. M., Doi, K.: Computer-Aided Diagnosis for Detection of Clustered Microcalcifications in Mammograms: Automated Optimization of Performance Based on Genetic Algorithm. Proceedings of IWCAD 1997, (Elsevier Science B.V., The Netherlands) (1997)
6. Ema, T., Doi, K., Nishikawa, R. M., Jiang, Y., Papaioannou, J.: Image feature analysis and Computer Aided Diagnosis in digital radiography: reduction of false-positive clustered microcalcifications using local edge-gradient analysis. Med. Phys. **22** (1995) 161–169
7. Cantú-Paz, E.: A survey of Parallel Genetic Algorithms. Report No. 97003, (Univ. of Illinois, Urbana, 1997) (1997)
8. Bevilacqua, A.: A dynamic load balancing method on a heterogeneous cluster of workstations. Informat. **23** (1999) 49–56

Dynamic Flies: Using Real-Time Parisian Evolution in Robotics

Amine M. Boumaza[1] and Jean Louchet[2]

[1] INRIA, projet FRACTALES, Domaine Voluceau BP 105,
78153 Le Chesnay cedex, France
amine.boumaza@inria.fr
[2] ENSTA, 32 Boulevard Victor, 75739 Paris cedex 15, France
louchet@ensta.fr

Abstract. The Fly algorithm is a Parisian evolution strategy devised for parameter space exploration in computer vision applications, which has been applied to stereovision. The resulting scene model is a set of 3-D points which concentrate upon the surfaces of obstacles. In this paper, we present how the evolutionary scene analysis can be continuously updated and integrated into a specific real-time mobile robot navigation system. Simulation-based experimental results are presented.

1 Introduction

The Fly algorithm [1] is an application of the individual (or "Parisian") approach [2] to artificial evolution: it treats each individual as a part of the solution of the problem rather than as a complete potential solution. Its basic principle is to evolve a population of 3-D points (the 'flies') using a fitness function such that the flies concentrate into the objects surfaces in the scene. The fitness function is a pixel-based calculation: to evaluate a fly, the fitness correlates the neighbourhoods of the fly's projections on each image. High correlation values correspond to similar neighbourhoods and therefore to flies located on the visible surface of an object in the scene.

 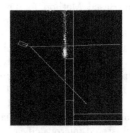

Synthetic images
(from the stereo pair)

Top view showing the walls
the flies and the robot

Classical "evolution strategy" - style genetic operators (random mutation, barycentric crossover) and a specific sharing operator are used to evolve the fly population.

E.J.W. Boers et al. (Eds.) EvoWorkshop 2001, LNCS 2037, pp. 288–297, 2001.

The aim of this paper is to present this method as the basis of a real-time vision system for a mobile robot. To this end we will describe more in detail the fly algorithm and show some of its interesting properties, concerning the processing of time-varying image sequences and real-time capabilities; then we describe the specific robot navigation algorithm which uses fly data as an input, and show some examples of navigation and obstacle avoidance using a robot simulator.

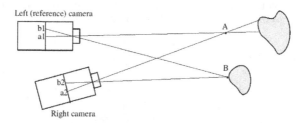

Fig. 1. Pixels b_1 and b_2, calculated projections of fly B, have highly correlated neighbourhoods, unlike pixels a_1 and a_2, projections of fly A, which receive their illumination from two different physical points on the surface of the object.

2   Fly-Based Robot Vision

2.1   Evolutionary Operators

Geometry and Fitness Function. A fly's chromosome is defined as the coordinates (x, y, z) of a 3-D point. As we are using two cameras, the coordinates of its projections are (x_L, y_L) in the image given by the left camera and (x_R, y_R) for the right camera. The positions of cameras are known, therefore x_L, y_L, x_R, y_R may be easily calculated from x, y, z using classical projective geometry [3] [4] formulas.

The fitness function evaluates the degree of similarity of the pixel neighbourhoods of the projections of the fly onto each image. It gives high fitness values to the flies lying on the surface of an object :

$$fitness(fly) = \frac{G}{\sum_{colours} \sum_{(i,j) \in N} (L\,(x_L + i,\, y_L + j) - R\,(x_R + i,\, y_R + j))^2}$$

where:

- (x_L, y_L) and (x_L, y_L) are the coordinates of the left and right projections of the current individual (see Fig. 1),
- $L\,(x_L + i,\, y_L + j)$ is the grey value of the left image at pixel $(x_L + i,\, y_L + j)$, similarly with R for the right image.
- N is a small neighbourhood of the origin.

On colour images, square differences are calculated on each colour channel.

The normalizing factor G, which is a local contrast measurement based on image gradient, reduces the fitness of flies over unsignificant regions. Thus, highest fitness values are obtained for flies whose projections have similar and *significant* (*non-uniform*) pixels.

Artificial Evolution Operators. An individual's chromosome is the triple (x, y, z) which contains the fly's coordinates in the coordinate system.

The population is initialised randomly inside the intersection of the cameras fields of view, from a given distance (clipping line) to infinity (fig. 2).

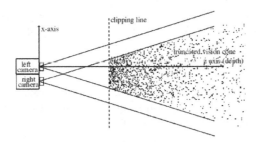

Fig. 2. The fly population is initialised inside the intersection of the cameras 3-D fields of view.

The statistical distribution is chosen in order to obtain a uniform distribution of their projections in the left image. In addition, their depth is distributed from an arbitrary clipping line to infinity using a uniform distribution of the values of z^{-1}: the flies' probability density is thus lower at high distances.

Selection : is elitist and deterministic. It ranks the flies according to their fitness values and retains the best individuals (around 50%).

2-D sharing : reduces the fitness values of flies located in crowded areas to prevent them from getting concentrated into a small number of maxima. It reduces each fly's fitness by $K \times N$, where K is a "sharing coefficient" and N the number of flies which project into the left image within a distance R ("sharing radius") to the given fly. The sharing radius is given by [1]:

$$R \approx \frac{1}{2} \left(\sqrt{\frac{N_{pixels}}{N_{flies}}} - 1 \right)$$

Mutation : allows exploration of the search space. It uses an approximation of a Gaussian random noise added to the parent fly's chromosome parameters (x, y, z). Standard deviations σ_x, σ_y, σ_z are the same order of magnitude as R.

The barycentric crossover operator : builds an offspring randomly located on the line segment between its parents: the offspring of two flies F_1 (x_1, y_1, z_1) and F_2 (x_2, y_2, z_2) is the fly F_3 (x_3, y_3, z_3) defined by :

$$\overrightarrow{OF} = \lambda \overrightarrow{OF_1} + (1 - \lambda) \overrightarrow{OF_2}$$

The weight is chosen using a uniform[1] random law in $[0, 1]$. The crossover may be interpreted as an exploration of potential straight or planar primitives in the scene.

2.2 Integrating Velocities: An Asynchronous Algorithm for Dynamic Flies

The "standard" fly algorithm presented above is dedicated to static images. Experiments showed that if the changes in the scene are slow enough, using the set of flies obtained with the last stereo pair to initialise the current population of flies speeds up convergence significantly. It is even beneficial to consider that the flies evolve continuously [5] and to use, for each evaluation, the most recent image data available from the cameras - provided that pixels from the right and left cameras are synchronous. This is an interesting feature, as modern CMOS technology imagers allow asynchronous random access to pixels: this enables to use constantly updated image data and to get a potentially faster reaction of the system to external events.

Conversely, unlike current stereo analysis algorithms, the fly algorithm is able to continuously update its results, which allows the robots navigation system to use constantly updated scene descriptions.

In order to better exploit image data coming from moving cameras, we enriched the chromosome representation of the flies by adding velocities: the chromosome of a "dynamic fly" is now $(x, y, z, \dot{x}, \dot{y}, \dot{z})$ where $\dot{x}, \dot{y}, \dot{z}$ are the velocity components of the fly.

Each time a new image frame is used, the coordinates of the flies are updated using their velocities. The fitness function does not use the velocities, but flies with correct velocities will survive better through generations. The mutation and crossover operators are extended to velocities, using the same weights. An additional genetic operator is added, *immigration*, which introduces totally new randomly generated flies. This operator ensures exploration of space and faster detection of new obstacles in the field of view.

Once the algorithm has converged on the first images, convergence on a new image is obtained in about 10 generations, depending on how much the scene has moved since the last frame.

---

[1] The goal of the crossover operator is to fill in surfaces whose contours are detected, rather than to extend them. It is therefore not desirable to use coefficients outside $[1, 0]$, otherwise it would allow the center of gravity to lie outside of the objects boundary.

3   Application to Obstacle Avoidance

3.1   Using Artificial Potential Fields

The obstacle avoidance methods which can be found in literature use results of conventional stereo analysis algorithms as an input (i.e. in general polyhedral models). Due to the novelty of the "fly" approach, there is as far as we know, no existing navigation and obstacle avoidance method able to use the fly representation. We chose a method based on artificial potential fields, because of their simplicity and also to respect real time constraints.

The potential field navigation methods proposed by O. Khatib [6] is among the most widely used in robot navigation and local planning. It does not require a lot of computation time, which allows its implementation in real time.

The principle of this method is inspired form Physics. The environment is modelled as a potential field and the robot as a charged particle under the influence of this field. The potential field includes an attractive and a repulsive potential. The attractive potential constantly attracts the robot to the goal. The role of the repulsive one is to repel the robot from the obstacles present in the environment.

In practice the robot follows the force induced from the potential field, easily translated into a steering command i.e. a heading and a velocity proportional to the force. The force $\vec{F}$ is the sum of an attractive and a repulsive force.

$$\vec{F} = \vec{F_a} + \vec{F_r}$$

The attractive force $\vec{F_a}$ has a constant amplitude except in a small neighbourhood around the goal. The repulsive force $\vec{F_r}$ must take into account all the flies which represent obstacles in the environment. However, involving each fly individually in the potential would induce too many calculations. Thus we use a discrete representation of the environment.

3.2   Robot's Internal Representation

As shown in figure 3, the robot uses an internal representation of the world which consists of a 2-D grid of cells centred on it. Each cell concentrates the information coming from the flies it contains and generates its own contribution to $\vec{F_r}$. This is similar to the "evidence grids[2]" proposed by [9].

Once the space around the robot is discretized, we associate to each cell a value ("confidence weight") computed according to the fitness of the flies that lie within the cell.

$$val(c_i) = \sum_{m \in c_i} fitness(m)$$

where c_i is a given cell of the grid. The value of $val(c_i)$ is then normalized between [0,1] to keep the cell values bounded. This coefficient indicates if the fly is well positioned on a surface of an object or not. Therefore we define the resulting repulsive force as:

---

[2] Similar approaches were used in [7], [8] to model sensory data issued from sonars.

Fig. 3. The discretized environment of the robot. The grey cone shows the robot cameras' field of view, the square grid its internal memory.

$$\vec{F_r} = \sum_{c_i \in grid} \vec{F_{c_i}}$$

where the amplitude of $\vec{F_{c_i}}$ is proportional to $val(c_i)$.

Appendix A shows the potential functions and forces used.

3.3 Escaping Local Minima

Local minima are a considerable weakness of potential-based navigation methods. These are configurations where the magnitude of the total force is zero (figure 5a, c). The methods[3] we use to solve this problem rely on altering the attractive force to get the robot out of the minimum.

Random walk. This is a method often used in gradient descent algorithms, not requiring a lot of resources. We implemented it by generating a virtual goal (attractive potential) in a random place to deviate the trajectory of the robot from the local minimum. The virtual goal is generated so that there are no or few obstacles between it and the robot or the global goal. We translated this into a cost function defined in appendix B. Among several randomly generated virtual goals, we keep the one with minimum cost.

Wall following. In the case the robot is stuck behind an obstacle, an alternative method is to deviate its trajectory by changing the direction of the attractive force without using a virtual goal, but introducing a new attractive force with a direction at an obtuse[4] angle with the repulsive force so that the new total force is roughly parallel to the obstacle edge. This way we expect the robot to follow the edges of the obstacle until it finds an obstacle-free path to the goal. This method is used in the example shown on figure 5.

---

[3] [10] [11] propose heuristic methods to solve situation of local minima, and [12] define generalized potential functions that avoid local minimum configurations.

[4] [13] [14] propose similar approaches with angles of $145°$ and $120°$

3.4   Robot Simulator Architecture

The simulator is divided into four parts (figure 4). The sensing system is modeled as an image synthesis algorithm which creates stereo image pairs as seen by the robot. The vision system uses this input to build a fly-based representation of space. The navigation system uses this fly system to create commands transmitted to the robot simulator.

4   Experimental Results

On figure 5, images (a) and (c) show the robot getting trapped into local minima. Images (b) and (d) show how these situations are resolved using wall following. Images (c) and (f) show other examples of complex trajectories generated by wall following (e) and random walk (f). On the latter example, the robot has been able to find the door entrances to reach its destination.

5   Conclusion

As a consequence of the fact that the fly algorithm delivers a scene description which consists of a set of 3-D points, it is not possible to compare directly its performance with conventional stereo analysis methods [15]. In this work, we developed a first platform to develop and test the fly algorithm in a mobile robotics context. This is an important step into this real-world application of the Parisian approach.

Let us now examine the speed point of view. Standard synchronous vision systems, based on image segmentation and stereo matching with a typical $25\,Hz$ ($40\,ms$) frame rate, generally use all the interval between two frames to make their calculations and (in the best cases) deliver their result to the navigation system with an $80\,ms$ delay.

In our approach, the combination of random access cameras[5] and Parisian evolution allows to get a proper convergence in 10 generations: this takes less than $40\,ms$ with a 5000 individual population, using non-optimised C code with a conventional $300\,MHz$ industry-standard processor. Thus this combination should be at least twice as fast as conventional methods. Using a more powerful processor would allow to increase the number of generations and get proportionally faster convergence and shorter reactions to new events in the scene.

Future work will include experiments on real wheeled robots, studying self-adaptation of genetic parameters in the flies chromosomes and fly speciation, so that flies can fulfil different specialised roles in the analysis process.

A   The Potential Function

The potential function used in this paper is a mix of conic and parabolic wells defined as follows:

---

[5] Stereo algorithms based on conventional image processing need to use complete bitmap images as inputs, and therefore cannot easily take advantage of random access camera technology.

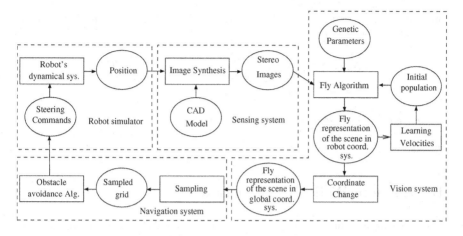

Fig. 4. The architecture of the simulator.

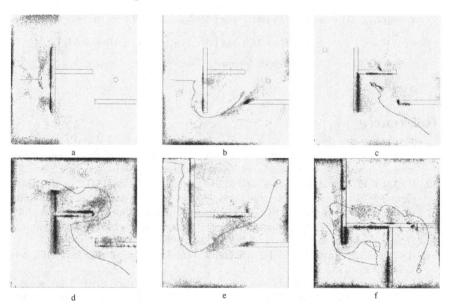

Fig. 5. Examples of trajectories.

$$U_a(q) = \begin{cases} k_a\, dist_{goal}(q)^2 & if\ dist_{goal}(q) \le d_0 \\ d_0\, k_a\, dist_{goal}(q) & if\ not \end{cases}$$
$$U_r(q) = k_r\, \frac{1}{dist_{obs}(q)}$$

where U_a and U_r are respectively the attractive and repulsive potentials, k_r, k_a are positive constant, $dist_{goal}$, $dist_{obs}$ respectively the distances from the goal and from the obstacles, and d_0 a fixed distance from the goal. These potentials generate attractive and repulsive forces so that:

$$\vec{F_a}(q) = -\vec{\nabla} U_a(q) = \begin{cases} -k_a(q - q_{goal}) \, \vec{u_a} \; if \; dist_{goal}(q) \le d_0 \\ -d_0 \, k_a \, \vec{u_a} \qquad\quad if \; not \end{cases}$$

$$\vec{F_r}(q) = -\vec{\nabla} U_r(q) = k_r \, \frac{1}{dist_{obs}(q)^2} \, \vec{u_r}$$

where q_{goal} is the goal position, $\vec{u_a}$ a unit vector directed toward q_{goal} from q. since $\vec{F_r} = \sum_{c_i \in grid} \vec{F_{c_i}}$ then $\vec{F_{c_i}} = val(c_i) \, \frac{1}{dist_{c_i}(q)^2} \, \vec{u_{c_i}}$. where $\vec{u_{c_i}}$ is a unit vector directed from the centre of the cell c_i to q.

B   Random Walk Cost Function

The following cost function is used in the "random walk" process to choose which virtual goal will be used by the robot to get out of a local minimum.

$$Cost(G_{virtual}) = w_1 \left(\# \, flies \; M \mid \cos \left(\overset{\frown}{\overrightarrow{MG}_{virtual}, \, \overrightarrow{MR}} \right) < threshold_1 \right) +$$
$$w_2 \left(\# \, flies \; M \mid \cos \left(\overset{\frown}{\overrightarrow{MG}_{real}, \, \overrightarrow{MR}} \right) < threshold_2 \right) \quad +$$
$$w_3 \, dist(G_{real}, \, G_{virtual})$$

References

1. J. Louchet. Using an individual evolution strategy for stereovision. *Genetic Programming and Evolvable Machines. Kluwer*, to appear, 2001.
2. P. Collet, E. Lutton, F. Raynal, and M. Schoenauer. Individual GP: an alternative viewpoint for the resolution of complex problems. In *Genetic and Evolutionary Computation Conference GECCO99*. Morgan Kaufmann,San Francisco, CA, 1999.
3. R. M. Haralick. Using perspective transformations in scene analysis. *Computer Graphics and Image Processing*, (13):191–221, 1980.
4. R.C. Jain, R. Kasturi, and B.G. Schunck. *Machine Vision*. McGraw-Hill, New York, 1994.
5. R. Salomon and P. Eggenberger. Adaptation on the evolutionary time scale: a working hypothesis and basic experiments. In *Springer Lecture Notes on Computer Science*, number 1363, pages 251–262. Springer-Verlag, Berlin, 1997.
6. O. Khatib. Real time obstacle avoidance for manipulators and mobile robots. *The International Journal of Robotics Research*, 5(1):90–99, Spring 1986.
7. I. Ulrich and J. Borenstein. Vfh+: Reliable obstacle avoidance for fast mobile robots. In *Procedings of the IEEE Conference On Robotics and Automation, ICRA'98*, pages 1572–1577, Leuven, Belgium, May 1998.
8. J. Van Dam and B. Krose F. Groen. Transforming occupancy grids under robot motion. In S. Gielen and B. Kappe, editors, *Artificial neural networks*, page 318. Springer-Verlag, 1993.
9. M. C. Martin and H. P. Moravec. Robots evidence grid. Technical report, The Robotics Institute, Carnegie Mellon University, March 1996.

10. Y.Koren and J. Borenstein. Potential field methodes and their inherent limitations for mobile robot navigation. In *Procedings of the IEEE Conference On Robotics and Automation, ICRA'91*, pages 1398–1404, Sacramento, California, April 7-12 1991.

11. John S. Zelek. Complete real-time path planning during sensor-based discovery. In *IEEE/RSJ International Conference on Intelligent Robots and systems*, 1998.

12. B. H. Krogh. A generalized potential field approach to obstacle avoidance control. *Robotics Research*, August 1984.

13. J. Borenstein and Y.Koren. Real-time obstacle avoidance for fast mobile robots. In *IEEE Transaction on System, Man, and Cybernetics*, volume 19, pages 1179–1187, Sept./Oct. 1989.

14. R. A. Brooks and J. H. Connel. Asynchronous distributed control system for a mobile robot. In *SPIE86Cambridge Symposium on Optical and Optoelectronic Engineering, Cambridge, MA*, pages 77–84, October 1986.

15. CUMULI project: Computational understanding of multiple images, http://www.inrialpes.fr/cumuli.

16. I. Rechenberg. Evolution strategy. In J.M. Zurada, R.J. MarksII, and C.J. Robinson, editors, *Computational Intelligence imitating life*, pages 147–159. IEEE Press, Piscataway, NJ, 1994.

17. P.K. Ser, S. Clifford, T. Choy, and W.C. Siu. Genetic algorithm for the extraction of nonanalytic objects from multiple dimensional parameter space. *Computer Vision and Image Understanding, vol. 73 no. 1, Academic Press: Orlando, FL*, pages 1–13, 1999.

18. E. Lutton and P. Martinez. A genetic algorithm for the detection of 2D geometric primitives in images. In *The Proceedings of the International Conference on Pattern Recognition, ICPR'94*, pages 526–528, Los Alamitos, CA, October 9-13 1994. IEEE Computer Society.

ARPIA: A High-Level Evolutionary
Test Signal Generator

Fulvio Corno, Gianluca Cumani, Matteo Sonza Reorda, and Giovanni Squillero

Politecnico di Torino
Dipartimento di Automatica e Informatica
Corso Duca degli Abruzzi 24 I-10129, Torino, Italy
http://www.cad.polito.it/

Abstract. The integrated circuits design flow is rapidly moving towards higher description levels. However, test-related activities are lacking behind this trend, mainly since effective fault models and test signals generators are still missing. This paper proposes ARPIA, a new simulation-based evolutionary test generator. ARPIA adopts an innovative high-level fault model that enables efficient fault simulation and guarantees good correlation with gate-level results. The approach exploits an evolutionary algorithm to drive the search of effective patterns within the gigantic space of all possible signal sequences. ARPIA operates on register-transfer level VHDL descriptions and generates effective test patterns. Experimental results show that the achieved results are comparable or better than those obtained by high-level similar approaches or even by gate-level ones.

1 Background

In recent years the application specific integrated circuit (ASIC) design flow experienced radical changes. Deep sub-micron integrated circuit (IC) manufacturing technology is enabling designers to put millions of transistors on a single integrated circuit. Following Moore's law, design complexity is doubling every 12-18 months. In addition, there is an ever-increasing demand on reducing time to market. With complexity skyrocketing and such a competitive pressure, designing at high levels of abstraction has become more of a necessity than an option.

At the present time, exploiting design-partitioning techniques, register-transfer level (RT-level) automatic logic synthesis tools can be successfully adopted in many ASIC design flows. However, not all activities have already migrated from gate- to RT-level and are not yet mature enough to.

High-level design for testability, testable synthesis and test pattern generation are increasing their industrial relevance [1]. During the development of ASIC, designers would like to be able to foresee its testability before starting the logic synthesis process. Furthermore, RT-level automatic test signals generators are expected to exploit higher-level compact information about design structure and behavior, and to be able to produce more effective sequences within reduced CPU time. The test signals may possibly be completed after synthesis by ad-hoc gate-level tools.

E.J.W. Boers et al. (Eds.) EvoWorkshop 2001, LNCS 2037, pp. 298-306, 2001.

However, despite the big effort of electronic design automation (EDA) industries, tackling test issues at high levels is still an unsolved problem. The lack of a common fault model is probably the hardest theoretical barrier. Over the years, the computer aided design (CAD) community agreed on some well-defined gate-level fault models. The most popular is the permanent single stuck-at fault, and all commercial products are able to use it. On the contrary, up to now it was unable to agree on any high-level fault model.

The most important technical barrier, on the other side, is the lack of efficient algorithms to generate effective test signals. As a matter of fact, at the present time even *simulating* RT-level test signals is a challenging task. Fault simulation algorithms for RT-level designs are known since more than a decade [2], but commercial tools usually don't include these capabilities. Furthermore, classical algorithms are difficult to integrate in simulators, mainly due to the complexity and to the several peculiarities of hardware description languages. Some prototypical fault simulators were proposed [3] [4], but until some fault model becomes widely accepted EDA industries have no good reason to invest and this situation is not likely to change.

Even so, researchers and pioneering design groups already need test signals on their RT-level designs. Many generators were proposed. Nevertheless, since any attempt of backward justification must take into account all structural, behavioral and timing specifications [5], traditional algorithms are almost unusable. Researchers sometimes achieved good results, but they were generally limited to specific classes of circuits. However, interesting successful results have been reported using evolutionary algorithms [6]. These approaches exploit natural evolution principles to drive the search of effective patterns within the gigantic space of all possible signal sequences. Evolutionary heuristics begin to appear a reasonable alternative to traditional techniques.

This paper presents ARPIA, a high-level evolutionary automatic test signals generator. Experimental results gathered using the prototypical implementation is remarkable. The effectiveness of the generated test signals is at least comparable with (and in several cases higher than) that of previously proposed approaches. Additionally, thanks to the evolutionary algorithm and to a fault dropping mechanism, computational requirements of the new system are lower.

Next Session details ARPIA. An analysis of the proposed approach is illustrated in Section 4. Section 5 draws some conclusions.

2 ARPIA

ARPIA is a simulation-based evolutionary test signals generator. Being an evolutionary algorithm, it evolves a population seeking fitter individuals. But, since individuals are test sequences for a digital circuit, the fitness measures the sequence ability to detect faults in the design. And it is computed by simulation. Given a fault model, the *fault coverage* is defined as the percentage of faults that the test sequence is able to detect. Thus, the goal of ARPIA can be rephrased as "generate a sequence of signals that attains maximum fault coverage."

ARPIA shares the same philosophy with [6]. They are both simulation-based approaches, and individuals are evaluated resorting to an RT-level fault simulator.

However, the two methodologies exploit different fault models, different fault simulation techniques and different evolutionary algorithms. Next Sections detail these three key points.

2.1   Fault Model

Many fault models have been proposed in the past, mainly by borrowing from software-testing metrics [7]. While software derived metrics are well known and quite standardized, and they are already implemented in some commercial tools, their usefulness for hardware testing is quite low. In particular, software metrics neglect observability issues and the effects of logic synthesis, and thus are not an accurate indicator of circuit testability.

One successful proposal of a hardware-related fault model is *Observability-Enhanced Statement Coverage* [8] [4]: it introduces the concept of *tag* as the possibility that an incorrect value is computed at a given location, thus approximating the effects of fault propagation. Since this fault model does not assume any specific fault effect, its generality prevents explicit fault simulation.

An extension of observability-enhanced statement coverage was first proposed in [9] and then refined in [10], where explicit *RT-level assignment single-bit* stuck-at's are used instead of generic tags. An RT-level assignment single-bit stuck-at fault is defined as a single-bit stuck-at in the effect of an RT-level assignment operation: when a fault is present, the affected object (signal or variable target of an assignment statement) loads the correct value, except for one bit that is forced to 0 or 1.

Fig. 1 shows the example of a stuck-at fault. The fault affects the third leftmost bit of the assignment operation, and modifies the result of the expression, after it has been computed and before it is assigned to the target signal. The faulty signal is updated as usual, according to propagation rules, but with a faulty value. Other assignments of the same signal are assumed to be fault-free, since stuck-at faults on the same signal but on different statements are considered different. More details about the fault model can be found in [9].

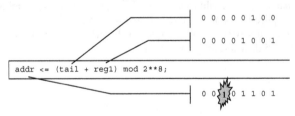

Fig. 1. Stuck-At Fault Example

The current fault model exploits several heuristics to filter out unrepresentative RT-level stuck-at. This elimination enables the RT-level assignment single-bit stuck-at fault coverage to be highly correlated with gate-level fault coverage: the correlation coefficient R, computed over a set of 10 different circuits by simulating the test sequences generated in three different ways, reaches R=0.77 [10].

2.2   Fault Simulation Technique

Fault simulation is made possible by creating routines that *change* the target signal/variable bit value during simulation, using the simulator scripting language (Tcl), when a given target assignment instruction is executed. The fault injection procedures, presented in [9], must face various issues derived from the fault model, from VHDL semantics and from the simulator itself.

This fault simulation environment allows us to compute fault coverage figures at the RT-level with a minimal CPU time overhead, since the VHDL model of faulty circuits is simulated at the same speed as the fault-free model. The main time penalty is at fault activation time instants, where some breakpoints are set and TCL commands to modify values are executed.

2.3   Algorithm

A fault can be marked as *tested* only when it is both *excited* and *observed*. Given a fault, the target of the whole process is first to force the corresponding bit to a value that make the fault visible (excitation), then to propagate the fault effects to some primary output (observation).

```
current_sequence = random_sequence(L);
steady_state_factor = 0;
do
{
  new_sequence = mutate(current_sequence);
  if(excited_faults(new_sequence) > excited_faults
                                  (current_sequence)) {
    current_sequence = new_sequence;
    steady_state_factor = 0;
  } else {
    increase(steady_state_factor);
  }
} while(excited_faults(current_sequence) == 0 &&
    steady_state_factor < steady_state_limit)
```

Fig. 2. Phase One Pseudo Code

The two problems are tackled separately, with two different strategies. Indeed, test generation is performed in three phases. The first is aimed at exciting faults, the second tackle observation, while the third dynamically optimize the fault list.

The goal of the first phase is to produce a set of signals able to excite an untested fault. The first phase implements a simple first-improvement hill climber (**Fig. 2**). The local search procedure starts with a random sequence of given length L. In each step, a new sequence is generated by randomly mutating the current one. If the new sequence excites a larger number of untested faults, it becomes the current one. Otherwise it is discarded. The process ends whenever the current sequence is able to excite at least k fault, or after a predefined amount of useless steps. It is worth noting that the number of excited faults is computed over the untested faults only.

Three mutations are currently implemented by ARPIA: *add*, *delete* and *change*. The first two respectively add and delete a vector of input signals from the test sequence. The last one randomly changes a vector of input signals in the sequence.

When a test sequence able to excite a sufficient number of faults is found, it is transferred to the second phase together with the excited faults. The goal of the second phase is to observe each single fault in the excited set. This stage exploits an evolutionary algorithm similar to an evolution strategy [11].

First, a target fault f_t is selected from the excited set. Then, a population of P sequences is created by mutating the original sequence and evolved using a $(P+P)$ strategy. In every evolution step, P new individuals are generated by mutating the P original ones (each sequence generates exactly one new sequence). The P fitter individuals are selected for survival among the $2P$. The same three mutation operators of the first phase are adopted.

In the second phase, given a target fault f_t, the fitness measures how far a sequence is able to propagate f_t effects. More precisely, it is the maximum number of differences caused by the fault during the application of a single vector of signals of the test sequence (1).

$$evaluation\_two(S, f_t) = \underset{v\ S}{MAX}\ \underset{objects}{different\_bits(object)} \tag{1}$$

The evolution is halted whenever f_t is detected or after a certain amount of generations. The pseudo code is shown in **Fig. 3**.

The second phase is iterated until all faults in the excited set have been tested or aborted.

The evolution-strategy approach was chosen because of the complexity of the encoding. Individuals are sequence of vectors. Each vector of signals is simulated in a clock cycle. Circuits contain memory elements, thus the behavior in a clock cycle depends both on current input signals and previous ones. The effect of a traditional recombination operator, like the uniform crossover, can be very similar to a complete random mutation at phenotypic level. We shun any risk and exploit an algorithm that "omits recombination since its philosophy relies on species as evolving entities" [12].

After each successful second phase, an optimization mechanism called *fault dropping* is activated. All still untested faults are simulated with the new sequence, seeking if any additional fault is detected by it. This is more of a possibility than an expectation, since the starting sequence found in the first phase is required to be able to excite more than one fault. The fault dropping mechanism greatly enhances overall algorithm performance.

3   Experimental Analysis

In order to practically evaluate the effectiveness of the proposed approach, we implemented a prototype. The generator is composed of about 1,500 lines in ANSI C and interacts with V-System 5.3 VHDL simulator by Model Technology. Special techniques are adopted to speed-up fault simulation [9]. During experiments we adopt the following parameter values:

```
for t = 1 to P {
   population[t] = mutate(starting_sequence);
}
generations = 0;
success = 0;

do
{
   generations = generations + 1;
   for t = 1 to P {
      population[P + t] = mutate(population[t]);
      if(tested(f_t, pupulation[t])
         success = 1;
   }
   sort(population);
} while(not success && generations < generations_limit);
```

Fig. 3. Phase Two Pseudo Code

- first-phase initial sequence length of 50 clock cycles ($L = 50$)
- first phase sequence required to excite 5 different faults ($k = 5$)
- second phase population of 10 individuals ($P = 10$).

Table 1. ARPIA Result

Circuit	CPU time [s]	RT Faults			Gate Faults		
		Tot	Det	FC%	Tot	Det	FC%
b01	78.80	81	81	100.00%	258	258	100.00%
b02	39.19	43	39	90.70%	150	149	99.33%
b03	1,089.96	213	145	68.08%	822	615	74.82%
b04	1,627.86	424	353	83.25%	3,356	3,035	90.44%
b05	1,932.27	778	244	31.36%	5,552	1,856	33.43%
b06	200.31	110	82	74.55%	5,552	5,387	97.02%
b07	9,297.26	289	146	50.52%	2,404	1,401	58.28%
b08	2,832.38	154	118	76.62%	918	839	91.39%
b09	4,970.53	240	196	81.67%	900	768	85.33%
b10	778.35	172	127	73.84%	1,054	961	91.18%
b11	34,837.11	381	263	69.03%	2,868	2,614	91.14%
b12	7,890.20	870	115	13.22%	5,280	1,105	20.92%
b13	2,801.85	284	224	78.87%	1,818	1,501	82.56%
b14	473,741.90	10,493	9,114	86.86%	28,990	23,708	81.78%
b15	590,611.31	4,900	2,026	41.35%	55,568	18,060	32.50%

Table 1 reports the experiments performed on the ITC99 RT-level benchmarks. These benchmarks are representative of typical circuits, or circuit parts, that can be automatically synthesized as a whole with current tools and are described in [13]. Experiments have been run on a Sun Enterprise 250 running at 400 MHz and equipped with 2 Gbytes of RAM.

The first column of Table 1 reports the name of the benchmark, while the CPU time required to generate the test signal sequence is shown in the second column. RT-level fault figures are reported in the next column block in terms of: total number of faults [Tot], number of detected faults [Det] and percent fault coverage [FC%]. The next column block shows gate-level figures: total number of gate-level faults [Tot], number of detected [Det] and respective fault coverage [FC%].

Results show that ARPIA is able to generate test sequences that are highly effective both at RT-level and at gate-level, within an acceptable CPU time. However, to better analyze the tool performance, we need to compare it with different approach (**Table 2**).

Table 2. Comparison with ARTIST and ARPIA without Evolutionary Algorithm

Circuit	ARPIA (no ES)		ARTIST		ARPIA	
	RT	Gate	RT	Gate	RT	Gate
b01	100.00%	99.61%	100.00%	100.00%	100.00%	100.00%
b02	90.70%	99.33%	90.70%	99.33%	90.70%	99.33%
b03	56.81%	69.10%	68.08%	74.82%	68.08%	74.82%
b04	69.34%	69.19%	83.79%	91.03%	83.25%	90.44%
b05	11.31%	5.42%	31.36%	33.50%	31.36%	33.43%
b06	70.00%	93.38%	74.80%	97.35%	74.55%	97.02%
b07	47.75%	56.49%	50.52%	58.28%	50.52%	58.28%
b08	52.60%	28.00%	60.10%	71.68%	76.62%	91.39%
b09	62.50%	48.89%	77.84%	81.33%	81.67%	85.33%
b10	46.51%	65.84%	73.69%	90.99%	73.84%	91.18%
b11	43.57%	58.79%	68.64%	90.62%	69.03%	91.14%
b12	2.76%	4.36%	29.06%	45.99%	13.22%	20.92%
b13	31.69%	31.19%	n/a	n/a	78.87%	82.56%
b14	11.57%	37.91%	n/a	n/a	86.86%	81.78%
b15	14.69%	12.75%	n/a	n/a	41.35%	32.50%

The first column block of **Table 2** reports data for the first prototype of ARPIA, where a simple hill-climber was exploited instead of the evolution strategy. The gap between the two RT-level figures shows the fundamental role played by the evolutionary algorithm. The gap between gate-level fault coverage statistics is a mere consequence.

In the second column group of **Table 2** we reported results attained by ARTIST [6], a highly-optimized tool exploiting a genetic algorithm. The difference between these two tools can be explained resorting to both the fault model and the new evolutionary mechanism. A deeper comparison between the results of ARPIA and of ARTIST (complete data can not be reported here for lack of space) shows that the former is characterized by a higher efficiency (i.e., it requires a lower CPU time), thanks to fault dropping and a higher compactness of the generated sequences. Indeed, ARTIST was not able to tackle some of the benchmarks due its lower efficiency (marked with "n/a" in the table).

The effectiveness of the evolutionary algorithm can also be seen in Table 3, where the number of excited faults is shown in column [Exc] together with the number of detected faults [Det]. The effectiveness of the evolutionary algorithm can be defined as its ability to observe (i.e., to detect) excited faults, not considering faults that are certainly unobservable due to incorrect design ([Err] column).

It should be noted that phase two effectiveness is quite high also for b12, a problematic circuit where ARPIA only manages to get 13.22 %   RT-level fault coverage. Thus, primarily the first phase can be hold responsible for the low performance.

Table 3. Phase Two Effectiveness

Circuit	RT-Level Faults				Phase2
	Tot	Exc	Det	Err	Efficacy
b01	81	81	81	0	100.00%
b02	43	43	39	4	100.00%
b03	213	148	145	3	100.00%
b04	424	356	353	3	100.00%
b05	778	340	244	19	76.01%
b06	110	87	82	5	100.00%
b07	289	240	146	20	66.36%
b08	154	118	118	0	100.00%
b09	240	196	196	0	100.00%
b10	172	139	127	12	100.00%
b11	381	289	263	26	100.00%
b12	870	130	115	1	89.15%
b13	284	266	224	16	89.60%
b14	10,493	10,165	9,114	0	89.66%
b15	4,900	2,244	2,026	24	91.26%

3   Conclusions and Future Works

Due to the wide adoption of logic synthesis tools, high-level techniques are increasingly necessary in order to shift test-related activities towards the description level adopted by designers. A crucial point for developing effective high-level test signal generator lies in the identification of a suitable fault model. Another crucial point is the availability of a suitable algorithm for test generation.

In this paper we exploit a recent fault model to propose a methodology for generating high quality test signals. The approach is based on an evolution strategy. Experimental data show the potential of the proposed approach with regard to similar techniques, the effectiveness of the fault model, and the convenience of the evolutionary algorithm.

We are currently improving ARPIA in two directions. First, we are trying to replace the hill-climber in the first phase with sharper algorithms, solving the issues stemming from the lack of information concerning the system behavior before fault excitation. We are also considering the adoption of an evolutionary core similar to the one exploited in [14]. Secondly, we are tuning the evolution strategy adopted in the second phase, experimenting more complex selection schemes. We are pondering to change the algorithm, adding back recombination in some constrained form.

4   References

[1]   "High Time for High-Level Test Generation," Panel at the *IEEE International Test Conference*, 1999, pp. 1112-1119

[2]   M. Abramovici, M. A. Breuer, A. D. Friedman, *Digital systems testing and testable design*, Computer Science Press, 1990

[3]   A. Fin, F. Fummi, "A VHDL Error Simulator for Functional Test Generation," *IEEE European Design, Automation and Test Conference*, 2000, pp. 390-395

[4]   F. Fallah, S. Devadas, K. Keutzer, "OCCOM: Efficient Computation of Observability-Based Code Coverage Metrics for Functional Verification," *DAC98: 34th Design Automation Conference*, 1998

[5]   F. Ferrandi, F. Fummi, L. Gerli, D. Sciuto, "Symbolic Functional Vector Generation for VHDL Specifications," *DAC99: 35th Design Automation Conference*, 1999, pp. 442-446

[6]   F. Corno, M. Sonza Reorda, G. Squillero, "High-Level Observability for Effective High-Level ATPG," *VTS2000: 18th IEEE VLSI Test Symposium*, May 2000, pp. 411-416

[7]   B. Beizer, *Software Testing Techniques* (2nd ed.), Van Nostrand Rheinold, 1990

[8]   F. Fallah, P. Ashar, S. Devadas, "Simulation Vector Generation from HDL Descriptions for Observability-Enhanced Statement Coverage," *DAC99: 35th Design Automation Conference*, 1999, pp. 666-671

[9]   F. Corno, G. Cumani, M. Sonza Reorda, Giovanni Squillero, "RT-level Fault Simulation Techniques based on Simulation Command Scripts," *DCIS 2000: XV Conference on Design of Circuits and Integrated Systems*, November 21-24, 2000

[10] F. Corno, G. Cumani, M. Sonza Reorda, G. Squillero, "An RT-level Fault Model with High Gate Level Correlation," *IEEE International High Level Design Validation and Test Workshop*, November 8-10, 2000

[11] T. Bäck, F. Hoffmeister, and H.-P. Schwefel, "A survey of evolution strategies." *Proceedings of the Fourth International Conference on Genetic Algorithms*, 1991, pp. 2-9

[12] H.-P. Schwefel, F. Kursawe, *On Natural Life's Tricks to Survive and Evolve*, Proceedings of the 1998 IEEE International Conference on Evolutionary Computation, 1998, pp. 1-8

[13] F. Corno, M. Sonza Reorda, G. Squillero, "RT-Level ITC 99 Benchmarks and First ATPG Results," *IEEE Design & Test, Special issue on Benchmarking for Design and Test*, July-August 2000, pp. 44-53

[14] F. Corno, M. Sonza Reorda, G. Squillero, "Automatic Validation of Protocol Interfaces Described in VHDL," *EvoTel2000: European Workshops on Telecommunications*, Edinburgh (UK), May 2000, pp. 205-213

A Pursuit Architecture for Signal Analysis

Adelino R. Ferreira da Silva

Universidade Nova de Lisboa,
Dept. de Eng. Electrotécnica,
2825 Monte de Caparica, Portugal
afs@mail.fct.unl.pt

Abstract. One of the main goals of signal analysis has been the development of signal representations in terms of elementary waveforms or atoms. Dictionaries are collections of atoms with common parameterized features. We present a pursuit methodology to optimize redundant atomic representations from several dictionaries. The architecture exploits notions of modularity and coadaptation between atoms, in order to evolve an optimized signal representation. Modularity is modeled by dictionaries. Coadaptation is promoted by introducing self-adaptive, gene expression weights associated with the genetic representation of a signal in a proper dictionary space. The proposed model is tested on atomic pattern recognition problems.

1 Introduction

An all pervasive scientific methodology to describe and analyze complicated phenomena, is to represent them as a superposition of simple, well-understood objects. In physics and engineering, an important aspect of many of these representations is the chance to extract relevant information hidden in the underlying process or signal representation. For example, linear transformations are often applied with the aim of simplifying signal analysis. Such transformations are used for many diverse tasks such as reconstruction, classification, compression, coding, and diagnostics.

One of the main goals of signal analysis in recent years has been the development of signal representations in terms of elementary waveforms well localized in time and frequency, called time-frequency atoms. Collections of waveforms with common parameterized features are called dictionaries. Several alternative time-frequency dictionaries have been developed. For instance, wavelet dictionaries use translations and dilations of a basic mother waveform or wavelet [1]. A wavelet packet dictionary includes the standard wavelet dictionary and a collection of oscillating waveforms spanning a range of frequencies and durations. A local cosine packet dictionary contains the standard orthogonal Fourier dictionary and a variety of sinusoids of various frequencies weighted by windows of various widths and locations. A more general framework for signal representation is provided by atomic decomposition analysis [2]. In this case, the representation is extended with the introduction of non-orthogonal signal decompositions.

E.J.W. Boers et al. (Eds.) EvoWorkshop 2001, LNCS 2037, pp. 307–316, 2001.
© Springer-Verlag Berlin Heidelberg 2001

Non-orthogonal bases produce highly overcomplete dictionaries. The freedom of choice provided by non-orthogonal representations leads to a considerable combinatorial explosion. For general dictionaries, finding the optimal expansion over a redundant dictionary of waveforms is NP-complete [3]. On the other hand, the non-orthogonality between atoms introduce complex signal interactions which cannot be exploited by standard signal processing algorithms.

In this paper we propose a pursuit methodology to optimize redundant, non-linear approximations from several dictionaries. The proposed methodology, referred to as evolutionary pursuit (EP), relies on evolutionary computation techniques [4] to select well-adapted approximations. In EP, finding the best approximation for a given signal is viewed as an evolutionary learning process. The EP approach reduces the computational complexity of atomic decomposition analysis by subdecomposing the fitness optimization process. One critical requirement for atomic decomposition analysis is cooperative coevolution. We need an evolutionary algorithm to evolve and sustain a diverse and cooperative population in order to promote interaction between atoms in different decompositions. Simple genetic algorithms are not entirely adequate for this purpose, since they have difficulty in maintaining diversity, thus leading to strong convergent properties. We introduce an architecture for evolving coadapted sets of atomic decompositions. The architecture exploits explicit notions of modularity and coadaptation between atoms, in order to evolve an optimized signal representation. Modularity is modeled by considering a pursuit model of decompositions coming from separate dictionaries. Considering a molecule as a set of atoms, finding an optimal atomic decomposition amounts to optimizing the best combination of molecules from separate dictionaries. Coadaptation is promoted in EP by introducing self-adaptive, gene expression weights associated with the genetic representation of a signal in a proper dictionary space. Each individual is modeled after a composite genome with a genome representation for each component part. More specifically, each individual is able to generate signal representations for each dictionary within the set of all dictionary spaces. Gene expression weights are evolved for each component separately, and used to promote fitness adaptation between molecules from different libraries in the pursuit model.

The paper is organized as follows. In Sect. 2, we review the basic formulation used to characterize signal approximations. Section 3, formulates signal approximations in terms of coadapted decompositions. In Sect. 4 we propose an architecture to support the evolution of adapted decompositions. Section 5 presents the results of the application of the proposed methodology to atomic pattern recognition problems. Section 6 summarizes the conclusions.

2 Signal Approximations

In this section we review some relevant aspects of signal approximation theory. For more details we refer to [1]. Let f be a discrete digital signal. A dictionary $\mathcal{D}$ is a collection of waveforms $(\phi_\gamma)_{\gamma \in \Lambda}$, with γ a parameter [2]. We envision an optimal superposition of dictionary elements, in the sense that the residue R_M

in an approximate signal expansion

$$f = \sum_{m=0}^{M-1} \alpha_{\gamma_m} \phi_{\gamma_m} + R_M$$

is minimized, being α_{γ_m} appropriate expansion coefficients. We are mainly interested in dictionaries that are overcomplete, because they give us the possibility of adaptation, i.e., choosing among many representations the one that is best suited to our purposes. This adaptation capability draws from the fact that overcomplete dictionaries yield non-unique decompositions. Following a terminology close to the one proposed in [1], let $\mathcal{B} = \{g_m\}_{m \in \mathbb{N}}$ be an orthonormal basis of a Hilbert space $\mathcal{H}$. Any $f \in \mathcal{H}$ can be decomposed in a basis $\mathcal{B}_b$ indexed by $b \in \mathbb{N}$,

$$f = \sum_{m=0}^{+\infty} \langle f, g_{bm} \rangle g_{bm}.$$

For an approximation in basis $\mathcal{B}_b$, instead of representing f by all inner products $\{\langle f, g_{bm} \rangle\}_{m \in \mathbb{N}}$ we may use only the best M components. The approximation error in basis $\mathcal{B}_b$ is then

$$\epsilon_b[M] = \|f - \tilde{f}_{bM}\|^2 = \sum_{m=M}^{+\infty} |\langle f, g_{bm} \rangle|^2. \tag{1}$$

Using the simplified notation $\alpha_b[i] = \langle f, g_{bm_i} \rangle$ for expressing the coefficient of rank i, the best non-linear approximation in basis $b = \mathcal{B}_1$ is

$$\tilde{f}_{bM} = \sum_{i=0}^{M-1} \alpha_b[i] g_{bm_i}.$$

For our purposes, a dictionary $\mathcal{D}^k, k \in \mathbb{Z}^+$, is a union of orthonormal bases in a single space of finite dimension N, $\mathcal{D}^k = \cup_{b \in \mathbb{N}} \mathcal{B}_b$. To optimize non-linear signal approximations in a single dictionary, we will adaptively select a "best" basis from a dictionary of bases by minimizing a cost functional.

3   Formulation of Coadapted Decompositions

To improve the approximation of complex signals we need to consider general non-orthogonal signal decompositions, and multiple dictionaries. The non-orthogonality between atoms introduce complex signal interactions which cannot be fully exploited by standard signal processing techniques. Evolutionary algorithms rely on a population of individuals, each of which represents a search point in the space of potential solutions. To construct well-adapted and parsimonious signal decompositions, EP evolves populations of bases taken from several dictionaries. In addition, the number of atoms to retain in the final approximation from each dictionary may be self-adaptively selected by the evolutionary algorithm

[4] in order to minimize the final signal residue. The self-adaptive procedures in EP follow the standard evolution strategies approach [5]. The particular choice of which dictionaries to use depends upon the application.

Consider a signal of dimension L. Let K be the number of dictionaries $\mathcal{D}^k, k = \{1, \ldots, K\}$ from which bases are to be selected. For each dictionary $\mathcal{D}^k$, select an initial population $\mathcal{P}^{k,0}$ of μ bases,

$$\mathcal{B}_b^k = \{g_{bm}^k\}_{\substack{b=\{1,\ldots,\mu\}, \\ 1 \leq m \leq N_k}},$$

where N_k is the number of vectors in basis $\mathcal{B}_b^k$ for dictionary $\mathcal{D}^k$. For each generation g, we evolve the decomposition associated with each dictionary $\mathcal{D}^k$, and extract the residue at each step. Thus, for dictionary $\mathcal{D}^1$,

$$R_b^{1,g} = f - \sum_{i=1}^{N_1} \alpha_b^{1,g}[i] g_{bm_i}^{1,g}. \tag{2}$$

The application of the genetic algorithm to the population of residues at generation g for dictionary $\mathcal{D}^1$, $R_b^{1,g}, b = \{1, \ldots, \mu\}$, yields a local best basis $g_{\beta_1 m}^{1,g}$, and a local minimum residue $R_{\beta_1}^{1,g}$. This residue is the signal to be optimized by bases taken from the next dictionary $\mathcal{D}^2$, in the current generation. A similar optimization process is applied to all dictionaries in sequence.

The evolutionary procedure (2) for calculating the residues is unsuited to deal with non-orthogonal decompositions. Since, in general, we are dealing with the superposition of non-orthogonal atoms, the best-adapted coefficients $\alpha_b^{k,g}[i]$ are not specified by the bases $g_{bm}^{k,g}$ alone, as in the single dictionary case with orthogonal bases, but have to be evolved to achieve improved signal approximations. Therefore, we reformulate (2) in terms of weight factors $w_i^{1,g}$ to be applied to the coefficients $\alpha_b^{1,g}[i]$ at each generation g,

$$R_b^{1,g} = f - \sum_{i=1}^{N_1} w_i^{1,g} \alpha_b^{1,g}[i] g_{bm_i}^{1,g}. \tag{3}$$

The coefficients $\alpha_b^{1,g}[i]$ are evaluated directly from the population of μ bases, as in the single dictionary case. The weights $w_i^{1,g}$ are evolved by EP. There is a different set of weights for each dictionary. The objective is to evolve a set of bases, weights and number of atoms for each dictionary, such that the final residue at generation $g = G$, $R^G = \|f - \tilde{f}^G\|$ is minimized. The final signal approximation $\tilde{f}^G$ is synthesized from the best evolved bases $\mathcal{B}_\beta^k$ among all K dictionaries at the final generation G,

$$\tilde{f}^G = \sum_{k=1}^{K} \sum_{i=1}^{N_k} w_i^{k,G} \alpha_\beta^{k,G}[i] g_{\beta m_i}^{k,G}. \tag{4}$$

Each best evolved basis per dictionary contributes with the best $N_k \geq 0$ coefficients to build the final (non-linear) signal approximation.

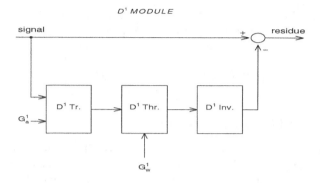

Fig. 1. Pursuit architecture: first stage

4   A Pursuit Architecture of Coadapted Dictionaries

The formulation of coadapted decompositions introduced in Sect. 3 may be expressed in terms of a pursuit architecture of coadapted dictionaries. The proposed pursuit architecture supports the emergence of adapted decompositions. The architecture exploits modularity and coadaptation between molecules in order to evolve an optimized signal representation. The first stage of the pursuit model associated with dictionary $\mathcal{D}^1$ is depicted in Fig. 1. The model is a cascade of K stages, following the pattern represented in Fig. 1. There is one stage for each dictionary in the model. Each stage is composed of a transformation unit (Tr.), a thresholding and weighting unit (Thr.), and an inversion unit (Inv.). The exact transformation to be applied at stage k, is governed by the genetic material $\mathcal{G}_a^k$ for dictionary $\mathcal{D}^k$. Likewise, the type of weighting mechanism to use is governed by the gene expression weights $\mathcal{G}_w^k$. At the end of each stage, a residue is extracted to be processed by the next stage.

4.1   Modularity

As far as modularity is concerned, dictionaries $\mathcal{D}^k, k \in \mathbb{Z}^+$, are the basic modular units to be considered. For computational efficiency reasons, we consider dictionaries for which fast transforms are available. A one-dictionary molecule $\mathcal{M}^k$, is defined as a signal composed by the superposition of atoms taken from dictionary $\mathcal{D}^k$. In the current implementation, we have used a maximum of four dictionaries, $\mathcal{D}^k, \{1 \leq k \leq 4\}$: wavelet packets, local cosine packets, cosine transforms and Dirac signals [1]. At each generation, each unit contributes with a one-dictionary molecule for the adaptation process (molecules may be empty, i.e., have zero atoms). The connections between modules implement the residue extraction specification formulated in Sect. 3. These connections represent a feature extraction mechanism for atomic recognition, updated at each stage $\mathcal{D}^k$. The superposition of the evolved one-dictionary molecules M^k at each stage, gives the signal approximation $\tilde{f} = \sum_k M^k$, to be evaluated for fitness at each generation. Fitness has been evaluated in terms of the ℓ_2-error,

$$\ell_2 = \|f - \tilde{f}\|_2 = \left[\sum_{i=1}^{L}[f(x_i) - \tilde{f}(x_i)]^2\right]^{\frac{1}{2}}. \tag{5}$$

4.2   Intra-dictionary Adaptation

The second fundamental pursuit mechanism deals with coadaptation. However, before describing the coadaptation mechanism used in the pursuit architecture, we have to specify the underlying genetic representation, and the genotype-to-phenotype mapping used for the purpose of fitness evaluation. Both of these aspects, the representation and the fitness evaluation, have been described more fully in previous work when dealing with atomic decompositions for wavelet packet and cosine packet dictionaries [6,7]. In the following, we will refer to the construction used in those references as the intra-dictionary genetic algorithm. The fundamental aspects of the algorithm are as follows. For each dictionary $\mathcal{D}^k$ there is a genotype sequence $\mathcal{G}_a^k$, which defines how some discrete function f of L sample values is transformed to obtain its representation in the transformed library space. For instance, the wavelet packet dictionary may be organized as a tree data structure. Therefore, the genotype sequence $\mathcal{G}_a^k$ which defines the wavelet packet decomposition may be specified by a breadth-first sequence [6]. Alternatively, a tree genotype or some other valid representation could be used for the same purpose. The genotype sequence $\mathcal{G}_a^k$ guides the decomposition process and constructs a phenotypic representation of f in the transformed space, F^k, for dictionary $\mathcal{D}^k$. For fitness evaluation purposes, the best terms of the F^k representations for each dictionary $\mathcal{D}^k$ are used to reconstruct the approximation $\tilde{f}$. When dealing with single dictionaries, it has been shown that the intra-dictionary genetic algorithm is robust for atomic pattern recognition [6,7].

4.3   Coadaptation: Inter-dictionary Adaptation

When considering multiple dictionaries, each dictionary $\mathcal{D}^k$ contributes with N_k evolved coefficients from the transformed space F^k, or equivalently N_k atoms, for the reconstructed signal $\tilde{f}$. We suppose that we know which dictionaries are available. Furthermore, we suppose that the total number of atoms, N, used to build the reconstructed signal $\tilde{f}$ is specified. For multiple dictionaries, the atomic decomposition problem amounts to address the following two questions. First, which coefficients should be selected from each dictionary ? Second, how many coefficients should be selected from each dictionary ? In general the best representation for a given signal involves interaction between atoms from different dictionaries. Therefore, we are left with a coadaptation problem for atoms from different dictionaries, or inter-dictionary adaptation. For atoms from the same dictionary, adaptation is solved by the intra-dictionary evolutionary algorithm outlined in Sect. 4.2.

Coadaptation is addressed by defining, for each dictionary, a new genome component $\mathcal{G}_w^k$ which weights the intermediate phenotypic representation generated by the component $\mathcal{G}_a^k$ specified in Sect. 4.2. Each gene $g_a[i]$ of $\mathcal{G}_a^k$ has an associated gene $g_w[i]$ of $\mathcal{G}_w^k$. The genes $g_w[i]$ are used to weight the expression of the genes $g_a[i]$ in transformed space. Fitness is evaluated by the ℓ_2-error $\|f - \tilde{f}\|_2$, with $\tilde{f} = \sum_{k=1}^{4} \tilde{f}^k$, where each component $\tilde{f}^k$ is obtained through the inverse transform associated with dictionary $\mathcal{D}^k$. For each dictionary $\mathcal{D}^k$, N_k transformed coefficients F^k are evolved. It is worth noting that the genetic weights $\mathcal{G}_w^k$ are not applied to the fitness function directly, but are used as regulatory entities applied to intermediate genetic transcriptions (see Fig. 1). Although we do not elaborate on this, this mechanism has features in common with gene expression mechanisms as proposed for instance in [8]. However, the genetic weights are not used simply as gene activation entities, but rather as regulatory mechanisms within the decomposition and reconstruction (transcription) process. The determination of the contribution each vector of coefficients is making to the solution as a whole (credit assignment), is regulated by the genes $\mathcal{G}_w^k$. This regulation fosters the maintenance of genetic diversity. In fact, even if the vectors of coefficients F^k are correctly identified for each dictionary $\mathcal{D}^k$, the expression weights $\mathcal{G}_w^k$ may degrade the fitness value assigned to an individual. Therefore, individuals representing less than exact decompositions will have an increased chance of surviving, if their gene expression weights promote well-fitted combinations of molecules among dictionaries. The weights are evolved in a self-adaptive manner [5], in close connection with the evolved dictionary representation.

5 Experimental Results

Following the methodology outlined in Sect. 3, a series of statistical tests were conducted to evaluate the discriminating capabilities of the evolutionary algorithm in identifying the atoms and the time-frequency coefficients in molecular test signals. The evolutionary program used in the tests is a hybrid implementation based on the steady-state genetic algorithm as implemented in [9], and the evolutionary algorithm with self-adaptive Gauss-Cauchy mutation [10,4]. We have used the following genetic parameters: population size: 50; number of generations: 200; crossover probability: 0.95; mutation probability: 0.01; replacement probability: 0.9. The basic operators used to drive the EP program were as follows: self-adaptive Gauss-Cauchy mutation for real genomes [10]; uniform mutation with constraints for admissible tree genomes [11]; blend crossover (BLX) for real genomes [9]; uniform crossover for integer genomes; tournament selection as the selection method.

In this work, we consider synthetic signals composed of atoms taken from a mixture of dictionaries. The atoms used in the construction of the synthetic signals were randomly taken from four dictionaries [1]: wavelet packets (WP), local cosine packets (CP), discrete cosines (DC), and diracs (DI). As an example, Fig. 2 shows a randomly generated test signal constructed with 4 wavelet packet atoms, 4 cosine packet atoms, 2 cosines and 2 diracs. The signal has length

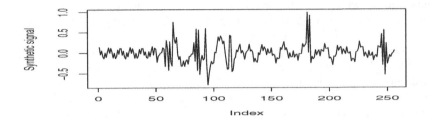

Fig. 2. Example of a synthetic test signal

$L = 256$. Table 1 specifies the atomic composition of the molecular signals used in the test suite, in terms of the dictionaries WP, CP, DC, and DI. The Daubechies-8 filter [1] was used to generate the atoms from the wavelet packet dictionary. Molecular tests signals were randomly generated with a total number of atoms in the range $r_a = \{6, 8, 12, 16, 20\}$, for different molecular compositions.

To evaluate the ability of the evolutionary algorithm to correctly identify the elementary waveforms in the test signal f, we have used two error measures: one in transformed space, the ℓ_0-error, and the other in signal space, the ℓ_2-error. As explained in Sect. 4.2, the set of evolved coefficients F^k ,$\{1 \leq k \leq 4\}$, in transformed space are used to reconstruct the signal approximation $\tilde{f}$. There are two characteristics associated with the F^k coefficients: their position in the transformed signal representation, and their numerical value. The ℓ_0-error gives an indication of the positioning of the coefficients in the vectors F^k, and measures the coefficients' misidentification error rate. More specifically, the ℓ_0-error measure is defined as follows. Suppose that each coefficient is normalized so that $\|F_i^k\| = 1$. Let C be the union of the sets of normalized coefficients C^k from all dictionaries $k = \{1, \ldots, 4\}$, $C = \cup_k C^k$. Likewise, let E be the union of the sets of normalized evolved coefficients E^k from all dictionaries $k = \{1, \ldots, 4\}$, $E = \cup_k E^k$. The ℓ_0-error measure is defined by the number of elements in the asymmetric difference of the two sets, $\ell_0 = \#\{C - E\} = \#\{C - (C \cap E)\}$.

Table 1. Atomic composition of test signals and errors

Test	Atoms	WP	CP	DC	DI	ℓ_0-error	ℓ_2-error
1	6	4	0	1	1	0	2.7e-7
2	8	2	2	2	2	0	7.3e-4
3	12	4	4	2	2	0	0.017
4	12	6	6	0	0	0	0.018
5	16	4	4	4	4	0	0.044
6	16	8	8	0	0	0	0.092
7	20	10	10	0	0	0	0.20
8	20	8	8	4	0	0	0.23

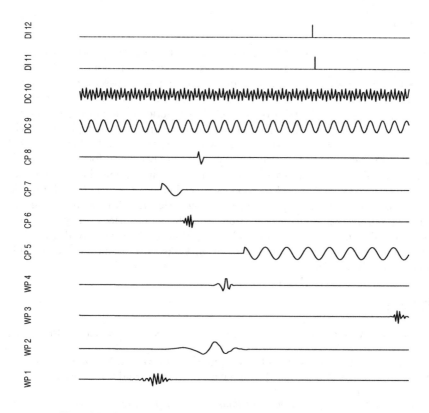

Fig. 3. Evolved atomic components of the signal depicted in Fig. 2

For each test 30 samples were generated. Each new sample generates a new random atomic composition, with the exact number of atoms from each dictionary specified by the corresponding test number in Table 1. For wavelet packets and cosine packets the decomposition tree used to synthesize the signal was randomly generated anew for each sample as well. Table 1 presents the median values of the ℓ_2-errors, between the test signals referenced in Table 1, and the signals reconstructed with the atoms found by evolutionary pursuit. The median values of the misidentification errors for the atomic coefficients, as measured by the ℓ_0-error, are also reported in the same table. The specification of the number of atoms to take from each dictionary was used to run the tests, and obtain the results presented in Table 1. The median ℓ_0-errors in Table 1 show that all the atoms were correctly identified in most of the stochastic tests, for the number and mixture of atoms considered in the suite of tests. The low ℓ_2-errors indicate that a good visual reconstruction of the original signal has been achieved. Moreover, the ℓ_2-errors show a nice degradation of the reconstructed signal in relation to the original signal, as the number of atoms increases. Fig. 3 depicts the evolved atomic components of the signal shown in Fig. 2.

6   Conclusion

The problem with overcomplete systems in general, is that they lead to large scale optimization problems with highly non-convex objective functions. We presented a modular architecture for signal decomposition analysis, based on dictionaries for which fast transforms are available. Intra-dictionary adaptation, operates on overcomplete systems of orthogonal bases. Inter-dictionary adaptation, operates on sequences of intra-dictionary approximations and require coadaptation mechanisms. The tests with synthetic signals used in the paper show that the proposed pursuit approach is able to perform component analysis with low misidentification errors. The tests were carried out under the supposition that the number of atoms to extract from each dictionary were known *a priori*. This limitation may be overcome, by incorporating in the genetic algorithm a mechanism to learn the number of atoms to take from each dictionary. However, preliminary tests have shown that this additional adaptation mechanism may degrade the misidentification rate for atomic recognition tasks. We intend to research this topic further in the future.

References

1. Stéphane Mallat. *A Wavelet Tour of Signal Processing.* Academic Press, San Diego, 1998.
2. S. S. Chen. *Basis Pursuit.* PhD thesis, Stanford University, November 1995.
3. G. M. Davis, S. Mallat, and M. Avelanedo. Greedy adaptive approximations. *J. Constr. Approx.*, 13:57–98, 1997.
4. Thomas Bäck. *Evolutionary Algorithms in Theory and Practice: Evolution Strategies, Evolutionary Programming, Genetic Algorithms.* Oxford University Press, New York, 1996.
5. H.-G. Beyer. Towards a theory of evolution strategies: Self-adaptation. *Evolutionary Computation*, 3(3):311–347, 1996.
6. A. F. da Silva. Genetic algorithms for component analysis. In D. Whitley et. al., editor, *Proceedings of the 2000 Genetic and Evolutionary Computation Conference, GECCO-2000, Las Vegas, Nevada*, pages 243–250. Morgan Kaufmann Publishers, San Francisco, 2000.
7. A. F. da Silva. Evolutionary time-frequency analysis. In A. Zalzala et. al., editor, *Proceedings of the 2000 Congress on Evolutionary Computation, CEC2000, La Jolla, California*, pages 1102–1109. IEEE Press, 2000.
8. H. Kargupta. The genetic code and the genome representation. In Annie S. Wu, editor, *Proceedings of the 2000 Genetic and Evolutionary Computation Conference Workshop Program*, pages 179–184, 2000.
9. M. Wall. *GAlib: A C++ Library of Genetic Algorithm Components.* Mechanical Engineering Department, Massachusetts Institute of Technology, August 1996.
10. K. Chellapilla. Combining mutation operators in evolutionary programming. *IEEE Trans. on Evolutionary Computation*, 2(3):91–96, September 1998.
11. A. F. da Silva. Evolutionary wavelet bases in signal spaces. In S. Cagnoni et al., editor, *Real-World Applications of Evolutionary Computing*, volume 1803 of *Lecture Notes in Computer Science*, pages 44–53. Springer-Verlag, 2000.

Genetic Algorithm Based Heuristic Measure for Pattern Similarity in Kirlian Photographs

Mario Köppen[1], Bertram Nickolay[1], and Hendrik Treugut[2]

[1] Fraunhofer IPK Berlin
Pascalstr. 8-9, 10587 Berlin, Germany
{mario.koeppen|bertram.nickolay}@ipk.fhg.de
[2] Stauferklinik Schwäbisch Gmünd
Wetzgauer Str. 85, 73557 Mutlangen, Germany

Abstract. This paper presents the use of a genetic algorithm based heuristic measure for quantifying perceptable similarity of visual patterns by the example of Kirlian photographs. Measuring similarity of such patterns can be considered a trade-off between quantifying strong similarity for some parts of the pattern, and the neglection of the accidental abscense of other pattern parts as well. For this reason, the use of a dynamic measure instead of a static one is motivated. Due to their well-known schemata processing abilities, genetic algorithm seem to be a good choice for "performing" such a measurement. The results obtained from a real set of Kirlian images shows that the ranking of the proposed heuristic measure is able to reflect the apparent visual similarity ranking of Kirlian patterns.

1 Introduction

Kirlian photography was invented in the former Soviet Union in 1939 by Semyon Kirlian [1]. Due to the fact that images obtained through the Kirlian process seems to reveal aura-like features around living objects, it has a long time fascinated scientists and laymen, but charlatans as well. The Kirlian process and a number of processes derived later based on the original one all use high frequency, low current electricity to create an electrical "corona" around the object to be photographed. Actually, the discharge is photographed. It has been suspected that Kirlian photographs reveal relevant medical information about the "energetic" state of the object, esp. the human body or parts of it, and can be used for the diagnosis of disease [2] [3] [4]. The medical term for such a diagnostic method is electroradiography. One well-known approach to a therapy based on Kirlian photography was introduced by Mandel [5], but there are more [6].

In Fig. 1 the Kirlian photographs of four different subject's fingertips can be seen. There are typical features of such photographs, as protuberances, gaps and discharge marks. All of them give the human observer of such images the impression of a unique pattern category. However, a serious controversy is raging among researchers on the "non-random" nature of Kirlian images, especially the

E.J.W. Boers et al. (Eds.) EvoWorkshop 2001, LNCS 2037, pp. 317–324, 2001.

question, whether the Kirlian images of one and the same person show similar features or not. In some cases, as it is shown in Fig. 2, a visual similarity of images taken from the same person at different moments can be clearly seen. Already this fact urges an explanation.

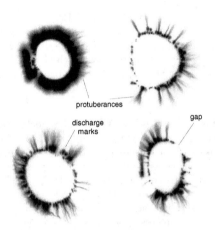

Fig. 1. Kirlian images of fingertips of four different subjects.

This paper presents an approach to the quantification of the visual similarity of Kirlian images. Due to the different sources of influence on the picturing process, standard statistical methods will not give very much qualitative insight into this issue (see [7] [8] for a study using conventional statistics). Consider for example the presence of gap features in the pattern. Gaps could appear for two different reasons: as partly belonging to the suspected person-specificity of the Kirlian image, or as being caused by a random influence on the individual picturing process. Hence, gaps could stand for visual similarity or not. Also, the local distribution of discharge marks could appear differently. Of more importance is the *partial* high similarity, thus giving a contradictory goal for similarity measurement. A means for reflecting *schematical* similarity has to be used instead of a direct computation from image data.

In order to solve this problem, a dynamic measure is used instead of a static one. The similarity is ranked by the degree, to which it is possible for an adaptive procedure to re-establish that similarity. Genetic algorithms are proposed for fulfilling this task.

In section 2, the necessary preprocessing of the Kirlian images and the used material are presented. Section 3 discusses the problem of processing measures and emphasizes the use of a genetic algorithm based heuristic measure for the purpose of this study. Finally, in section 4 the results on the Kirlian images, which were achieved with this approach, are presented and discussed.

Fig. 2. Kirlian images of the same fingertip of a subject, taken at different moments.

2   Material and Method

From 30 adult subjects, which suffered from several diseases, 120 Kirlian images were taken, four images of the hands and feets of each subject, with 15 minutes time delay between two photographs. Out of those 120 examinations, 32 were selected according to an apparent high visual similarity of at least three finger-tips. The goal was to provide a quantitative measure that reflects this visual similarity, and which is able to separate the class of Kirlian images from random patterns and may hint on person-specificity. The goal was not to relate the Kirlian images to the special diseases of the subjects.

The Kirlian images were digitized using a resolution of 400dpi. As a result, there were 96 digital images (four images of three fingers of eight subjects) prepared for further examination.

In order to get a comparable representation for all those images (which are differing in size, quality and presence of features), an unrolling procedure was applied, with an interactively set center. Thus, the nearly circular structure of the Kirlian patterns is reflected. For 256 equally-angled directions, the distance between the first and last group of pixels[1], which was meet along that directions, was taken and plotted against the direction angle. This "coarse polar transformation" gives a pattern, which will be referred to as signature of the Kirlian image in the following (see Figure 3). When there was no group of pixels found, the value 0 was assigned to the corresponding direction.

The reason for unrolling the Kirlian images were

- normalization of the results,
- reduction to the one-dimensional case,
- first quantification of the image data,
- noise reduction,
- accounting for shape irregularity and
- special treatment of gaps.

The unrolled Kirlian image signatures were used for further processing.

---

[1] At least three pixels in a sequence are black.

3   Genetic Algorithms for Heuristic Measuring

In Fig. 4, four Kirlian signatures of two different subjects can be seen. While there is still an apparent schematic similarity of the signatures of the same subject, and an obvious dissimilarity for signatures of different subjects, the provision of a suitable measure to measure this is not so easy. The idea was to use an adaptive procedure for such a measuring. The adaptive task here is given by the goal to design a signature, which *is equal* to all four signatures at once. Of course, this task has no exact solution, but there will be an optimal one. The matching quality can be used as a measure of similarity. Instead of being a static measure, such a procedure can be considered a dynamic measure. In artificial intelligence research, such measures are infrequently used and referred to as heuristic measures.

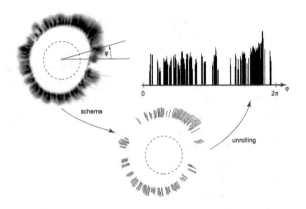

Fig. 3. Unrolling of a Kirlian image.

To provide an analogy for the heuristic measure approach: consider the task to get to know, whether a given lake is deep or shallow. An easy way would be to measure the depth of the lake at every point, then to compute the average depth and decide from the obtained value. However, the effort might be too high for getting the depth everywhere. So, the other approach is to instruct a subject to swim to a shallow position of the lake, starting of from a random position. The subject may use all of her skills to solve this task. Then, the subject is repeatedly placed anywhere in the lake, and a couple of minutes is given to her to find to a shallow place. Now, it is counted how often the subject was able to find a shallow place among all trials. From this frequency, the depth of the lake can be judged. This gives a *heuristical* measure for the depth of the lake.

It has to be noted that the actual size of the measure is not of interest, since this value depends on the method used for the adaptation. What counts is the relative ranking of those values for different objects, say different lakes. It follows that for a reasonable use of such a measuring, the measure needs to be calibrated to the applied method, e.g. on lakes, which are known to be very deep or very shallow. Then, the obtained values can be related to the wanted information.

Back to the Kirlian images, as a prominent heuristical search procedure, genetic algorithms (GA) were chosen. GAs are well-known for their inherent building-block (or schemata) processing capabilities, which exposes them among other optimization procedures [9] [10].

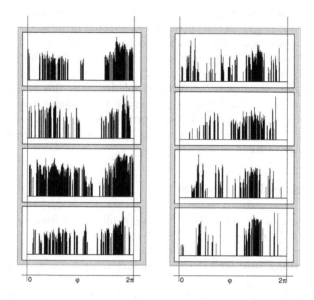

Fig. 4. Unrollings of two sets of Kirlian images.

GA maintain a population of *bitstrings*, i.e. vectors composed of the elements 0 and 1. A *schema* is a bitstring containing "wildcards", i.e. a bitstring with some unspecified positions. In this sense, e.g. 1011 is a realization of the schema 1**1, and 0111 not. During the GA run, not only the space of bitstrings (so-called *searchspace*) is searched for better bitstrings (solutions), but the space of all schemata as well. This was a long time suspected the "hidden force" that drives GA towards successful solutions, and it was initially formulated as [11]:

Short, low order, above average schemata receive exponentially increasing trials in subsequent generations.

However, there were prominent failures of GAs as well, and the question about the importance of building blocks and their features for genetic search is a still ongoing research topic by itself (see [12] for a newer work in this field).

From this, the possible ability of a GA to find building blocks (at least reflected by the obtained results) that match a set of different bitstrings as close as possible can be expected. The fitness of a bitstring is the average hamming distance of the bitstring to the given set of bitstrings. The average fitness obtained from the best individuals of several runs of a GA on the bitstring matching problem is a heuristic measure for schematic similarity of the given set of bitstrings.

It has to be noted that the use of standard statistical measures for Kirlian images was considered in [7] and [8], yielding no insight into the similarity of the patterns. This is no wonder, since the possible absence of features, despite of a present high similarity in all other parts of the pattern, may deceive direct computations.

A similar procedure for solving a pattern recognition problem was presented in [13]. There, in a system for automatic processing of invoices, the schematic similarity (but not equality) of invoice table rows was detected by running a genetic algorithm for a few generations, with the task of generating a bitstring pattern for *all* text rows of the document at once. The GA was able to detect the group of rows comprising an invoice table by using this method. However, when there was no invoice table present within the document image at all, the achieved fitness value remained below a certain threshold. This can be considered a heuristically measuring based extraction of invoice table from the invoice document image as well.

4   Results and Discussion

For heuristicically measuring the schematic similarity of four signatures, those signatures were represented as four bitstrings of length 256. The bitstrings at position i were set to 1, if the value of the signature at the i-th angle (i.e. at angle $2\pi i/256$) was over 50% of the maximum value of the signature, and to 0 if this was not the case.

For the GA, a population of 10 bitstrings of length 256 was used. The genetic operations were roulette-wheel selection, one-point crossover, pointwise mutation and elitist selection. The fitness measure for a bitstring was the average hamming distance of the bitstring to the four bitstrings, which were derived from the four sigtnatures. However, for the hamming distance only positions of the signature bitstrings were regarded, for which the left and right neighbors carried the same bitvalue. In each generation, 30 children were generated, and 200 generation cycles were performed at all. The average fitness of the best individuals, taken over ten runs of the GA, gave the heuristical measure for signature similarity.

In order to "calibrate" the GA-based heuristical measure, two experiments were performed in advance. In one experiment, the procedure was applied to 4-tupel of random patterns, i.e. random sets of 256 values from $[0,1]^2$. The average best fitness values of those runs were between 0.2578 and 0.2930.

In a second experiment, the same was done with four randomly selected Kirlian images (i.e. Kirlian images of different subjects). This time, values between 0.2815 and 0.3164 were obtained, thus notably higher values as for the random patterns.

---

[2] Note, that the order of magnitude of the values does not influence the heuristical measure.

Table 1. Result of the Genetic Algorithm based heuristic measure for Kirlian images of the same person.

Finger	Measure	Finger	Measure
K3-l2	0.503625	K13-l3	0.437500
K3-l3	0.410156	K13-r4	0.351563
K3-r3	0.476562	K13-l1	0.445312
K5-r2	0.398437	K15-l3	0.347656
K5-l3	0.503906	K15-r3	0.363281
K5-r3	0.500000	K15-l4	0.441406
K6-r2	0.496094	K18-l4	0.488281
K6-l4	0.406250	K18-l1	0.515625
K6-r1	0.441406	K18-r1	0.523437
K11-l2	0.523437	K19-l2	0.394531
K11-r2	0.429687	K19-r2	0.441406
K11-l3	0.371094	K19-l3	0.386719

Table 1 gives the results for the final experiment with the test subjects. All measures are above 0.34 and can be considered to be clearly distinct from random patterns and from pattern groups of several subjects.

The study shows a clear ranking: at the lowest level, there are random patterns; at the next level, there are inter-subject Kirlian patterns; and on the highest level there are intra-subject Kirlian patterns. The heuristic measure correctly describes the visual similarity of the intra-subject Kirlian photographs.

The question, whether this hints on a person-specificity of Kirlian images at all should be discussed with care. At least, there is person-specificity for the selected patterns. According to the heavy scientific attacks on electroradiographic therapies in general, this reflects the fact that there are at least some circumstances, where Kirlian images of the same person are more similar to each other than to Kirlian images of different persons. From a physical point of view, this is an astonishing fact. While the basic physical mechanisms of the corona discharge are fully understood, there is no proper model for the appearance of the typical features of a Kirlian image for the same person. However, two applications of even low person-specificity of Kirlian images are nevertheless of high interest: for biometrical applications, and for clinical therapies based on the feedback of a subject to its own Kirlian image.

Acknowledgment. This work was supported by the Karl and Veronica Carstens Stiftung. The authors would also like to thank the unknown reviewers for their helpful suggestions.

References

1. S.D. Kirlian and V.K. Kirlian. Photography and visual observations by means of high frequency currents. *J.Sci.Appl.Photography*, 6:397–403, 1964.

2. R.S. Chouhan. Towards a biophysical explanation of the coronal formations obtained in kirlian photography in relation to cancer. In *Proc. of the 3rd Intl. Conf. for Medical and Applied Bioelectrography, Helsinki, Finlandia*, pages 19–21, 1996.

3. L. Konikiewicz. Kirlian photography in theory and clinical applications. *J.Biol.Photogr.Assoc.*, 45:115–134, 1997.

4. Y. Omura. Acupuncture, infra-red thermography and kirlian photography. *Acupunct.Electrother.Res.*, 2:43–86, 1977.

5. Peter Mandel. *Energetische Terminalpunkt-Diagnose*. Synthesis-Verlag, 1983. (in German).

6. J. Pehek, H. Kyler, and D. Faust. Image modulation in corona discharge photography. *Science*, 194:263–270, 1976.

7. H. Treugut. Kirlian Fotographie: Reliabilität der Energetischen Terminalpunktdiagnose (ETD) nach Mandel bei gesunden Probanden. *Forsch.Komplementärmed.*, 4:210–217, 1997. (in German).

8. H. Treugut. Kirlian Fotographie: Reliabilität der Energetischen Terminalpunktdiagnose (ETD) nach Mandel bei Kranken. *Forsch.Komplementärmed.*, 5:224–229, 1998. (in German).

9. John H. Holland. *Adaptation in Natural and Artificial Systems*. University of Michigan Press, 1975.

10. David E. Goldberg. *Genetic Algorithms in Search, Optimization & Machine Learning*. Addison-Wesley, Reading MA, 1989.

11. Shumee Baluja. Population-based incremental learning: A method for integrating genetic search based function optimization and competitive learning. Technical Report CMU-CS-94-163, Computer Science Department, Carnegie Mellon University, Pittsburgh, PA, 1994.

12. Chris Stephens and Henri Waelbroeck. Schemata evolution and building blocks. *Evolutionary Computation*, 7:109–124, 1999.

13. Mario Köppen, Dörte Waldöstl, and Bertram Nickolay. A system for the automated evaluation of invoices. In Jonathan H. Hull and Suzanne L. Taylor, editors, *Document Analysis Systems II*, pages 223–241. World Scientific, Singapore a.o., 1997.

Evolutionary Signal Enhancement Based on Hölder Regularity Analysis

Jacques Lévy Véhel and Evelyne Lutton

Projet Fractales — INRIA, B.P. 105, 78153 Le Chesnay cedex, France,
Jacques.Levy_Vehel@inria.fr, Evelyne.Lutton@inria.fr,
http://www-rocq.inria.fr/fractales

Abstract. We present an approach for signal enhancement based on the analysis of the local Hölder regularity. The method does not make explicit assumptions on the type of noise or on the global smoothness of the original data, but rather supposes that signal enhancement is equivalent to increasing the Hölder regularity at each point. The problem of finding a signal with prescribed regularity that is as near as possible to the original signal does not admit a closed form solution in general. Attempts have been done previously on an analytical basis for simplified cases [1]. We address here the general problem with the help of an evolutionary algorithm. Our method is well adapted to the case where the signal to be recovered is itself very irregular, e.g. nowhere differentiable with rapidly varying local regularity. In particular, we show an application to SAR image denoising where this technique yields good results compared to other algorithms. The implementation of the evolutionary algorithm has been done using the EASEA (EAsy specification of Evolutionary Algorithms) language.

1 Introduction

A large number of techniques have been proposed for signal enhancement. The basic frame is as follows. One observes a signal Y which is some combination $F(X, B)$ of the signal of interest X and a "noise" B. Making various assumptions on the noise, the structure of X and the function F, one then tries to derive a method to obtain an estimate $\hat{X}$ of the original signal which is optimal in some sense. Most commonly, B is assumed to be independent of X, and, in the simplest case, is taken to be white, Gaussian and centered. F usually amounts to convoluting X with a low pass filter and adding the noise. Assumptions on X are almost always related to its regularity, e.g. X is supposed to be piecewise C^n for some $n \geq 1$. Techniques proposed in this setting resort to two domains: functional analysis and statistical theory. In particular, wavelet based approaches, developed in the last ten years, may be considered from both points of view [2, 3].

Our approach in this work is different from previous ones in several respects. First, we do not make explicit assumptions on the type of noise and the coupling between X and B through F. However, if some information of this type

E.J.W. Boers et al. (Eds.) EvoWorkshop 2001, LNCS 2037, pp. 325–334, 2001.

is available, it can readily be used in our method. Second, we do not require that X belong to a given global smoothness class but rather concentrate on its *local* regularity. More precisely, we view enhancement as equivalent to increasing the Hölder function α_Y (see next section for definitions) of the observations. Indeed, it is generally true that the local regularity of the noisy observations is smaller than the one of the original signal, so that in any case, $\alpha_{\hat{X}}$ should be greater than α_Y. We thus define our estimate $\hat{X}$ to be the signal "closest" to the observations which has the desired Hölder function. Note that since the Hölder exponent is a local notion, this procedure is naturally adapted for signals which have sudden changes in regularity, like discontinuities. From a broader perspective, such a scheme is appropriate when one tries to recover signals which are highly irregular and for which it is important that the restauration procedure yields the right regularity structure (i.e. preserves the evolution of α_X along the path). An example of this situation is when denoising is to be followed by image segmentation based on textural information: Suppose we wish to differentiate highly textured zones (appearing for instance in MR or radar imaging) in a noisy image. Applying an enhancement technique which assumes that the original signal is, say, piecewise C^2, will induce a loss of the information which is precisely the one needed for segmentation, since the denoised image will not contain much texture. The same difficulty occurs in other situations such as change detection from noisy sequences of aerial images or automatic monitoring of the evolution of lung diseases from scintigraphic images: in such cases, the criterion for a change is often based on a variation of the irregularity in certain regions, and one needs to preserve this information.

In addition to the examples given above, other situations where such conditions occur are turbulence data analysis and characterization of non-voiced parts of voice signals. Since the method is highly non linear and quite complex, an analytic solution is not possible. We thus resort to a stochastic optimisation method based on evolutionary altgorithms.

The remaining of this paper is organized as follows. Section 2 recalls some basic facts about Hölder regularity analysis, which is the basis of our approach. The denoising method is explained in section 3. The evolutionary implementation based on EASEA is then detailed in section 4, finally numerical results on both 1D and 2D signals are displayed in section 5.

2   Hölder Regularity Analysis

A popular way to measure the local irregularity of signals is to consider Hölder spaces. We will focus in this paper on the Hölder characterizations of regularity. To simplify notations, we assume that our signals are nowhere differentiable. Generalisation to other signals simply requires to introduce polynomials in the definitions [4].

Let $\alpha \in (0,1)$, $\Omega \subset \mathbf{R}$. One says that $f \in C_l^\alpha(\Omega)$ if:

$$\exists\, C : \forall x, y \in \Omega : \frac{|f(x) - f(y)|}{|x - y|^\alpha} \le C$$

Let: $\alpha_l\left(f, x_0, \rho\right) = \sup\left\{\alpha : f \in C_l^{\alpha}\left(B\left(x_0, \rho\right)\right)\right\}$ $\alpha_l\left(f, x_0, \rho\right)$ is non increasing as a function of ρ.

We are now in position to give the definition of the local Hölder exponent :

Definition 1 *Let f be a continuous function. The* local Hölder exponent *of f at x_0 is the real number:*$\alpha_l\left(f, x_0\right) = \lim_{\rho \to 0} \alpha_l\left(f, x_0, \rho\right)$

Since α_l is defined at each point, we may associate to f the function $x \to \alpha_l(x)$ which measures the evolution of its regularity.

This regularity characterization is widely used in fractal analysis because it has direct interpretations both mathematically and in applications. It has been shown for instance that α_l indeed corresponds to the auditive perception of smoothness for voice signals. Similarly, simply computing the Hölder exponent at each point of an image already gives a good idea of its structure, as for instance its edges [5]. More generally, in many applications, it is desirable to model, synthesize or process signals which are highly irregular, and for which the relevant information lies in the singularities more than in the amplitude. In such cases, the study of the Hölder functions is of obvious interest.

In [6], a theoretical approach for signal denoising based on the use of a Hölder exponent and the associated multifractal spectrum was investigated. We develop here another enhancement technique that uses the information brought by the local Hölder function, which is simple from an algorithmic point of view, and yields good results on several kind of data.

3   Signal Enhancement

We adopt in this paper a functional analysis point of view. This means that we do not make any assumption about the noise structure, nor the way it interacts with the data. Rather, we seek a regularized version of the observed data that fulfills some constraints. A statistical approach, classically based on risk minimization, will be presented elsewhere [1].

Let X denote the original signal and Y the degraded observations. We seek a regularized version $\hat{X}$ of Y that meets the following constraints: a) $\hat{X}$ is close to Y in the L^2 sense, b) the (local) Hölder function of $\hat{X}$ is prescribed.

If α_X is known, we choose $\alpha_{\hat{X}} = \alpha_X$. In some situations, α_X is not known but can be estimated from Y. Otherwise, we just set $\alpha_{\hat{X}} = \alpha_Y + \delta$, where δ is a user-defined positive function, so that the regularity of $\hat{X}$ will be everywhere larger than the one of the observations. We must solve two problems in order to obtain $\hat{X}$. First, we need a procedure that estimates the local Hölder function of a signal from discrete observations. Second, we need to be able to manipulate the data so as to impose a specific regularity. A third difficulty arises from the following analysis: Assume the simplest case of an L^2 signal corrupted by independent white Gaussian noise. It is easy to check that almost surely $\alpha_Y = -\frac{1}{2}$ everywhere, because a) $-\frac{1}{2}$ is the regularity of the noise, b) $\alpha_X \geq 0$ since $X \in L^2$, c) the regularity of the sum of two signals which have everywhere different Hölder exponents is the min of the two regularities. Thus α_Y does not depend on X, and one cannot go back from α_Y to α_X. This fact casts doubts on the efficiency on the whole approach, since the information it is based on is degenerate in this case.

All these problems are solved once one realizes that the mathematical notion of Hölder regularity is an abstraction that makes sense only asymptotically. One needs to analyze carefully how it should be adapted to a finite setting, much in the same way as what is done for abstract white noise. In particular, we are interested in a perceptual notion of regularity: If two 2D functions A and B are such that $\alpha_A < \alpha_B$, but an imaging of A and B at a given resolution yields the contrary visual impression that A is smoother than B, then of course our algorithm should go with the perceptual information. In other words, in practical applications we are not interested in the asymptotic behavior, but in the scales which are really in the image, and our estimate of α should reflect this fact. This means precisely one thing: the estimation procedure should yield results in agreement with what is perceived, and not care for the "true" α, which may or may not be accessible from the finite data. To go back to the example above, while it is true that at infinite resolution the sum "signal + white noise" would look much the same as white noise as far as regularity is concerned, this is not the case at finite resolution, where the influence of the signal is still perceptible. In addition, since our procedure is differential, i.e. we wish to impose $\alpha_{\hat{X}} = \alpha_Y + \delta$ or $\alpha_{\hat{X}} - \alpha_X = 0$, for *estimated* α_X and α_Y, we do not care about constant bias.

We will use a wavelet based procedure for estimating and controlling the Hölder function. This is made possible by results in [4] and [7] which imply that:

Proposition 1 *Let* $\{\psi_{j,k}\}_{j,k}$ *be an orthonormal wavelet basis, where as usual j denotes scale and k position, and assumes that ψ is regular enough and has sufficiently many vanishing moments. Then, X has local Hölder exponent α at t if and only if for all (j,k) such that t belongs to the support of $\psi_{j,k}$,*

$$|c_{j,k}| \le C2^{-j(\alpha+\frac{1}{2})} \tag{1}$$

where C is a constant and $c_{j,k}$ is the wavelet coefficient of X.

Although (1) is only an inequality, it suggests that one may estimate $\alpha_X(t)$ by linear regression of $\log(|c_{j,k}|)$ w.r.t. to the scale j (log denotes base 2 logarithm) considering those indices (j,k) such that the support of $\psi_{j,k}$ is centered above t. Of course this will be only approximate, but since (1) is a necessary and sufficient condition, and if enough wavelet coefficients are "large", we may hope to obtain results sufficient for our purpose.

Two points are essential in this estimation procedure:

- The estimation is obtained through a regression on a finite number of scales, defined as a subset of the scales available on the discrete data. This avoids the pathologies described above concerning the regularity of the sum of two signals. In particular, it is possible to express the Hölder function of the noisy signal $Y = X +$ Gaussian white noise as a function of α_X, and thus to estimate conversely α_X from α_Y [1].

– The use of (orthonormal) wavelets allows to perform the reconstruction in a simple way: Starting from the coefficient $(d_{j,k})$ of the observations, we shall define a procedure that modifies them to obtain coefficients $(c_{j,k})$ that verify (1) with the desired α, and then reconstruct $\hat{X}$ form the $(c_{j,k})$.

We may now reformulate our program as follows: For a given set of observations $Y = (Y_1, \ldots, Y_{2^n})$ and a target Hölder function α, find $\hat{X}$ such that $||\hat{X} - Y||_{L^2}$ is minimum and the regression of the logarithm of the wavelet coefficients of $\hat{X}$ above any point i w.r.t. scale is $-(\alpha(i) + \frac{1}{2})$. Note that we must adjust the wavelet coefficients in a global way. Indeed, each coefficient at scale j subsumes information about roughly 2^{n-j} points. Thus we cannot consider each point i sequentially and modify the wavelet coefficients above it to obtain the right regularity, because point $i + 1$, which shares many coefficients with i, requires different modifications. The right way to control the regularity is to write the regression contraints simultaneously for all points. This yields a system which is linear in the logarithm of the coefficients:

$$\Delta L = A$$

where Δ is a $(2^n, 2^{n+1} - 1)$ matrix of rank 2^n, and

$$L = (\log |c_{1,1}|, \log |c_{2,1}|, \log |c_{2,2}|, \ldots \log |c_{n,2^n}|),$$
$$A = -\frac{n(n-1)(n+1)}{12} \left(\alpha(1) + \frac{1}{2}, \ldots, \alpha(2^n) + \frac{1}{2} \right)$$

Since we use an orthonormal wavelet basis, the requirements on the $(c_{j,k})$ may finally be written as:

minimize $\sum_{j,k} (d_{j,k} - c_{j,k})^2$ subject to: $\forall\, i = 1, \ldots, 2^n$,

$$\sum_{j=1}^{n} s_j \log(|c_{j,E((i-1)2^{j+1-n})}|) = -M_n(\alpha(i) + \frac{1}{2}) \qquad (2)$$

where $E(x)$ denotes the integer part of x and the coefficients $s_j = j - \frac{n+1}{2}$, $M_n = \frac{n(n-1)(n+1)}{12}$ and equation (2) are deduced from the requirement that the linear regression of the wavelet coefficients of $\hat{X}$ above position i should equal $-(\alpha(i) + \frac{1}{2})$.

Finding the global solution to the above program is a difficult task; in particular, it is not possible to find a closed form formula for the $c_{j,k}$. In the following, we show how this problem can be addressed with an evolutionary algorithm.

4   Evolutionary Signal Enhancement with EASEA

An evolutionary technique seems to be appropriate for the optimisation problem described in equation (2): a large number of variables are involved, and the function to be optimised as well as the contraint are non linear. We describe in this section an implementation based on the EASEA [8] language and compiler.

EASEA (EAsy Specification of Evolutionary Algorithms) is a language dedicated to evolutionary algorithms. Its aim is to relieve the programmer of the task of learning how to use evolutionary libraries and object-oriented programming by using the contents of a .ez source file written by the user.

EASEA source files only need to contain the "interesting" parts of an evolutionary language, namely the fitness function, a specification of the crossover, the mutation, the initialisation of a genome plus a set of parameters describing the run. With this information, the EASEA compiler creates a complete C++ source file containing function calls to an evolutionary algorithms library (either the GALIB or EO for EASEA v0.6). Therefore, the minimum requirement necessary to write evolutionary algorithms is the capability of creating non-object-oriented functions, specific to the problem which needs to be solved.

In our case, the evolutionary optimisation involved to enhance a signal (1D or 2D) was implemented using a simple structure on which genetic operators were defined. We used GALib [9] as the underlying evolutionary library.

We describe below the implementation for 1D signals, an implementation for image denoising was also produced based on the same principle, and results for 1D and 2D data are presented in the next section.

The Haar wavelet transform has been used to produce the $d_{j,k}$ associated to the observed signal Y. We also suppose that we know the desired Hölder exponents $\alpha(i)$ (either $\alpha(i) = \alpha_Y(i) + \delta$ where the $\alpha_Y(i)$ are the Hölder exponents of Y and δ is a user defined regularisation factor, or $\alpha(i)$ is set a priori).

The problem is to find some multiplication factor $u_{j,k}$ such that $c_{j,k} = u_{j,k} * d_{j,k}, j \in [0..n-1], k \in [0..2^j - 1]$. As is usual in wavelet denoising, we let unchanged the first l levels and seek for the remaining $u_{j,k}$ in $[0, 1]$. The genome is made of the $u_{j,k}$ coefficients, for $j \in [l..n-1]$ and $k \in [0..2^j - 1]$. These coefficients are encoded as a real numbers vector of size SIZE_MAX = $2^n - 2^l$, which can be written using EASEA syntax as:

```
GenomeClass { double     U[SIZE_MAX]; }
```

The EASEA Standard functions sections contain the specific genetic operators, namely:

1. **The initialisation function:** Each $u_{j,k}$ coefficient is randomly set to a value in $[0, 1]$. Two initial solutions are also put in the initial population: $u_{j,k} = 1$. and $u_{j,k} = 2^{-k\delta}$.
2. **The crossover function:** a barycentric crossover has been easily defined as follows: Let parent1 and parent2 be the two genomes out of which child1 and child2 must be generated, and let alpha be a random factor:

```
\GenomeClass::crossover:
    double alpha = (double)random(0.,1.);
    if (&child1) for (int i=0; i<SIZE_MAX; i++)
        child1.U[i] = alpha*parent1.U[i] + (1.-alpha)*parent2.U[i];
    if (&child2) for (int i=0; i<SIZE_MAX; i++)
        child2.U[i] = alpha*parent2.U[i] + (1.-alpha)*parent1.U[i];
\end
```

3. **The mutation function:** Mutation is a random perturbation of radius $SIGMA = 0.01$, applied with probability PMut on each gene.

```
\GenomeClass::mutator://Must return the number of mutations as an int
    int NbMut=0;
    for (int i=0; i<SIZE_MAX; i++)
if (tossCoin(PMut)){    NbMut++;
Genome.U[i]+=SIGMA*(double)random(-1.,1.);
Genome.U[i] = MIN(1.,Genome.U[i]); Genome.U[i] = MAX(0.,Genome.U[i]);}
    if (NbMut==0) identicalGenome=true;   // saves evaluation time
    return NbMut;
\end
```

4. **The evaluation function:** The fitness function has two aims: minimise $\sum((1 - u_{j,k}) * c_{j,k})^2$, making sure constraint (2) is satisfied, i.e. the Hölder exponents are the ones we want. The constraint is integrated to the fitness function using a high penalisation factor W:

$$Fitness = \sum_{j,k}((1 - u_{j,k}) * c_{j,k})^2 + W * \sum_i |\alpha_u(i) - \alpha(i)|$$

We use the GALib steady state genetic engine with replacement percentage of 60% and a selection by ranking. Crossover and mutation probabilities are fixed respectively to 0.9 and 0.001. Genome size, Population size, and number of generations are fixed for each experiment, see section 5.

5   Numerical Experiments

We first show results of enhancement on synthetic 1D data. The original signal is a generalised Weierstrass function [10] with $\alpha_X(t) = t$ for all t, (i.e. $X(t) = \sum_{n=0}^{\infty} 2^{nt} sin(2^{nt})$, $t \in [0.1]$) which has been corrupted by white Gaussian noise. Figure 1 shows the original signal, the noisy one, and the result of the enhancement procedure. For comparison, a denoising using a classical wavelet shrinkage is also displayed. For both procedures, the parameters were set so as to obtain the best fit to the known original signal. It is seen that, for such irregular signals, the Hölder regularity based enhancement yields more satisfactory results. The constraint was to find Hölder exponents that verify $\alpha(t) = t$ for all t. Parameters of the evolutionary algorithm were as follows:

Genome Size	SIZE_MAX = 496
Population Size	25
Number of generations	50000
Computation time	1365.96 seconds for 293506 evaluations

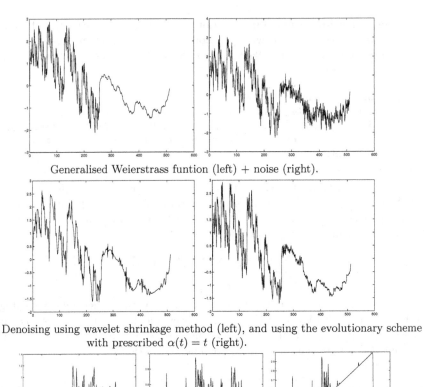

Generalised Weierstrass funtion (left) + noise (right).

Denoising using wavelet shrinkage method (left), and using the evolutionary scheme with prescribed $\alpha(t) = t$ (right).

Left: Estimated Hölder exponents of the original function (left), of the function + noise (middle) and of the reconstructed function (right).

Fig. 1. Results on a generalised Weierstrass function $\alpha(t) = t$

Our second example deals with a synthetic aperture radar (SAR) image. A huge literature has been devoted to the difficult problem of enhancing these images, where the noise is non Gaussian, correlated and multiplicative. A fine analysis of the physics of the speckle suggests that it follows a K distribution [11]. Classical techniques specifically designed for SAR image denoising include geometric filtering and Kuan filtering. Wavelet shrinkage methods have also been adapted to this case [12]. Figure 2 show an original SAR image, its denoising with the Hölder method and with soft thresholding. Notice how the river, with appears with a "Λ" shape in the middle of the image is nicely uncovered by the regularity based enhancement. As no a priori knowledge about

Hölder exponents of the signal was available, the constraint was to find Hölder exponents that verify $\alpha_{denoised}(t) = \alpha_{original}(t) + \delta$ for all t. Parameters of the evolutionary algorithm were as follows:

Genome Size	SIZE_MAX = 21845
Population Size	50
Number of generations	100
Computation time	1702.46 seconds for 3051 evaluations
Regularisation factor δ	0.5 and 0.7

Left: Original SAR image. Right: Image denoised using soft thresholding.

Image denoised using the Hölder regularity scheme. Left: $\delta = 0.5$. Right: $\delta = 0.7$

Fig. 2. Experiments on a 256x256 area of a SAR image (courtesy IRD)

6   Conclusion

We have shown in this paper how an evolutionary algorithm can be applied to a signal or image enhancement technique based on a "fractal" analysis. Good results have been obtained as well as on 1D or 2D signal in comparison to other techniques in an affordable computation time.

Quality of the results is also very dependent from the quality of Hölder exponents estimation. Much better estimate of $\alpha_X(t)$ can be obtained by measuring the *oscillations* of X in balls centered at t and of radii ϵ_k, and then regressing the logarithm of these oscillations w.r.t. the logarithm of the ϵ_k. However, this procedure does lead to a more complex inverse problem, i.e. obtaining a signal with prescribed regularity. Future work will be devoted to an evolutionary formulation of this problem.

References

1. J. Lévy Véhel, *Statistical denoising of irregular signals,* INRIA internal report.
2. R.A. Devore, B. Lucier *Fast wavelet techniques for near optimal image processing.* 1992 IEEE Military Communications Conference, 2–12 (1992).
3. D.L. Donoho, *De-noising by soft-thresholding.* IEEE Trans. Inf. Theory 41, No. 3, 613–627 (1994).
4. S. Jaffard, *Pointwise smoothness, two-microlocalization and wavelet coefficients.* Publ. Mat. 35, No. 1, 155–168 (1991).
5. J. Lévy Véhel, *Fractal Approaches in Signal Processing,* Fractals, **3** (4), pp 755-775, 1995.
6. J. Lévy Véhel, B. Guiheneuf "Multifractal Image Denoising," *SCIA,* 1997.
7. B. Guiheneuf, J. Lévy Véhel, *2 micro-local analysis and applications in signal processing,* Int. Conf. on Wavelet, Tangier, 1997.
8. Pierre Collet, Evelyne Lutton, Marc Schoenauer, Jean Louchet, "Take it EASEA," Parallel Problem Solving from Nature VI, vol 1917, Springer Verlag pp 891-901, Paris, septembre 2000. *EASEA home page:* http://www-rocq.inria.fr/EASEA/
9. M. Wall, *GAlib home page:* http://lancet.mit.edu/ga/, MIT.
10. K. Daoudi, J. Lévy Véhel, Y. Meyer, *Construction of functions with prescribed local regularity,* Constructive Approximation, 1989.
11. C.J. Oliver, *Information from SAR images,* J. Phys. D, **24**, 1493-15144, 1991.
12. L. Gagnon, F. Drissi Smaili, *Speckle noise reduction of airborne SAR images with symetric Daubechies wavelets,* Signal and Data Processing of Small Targerts, Proc. SPIE 2759, 1996.

Building ARMA Models with Genetic Algorithms

Tommaso Minerva[1] and Irene Poli[2]

[1] Faculty of Economics, University of Modena and Reggio Emilia,
via Berengario 51, 41100 Modena, Italy
minerva@unimo.it
[2] Department of Statistics, Università Ca' Foscari, Venezia (Italy)
irenpoli@unive.it

Abstract. The current state of the art in selecting ARMA time series models requires competence and experience on the part of the practitioner, and sometimes the results are not very satisfactory. In this paper, we propose a new automatic approach to the model selection problem, based upon evolutionary computation. We build a genetic algorithm which evolves the representation of a predictive model, choosing both the orders and the predictors of the model. In simulation studies, the procedure succeeded in identifying the data generating process in the great majority of cases studied.

1 Introduction

Modelling linear and stationary time series, one frequently chooses the class of ARMA models because of its high performance and robustness. The selection of a particular ARMA model, however, is neither easy nor without implications for the goal of the analysis ([1],[2],[3],[4],[5],[6],[7],[8]).

The most common approach to the model selection problem is the well-known Box-Jenkins procedure ([9], [10]), which involves a pattern recognition analysis, a model parameter estimation, and diagnostic checking based on the estimated errors. The result of the procedure depends greatly on the competence and experience of the investigators and is affected by a strong path-dependence. The pattern recognition analysis, in particular, employs a wide set of correlation measures whose values are supposed to provide insight into the optimal orders p and q of the autoregressive and moving average components. Unfortunately, one frequently observes that different models have similar estimated correlation patterns, and the choice among competitive models can be quite arbitrary. Some concern derives also from comparative studies in which experts, asked to identify a number of series, frequently reach different conclusions ([11], [12]). A further objection to the Box-Jenkins procedure for model selection is related to the time required to develop the identification model, which sometimes can be excessively high.

In this paper we introduce a computational method for selecting a model within the ARMA class models. We adopt the evolutionary computational approach based on the idea that certain algorithms based upon biological evolution can be reproduced in a

E.J.W. Boers et al. (Eds.) EvoWorkshop 2001, LNCS 2037, pp. 335-342, 2001.
© Springer-Verlag Berlin Heidelberg 2001

computer and can be used as a powerful search and optimisation tool. The evolution of natural systems has been studied for many years in the field of artificial intelligence and artificial life to solve optimisation problems characterised by a high level of complexity. The idea of studying evolution to build computational procedures for solving problems was first introduced by Box ([9]) but developed and algorithmically implemented by Holland ([13], [14],[15],[16],[17], [18]).

Evolutionary computation usually works by randomly generating a population of candidate solutions to a problem. This population then evolves, through a set of operators inspired by genetic variation and natural selection, towards the "best" population of solutions to the problem. Evolution is an adaptive process whereby the structures change with the action of the genetic operators. The change is stochastic and controlled by some probability law. It is also parallel, since the structures are transformed simultaneously.

For the ARMA model selection problem we build a genetic algorithm which evolves step by step a population of candidate models, by using a set of operators tailored to the specific problem. The procedure suggests how many predictors should be included in the model and chooses the predictors from a larger set of possible predictors. The performance of the procedure is then tested in a simulation study, where the genetic algorithm is asked to identify a number of time series generated by different processes. The results are very satisfactory: we reach a correct identification of the data generating process for the majority of the series (or the best predictive model) with a low number of generations and a low number of individuals in the algorithm population.

2   The EvoARMA Models

The usual representation of the autoregressive moving-average, $ARMA(p,q)$, model for a time series X_t , where p is the order of the autoregressive form and q the order of the moving average form, is given by

$$\phi(B)X_t = \theta(B)\varepsilon_t \tag{1}$$

where $\phi(B)=1- \phi_1 B_1 -....- \phi_p B_p$ represents the autoregressive terms, $\theta(B)= 1- \theta_1 B_1 -...- \theta_q B_q$, represents the moving average term, B is the backshift operator, and $\varepsilon(t)$ is a sequence of uncorrelated random variables with means 0 and variance $\sigma^2 >0$. The sequence $\varepsilon(t)$ is the so-called white noise process. The model is supposed to be stationary and invertible, i.e., the equations $\phi(z)=0$ and $\theta(z)=0$ have the roots outside the unit circle. Moreover the model coefficients $\phi_1...\phi_p$, $\theta_1...\theta_q$ and the white noise variance σ^2 are supposed to be constant, i.e. they do not depend on time. Modelling time series within the class of ARMA models deals with the selection of both the autoregressive order, p, and the moving average order, q, as well as the relevant (non zero) predictors.

Our concern in this paper is with choosing both the orders p and q of the model and the relevant predictors to be considered in the representation according to the evolutionary paradigm of genetic algorithms according to previous preliminary works ([19], [20], [21]).

We encode the ARMA models as vectors ("chromosomes") in binary digits to represent different orders and different predictors. Each chromosome is composed of various fields to identify: the number p of autoregressive predictors, the number q of moving average predictors, the predictors X_{t-i}, $i=1,...,p$, and the predictors ε_{t-j}, $j=1,...,q$. In this study we assume that the first two fields of the chromosome are represented by 4 binary digits each, encoding an integer running from 1 to 15 that selects respectively the number of autoregressive terms (p) and the number of moving average terms (q). The following $p+q$ fields are each represented by 5 binary digits. Each such field encodes an integer running from 1 to 31, which represents the respective autoregressive or moving average predictor.

For example the chromosome in the binary alphabet

$$0010 \quad 0001 \quad 00001 \quad 10110 \quad 01000$$

whose real representation is

$$2 \quad 1 \quad 1 \quad 22 \quad 8$$

denotes a model for the time series X_t with 2 AR predictors and 1 MA term, which are specified by X_{t-1}, X_{t-22}, and ε_{t-8}.
The resulting model is

$$X_t - \phi_1 X_{t-1} - \phi_{22} X_{t-22} = \varepsilon_t - \theta_8 \varepsilon_{t-8}$$

To build the genetic algorithm we randomly generate a population of candidate ARMA models to represent the dynamics of the particular time series.

For each individual model in this population, we then proceed with the statistical estimate of the parameters using a set of data called the training or learning set. We next evaluate each individual with an objective function which measures the optimality of each model with respect to the prediction problem. A different set of data, the validation set, is then chosen to evaluate the objective function that we define as a modified form of Akaike's information criterion. This objective function takes the form:

$$f(m) = N_v \ log \ (1/N_v \ (\ \Sigma \left(y_{0+l} - \hat{y}(l)\right)^2 \ )+ 2M \qquad (2)$$

where N_v is the number of data in the validation set, $l = 1, 2, ...$ is the prediction lead time, $\left(y_{0+l} - \hat{y}(l)\right)$ is the l-step ahead prediction error, using model m, and M is the number of the parameters in the model m.

According to these values a stochastic selection of the individuals to become candidates for the new generation is performed. Selection is based on ranking the individuals by the f values and then assigning each individual a transition probability which is proportional to the f values. We select the individuals by sampling with replacement from the population, so that each individual can be selected more than once to become a member of the next generation. This allows the good models to be reproduced in more than one copy in next generation.

The *crossover* operator is performed by randomly choosing pairs of individuals with probability p_c and crossing over parts of these individuals at randomly chosen points. We perform a multi-crossover operator because of the field representation of the individuals. Crossover works by recombining "building blocks" in the population and designing a (hopefully) improved set of solutions. Recombination, however, may result in the loss of some good building blocks that can include relevant variables for the prediction problem. A *multimutation* operator is then introduced and applied with probability p_m to recreate good predictive components. In our experiment, whose results will be presented in following section, we also assume an elitistic transition which involves copying the best 10% of the individuals in the population into the next one to replace the lowest fitness individuals. We stop the procedure after a fixed number N_G of generations. The resulting best ARMA model is called the EvoARMA model.

3 Simulation Studies

The proposed procedure is now applied to identify a set of time series generated by known stochastic processes. We consider a large class of simulated ARMA time series that includes a diversity of orders for the autoregressive and moving average time series. In this study *ARMA(p,q)* denotes p autoregressive predictors and q moving average predictors which can be selected from a much larger set of predictors We generate 400 time series with 500 data each (using the System Identification Matlab Toolbox routines).

The implementation of the genetic algorithm follows these steps:

[1] randomly generate an initial binary population of individual m_i, $i = 1,....M$. In some experiments M is 100, in others 250, as denoted in the results. Each individual defines an ARMA model, that is the orders of the autoregressive and moving average representations, and the identified predictors;

[2] decode each chromosome; estimate the model parameters using the learning set of data given by 400 values out of the 500 of the data set; formulate predictions for the validation data set (100 values); set the number of generations N_g (50, 100, or 500, depending on the experiment).

[3] compute the fitness function values $f(m_i)$;

[4] rank each model according to the values $f(m_i)$;

[5] perform the elitism transition;

[6] select the values of the procedural parameters for selection, multi-crossover and multi-mutation, and perform the genetic operators as defined above;

[7] set $g = g + 1$; if $g > N_g$ then stop, otherwise return to 3.

The first results of these simulations are presented in Table 1. The algorithm has been applied to different model structures belonging to the ARMA class, and evaluated for two sizes of the genetic algorithm population, namely 100 and 250 individuals. To determine the effect of the number of generations admitted for the evolution we have considered three possible values: 50, 100, 500. For each ARMA structure we then generated 40 different models. Applying the genetic algorithm we evolved the populations through the generations to identify the generating process of the data. The values inside of the Table 1 represent the rate (%) of correct identification of the series randomly generated from the models for two different predictive horizons: 1 time unit and three time units. We notice the very high values of the rate of correct identification, in particular for series with low orders. This result holds for different size of the algorithm population and different lead prediction time; the rate of correct identification increases for increasing sizes of the population and increasing number of generations before stopping.

Table 1 Rate (%) of correct model identification for different predictive lead time varying the number of generation while keeping constant the population size. For each structure we tested 40 simulated time series.

Lead time	1 time unit			3 time units		
Generations	50	100	500	50	100	500
Individuals	250	250	250	250	250	250
ARMA(1,0)	100	100	100	100	100	100
ARMA(2,0)	100	100	100	95	100	100
ARMA(3,0)	95	100	100	95	100	100
ARMA(4,0)	95	95	100	95	95	100
ARMA(5,0)	90	95	97	95	90	97
ARMA(0,1)	95	100	97	85	90	97
ARMA(0,2)	90	100	97	80	90	92
ARMA(1,1)	95	95	95	85	85	90
ARMA(1,2)	90	92	95	75	85	87
ARMA(2,1)	90	90	95	80	85	90
ARMA(2,2)	85	90	95	70	80	85

In Table 1 we maintain a fixed population size increasing the allowed number of generation. We consider the models as identified only if the GA correctly indicated the AR and MA orders and the AR and MA terms of the generating process of the data in a single run. We notice that as the number of generations increases for a fixed number of individuals (250), the rate of correct model identification increases.

Table 2 Rate (%) of correct model identification for different predictive lead time varying the population size while keeping constant the number of generation. For each structure we tested 40 simulated time series.

Lead time	1 time unit		3 time units	
Generations	500	500	500	500
Individuals	100	250	100	250
ARMA(1,0)	100	100	100	100
ARMA(2,0)	100	100	100	100
ARMA(3,0)	100	100	97	100
ARMA(4,0)	95	100	95	100
ARMA(5,0)	95	97	95	97
ARMA(0,1)	97	97	97	97
ARMA(0,2)	95	97	92	92
ARMA(1,1)	95	95	90	90
ARMA(1,2)	95	95	87	87
ARMA(2,1)	92	95	82	85
ARMA(2,2)	90	95	82	82

In Table 2 we maintain a fixed number of generations and we increase the number of individuals in each generation: the rate of correct model identification increases with the number of individuals in each population.

In the case of an incorrect model identification the algorithm can, however, give good answers identifying the best predictive model on the validation data set which can be different from the model of the generating process of the data.

Moreover, the genetic algorithm we proposed has been able to identify the correct model also in cases where the gaussian white noise has comparable amplitude with the deterministic part of the time series.

Finally, a study on the behaviour of the fitness function with respect to the generation iteration shows that the convergence is reached in the majority of the cases after few generations. It was found that more generations can help in identify the data generating process without improving the predictive capabilities.

The GA has been developed within the Matlab environment because it provides a full programmable and flexible language and a large set of specialized libraries

(System Identification Toolbox, Statistics Toolbox, etc...). We implemented the algorithm on a biprocessors SUN Sparc Ultra II where the required CPU time (in seconds) to evaluate each generation ranges from 5 to 80 seconds depending on the population size.

In conclusion we regard these first results as quite satisfactory. The GA seems to perform quite well in model identification, with a high rate of correct identification. In addition, as will be reported later, when the model is not correctly identified or when the gaussian noise is quite high, the procedure is still able to generate good predictions. Moreover, the EvoARMA models seem to solve the problems of the Box-Jenkins identification procedure, since they do not require the competence and experience of a researcher to identify the generating process of the data, but represent an automatic and fast search of the time series model from the data.

References

1. Chen, C.W.S, Subset Selection of Autoregressive Time Series Models, Journal of Forecasting, 18, 505-516, (1999)
2. Choi, B.S. Arma Model Identification, Springer Series in Statistics, Springer-Verlag, (1992)
3. Hamilton, J.D., Time Series Analysis, Princeton University Press, (1994).
4. Toscano, E.M.M., Reisen, V.A., The use of Canonical Correlation Analysis to Identify the Order of Multivariate ARMA Models: Simulation and Application, Journal of Forecasting, 19, 441.456, (2000).
5. Mills, T.C., The econometric Modelling of Financial Time Series, Cambridge University Press. (1999)
6. Shah, C., Model Selection in Univariate Time Series Forecasting Using Discriminant Analysis, International Journal of Forecasting, 13, 489-500, 1997.
7. Tong, H., Non Linear Time Series: a Dynamical System Approach, Oxford University Press, 1990.
8. Weigend, A.S. and Gershenfield, N.A. (editors) Time Series Prediction: Forecasting the Future and Understanding the Past, Santa Fe Institute Studies in the Sciences of Complexity XV; Proceedings of the NATO Advanced Research Workshop on Comparative Time Series Analysis (Santa Fe, NM, May 1992.) Reading, MA: Addison-Wesley. Time Series Prediction, (1994).
9. Box, G.E.P., Evolutionary Operation: a method for increasing industrial productivity. Journal of the Royal Statistical Society, C, 6, 2, 81-101, (1957).
10. Box, G.E.P, Jenkins, G.M., Reinsel, G.C. Time Series Analysis. Forecasting and Control, San Francisco: Holden-Day, (1994).
11. Beveridge, S. and Oickle, C., A comparison of Box-Jenkins and objective methods for determining the order of non-seasonal ARMA model, *Journal of Forecasting*, Vol. 13, 419-34, (1994).
12. Lusk, E. J. and Neves, J. S., A comparative ARIMA analysis of the 111 series of the Makridakis competition, Journal of Forecasting, 3, 329-32, (1984).
13. Fogel, D.B., Evolutionary Computation, IEEE Press, New York, (1995)

14. Forrest, S., Genetic Algorithms: Principles of natural selection applied to computation, Science, 261, 872-878, (1993).
15. Holland, J.H. Adaptation in natural and artificial systems, University of Michigan Press, (1975)
16. Holland, J.H. Adaptation in natural and artificial systems, the MIT Press, Cambridge Mass, (1992).
17. Mitchell, M., An introduction to genetic algorithms, The MIT Press, Cambridge Mass. 1996.
18. Goldberg, D.E. Genetic Algorithms in Search, Optimization and Machine Learning, Reading: Addison Wesley, (1989).
19. Minerva, T., Paterlini S. and Poli I., Algoritmi ibridi per l'analisi di serie storiche finanziarie, Scienza e Business, 1, **3-4,** 56-77 (in italian), (1999).
20. Minerva T., Poli I., Genetic Algorithms to Identify Time Series Models, Department of Economics of the University of Modena Working Papers, 229, (1997)
21. Minerva, T. and Poli, I., A Neural Net Model to Predict High Tides in Venice, in S. Borra, R. Rocci, M. Vichi e M. Schader "Studies in Classification, Data Analysis and Knowledge Organization" Springer, 2001.

Evolving Market Index Trading Rules Using Grammatical Evolution

Michael O'Neill[2], Anthony Brabazon[1], Conor Ryan[2], and J.J. Collins[2]

[1] Dept. Of Accountancy, University College Dublin, Ireland.
`Anthony.Brabazon@ucd.ie`
[2] Dept. Of Computer Science And Information Systems,
University of Limerick, Ireland.
`{Michael.ONeill|Conor.Ryan|J.J.Collins}@ul.ie`

Abstract. This study examines the potential of an evolutionary automatic programming methodology to uncover a series of useful technical trading rules for the UK FTSE 100 stock index. Index values for the period 26/4/1984 to 4/12/1997 are used to train and test the model. The preliminary findings indicate that the methodology has much potential, outperforming the benchmark strategy adopted.

1 Introduction

The objective of this study is to determine whether an evolutionary automatic programming methodology, Grammatical Evolution, is capable of uncovering useful technical trading rules for the UK FTSE 100 index.

The paper is organised as follows. Section two discusses the background to the technical indicators utilised in this study. Section three describes the evolutionary algorithm adopted, Grammatical Evolution [1] [2]. Section four outlines the data and function sets used. The following sections provide the results of the study followed by a discussion of these results and finally a number of conclusions are derived.

1.1 Technical Analysis

A market index is comprised of a weighted average measure of the price of individual shares which make up that market. The value of the index represents an aggregation of the balance of supply and demand for these shares. Some market traders, known as technical analysts, believe that prices move in trends and that price patterns repeat themselves [3]. If we accept this premise, that there are rules, although not necessarily static rules, underlying price behaviour it follows that trading decisions could be enhanced through use of an appropriate rule induction methodology such as Grammatical Evolution (GE). Although controversy exists amongst financial theorists regarding the veracity of the claim of technical analysts, recent evidence has suggested that it may indeed be possible to uncover patterns of predictability in price behaviour. Brock, Lakonishok and

E.J.W. Boers et al. (Eds.) EvoWorkshop 2001, LNCS 2037, pp. 343–352, 2001.

LeBaron [4] found that simple technical trading rules had predictive power and suggested that the conclusions of earlier studies that technical trading rules did not have such power were "premature". Other studies which indicated that there may be predictable patterns in share price movements include those which suggest that markets do not always impound new information instantaneously [5] [6], that stock markets can overreact as a result of excessive investor optimism or pessimism [7], that returns on the market are related to the day of the week [8] or the month of the year [9]. The continued existence of large technical analysis departments in international finance houses is consistent with the hypothesis that technical analysis has proven empirically useful.

1.2   Potential for Application of Evolutionary Automatic Programming

As noted by Iba and Nikolaev [10] there are a number of reasons to suppose that the use of an evolutionary automatic programming (EAP) approach can prove fruitful in the financial prediction domain. EAP can conduct an efficient exploration of the search space and can uncover dependencies between input variables, leading to the selection of a good subset for inclusion in the final model. Additionally, use of EAP facilitates the utilisation of complex fitness functions including discontinuous, non-differentiable functions. This is of particular importance in the financial domain as the fitness criterion may be complex, usually requiring a balancing of return and risk. EAP, unlike for example basic neural net approaches to financial prediction, does not require the ex-ante determination of optimal model inputs and their related transformations. Another useful feature of EAP is that it produces human-readable rules that have the potential to enhance understanding of the problem domain.

1.3   Motivation for Study

This study was motivated by a number of factors. Much of the existing literature concerning the application of genetic algorithms (GA) or GP to the generation of technical trading rules [11] [12] [13] [14] [15] concentrates on the US and to a lesser extent the Japanese stock markets. Published research on this area is both incomplete and scarce. To date, only a limited number of GA / GP methodologies and a limited range of technical indicators have been considered. This study addresses these limitations by examining index data drawn from the UK stock market and by adopting a novel evolutionary automatic programming approach.

2   Background

As with any modelling methodology, issues of data pre-processing need to be considered. Rather than attempting to uncover useful technical trading rules for the FTSE 100 index using raw current and historical price information, this information is initially pre-processed into technical indicators. The objective of these pre-processing techniques is to uncover possible useful trends and other information in the time series of the raw index data whilst simultaneously reducing the noise inherent in the series.

2.1 Technical Indicators

The development of trading rules based on current and historic market price information has a long history [16]. The process entails the selection of one or more technical indicators and the development of a trading system based on these indicators. These indicators are formed from various combinations of current and historic price information. Although there are potentially an infinite number of such indicators, the financial literature suggests that certain indicators are widely used by investors [4][3][17].

Four groupings of indicators are given prominence in prior literature:

   i. Moving average indicators
  ii. Momentum indicators
 iii. Trading range indicators
  iv. Oscillators

Given the large search space, an evolutionary automatic programming methodology has promise to determine both a good quality combination of, and relevant parameters for, trading rules drawn from individual technical indicators.

We intend to use of each of these groupings as our model is developed, but in our preliminary investigation, we have limited our attention to moving average indicators.

Moving Average Indicators. The simplest moving average systems compare the current share price or index value with a moving average of the share price or index value over a lagged period, to determine how far the current price has moved from an underlying price trend. As they smooth out daily price fluctuations, moving averages can heighten the visibility of an underlying trend. A variation on simple moving average systems is to use a moving average convergence divergence (MACD) oscillator. This is calculated by taking the difference of a short run and a long run moving average. In a recursive fashion, more complex combinations of moving averages of values calculated from a MACD oscillator can themselves be used to generate trading rules. For example, a nine day moving average of a MACD oscillator could be plotted against the raw value of that indicator. A trading signal may be generated when the two plotted moving averages cross. Moving average indicators are trend following devices and work best in trending markets. They can have a slow response to changes in trends in markets, missing the beginning and end of each move. They tend to be unstable in sideways moving markets, generating repeated buy and sell signals (whipsaw) leading to unprofitable trading. Trading systems using moving averages trade-off volatility (risk of loss due to whipsaw) against sensitivity. The objective is to select the lag period which is sensitive enough to generate a useful early trading signal but which is insensitive to random noise.

A description of the evolutionary automatic programming system used to evolve trading rules now follows.

3   Grammatical Evolution

Grammatical Evolution (GE) is an evolutionary algorithm that can evolve computer programs in any language. Rather than representing the programs as parse trees, as in traditional GP [18], a linear genome representation is adopted. A genotype-phenotype mapping process is used to generate the output program for each individual in the population. Each individual, a variable length binary string, contains in its codons (groups of 8 bits) the information to select production rules from a Backus Naur Form (BNF) grammar. The BNF is a plug-in component to the genotype-phenotype mapping process, that represents the output language in the form of production rules. It is comprised of a set of non-terminals that can be mapped to elements of the set of terminals, according to the production rules. An example excerpt from a BNF grammar is given below. These productions state that S can be replaced with either one of the non-terminals expr, if-stmt, or loop.

```
S ::= expr      (0)
    | if-stmt   (1)
    | loop      (2)
```

The grammar is used in a generative process to construct a program by applying production rules, selected by the genome, beginning from the start symbol of the grammar.

In order to select a rule in GE, the next codon value on the genome is generated and placed in the following formula:

$$Rule = Codon\ Value\ MOD\ Num.\ Rules$$

If the next codon integer value was 4, given that we have 3 rules to select from as in the above example, we get $4\ MOD\ 3\ =\ 1$. S will therefore be replaced with the non-terminal if-stmt.

Beginning from the left hand side of the genome codon integer values are generated and used to select rules from the BNF grammar, until one of the following situations arise:

i. A complete program is generated. This occurs when all the non-terminals in the expression being mapped, are transformed into elements from the terminal set of the BNF grammar.

ii. The end of the genome is reached, in which case the *wrapping* operator is invoked. This results in the return of the genome reading frame to the left hand side of the genome once again. The reading of codons will then continue unless an upper threshold representing the maximum number of wrapping events has occurred during this individual's mapping process. This threshold is currently set to ten events.

iii. In the event that a threshold on the number of wrapping events is exceeded and the individual is still incompletely mapped, the mapping process is halted, and the individual assigned the lowest possible fitness value.

GE uses a steady state replacement mechanism, such that, two parents produce two children the best of which replaces the worst individual in the current population if the child has a greater fitness. The standard genetic operators of point mutation, and crossover (one point) are adopted. It also employs a duplication operator that duplicates a random number of codons and inserts these into the penultimate codon position on the genome. A full description of GE can be found in [1] [2].

4   Problem Domain & Experimental Approach

We describe an approach to evolving trading rules using GE. This study uses daily data for the UK FTSE 100 stock index drawn from the period 26/4/1984 to 4/12/1997. The training data set was comprised of the first 440 trading days of the data set. The remaining data was divided into five hold out samples totaling 2125 trading days. The division of the hold out period into five segments was undertaken to allow comparison of the out of sample results across different market conditions in order to assess the stability and degradation characteristics of the developed model's predictions. The extensive hold out sample period helps reduce the possibility of training data overfit. The rules evolved by GE are used to generate one of three signals for each day of the training or test periods. The possible signals are *Buy, Sell,* or *Do Nothing.* Permitting the model to output a Do Nothing signal reduces the hard threshold problem associated with production of a binary output. This issue has not been considered in a number of prior studies. A variant on the trading methodology developed in Brock et al. [4] is then applied. If a buy signal is indicated, a fixed investment of $1,000 (arbitrary) is made in the market index. This position is closed at the end of a ten day (arbitrary) period. On the production of a sell signal, an investment of $1,000 is sold short and again this position is closed out after a ten day period. This gives rise to a maximum potential investment of $10,000 at any point in time (the potential loss on individual short sales is in theory infinite but in practice is unlikely to exceed $1,000). The profit (or loss) on each transaction is calculated taking into account a one-way trading cost of 0.2% and allowing a further 0.3% for slippage. The total return generated by the developed trading system is a combination of its trading return and its risk free rate of return generated on uncommitted funds.

The rate adopted in this calculation is simplified to be the average interest rate over the entire data set (8.5%).

The only technical indicator that we adopt for these experiments is the moving average, where the period is determined by evolution. We choose to do this for the sake of simplicity in these preliminary experiments.

As well as the moving average the grammar also allows the use of the binary operators f_and, f_or, and the standard arithmetic operators, and the unary operator f_not. The operations f_and, f_or, and f_not are fuzzy logic operators returning the minimum, maximum, of the arguments, and 1 - the argument, respectively. We are therefore getting a mix of types for free, through the grammar and the genotype-phenotype mapping process of GE.

The signals generated for each day, Buy, Sell, or Do Nothing, are post-processed using fuzzy logic. The trading rule, a fuzzy trading rule, returns values in the range 0 to 1. We use pre-determined membership functions, in this case, to determine what the meaning of this value is. The membership functions adopted were as follows:

$$Buy = 0.0 >= Value < .33$$

$$Sell = .33 >= Value < .66$$

$$DoNothing = .66 >= Value <= 1.0$$

4.1   Data Preprocessing

The value of the FTSE 100 index increased substantially over the training and testing period, rising from 1130.9 to 5082.3. Before the trading rules were constructed, these values were normalised using a two phase preprocessing. Initially the daily values were transformed by dividing them by a 75 day lagged moving average. These transformed values are then normalised using linear scaling into the range 0 to 1. This procedure is a variant on that adopted by Allen and Karjalainen [11]and Iba and Nikolaev [10].

4.2   Selection of Fitness Function

A key decision in applying a GP methodology to construct a technical trading system is to determine what fitness measure should be adopted. A simple fitness measure such as the profitability of the system both in and out of sample is inadequate as it fails to consider the risk associated with the developed trading system. The risk of the system can be estimated in a variety of ways. One possibility is to consider market risk, defined here as the risk of loss of funds due to a market movement. A measure of this risk is provided by the maximum drawdown (maximum cumulative loss) of the system during a training or test period. This measure of risk can be incorporated into the fitness function in a variety of formats including: (return / maximum drawdown) or return - 'x'(maximum drawdown), where 'x' is a pre-determined constant dependent on an investor's psychological risk profile. For a given rate of return, the system generating the lowest maximum drawdown is preferred.

This study incorporates drawdown in the fitness function by subtracting the maximum cumulative loss during the training period from the profit generated during that period. This is a conservative approach which will encourage the evolution of trading systems with good return to risk characteristics. This will provide a more stringent test of trading rule performance as high risk / high reward trading rules will be discriminated against. The adoption of a risk conservative approach will facilitate the comparison of the final results with those of a benchmark buy and hold trading strategy.

5   Results

The results from our preliminary experiments are now given. Runs were conducted with a population size of 500 for 100 generations. Trading rules were evolved with a performance superior to that of a benchmark buy and hold strategy. Under this benchmark, an amount of $10,000 is invested in the market at the beginning of each of the test periods. The gain on this investment to the end of each period is then calculated. The best individual (set of trading rules) found to date made a profit of US$2491 over the training period.

When tested on the 5 out of sample periods following the training data set we find that this individual was consistently profitable, with the exception of a small loss in test period 4. It is noteworthy that the performance of this individual showed no significant evidence of degradation in succeeding out of sample test periods. In some cases the individual performed better out of sample than in the training period. This individual demonstrated robust performance, showing an ability to adapt to a period of crisis in the market in the second test period caused by the market collapse in Oct 1987. Plots of the index over each of the test periods and the training period can be seen in Fig. 1.

To facilitate assessment of these results, they are compared with those of the benchmark buy and hold strategy. The results of this buy and hold strategy can be seen in table 1.

Table 1. A comparison of benchmarks with the best of run individual.

Trading Period (Days)	Buy & Hold Profit (US$)	Best-of-run Profit(US$)	Best-of-run Avg. Daily Investment
Test 1 (440 to 805)	5244	1190	7959
Test 2 (805 to 1170)	-1376	5459	4356
Test 3 (1170 to 1535)	1979	2122	6973
Test 4 (1535 to 1900)	1568	-595	7109
Test 5 (3196 to 3552)	3852	10143	6315
Total	11267	18319	

In assessing these results, the market risk profile of each trading strategy should be considered. The buy and hold strategy maintains an investment of

$10,000 in the market at all times whereas the maximum investment of the developed trading system, ignoring drawdown, is $10,000. Looking at table 1 we can see the average daily investment made by the best of run individual for each test period. Averaged over all 5 test periods the developed system has an investment of $6542 in the market.

There is no clear evidence that the trading system has higher market risk than the buy and hold strategy.

6   Discussion

In evaluating the performance of any market predictive system, a number of caveats must be borne in mind. Any trading model constructed and tested using historic data will tend to perform less well in a live environment than in a test period for a number of reasons. Live markets have attendant problems of delay in executing trades, illiquidity, interrupted / corrupted data and interrupted markets. The impact of these issues is to raise trading costs and consequently to reduce the profitability of trades generated by any system. An allowance for these costs ("slippage") has been included in this study but it is impossible to determine the scale of these costs ex-ante with complete accuracy. In addition to these costs, it must be remembered that the market is competitive. As new computational technologies spread, opportunities to utilise these technologies to earn excess risk-adjusted profits are eroded. As a result of this technological "arms-race", estimates of trading performance based on historical data may not be replicated in live trading as other market participants will apply similar technology. This study ignores impact of dividends. Although a buy-and-hold strategy will generate higher levels of dividend income than an active trading strategy, the precise impact of this factor is not determinable ex-ante. It is notable that the dividend yield on most stock exchanges has fallen sharply in recent years and that the potential impact of this factor has lessened.

7   Conclusions & Future Work

GE was shown to successfully evolve trading rules with a performance superior to the benchmark buy and hold strategy. These preliminary results, with regard to the potential utility of technical analysis, are more positive than those reported in some earlier studies. Allen and Karjalainen [11] found that after transaction costs, the technical trading rules developed in their study, using a more traditional GP methodology, did not produce excess returns. However, the scope of their finding is limited as the methodology adopted in the study did not compare returns with a similar risk profile. The risk of the benchmark buy-and-hold portfolio exceeded that of the portfolio generated by the technical trading rules because an investor following the technical trading system was only invested in the market 57% of the time.

There is notable scope for further research utilising GE in this problem domain. Our preliminary methodology has included a number of simplifications, for example, we only considered moving averages, a primitive technical indicator. The incorporation of additional technical may further improve the performance of our approach.

References

1. O'Neill M., Ryan C. (2001) Grammatical Evolution. *IEEE Trans. Evolutionary Computation*. 2001.
2. Ryan C., Collins J.J., O'Neill M. (1998). Grammatical Evolution: Evolving Programs for an Arbitrary Language. *Lecture Notes in Computer Science 1391, Proceedings of the First European Workshop on Genetic Programming*, pages 83-95. Springer-Verlag.
3. Murphy, John J. (1999). Technical Analysis of the Financial Markets, New York: New York Institute of Finance.
4. Brock, W., Lakonishok, J. and LeBaron B. (1992). 'Simple Technical Trading Rules and the Stochastic Properties of Stock Returns', Journal of Finance, 47(5):1731-1764.
5. Hong, H., Lim, T. and Stein, J. (1999). 'Bad News Travels Slowly: Size, Analyst Coverage and the Profitability of Momentum Strategies', Research Paper No. 1490, Graduate School of Business, Stanford University.
6. Chan, L. K. C., Jegadeesh, N. and Lakonishok, J. (1996). 'Momentum strategies', Journal of Finance, Vol. 51, No. 5, pp. 1681 - 1714.
7. Dissanaike, G. (1997). 'Do stock market investors overreact?', Journal of Business Finance & Accounting (UK), Vol. 24, No.1, pp. 27-50.
8. Cross, F. (1973). 'The Behaviour of Stock prices on Friday and Monday', Financial Analysts' Journal, Vol. 29(6), pp.67-74.
9. DeBondt, W. and Thaler, R. (1987). 'Further Evidence on Investor Overreaction and Stock Market Seasonality', Journal of Finance, Vol. 42(3):pp.557-581.
10. Iba H. and Nikolaev N. (2000). 'Genetic Programming Polynomial Models of Financial Data Series', In *Proc. of CEC 2000*, pp. 1459-1466, IEEE Press.
11. Allen, F., Karjalainen, R. (1999) Using genetic algorithms to find technical trading rules. *Journal of Financial Economics*, **51**, pp. 245-271, 1999.
12. Colin, A. (1994). 'Genetic Algorithms for Financial Modelling', in Guido Deboeck (Editor) (1994). Trading on the edge: neural, genetic and fuzzy systems for chaotic and financial markets, New York: John Wiley & Sons Inc.
13. Bauer R. (1994). Genetic Algorithms and Investment Strategies, New York: John Wiley & Sons Inc.
14. Neely, C., Weller P. and Dittmar, R. (1997). 'Is technical analysis in the foreign exchange market profitable? A genetic programming approach", Journal of Financial and Quantitative Analysis, Vol. 32, No. 4, pp. 405 - 428.
15. Deboeck G. (1994). 'Using GAs to optimise a trading system', in Guido Deboeck (Editor) (1994). Trading on the edge: neural, genetic and fuzzy systems for chaotic and financial markets, New York: John Wiley & Sons Inc.
16. Brown, S., Goetzmann W. and Kumar A. (1998). 'The Dow Theory: William Peter Hamilton's Track Record Reconsidered', Journal of Finance, 53(4):1311-1333.
17. Pring, M. (1991). Technical analysis explained: the successful investor's guide to spotting investment trends and turning points, New York: Mc Graw-Hill Inc.
18. Koza, J. (1992). *Genetic Programming*. MIT Press.

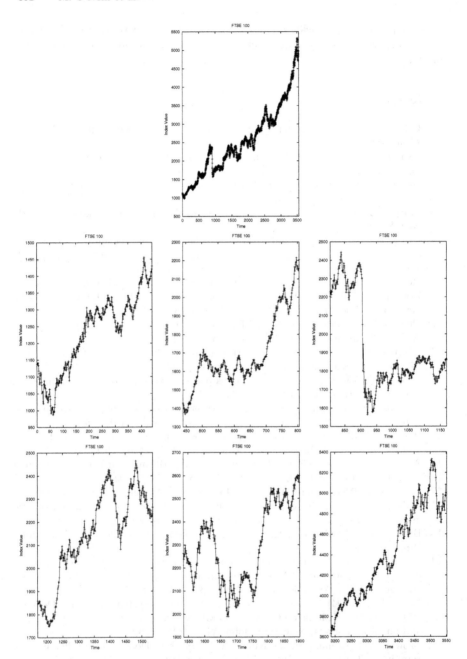

Fig. 1. A plot of the FTSE 100 over the entire data set (top), over the training period (middle-left), over the first two test periods. Days 365 to 730 (middle-center), and days 730 to 1095 (middle-right), and the third, fourth & fifth test periods (bottom row, from left to right).

Autonomous Photogrammetric Network Design Using Genetic Algorithms

Gustavo Olague

Departamento de Ciencias de la Computación, División de Física Aplicada, Centro de Investigación Científica y de Educación Superior de Ensenada, B.C. Km. 107 carretera Tijuana-Ensenada, 22860, Ensenada, B.C. México.
golague@cicese.mx

Abstract. This work describes the use of genetic algorithms for automating the photogrammetric network design process. When planning a photogrammetric network, the cameras should be placed in order to satisfy a set of interrelated and competing constraints. Furthermore, when the object is three-dimensional a combinatorial problem occurs. Genetic algorithms are stochastic optimization techniques, which have proved useful at solving computationally difficult problems with high combinatorial aspects. EPOCA (an acronym for "Evolving POsitions of CAmeras") has been developed using a three-dimensional CAD interface. EPOCA is a genetic based system that provides the attitude of each camera in the network, taking into account the imaging geometry, as well as several major constraints like visibility, convergence angle, and workspace constraint. EPOCA reproduces configurations reported in the photogrammetric literature. Moreover, the system can design networks for several adjoining planes and complex objects opening interesting new research avenues.

1 Introduction

Photogrammetric network design is the process of placing cameras in order to perform photogrammetric tasks. An important aspect of any close range photogrammetric system is to achieve an optimal spatial distribution of the cameras comprising the network. Planning an optimal photogrammetric network for some special purpose, such as for monitoring structural deformation or for determining the precise shape characteristics of an object demands special attention from the quality of the network design. Previous approaches to photogrammetric network design have attempted to identify the main stages in the process. Following the widely accepted classification scheme of Grafarend [1], network design has been divided into four design stages from which only the first three are used in close-range photogrammetry:

1. Zero Order Design (ZOD): This stage attempts to define an optimal datum in order to obtain accurate object point coordinates and exterior orientation parameters.

E.J.W. Boers et al. (Eds.) EvoWorkshop 2001, LNCS 2037, pp. 353–363, 2001.

2. First Order Design (FOD): This stage involves defining an optimal imaging geometry, which in turn determines the accuracy of the system.

3. Second Order Design (SOD): This stage is concerned with adopting a suitable measurement precision for the image coordinates. It consists usually in taking multiple images from each camera station.

4. Third Order Design (TOD): This stage deals with the improvement of a network through the inclusion of additional points in a weak region.

Photogrammetric measurement operations attempt to satisfy, in an optimal manner, several objectives such as precision, reliability and economy. The ZOD and SOD are greatly simplified in comparison to geodetic networks for which the four stages were originally developed. Indeed FOD, the design of network configuration or the sensor placement task needs to be comprehensively addressed for photogrammetric projects. This design stage must provide an optimal imaging geometry and convergence angle for each set of points placed over a complex object [2]. Photogrammetrists have acknowledged the degree of expertise needed to carry out a photogrammetric project. For example, Mason and Grün [3] developed a work called CONSENS that follows the expert system approach and uses multiple cameras in combination with optical triangulation. It outlines a method of overcoming the set of constraints and objectives presented in camera station placement. The method is based on the theory of generic networks, which constitutes compiled expertise, describing an ideal configuration of four camera stations that can be employed to provide a strong imaging geometry for the class of planar network design problems. Complex objects are divided into planes; each plane is evaluated through one of these networks and then connected with some additional cameras with the purpose of establishing just one common datum. However, the expert system approach has shown it unlikely that full automation of the network design process will be achieved, due in large part to human expert's extensive use of common-sense reasoning [2]. On the other hand, the Grafarend classification just presented serves the photogrammetric user by identifying what set of tasks needs to be implemented in designing a network. Despite the progress that photogrammetrists have made in understanding this design problem, the photogrammetric measurement technique has rarely been applied by other than experienced photogrammetrists. Although its definition seems simple, it reaches a high complexity mainly due to the numerous constraints and design decisions that need to be made. Photogrammetric network design is also difficult to obtain due to the unknown number of configurations all having very similar accuracy, but with a very different imaging geometry. Consequently, photogrammetric network design in many machine vision applications is often conducted in a trial-and-error fashion or using heuristic reasoning strategies [4]. These strategies fail at solving the problem for the case of complex objects. Moreover, the main question, how to obtain an initial configuration with an optimal imaging geometry, is unsolved and left as the responsibility of the designer. The motivation of this research is to reduce the cost of vision system design and to equip autonomous inspection systems with photogrammetric network capabilities, e.g., measurement robots used in flexible manufacturing, see Figure 1.

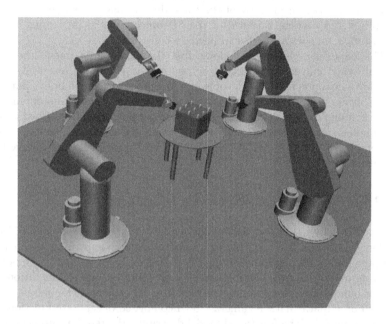

Fig. 1. Photogrammetric network simulation of four robots, each camera is mounted on the robot's hand, with the goal of measuring the box on the table.

Expert photogrammetrists regard simulation as a viable strategy to the problem of photogrammetric network design [2]. Computer simulation of close range photogrammetric networks has been successfully employed and, with the sophistication of computers, a considerable boost to interactive network design has been achieved. The process of photogrammetric network design optimization through computer simulation can follow a number of approaches. One traditional procedure is based on the stages ZOD, FOD and SOD. Given the criteria related to required triangulation precision, the initial step is to adopt a suitable observation and measuring scheme (FOD stage). This entails the selection of an appropriate camera format, focal length, and image measurement system, as well as a first approximation to suitable network geometry. Once this design stage is finished, the network is evaluated against the specified criteria. If the network fails to achieve the criteria, a new stage to diagnose and identify the problem is performed. FOD or SOD will be applied to produce the new solution. If both corrections are insufficient a completely new network will be proposed until a solution to the problem is achieved. In this way, network design is iterative in nature. The aim of this paper is to present a new simulation-based method for solving the most fundamental stage in network design. The problem is set in terms of a global optimization design [5,6], which is capable of managing the problem using an adaptive strategy. It explores the solution space using both non-continuous optimization and combinatorial search. The approach then is to minimize the uncertainty of the three dimensional measurements using as a criterion the average variance of the 3D object points, presuming that the optimization satisfies a number of primary constraints.

This paper is organized as follows: first the bundle adjustment, the mathematical model universally accepted by photogrammetrists, is reviewed in order to obtain a criterion useful to the optimization process. Then, a brief summary of the constraints on network design is presented. The problem of photogrammetric network design in terms of a stochastic global optimization is described together with implementation details about visibility and occlusion constraints related to the complexity of the search space. Finally, results are presented followed by a conclusion.

2   Photogrammetric Network Modeling

Brown originally developed the bundle method in a fully general form. Today, the bundle method is recognized as a critical factor in exploiting the mensuration potential of photogrammetry and is almost exclusively used in applications requiring high accuracy. The method accords simultaneous consideration to all sets (or bundles) of photogrammetric rays from all cameras. The bundle method is based on a mathematical camera model comprised of separate functional and stochastic models. The functional model describing the relationship between the desired and measured quantities consists of the well-known collinearity equations. The collinearity equations, derived from the perspective transformation, are based on the fundamental assumption that the perspective center, the ground point and its corresponding image point, all lie on a straight line. For each pair of image coordinates (x_{ij}, y_{ij}) observed on each image, the following pair of equations is written:

$$x_{ij} = x_p - f \left[\frac{m_{11}(X_j - X_i^c) + m_{12}(Y_j - Y_i^c) + m_{13}(Z_j - Z_i^c)}{m_{31}(X_j - X_i^c) + m_{32}(Y_j - Y_i^c) + m_{33}(Z_j - Z_i^c)} \right]$$

$$y_{ij} = y_p - f \left[\frac{m_{21}(X_j - X_i^c) + m_{22}(Y_j - Y_i^c) + m_{23}(Z_j - Z_i^c)}{m_{31}(X_j - X_i^c) + m_{32}(Y_j - Y_i^c) + m_{33}(Z_j - Z_i^c)} \right], \quad (1)$$

where (x_{ij}, y_{ij}) denote the coordinates of point j on photograph i, f and (x_p, y_p) are the camera constant and image coordinates of the principal point of the sensor defining the sensor's orientation, (X_j, Y_j, Z_j) are the object space coordinates of the corresponding point feature, (X_i^c, Y_i^c, Z_i^c) are the object space coordinates of the perspective center, and m_{kl} are elements of an orthogonal matrix which defines the rotation between the image and object coordinate systems. This system of equations assumes that light rays travel in straight lines, that all rays entering a camera lens system pass through a single point and that the lens system is distortion-less or, as is usual in highly accurate measurement, that distortion has been cancelled out after having been estimated. Due to the nature of the measurement process, observations are accompanied by errors. Because of random errors, as evidenced by the small differences between observations of the same quantity, observations can be regarded as random variables and their effects described by means of a stochastic model. Equation 1 can be linearized through the first order development using the Taylor series. A functional model can be given as

$$\mathbf{v} = \mathbf{Ay} - \mathbf{l}$$
$$\mathbf{C_1} = \sigma_0^2 \mathbf{P}^{-1}$$

where $\mathbf{l}$, $\mathbf{v}$ and $\mathbf{y}$ are the vectors of observations, residuals and unknown parameters, respectively; $\mathbf{A}$ is the design or configuration matrix; $\mathbf{C_1}$ the covariance matrix of observations; $\mathbf{P}$ the weight matrix; and σ_0^2 the variance factor. In situations where $\mathbf{A}$ is of full rank (i.e., where redundant or explicit minimal constraints are imposed), the parameter estimates $\mathbf{y}$ and the corresponding cofactor matrix $\mathbf{Q_y}$ and covariance matrix $\mathbf{C_y}$ are obtained as

$$\mathbf{y} = (\mathbf{A^T P A})^{-1} \mathbf{A^T P L} = \mathbf{Q_y A^T P L} , \tag{2}$$

and

$$\mathbf{C_y} = \sigma_0^2 \mathbf{Q_y} . \tag{3}$$

The ultimate aim of any photogrammetric measurement is the determination of triangulated object point coordinates along with estimates for their precision. The bundle method is simplified by considering two groups of parameters in the vector $\hat{\mathbf{y}}$: $\mathbf{y_1}$ comprising exterior orientation (self-calibration parameters were not considered for simplicity), and $\mathbf{y_2}$ containing object coordinate corrections. Equation 2, then assumes the form

$$\begin{pmatrix} \mathbf{y_1} \\ \mathbf{y_2} \end{pmatrix} = \begin{pmatrix} \mathbf{A_1^T P A_1} & \mathbf{A_1^T P A_2} \\ \mathbf{A_2^T P A_1} & \mathbf{A_2^T P A_2} \end{pmatrix}^{-1} \begin{pmatrix} \mathbf{A^T P L} \\ \mathbf{A^T P L} \end{pmatrix} ,$$

and the cofactor matrix $\mathbf{Q_y}$ can be written

$$\mathbf{Q_y} = \begin{pmatrix} \mathbf{Q_1} & \mathbf{Q_{1,2}} \\ \mathbf{Q_{2,1}} & \mathbf{Q_2} \end{pmatrix} .$$

The design optimization goal for precision is to achieve an optimal form of $\mathbf{Q_2}$ and therefore the covariance matrix of object point coordinates (X_j, Y_j, Z_j), considering the applicable design constraints. The criterion used in the minimization process was the average variance along the covariance matrix σ_c^2

$$\sigma_c^2 = \frac{\sigma_0^2}{3n} (trace \ \mathbf{Q_2}) .$$

Before dramatic improvements in computer processing power in recent years, a valid criticism of designing close range networks by simulation was the computation time required for a bundle adjustment after each design-iteration even for relatively small networks. As shown in [7], the covariance matrix can be obtained through the equation

$$\mathbf{Q_2} = \sigma_0^2[(\mathbf{A_2^T P A_2})^{-1} + \mathbf{K}] \,,$$

where

$$\mathbf{K} = \mathbf{M Q_1 M^T} \,,$$

and

$$\mathbf{M} = (\mathbf{A_2^T P A_2})^{-1} \mathbf{A_2^T P A_1} \,.$$

In this way, the determination of $\mathbf{Q_2}$ using this approach represents a rigorous approach that is termed Total Error Propagation (TEP.) On the other hand, it has been demonstrated [8] that for a wide range of convergent photogrammetric networks, $\mathbf{K} = \mathbf{0}$. This consideration is non-rigorous in that it implicitly assumes that exterior orientation parameters exhibit no dispersion and is called Limited Error Propagation (LEP). The perspective parameters are assumed to be error free and the variances in object point coordinates arise solely from the propagation of random errors in the image coordinate measurements. What is remarkable from a network design standpoint is that for strong networks (convergent networks) LEP is sufficiently accurate compared to TEP, causing considerable computation savings.

3   Constraints on Network Design

The problem of photogrammetric network design (PND) must deal with a series of constraints in order to propose an optimal camera distribution. The accuracy of the system is related to the imaging geometry (main objective in PND) as well as the convergence angle of each camera with respect to each object surface. In order to answer the most basic question of a favorable imaging geometry (FOD or the configuration problem) we must distinguish among the several constraints limiting the search space. Mason [9] has proposed a set of constraints and objectives that we separate into two parts:

3.1   Main Objective and Primary Constraints

Considering the constraints limiting the search space we identify the following main objective and three constraints due to the characteristics of the FOD problem:

- *Contribution to intersection angles or the imaging geometry.* Within a camera placement system the main objective is to know the contribution of each camera with respect to the others. Two fundamental questions need to be answered: how many cameras will be needed and where should they be placed. However, before answering the first question we need to answer the second one. Once we know where to place a given number of cameras, it is a trivial matter to decide on the number.

- *Convergence angle.* The reliability of image measurements from directions close to coplanar are difficult and even impossible to obtain. The minimum allowable incidence angle is dependent on the type of feature, its geometry and material. The accuracy of the measurement with respect to the convergence angle is a function of the viewing direction and the surface normal at the feature. In the case of circular targets the minimum convergence angle is about 20 to 30 degrees for the kind of retro-reflective targets that are normally used.
- *Working space constraint.* The workspace in which the photogrammetric survey is conducted can impose restrictions on the selection of an ideal imaging geometry. This constraint includes the walls of the room, any obstructions in the working environment, and the workspace of the robot where the camera could be mounted.
- *Visibility.* This constraint is related to the problem of obstructions in the environment. Viewpoints affected by occlusions caused by other objects in the workspace, or the object itself, should be avoided if possible. A ray tracing technique (POV-RAY, a free software package) was used in order to obtain visibility information of an object from different viewpoints. We created a database that was then used into our optimization process.

3.2   Secondary Constraints

Optical constraints such as field of view, depth of field, resolution, and image scale will not be taken into account when estimating a favorable imaging geometry. PND is mainly a function of the imaging geometry, as well as the convergence angle. Optical constraints lack significant importance once the camera observes the entire object. In this way, an optimal distance of the camera to the object can be defined a priori in order to measure the different object points. Thus, for the purpose here all object points appear within the field-of-view, in focus, at a given resolution and depth of field. In addition, in order to compute the exterior orientation parameters photogrammetrists affirm that the total number of points is irrelevant once a sufficient number of points are used during the simulation.

4   The Multi-cellular Genetic Algorithm

The multi-cellular genetic algorithm (MGA) then proceeds as follows:

1. An initial random population of N convergent networks that satisfy the environment constraints is chosen and is represented by (α_n, β_n), coded into a binary string representation.
2. Next, we evaluate each network, and store the corresponding maximum value of the diagonal of ΛP_n for each tree structure. This corresponds to the fitness value which says how good the network is, compared with other solutions in the population $P(t)$.

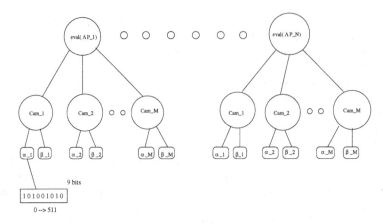

Fig. 2. The multi-cellular genetic algorithm is represented by a tree structure composed of a main node where the evaluation process is stored and several leaves corresponding to each camera. All cameras are codified in two parameters (α, β), which correspond to the cells of an artificial being. As network evaluation uses only the cells that satisfy the visibility constraints a combinatorial problem is then involved.

3. Then, we select a population of "good" networks by *tournament selection:* two networks are selected from $P(t)$ and are compared selecting the best individual according to its fitness, yielding the population $P(t + 1)$.

4. From this population, we recombine the binary strings (α_n, β_n) for each camera using the following operations:

   - Crossover, with a probability[1] Pc = 0.7. This operation was implemented using *one-cut-point*[2]. Let the two parents be:

$$\alpha_x = [\alpha_{x1}\ \alpha_{x2}\ \alpha_{x3}\ \alpha_{x4}\ \alpha_{x5}\ \alpha_{x6}\ \alpha_{x7}\ \alpha_{x8}\ \alpha_{x9}]\ ,$$

$$\alpha_y = [\alpha_{y1}\ \alpha_{y2}\ \alpha_{y3}\ \alpha_{y4}\ \alpha_{y5}\ \alpha_{y6}\ \alpha_{y7}\ \alpha_{y8}\ \alpha_{y9}]\ .$$

   If they are crossed after the random kth position = 4, the resulting offspring are:

$$\alpha'_x = [\alpha_{x1}\ \alpha_{x2}\ \alpha_{x3}\ \alpha_{x4}\ \alpha_{y5}\ \alpha_{y6}\ \alpha_{y7}\ \alpha_{y8}\ \alpha_{y9}]\ ,$$

$$\alpha'_y = [\alpha_{y1}\ \alpha_{y2}\ \alpha_{y3}\ \alpha_{y4}\ \alpha_{x5}\ \alpha_{x6}\ \alpha_{x7}\ \alpha_{x8}\ \alpha_{x9}]\ .$$

   - Mutation, with a probability Pm = 0.005. This operation alters one or more genes. Assume that the $\alpha_{y5} = 1$ gene of the chromosome α'_x is selected for a mutation. Since the gene is 1, it would be flipped into 0.

   These operations yield a new population, which we copy into $P(t)$.

5. Steps 2,3 and 4 are repeated until the optimization criterion stabilizes.

---

[1] The threshold values associated to Pc and Pm were adopted from to the literature.
[2] Due to the classification of the MGA this operation works like a *multiple-cut-point*.

Finally, this algorithm minimizes the maximum average variance along the co-variance matrix σ_c^2:

$$fitness = \min_{i=1...N} (\max \sigma_c^2) . \tag{4}$$

Thereby, the camera placement M_i relative to the world coordinate frame is optimized. Geometrically, each ΛP_i represents a hyper-ellipsoid, which changes its orientation and size as each sensor placement M_i does. Thus, an optimal placement solution is proposed, where the combined uncertainty of all points is minimal.

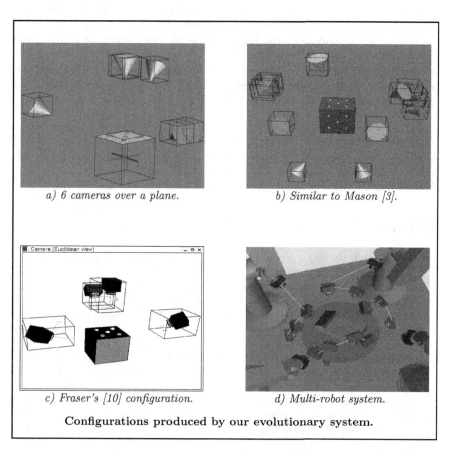

a) 6 cameras over a plane.   *b) Similar to Mason [3].*

c) Fraser's [10] configuration.   *d) Multi-robot system.*

Configurations produced by our evolutionary system.

Fig. 3. Configurations reported in the literature b) and c) were reproduced by EPOCA. Figure a) improves upon Fraser's configuration due to SOD operation, which is automatically generated. Moreover, EPOCA can be used in the case of complex objects, as can be appreciated from Figure d).

5   Examples and Conclusion

We have run a series of experiments to test the validity of our approach. We present select results in Figure 3, which show four configurations designed by EPOCA. The cameras are looking at a set of targets represented by their error ellipsoids aligned in one or two planes, as well as over a complex object. These configurations are a product of our evolutionary system. In fact, within a stochastic optimization process we cannot make conclusions from just one trial. Each configuration presented is the product of about 50 independent runs. Figure 3c illustrates a solution with four cameras looking at a planar surface. This solution is not the standard one used by the expert photogrammetrists: a photogrammetrist usually puts the four cameras at four-corners of a cube whose center contains the targets to be measured. In fact, Fraser [10] has already discussed our configuration; he noticed that this configuration is not atypical. Our experiments confirm Fraser's statement, hence the equivalence between both configurations [9].

Acknowledgements. This work was supported by CONACyT under project 35267-A. Prof. Roger Mohr contributed to the development of EPOCA with many useful ideas, criticisms and suggestions. I am also grateful to Dr. Scott Mason and Dr. Marc Schoenauer for his helpful comments and interest.

References

1. Grafarend, E.W., 1974, Optimization of Geodetic Networks, *Bollettino di Geodesia e Scienze Affini*, 33(4):351-406.
2. Fraser, C.S., 1996, Network Design, Close-Range Photogrammetry and Machine Vision, K.B. Atkinson, editor, Whittles Publishing, Chapter 9, pp. 256-281.
3. Mason, S.O. and A. Grün, 1995, Automatic Sensor Placement for Accurate Dimensional Inspection, *Computer Vision and Image Understanding*, 61(3):454-467.
4. Mason, S.O., 1997, Heuristic Reasoning Strategy for Automated Sensor Placement, *Photogrammetric Engineering & Remote Sensing*, 63(9):1093-1102, September.
5. Olague, G., 1998, Planification du placement de caméras pour des mesures 3D de précision, PhD Thesis, Institut National Polytechnique de Grenoble, France. ftp://ftp.imag.fr/pub/Mediatheque.IMAG/theses/98-Olague.Gustavo/notice-francais.html
6. Olague, G. and R. Mohr, 1998, Optimal Camera Placement to Obtain Accurate 3D Point Positions, In Proceedings of the *14th International Conference on Pattern Recognition*, Vol. 1, pages 8-10.
7. Brown, D.C., 1980, Application of Close-Range Photogrammetry to Measurements of Structures in Orbit, Vol. 1 and 2, GSI Technical Report No. 80-012, Melbourne.
8. Fraser, C.S., 1987, Limiting Error Propagation in Network Design, *Photogrammetric Engineering & Remote Sensing*, 53(5):487-493, May.
9. Mason, S.O., 1994, Expert System-Based Design of Photogrammetric Networks, PhD Thesis, Institut für Geodäsie und Photogrammetrie, Zurich.
10. Fraser, C.S., 1982, Optimization of Precision in Close-Range Photogrammetry, *Photogrammetric Engineering & Remote Sensing*, 48(4):561-570, April.

11. Olague, G., 2000, Design and Simulation of Photogrammetric Networks using Genetic Algorithms, In *American Society for Photogrammetry and Remote Sensing 2000, Annual Conference Proceedings*, 12 pages, Washington DC, USA, Copyright.

The Biological Concept of *Neoteny* in Evolutionary Color Image Segmentation – Simple Experiments in Simple Non-memetic Genetic Algorithms

Vitorino Ramos

CVRM – GeoSystems Centre, Technical Univ. of Lisbon (IST),
Av. Rovisco Pais, Lisboa, 1049-001, PORTUGAL
vitorino.ramos@alfa.ist.utl.pt

Abstract. *Neoteny*, also spelled *Paedomorphosis*, can be defined in biological terms as the retention by an organism of juvenile or even larval traits into later life. In some species, all morphological development is retarded; the organism is juvenilized but sexually mature. Such shifts of reproductive capability would appear to have adaptive significance to organisms that exhibit it. In terms of evolutionary theory, the process of *paedomorphosis* suggests that larval stages and developmental phases of existing organisms may give rise, under certain circumstances, to wholly new organisms. Although the present work does not pretend to model or simulate the biological details of such a concept in any way, these ideas were incorporated by a rather simple abstract computational strategy, in order to allow (if possible) for faster convergence into simple non-memetic Genetic Algorithms, i.e. without using local improvement procedures (e.g. via *Baldwin* or *Lamarckian* learning). As a case-study, the Genetic Algorithm was used for colour image segmentation purposes by using *K*-mean unsupervised clustering methods, namely for guiding the evolutionary algorithm in his search for finding the optimal or sub-optimal data partition. Average results suggest that the use of *neotonic* strategies by employing juvenile genotypes into the later generations and the use of linear-dynamic mutation rates instead of constant, can increase fitness values by 58% comparing to classical Genetic Algorithms, independently from the starting population characteristics on the search space.

1 Introduction

Evolution is carried out by a process dependent on mutation and natural selection. Expositions of this thesis, however, tend to overlook the fact that mutation occurs in the genotype, whereas natural selection acts only on the phenotype, the organism produced. It follows from this that the theory of evolution requires as one of its essential parts a consideration of the developmental or epigenetic processes (due to external not genetic influences) by which the genotype becomes translated into the phenotype. There are, of course, not only natural selective pressures that operate. Another evolutionary pressure strategy [1] has been to transfer the reproductive phase from the final stage of the life history to some earlier larval stage. If such a process is carried to its logical evolutionary conclusion, the final previously adult stage of the

E.J.W. Boers et al. (Eds.) EvoWorkshop 2001, LNCS 2037, pp. 364-373, 2001.
© Springer-Verlag Berlin Heidelberg 2001

life history may totally disappear, the larval stage of the earlier evolutionary form becoming the adult stage of the later derivative of it. It has been suggested that such process of *Neoteny* (i.e., the retention of juvenile characteristics in adulthood - the term was coined by *Kollman*) have played a decision role in certain earlier phases of evolution, evidence of which is now lost. More details on neoteny could be found in [2]. As his known, the process of diversity loss in genetic algorithms is often the cause of premature convergence, and as a consequence, the early convergence to an inferior local maximum. A large number of existing techniques are used to maintain diversity in *Genetic Algorithms* (GA, [3,4,5]). The present approach uses the concept of *Neoteny*. This last strategy was incorporated in simple non-memetic genetic algorithms, by simply preserving some individuals in the earlier generations (using elitism), and by randomly re-injecting this genetic material into the later generations, allowing for substantial increases in diversity, and (as it seems) for an appropriate balance between exploration and exploitation of the search space. Sections 4 and 5 are dedicated to detailed questions. Finally, and in order to study the impact of such abstract concept, yet computationally possible, a difficult combinatorial problem was chosen: color image segmentation. The aim of this paper is to study the impact of different mutation types (dynamic) and the use of older pseudo-solutions (neoteny) in the genetic implementation towards image segmentation.

2 Genetic Representation and Clustering in Image Segmentation

Image segmentation is a low-level image processing task that aims at partitioning an image into homogeneous regions [6]. How region homogeneity is defined depends on the application. A great number of segmentation methods are available in the literature to segment images according to various criteria such as for example grey level, colour, or texture. Recently, researchers have investigated the application of genetic algorithms [3,4,5] into the image segmentation problem. Probably the most extensive and detailed works are those from *Bhanu* [7] and *Pal* [8]. But we could find applications in parameter set optimization for these particular problems [7], in clustering using self-organized maps and GAs [9], within elastic-contour models [10], or in relaxation methods [11], which is and idea closely related to [12]. Finally, a fairly comprehensive review of other GA approaches in image processing is available in [13] - references include, animation, classification, feature extraction, filtering, image analysis, image processing, pattern recognition and naturally, image segmentation. As mentioned by *Andrey* [11], whether the GA is used to search in the parameter space of an existing segmentation algorithm [7], or in the space of candidate segmentations, an objective fitness function, assigning a score to each segmentation, has to be specified in both cases. However, evaluating a segmentation result is itself a difficult task. To date, no standard evaluation method prevails [14], and different measures may yield distinct rankings (the present author is nowadays developing image noise measures by MM [15], allowing for instance, their use in image filtering design by GA[s]). Another possible criterion is to think of homogeneous regions as the result of any appropriate and optimized clustering process, within the image feature space. Applications of GA[s] in clustering and grouping problems are intensively described in [16]. In the present approach, gray level intensities of RGB image channels are seen as feature vectors, and the k-mean clustering model proposed

366   V. Ramos

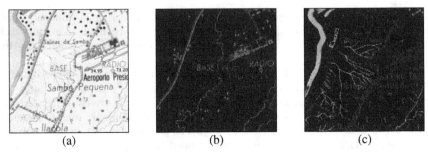

(a)   (b)   (c)

Fig. 1. (a) Original *Luanda* (Angola) Color Map ([8°50' S; 13°15' E], from N°89, 1:100.000, Aerial Photo – 1956 / Published – 1960) and as an example (b,c) the respective $3^{rd}/5^{th}$ color clusters (pointed by the GA) segmenting roads, some buildings, names, topographic lines (in red-b), and rivers, lagoons, and water lines (in blue-c).

by *J.MacQueen* in 1967, is then applied as a quantitative criterion (or GA fitness function), for guiding the evolutionary algorithm in his appropriate search. Since the *k*-mean clustering model allows to minimize the internal feature variance of each color cluster (or the maximization of the variances between different color clusters [17]), *natural* and homogeneous clusters can emerge if the GA is properly coded. In other words, the image segmentation problem is simply reformulated as an unsupervised clustering problem, and GAs are then used for finding the most appropriate and natural clusters. Since the clustering task cannot be successfully applied within the image 2D space itself (e.g., similar pixels can be very far apart) the problem is coded within another space - that one of their color features - in 3D (gray level intensities, for the three channels). By this reformulation, one can in fact guarantee that similar pixels will belong to the same color cluster. *Ramos* designed preliminary efforts in 1997 [17,18]. The standard minimization is then based on the different clustering combinations, of all points in the feature space. However, for high number of points in this 3D color space this minimization is hard to compute, since the combinatorial search space becomes very large. Nevertheless, the partition of this 3D histogram into different clusters must take in account the value of each point (that is, his frequency for a given RGB point). That is, the standard minimization described by *MacQueen*, suffers a little modification (the method becomes weighted by the frequency *f*, since the number of colors of any RGB point are an important information in the overall process [18]). Another important issue in the GA implementation is the problem's genetic coding. In order to do it appropriately, each chromosome codes the binary values of u_{ij} [2]. However, to improve the GA search time and since the number of different colors in one image can be high, each 3D feature color was submitted to a pre-partition. By this pre-procedure, the combinatorial search space is reduced, as also as the number of bits in each GA chromosome. That is, each 3D color cube (with side 256 - 8 bit images) that could represent up to 56^3 colors, was reduced to a maximum of 512 points (i.e., 512 small cubes with side 32). In other words, all RGB points that fall into a small cube are agglomerated, being the new point represented by the center of this small cube, and his frequency being equal to the sum of all frequencies of those points. Figure 1 shows some results applying this strategy.

3   Testing Dynamic Mutation Rates

Since the search space can be huge for similar applications (consider for example, satellite images or normal images at higher resolutions), and in order to speed-up the GA convergence (if possible), some experiments were conducted with dynamic mutation rates (i.e. time-dependent). As pointed by *Rudolph* [19] in 1994, one possible route to achieve global optimal convergence might be the introduction of time varying mutation *and* selection probabilities. *Rudolph* suggests using two simultaneous strategies instead of one, referring the work of *Davis* (1991, [20]), where it has been shown that the introduction of time varying mutation probabilities alone does not help. Anyway, all experiments were conducted in one-point crossover genetic algorithms (p_c= constant = 0.8), with 100 individuals (each pair of individuals selected via roulette wheel selection and windowing scaling, yields two new individuals), and within 3000 generations. Each individual was represented by a binary vector of length n = 531 (each 3 bit can code up to 8 color clusters, although only 6 are needed, since only 6 prominent colors are present in this maps / 177 color small cubes present). In these conditions, each generation g takes on average 0.0693 seconds (PENTIUM II - 333MHz / 128Mb RAM), which means about 3.5 minutes on 3000 generations (except for test #9, g_{max} = 6000 - see table 2 / image with 500^2 pixels and 214385 different colors - fig.1a). Then, 2 tests were run with constant mutation rates p_m = 0.15 (table 2 / column D=C), 8 with linear-dynamic mutation rates (column D=LD), and finally 25 with quadratic-dynamic mutation rates (column D=QD). C, LD and QD tests can be expressed by the following mutation rate expressions: • $C \Rightarrow p_m$ = 0.15; $g \in [0,3000]$; • $LD \Rightarrow p_m$ = 0.15 $(g=0)$ / p_m = 0.15/g ; $g \in [1,100]$ / p_m = 0.0015 ; $g \in [101,3000]$; • $QD \Rightarrow p_m$ = 0.15 $(g=0)$ / p_m = 0.15/g^2 ; $g \in [1,100]$ / p_m = 0.000015 ; $g \in [101,3000]$.

Many other functions were tried, some of them inspired on *Simulated Annealing* methods (SA, [21]) or in variants of it (e.g. *Adaptive Simulated Annealing, Re-Annealing, Quenching*, [22,23,24]), as the present problem seems similar [20]. In fact, both methods are applied in search-combinatorial-optimization problems, and both start from random points in the search space. Particularly interesting in the present case is that, the mutation rate in GA^s can be seen as the temperature parameter in SA^s (they both affect the convergence of the respective strategy and the balance between an appropriate exploring/exploiting character of the algorithm). Similarly, scheduling temperatures in SA^s (one of the most difficult problems to solve for this method) can be seen as the implementation of dynamic mutations on GA^s. Surprisingly (and even if several SA temperature scheduling rates were tried, generally of logarithmic or exponential nature [22,23,24,25,26,27]), the GA mutation settings that yields the best results were always the simplest ones (i.e. LD and QD - see table 1 for average results). Another fact, seems to be that the best dynamic rate should change with the starting population (compare for instance tests #2,11 and #3,12), suggesting that possibly the optimal mutation probability depends on the search landscape, the GA coding (introducing or not a multi-optimization problem and eventually several genotype mappings to the same phenotype), and finally on the objective function itself. All the previous results appear to be in some accordance with those from *Bäck* [28,29,30] and *Mühlenbein* [31]. As observed by *Bäck* [30] studying the objective function $f(x)=\sum_{i=1}^{n} x_i$ ("counting ones")   the optimal mutation probability depends strongly on the objective function value $f(x)$ and follows a hyperbolic law of the form

$$p_m(g) = \left[p_m^{-1}(0) + \frac{n - p_m^{-1}(0)}{T-1} \cdot g \right]^{-1} \tag{1}$$

$p_m = (2.(f(x)+1)-n)^{-1}$. In order to model the hyperbolic shape of the last equation, independently of the objective function, *Bäck* used a time-dependent mutation rate $p_m(g)$ (where n denotes the chromosome length, and T a given maximum of generations g). From the condition $p_m(T-1)=1/n$, the hyperbolic formulation $p_m = (a+b.g)^{-1}$ then yields Eq.1. There is however, at least one substantial difference. As mentioned by *Bäck* based on his own research and on *Muhlenbein*'s work, practical applications of genetic algorithms often favor larger or non-constant settings of the mutation rate, and the optimal mutation rate schedule analysis for a simple objective function provides a good confirmation of the usefulness of larger, varying mutation rates (in classical approaches they are generally $p_m \in [0.001,0.01]$, see [3,4]. For these reasons, *Bäck* imposed $p_m(0)=\frac{1}{2}$. However, comparing the GA efficiency based in *Bäck*'s function (Eq. 1 / with $p_m(0)= \frac{1}{2}$ or $p_m(0)=0.15$ / $T=3000$ / $n=531$), with the *LD/QD* functions, we come up with significant differences (tests #2,3,4,5). These results (although, they are statistically insufficient) probably point that optimal dynamic mutation rates should also be characterized in function of the problem's search landscape (which are manifestly different - "counting ones" *vs.* "K-means minimization function").

Bäck followed this same route [30], adapting the mutation rates according to the topology of the objective function, using the principle of strategy parameter self-adaptation as developed by *Schwefel* [29,32] for *Evolution Strategies* (ES), or similarly and independently by *Fogel* for *Evolutionary Programming* (EP, [32]). These models, however, were not applied or have been analyzed in the present framework; instead, a novel approach was considered: artificial *neoteny* (*aNeoteny*).

4   Implementing and Testing Artificial *Neoteny* (*aNeoteny*)

In order to implement *aNeoteny*, the preservation of older genotypes is a key-aspect. In general, this preservation was possible through capturing elitist individuals (*neotonic* individuals) from generations $g=0$ till $g=100$ (one per generation), and *throwing* them randomly into later generations (i.e. a randomly individual give his place in the population array, to one randomly chosen *neotonic* individual, generally for $g \in [1000,3000]$). Some questions however, are pertinent. For instance, at which period in the whole evolutionary process should this *neotonic* individuals be captured, how many should be thrown in (in the later generations), and when thus this *throwing* process should start? In order to answer these questions and to evaluate the possible contribution of *Neoteny* in the GA fitness convergence, several tests (38) were conducted (table 2). These tests can be roughly classified into six groups. The first group include tests #1 trough #8, and his purpose was to evaluate and compare the GA performance for different types of mutation with or without the implementation of neoteny (#7, 8) for the same random seed. The second group (tests #10-14) aims to performance (*J* value) in each generation, we can conclude that similar results can be evaluate the same effect but now for a different starting population (the nature of different starting populations can be analyzed, in terms of fitness, by table 3). The 3rd

Table 1. Analysis of different GA strategies with different starting populations (in the first column numeric values are from the test cardinality in table 2 / values for 3000 generations / best values for each random seed are in bold).

GA Strategy	R=9	R=7445	R=917	R=14	R=27	Average
C; 1,7.	201.61	191.15	183.36	205.07	201.52	196.54
LD; 2,8.	325.53	306.48	275.41	322.07	**314.29**	308.76
QD; 3,9.	312.69	**326.55**	286.61	290.89	270.74	297.50
LD/N; 4,10.	**326.43**	308.92	285.14	**323.07**	313.87	**311.49**
QD/N; 5,11.	314.13	321.01	**310.42**	281.46	279.56	301.32
LDN+R;35,37	323.25	290.81	284.94	317.57	312.83	305.88
QDN+R;36,38	315.93	326.28	292.87	297.81	305.70	307.72
Average	302.80	295.89	274.11	291.13	285.50	289.89

group (tests #15-25) was dedicated to evaluate if neotonic injection of genotypes could achieve the same results when incorporating that material at different generation intervals (i.e., at different evolutionary periods). Following the same concern, tests #26-30 (fourth group) analyses the effect on the average number of thrown neotonic individuals. The fifth group (tests #31-34) concerned the generation interval where neotonic individuals should be captured, and finally the sixth group (tests #35-38) analyses the effect of re-injecting one neotonic individual simultaneously with one complete random created individual. Average results, for different starting populations and strategies, can be found at table 1, and table 3 presents the random seed effect on some characteristics for these different starting populations used. The convergence of some GA strategies for each generation, in the present problem, are possible to visualize in [2].

5 Discussion and Future Work

The above strategy was applied in color maps (214385 different colors - see figure 1). For color skin mark segmentation and/or ornamental stone segmentation, see *Ramos* and *Muge*, [18]. Since the color maps have 6 prominent colors, the aim was to search for 6 color clusters. Overall results point to highly satisfactory results, namely for the segmentation of ornamental stones (29349 different colors) and for the case of human skin mark segmentation (303 colors). There is however some problems with the color map examples. The main reason is that the problem is by itself difficult (with large combinatorial search spaces), and that the pre-partition tends to reduce the discriminatory power of the overall strategy. This is mainly observed within pixels that form bounds of any important color object. Image acquisition with low resolutions interpolates somehow their gray level intensities into intermediate values (between inner and outer bounds), and the result (with pre-partition) is significantly altered, since similar pixels can belong to different small cubes (naturally with low probability). However if the number of these kind of pixels is high, the strategy tends to create himself another cluster. On the other hand, and by observing the GA achieved with half of the generations run (3000), since after this point, J values are

Table 2. Fitness for 38 GA runs (column **H**: $10^9/J$). **A**: Random seed; The number of generations for each test (#) was 3000, except for test#9 with 6000; Crossover probability was equal to 0.8; **D**: Mutation (C=constant=0.15, LD or QD / B = $Bäck$'s function with $p_m(0)$= ½ or $p_m(0)$=0.15); **E**: Average number of *Neotonic* individuals re-injected in the generation interval at column **G** (*one completely random individual is re-injected with one *Neotonic* individual); **F**: Generation interval where *Neotonic* individuals were captured (one for each generation).

#	A	D	E	F	G	H
1	9	C	0	-	-	201.611623
2	**9**	**LD**	**0**	**-**	**-**	**325.528410**
3	9	QD	0	-	-	312.694066
4	9	B [0.15]	0	-	-	203.964332
5	9	B [0.50]	0	-	-	180.236736
6	**9**	**LD**	**1**	**[1,100]**	**[1000,3000]**	**326.426236**
7	9	QD	1	[1,100]	[1000,3000]	314.125107
8	9	B [0.15]	1	[1,100]	[1000,3000]	207.823020
9	**9**	**LD**	**0**	**-**	**-**	**326.993288**
10	7445	C	0	-	-	191.146788
11	7445	LD	0	-	-	306.475341
12	**7445**	**QD**	**0**	**-**	**-**	**326.549272**
13	7445	LD	1	[1,100]	[1000,3000]	308.919431
14	7445	QD	1	[1,100]	[1000,3000]	321.010773
15	7445	QD	1	[1,100]	[500,3000]	319.063587
16	7445	QD	1	[1,100]	[350,3000]	320.481312
17	7445	QD	1	[1,100]	[320,3000]	316.784335
18	7445	QD	1	[1,100]	[300,3000]	322.772565
19	7445	QD	1	[1,100]	[285,3000]	316.366818
20	**7445**	**QD**	**1**	**[1,100]**	**[280,3000]**	**324.908299**
21	7445	QD	1	[1,100]	[279,3000]	317.947635
22	7445	QD	1	[1,100]	[277,3000]	318.843974
23	7445	QD	1	[1,100]	[275,3000]	319.100290
24	7445	QD	1	[1,100]	[200,3000]	316.244083
25	7445	QD	1	[1,100]	[150,3000]	316.556148
26	9	LD	2	[1,100]	[1000,3000]	319.990759
27	7445	QD	½	[1,100]	[280,3000]	312.452034
28	7445	QD	2	[1,100]	[280,3000]	311.933670
29	7445	QD	3	[1,100]	[280,3000]	303.136676
30	7445	QD	5	[1,100]	[280,3000]	297.281200
31	7445	QD	1	[100,200]	[1000,3000]	317.683124
32	7445	QD	1	[100,200]	[280,3000]	309.894177
33	7445	QD	1	[1,50]	[280,3000]	322.543241
34	7445	QD	1	[1,30]	[280,3000]	317.450920
35	9	LD	2*	[1,100]	[1000,3000]	323.254605
36	9	QD	2*	[1,100]	[1000,3000]	315.927842
37	7445	LD	2*	[1,100]	[1000,3000]	290.810651
38	**7445**	**QD**	**2***	**[1,100]**	**[1000,3000]**	**326.281866**

Table 3. Random seed effect on the initial population, in terms of fitness ($10^9/J$) for $g=0$ (100 chromosomes).

Random Seed	Best Fitness	Worst Fitness	Average Fitness	Std. Dev.	Sum
R=9	126.1	81.0	104.8	12.2	10484.0
R=14	125.6	78.4	100.7	11.8	10068.7
R=27	128.5	78.2	100.1	11.7	10005.4
R=917	132.9	79.0	101.3	12.3	10125.8
R=7445	128.5	79.3	101.9	12.1	10192.1

increasing very slow (compare tests #2,9 - a double value of generations adds around 0.45% in the fitness value, which is counter-productive). Future work includes three main lines. First, to study the cluster relations (clouds of points) for each segmentation problem. This can bring useful information into the GA approach, and simply geodesic neighborhood relations can be computed by using *Mathematical Morphology* [34] on the 3D color cube. Second, more relevant evaluating methods for image segmentation must be studied. In this topic, *Zhang*'s work [14] should be followed if possible. Significant improvements on the automatic design could also be achieved by using ISODATA models - since the number of clusters can be automatically chosen by the hybrid search model. Regarding the neotonic strategies, and by analyzing the results of tests #1 trough #14, it is clear that the strategy of implementing neotonic strategies and dynamic mutation rates can yield substantially (around 58%) the fitness values for the same number of generations, comparing to the use of constant mutation rates (table 1). The best result was achieved by using dynamic mutation rates and neotonic strategies (#6), although when we change the starting population the same result was only achieved by using non-neotonic implementations (test #12). It appears that starting populations with above-average individuals on it (see table 3 - random seeds $R=917$, $R=27$ and $R=7445$) do not need for higher exploring natures in the search space to achieve above-average fitness, either by incorporating a slowest decay in the mutation rate (e.g. *LD* versus *QD*) or by yielding the population diversity into the later generations via neotonic strategies. In fact, they appear to achieve good results simple by exploiting the above-average fitness and schema of their population. This is probably why, at constant mutation rates, the starting population with $R=9$ (#1) with greater average fitness, achieves better results than test#10 ($R=7445$). It appears also (#15-25) that under these circumstances, no optimal neotonic strategy can be found. In fact, throwing neotonic individuals at different temporal periods point that results can be different and only near fitness values could be found (#20). However, introducing diversity by neotonic implementations and simultaneously incorporating diversity into this diversity, by adding complete random created individuals (#35-38) could yield the fitness values to the same level, for $R=7445$. Apparently this last argument is in contradiction with the one of the last paragraph. However, is the author belief that for some starting populations (e.g., $R=7445$) the increase of diversity (increasing the exploring capabilities of the algorithm) by neotonic strategies cannot fulfill the exploiting power of simple genetic algorithms, unless, this diversity is himself increased. In other words, for a finite number of generations and for the precedent contexts, the best convergence could only be achieved either by increasing the exploring character of the algorithm, or by increasing his exploiting character, that is, renouncing for the suppose-to-be appropriate exploring/exploiting balance. This last point suggests that probably, a diversity *critical-mass* is needed within the evolutionary process, for some starting points in the search landscape. On the other hand, tests #26-30 suggest that no better results could be found by re-injecting more than one neotonic individual per generation. In fact, results decay when the number of neotonic individuals increases. Results also change if neotonic individuals are captured in different time-windows (tests #31-34 / column F - table 2). Why the interval [1,100] for capturing neotonic individuals, and the interval [1000,3000] for throwing them appear to be optimal, however, is hard to answer. Nevertheless, it appears to be important to give to the evolutionary search some time before re-injecting neotonic individuals, i.e. some evolutionary period where genetic exploitation should be processed in the classical way. Finally, a note about the neotonic strategy effect on the genetic image segmentation processing. In the case of color images, the differences between both techniques (classical *versus* neotonic) clearly affects the visual quality, namely at enhancing objects extracted (also) by the classical way.

Acknowledgements. The author wishes to thank to *David Fogel* (Natural Selection Inc. / USA), *Thomas Bäck* (Center for Applied Systems Analysis – CASA / Germany), and to *Rajeev Ayyagary* (Indian Statistical Inst. / India) for their useful references and comments on Dynamic Mutation Rates and Neoteny at COMP.AI.GENETIC (Jan.-Feb. 2000). *Ramos* work is possible under the PhD Research Fellow FCT-PRAXIS XXI (BD20001-99), PORTUGAL.

References:

1. Dawkins, Richard, 1976, *The Selfish Gene*, Oxford University Press, Oxford.
2. Ramos, V., "Artificial Neoteny in Evolutionary Image Segmentation", Proc. of SIARP'2000 - *5th IberoAmerican Symp. on Pattern Recognition*, F. Muge, M. Piedade & R. Caldas Pinto (Eds.), ISBN 972-97711-1-1, pp. 69-78, Lisbon, Portugal, 11-13 Sep. 2000.
3. Davis, L.D., 1991, *Handbook of Genetic Algorithms*, Van Nostrand Reinhold, New-York.
4. Goldberg, D.E., 1989, *Genetic Algorithms in Search, Optimisation and Machine Learning*, Addison-Wesley Reading, Massachusetts.
5. Michalewicz, Z., 1996, *Genetic Algorithms + Data Structures = Evolution Programs*, 3rd Ed., Springer-Verlag.
6. Duda, R. O. and Hart, P. E., 1973, *Pattern Classification and Scene Analysis*, John Wiley & Sons, New-York.
7. Bhanu, B. and Lee, S., 1994, *Genetic Learning for Adaptive Image Segmentation*, Kluwer Acad. Press.
8. Pal, S.K. and Wang P.P. (Eds.), 1996, *Genetic Algorithms for Pattern Recognition*, CRC Press.
9. Stern, J. M., 1992, "Simulated annealing with a temperature dependent penalty function", *ORSA Journal on Computing*, vol. 4, pp. 311-319.
10. Cagnoni, S., Dobrzeniecki, A.B., Poli, R. and Yanch, J.C., 1999, "Genetic Algorithm-based Interactive Segmentation of 3D Medical Images", *Image and Vision Computing* 17, pp. 881-895.
11. Andrey, P., 1999, "Selectionist Relaxation: Genetic Algorithms applied to Image Segmentation", *Image and Vision Computing* 17, pp. 175-187.
12. Ramos V., Almeida F., 2000, "Artificial Ant Colonies in Digital Image Habitats - A Mass Behaviour Effect Study on Pattern Recognition", Proceedings of ANTS'2000 – 2nd Inter. Workshop on Ant Algorithms (From Ant Colonies to Artificial Ants), M. Dorigo, M. Middendorf & T. Stüzle (Eds.), pp. 113-116, Brussels, Belgium, Sep. 7-9.

13. Bounsaythip, C. and Alander J.T., 1997, "Genetic Algorithms in Image Processing - A Review", *Proc. of the 3ʳᵈ Nordic Workshop on Genetic Algorithms and their Applications*, Metsatalo, Univ. of Helsinki, Helsinki, Finland, pp. 173-192.
14. Zhang, Y.J, 1996, "A Survey on Evaluating Methods for Image Segmentation*", Pattern Recognition* 29(8), pp. 1335-1246.
15. Ramos V., Muge F., 2000," On Image Filtering, Noise and Morphological Size Intensity Diagrams", RecPad'2000 – 11ᵗʰ Portuguese Conf. on Pattern Recognition, in A.C. Campilho and A.M. Mendonça (Eds.), ISBN 972-96883-2-5, pp. 483-491, Porto, Portugal, May 11-12.
16. Falkenauer, E., 1998, *Genetic Algorithms and Grouping Problems*, John Wiley & Sons, Boston.
17. Ramos, V., 1997, *Evolution and Cognition in Image Analysis*, MSc Thesis dissert. (in Portuguese), 230 pp., Instituto Superior Técnico - IST, Lisbon, Portugal, December.
18. Ramos V., Muge F., 2000, "Map Segmentation by Colour Cube Genetic K-Mean Clustering", Proc. of ECDL'2000 – 4ᵗʰ European Conference on Research and Advanced Technology for Digital Libraries, J. Borbinha and T. Baker (Eds.), *Lecture Notes in Computer Science*, Vol. 1923, pp. 319-323, Springer-Verlag , Heidelberg.
19. Rudolph, G., 1994, "Convergence Analysis of Canonical Genetic Algorithms", *IEEE Trans. on Neural Networks*, special issue on EP.
20. Davis, T.E., 1991*, Toward an Extrapolation of the Simulated Annealing Convergence Theory onto the Simple Genetic Algorithm*, PhD dissert., Gainesville: University of Florida.
21. Metropolis, N., Rosenbluth, A.W., Rosenbluth, M.N., Teller, A.H. and Teller, E., 1953; "Equation of State Calculations by Fast Computing Machines", *J. Chem. Phys.*, vol. 21, n. 6, pp. 1087 – 1092.
22. Ingber, Lester, 1989, "Very Fast Re-Annealing", *J. Mathl. Comput. Modelling*, v. 12, pp. 967-973.
23. Ingber, Lester, 1993, "Simulated Annealing: Practice versus Theory", *J. Mathl. Comput. Modelling*, v. 18, n. 11, pp. 29-57.
24. Ingber, Lester, 1995, "Adaptive Simulated Annealing (ASA): Lessons Learned", invited paper, Special issue (Simulated Annealing Applied to Combinatorial Optimization) of *Control and Cybernetics*.
25. Goldstein, J., 1988, "Mean Square Rates of Convergence in the Continuous Time Simulated Annealing Algorithm on {R}^d", *ADVAM: Adv. in Applied Mathematics*, vol. 9.
26. Rajasekaran, S., 1990, "On the Convergence Time of Simulated Annealing", *Computer and Information Science*, University of Pennsylvania.
27. Yoshimura, M. and Oe, S., 1999, "Evolutionary Segmentation of Texture Image using Genetic Algorithms towards Automatic Decision of Optimum Number of Segmentation Areas", *Pattern Recognition* 32, pp. 2041-2054.
28. Bäck, Th., 1992, "The Interaction of Mutation Rate, Selection, and Self-Adaptation within a Genetic Algorithm", in, Männer, R. and Manderick, B. (Eds.), *Parallel Problem Solving from Nature*, 2, pp. 85-94, Elsevier, Amsterdam.
29. Bäck, Th., Schwefel, H.-P., 1993, "An Overview of Evolutionary Algorithms for Parameter Optimization", *Evolutionary Computation*, 1(1), pp. 1-23.
30. Bäck, Th., Schütz, M., 1996, "Intelligent Mutation Rate Control in Canonical Genetic Algorithms", in, Ras, W. and Michalewicz, M. (Eds.): *Foundation of Intelligent Systems -* 9ᵗʰ Int. Syposium, ISMIS'96, pp. 158-167, Springer, Berlin.
31. Mühlenbein, H., 1992, "How Genetic Algorithms Really Work: I. Mutation and Hillclimbing", in, Männer, R. and Manderick, B. (Eds.), *Parallel Problem Solving from Nature*, 2, pp. 15-25, Elsevier, Amsterdam.
32. Schwefel, H.-P., 1981, *Numerical Optimization of Computer Models*, Chichester: Wiley.
33. Fogel, D.B., 1995, *Evolutionary Computation: Toward a New Philosophy of Machine Intelligence*, IEEE Press, Piscataway, NJ.
34. Serra, J., 1982, *Image Analysis and Mathematical Morphology*, Academic Press, London.

Using of Evolutionary Computations in Image Processing for Quantitative Atlas of Drosophila Genes Expression

Alexander V. Spirov[1], Dmitry L. Timakin[2], John Reinitz[3], and David Kosman[3]

[1] The Sechenov Institute of Evolutionary Physiology and Biochemistry, 44 Thorez Ave., St. Petersburg, 194223, Russia,
[2] Dept. of Automation and Control Systems, Computer Science and Engineering Faculty, St.-Petersburg State Technical University, 29 Polytechnic St, St.-Petersburg, 194064, Russia
[3] Dept. of Biochemistry and Molecular Biology, Box 1020 Mt. Sinai Medical School, One Gustave L. Levy Place, New York, NY 10029 USA

Abstract. It is well known, that organism of animal, consisting of many billions cells, is formed by consequent divisions of the only cell - zygote. In so doing, embryo cells are permanently communicating by means of biochemical signals. As a result, proper genes were being activated at proper time in proper cells of the embryo.

Modern confocal microscopes being equipped by lasers and computers give possibility to trace-through the cell fate of early embryo for such a classical model object, as fruit fly *Drosophila melanogaster*. By this approach, it is possible to retrace the detailed dynamics of activity of genes-controllers of development with the resolution on the level of individual nuclei for each of 4-6 thousand cells, composing early fly embryo. The final result of this analysis will be the quantitative atlas of *Drosophila* genes action (expression): http://www.iephb.nw.ru/ spirov/atlas. To achieve this aim we need to receive statistically authentic summary picture of detailed pattern dynamics proceeding from a large number of scanned embryos. This presupposes the elaboration of the methods of preprocessing, elastic deformation, registration and interpolation of the confocal-microscopy images of embryos.

For this purpose we apply modern heuristic methods of optimization to the processing of our images. Namely classic GA approach is used for finding a suitable elastic deformation, for registering the images and for finding a Fourier interpolation of concentration (gene-expression) surfaces. All GA programs considered are the developments of "evolution strategies program" from EO-0.8.5 C++ library (Merelo).

1 Introduction

1.1 Computer-Aided Analysis of Biological Images

The ongoing revolution in molecular genetics has progressed from the large scale automated characterization of genomic sequence to the characterization of the

E.J.W. Boers et al. (Eds.) EvoWorkshop 2001, LNCS 2037, pp. 374–383, 2001.

biological function of the genome. These investigations mark the beginning of the era of 'functional genomics' [1]. A key feature of the genomic scale approach is the automated treatment of large amounts of data. Both current and future work in this field is impossible without automated processing of images and updating of electronic image databases as well as computer-aided analysis of images [2].

Key aspects of the processing involve the segmentation of individual images, the registration of serial images, and the interpolation of 2D fields of concentrations of morphogenetic factors. Many problems of recognition, classification, segmentation, registration, and interpolation of images can be formulated as optimization problems. These optimization problems are typically difficult, involving multiple minima, grooves and a complex search space topology. Contemporary approaches based on evolutionary computations are a promising avenue for the solution of such problems.

Confocal laser scanning microscopy is a powerful tool for the imaging of gene expression in a developing embryo. Processing of images obtained by confocal scanning of stained embryos is a promising approach for acquisition of quantitative gene-expression data at the resolution of a single cell. Its advantage over other approaches is its ability to measure expression levels over embryo' space that gives 2D expression fields and makes possible 3D reconstruction of a whole embryo. But a concomitant disadvantage is its inability to measure more than a relatively small number of gene products from a given sample.

In this paper, we give a description of our computational tools by means of which observations of the expression of a small number of different genes in many samples can be synthesized into an integrated gene- expression dataset at cellular resolution (quantitative atlas of gene- expression). We are developing methods for the acquisition of such data for the segment determination genes in the fruit fly *Drosophila melanogaster*. It is well-known model organism for molecular biology studies and many biomedical problems of fundamental importance were first investigated in this organism.

The current technology of confocal scanning permits the observation of only three gene products simultaneously from an embryo. This means that a spatial expression map of all the segmentation genes must be synthesized from many observations, each made on a separate embryo. Although data can be taken from different combinations of three gene products, these patterns cannot be directly superimposed because of distortions caused by observational errors and implicit individual differences among the embryos treated.

If each embryo were stained for a common gene product and two others that vary among the dataset, it would be possible to make small coordinate transformations on each embryo, so that expression domains of the common gene product would be superimposed.

But it is impossible to perform this on a direct nucleus by nucleus basis because of individual differences among embryos and irregularity of the arrangement of nuclei, which do not lie on a rectangular or hexagonal grid. As a matter of fact, when we are making processing of such images, we have to deal with data-arrays with irregular spatial mesh. These coupled issues constitute the "registration-interpolation problem" for gene expression data. Here we describe our evolutionary computations-based approach to this registration-interpolation problem for quantitative atlas of Drosophila genes expression.

1.2   Elastic Deformations: "Stripe Straightening" Procedure

Early in the development the fruit fly embryo is shaped roughly like a hollow prolate ellipsoid, composed of a shell of nuclei which are not separated by cell membranes (Fig. 1).

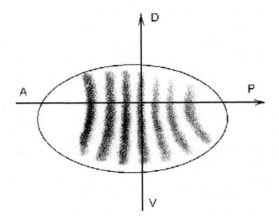

Fig. 1. Image of early (blastoderm stage) fly embryo with crescent-like stripes, in Cartesian physical coordinates. This is a confocally scanned image of an embryo stained by indirect fluorescence (immunostaining with polyclonal antisera against the *EVEN-SKIPPED* segmentation protein). Each small dot is an individual nucleus.

Fig. 1 shows that so called pair-rule stripes (early markers of the future segmental pattern [3]) are not parallel and straight, but have a crescent-like form. The curvature of the stripes is highest at the termini, and minimal at the central part. Each stripe specifies an anterior-posterior (A-P) location, and these stripes can be regarded as contours in an intrinsic coordinate system that is being created by the embryo itself. Another set of embryonic determinants exists for the dorsal-ventral (D-V) axis.

Our data processing begins with a smooth transformation of spatial coordinates. If the image is smoothly transformed such that the curvilinear coordinates are plotted orthogonally, the stripes appear straight, so the determination of these coordinates can be viewed as a "stripe straightening" procedure.

1.3   Registration of Serial Images

Our next procedure is registration of serial images. The registration cannot be performed on a direct nucleus by nucleus basis because of individual differences among embryos. Moreover, close inspection of the edge of a well-demarcated expression domain (see Fig. 1) shows irregularity due to the arrangement of nuclei, which do not lie on a rectangular or hexagonal grid. What is more, any two embryos of the same age can differ in size and form.

Our preliminary at hand computations demonstrated that the registration of *Drosophila* early images takes elastic deformations. So we can use practically the same approach, as is the case of the stripe straightening problem [4].

1.4   Interpolation

In a view of blastoderm nuclei don't form either regular square or hexagonal mesh, two-dimensional interpolation of expression patterns is non-trivial computational task. Hence following data-processing drastically depend on correct identification of interpolation function for such irregular meshes.

We used and compared several standard approaches for 2D interpolation. Particularly it was cubic spline and Fourier interpolations. However all these procedures require either regular mesh or are based on transition to a regular mesh. And it has appeared unacceptable for the level of precision, which is pursued in our project.

All this has motivated us to take advantage of interpolation by truncated two-dimensional Fourier polynomials. The power of series was chosen empirically. The Fourier coefficients were found by optimization techniques, while comparison of the interpolation result was performed on given irregular mesh of each image under treatment.

2   Methods and Approaches

The work reported here is part of a large scale project to construct a model of segment determination in the fruit fly *D. melanogaster* based on coarse-grained chemical kinetic equations [5]. The acquisition and mapping of gene expression data at a heretofore unprecedented level of precision is an integral part of this project. The current emphasis in our work is on the automated transformation of gene expression data in confocally scanned images into an electronic database of expression.

2.1   Images of Drosophila Genes Expression: The Dataset

In our experiments gene expression was measured using fluorescence tagged antibodies as described [6]. For each embryo a 1024×1024 pixel image with 8 bits of fluorescence data in each of 3 channels was obtained. To obtain the data in terms of nuclear location an image segmentation procedure was applied [7]. The segmentation procedure transforms the image into an ASCII table containing a series of data records, one for each nucleus. (About 2500- 3500 nuclei are described for each image.) Each nucleus is characterized by a unique identification number, the x and y coordinates of its centroid, and the average fluorescence levels of three gene-products.

At present over 1000 images were scanned and processed [6]. Our dataset contains data from embryos stained for 15 gene-products. Each embryo was stained for *EVE* (See Fig.1) and two other gene-products.

2.2   Technique of Genetic Algorithms

Following classical GA algorithm, the program generates a population of floating-point chromosomes. Initial chromosomes are randomly generated. After that the program evaluates every chromosome as described below; then, according to the truncation strategy, the average score is calculated. Copies of chromosomes with above average scores replace all chromosomes with a score less than average.

On the next step a predetermined proportion of the chromosome population undergoes mutation, so that one of the coefficients gets a small increment. This cycle is repeated: all chromosomes are consecutively evaluated, the average score is calculated and the winners' offspring substitutes for the losers in the process of reproduction.

All GA programs considered there are the developments of "evolution strategies program" from EO-0.8.5 C++ library [8]. This program implements a simple evolution strategy as defined by Rechenberg and Schwefel. It includes the so called *floating-point chromosomes* with increments, random *selection*, random *reproduction*, with *replacement* and *generational termination*.

Each chromosome is composed of genes, each gene is composed of values and standard deviations (*sigma*), used for mutation. The genetic operators used are *uniform crossover* and *mutation*. Mutation-operator acts on the gene values and sigmas: each gene in the chromosome has it's own sigma. Replacement takes the old and the new population, eliminate those that are not needed, and leave the rest for the other generations. Generational termination means that the program ends run after a fixed number of generations.

We developed our own *fitness functions* [9], suitable for our problems. Our initial populations is created by random variations of initial set of values from *init-file*. Our fitness functions read data from *input-file*. We had to modify the *ES.cpp* as well. All other files are taken from the initial "evolution strategies program" [8]. The program was compiled in the Microsoft Visual C++ 6.0.

2.3   Elastic Deformation Algorithm

Our goal here is to find the new, true A-P and D-V coordinates on the image (See Fig.1). We denote the true A-P coordinate by $\hat{x}$ and the true D-V coordinate by $\hat{y}$. For now, we can assume that $\hat{y} = y$.

We approximate the true coordinate system by a Taylor series. We expand in a Taylor series to third order around the origin (For details See[4]). After elimination all but three terms from the series, we found the model of image transformation as

$$\hat{x} = x + Axy^2 + Bx^2y + Cx^3. \tag{1}$$

The Fitness Function for Elastic Deformation. Our own fitness function for EO-0.8.5 C++ library performs evaluation of values of A, B and C parameters from eq.(1). The function applies the discussed Tailor transformation (1) to an image coordinates and then calculates *cost function* for the results of it's elastic deformation.

The Cost Function for Elastic Deformation. We use the following cost function. Each image under consideration was subdivided into a series of longitudinal strips. Then each strip is subdivided into boxes and the mean value of the brightness is calculated for each box. Each row of means gives the local profile of brightness along each strip. The cost function is computed by comparing each profile and summing the squares of differences between the strips. The task of the GA is to minimize this cost function.

2.4   Image Registration Algorithm

The stripe straightening procedure essentially simplified the organization of the patterns, so search of registration algorithm was simplified too. As a result, we had an opportunity to concentrate only on an A-P (X) coordinate to neglect the contribution of D-V (Y) one. Registration is implemented using pairs of digitized images. According to our approach, the spatial coordinates x of nuclei of one of the couple of embryos are transformed using the following polynomial expression (one-dimensional elastic transformation):

$$\hat{x} = \alpha x + \beta x^2 + \gamma x^3 + \delta x^4. \tag{2}$$

Now the problem is reduced to the determination of factors α, β, γ and δ.

The Fitness Function for Registration. Our own fitness function for EO-0.8.5 C++ library performs evaluation of the parameters of elastic deformation α, β, γ and δ from eq.(2). The function (2) applies the discussed polynomial transformation of coordinates to an image and then calculates cost function for the result of elastic deformation.

The Cost Function for Registration. We use the cost function similar to the previous one. We take the longitudinal strip from equatorial part of each embryo image. Then the strip is subdivided into boxes and the mean value of brightness is calculated for each box. Each row of means gives the local profile of the brightness along each embryo. The cost function is computed by comparing the profiles of pair of registered images and summing the squares of differences between the profiles.

2.5   Interpolation Algorithm

Our preliminary investigations show that all available mathematical packages and libraries give no possibility to achieve perfect Fourier interpolations for our expression surfaces with the irregular mesh. That is why we had to work out our own approach for our 2D-interpolation problem.

We found empirically that following truncated two-dimensional Fourier series are appropriate for solving of the problem:

$$C_{fourier} = C_0 +$$
$$\left[C_1 \cos\left(\tfrac{x}{4}\right) + C_2 \sin\left(\tfrac{x}{4}\right)\right] + ... + \left[C_{35} \cos\left(16 \cdot x\right) + C_{36} \sin\left(16 \cdot x\right)\right] + \quad (3)$$
$$\left[C_{37} \cos\left(\tfrac{y}{2}\right) + C_{38} \sin\left(\tfrac{y}{2}\right)\right] + ... + \left[C_{43} \cos\left(3 \cdot x\right) + C_{44} \sin\left(3 \cdot x\right)\right],$$

where $x = 2Pi \cdot x^0$, $y = 2Pi \cdot y^0$ and x^0, y^0 are raw co-ordinates of nuclei.

The power of the Fourier series was chosen empirically. The suitable power for x-coordinate is 36 and for y-coordinate is 8. The problem is to find all 45 coefficients for this series.

The Fitness Function for Interpolation. Our own fitness function for EO-0.8.5 C++ library performs evaluation of 45 coefficients for truncated 2D Fourier series (3). The function applies the Fourier transformation (3) to an image and then calculates cost function for the results of image interpolation.

The Cost Function for Interpolation. The cost function in this task is simply the sum the squares of differences between initial and calculated values of expression for all 2500-3500 nuclei of an embryo under treatment.

3   Results and Discussion

3.1   Search Spaces Features

The above-described task of image elastic deformations turned out to be a difficult numerical problem. This is caused first of all by the unusual geometry of search space with grooves, the bottom of which have several local minima [4].

On the contrary, owing to quasi-unidimensional form of the straightened images, the search space for our registration tasks turned to be smooth with unique extemum. Gradient methods are faster, then evolutionary search, but not robust enough. Large scale mis-registrations (e.g. where stripe 1 is mis-registered to stripe 2) is a problem for simplex method.

We fail to find any unusual features doing cross-sections through search space in case of our 2D Fourier interpolation for irregular mesh. Apparently the only serious difficulty is the number of coefficients under estimation.

3.2   Construction of a Quantitative Atlas of Gene-Expressions

One of the main objectives of our project is to construct a map of all relevant expression domains at each stage of early Drosophila embryo development. The purpose of registration-interpolation approach, considered there, is to construct a map of all segmentation gene products from a set of embryos of the same age.

The early fruit fly embryo has a form of prolate ellipsoid of revolution. The morphogenetic factors of the early Drosophila embryo are distributed along it's surface and have features along the principal axis of ellipsoid (i.e., anterior-posterior axis) as well as along the transverse axis (dorsal-ventral axis). These differences explain the necessity construction of the bank of these 2D expression patterns followed by 3D reconstruction of an embryo. Construction of an integrated dataset of the representative one-dimensional expression profiles can be considered as a first approximation of the real bank of patterns [10].

By now, we have got a representative surfaces of concentrations for 14 segmentation factors, by 8 temporal classes. These materials are represented on the corresponding pages of the Web database HOX Pro: http://www.iephb.nw.ru/ hoxpro and specialized Web base DroAtlas: http://www.iephb.nw.ru/ spirov/ atlas

General view of expression gradient (surface of expression) of primary morphogen *bicoid* (*bcd*) is shown in Fig.2 in the form of 3D diagram. There are good reasons to suppose that the early embryo organization along the anterior-posterior axis is under primary control of this gradient [3]. In turn, the different segmentation genes are activated at different concentration values of *bcd*. The superposition of the *bcd* gradient and domains of activity of four primary segmentation genes is shown in Fig.3.

Fig. 2. 2D morphogenetic exponential gradient of *bicoid*.

Fine spatial and temporal details of segmentation patterns are very important for the understanding of general mechanisms of control of embryo development. But, the revealing of these details is possible only by using of rather sophisticated

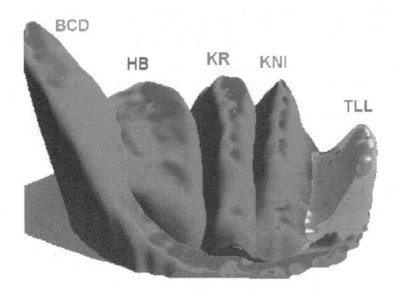

Fig. 3. Superposition of surfaces of expressions of *bcd* gradient and four zygotic genes from gap group.

methods of computer processing, classification and image analysis supported by the development and maintenance of data bank of these patterns.

4   Conclusions

The present state-of-art in the developmental genetics urgently faced with the problem of the analysis of the genes-controllers activity simultaneously (in parallel) for all cells of developing embryo. This actual problem produces new higher demands both to the experimental approaches and to the computer processing of the experimental data.

The variability of gene activity under the experimental resolution on the level of individual nuclei still stay high. The embryo cells does not form an ideal mosaic, though, they are arranged in layers close to the hexagonal packing. High variability of gene activity and the geometry of the cell-layers arrangement make the problem of unification of large arrays of expression-data very actual. In so doing, the important problem is not to lose the valuable details of the experimental data.

The using of the heuristic procedures of optimization is the most natural method of comparing, registration and interpolation of these data sets without losing of important details.

Among the all modern heuristic approaches we have chosen the Genetic Algorithms technique applied per se as well as in combination with the simplex method. Using of this technique in framework of the project gave possibility to perform the full range of procedures of processing of embryo confocal images, necessary for the construction of the computer data bank and atlas of unified gene activity patterns.

Acknowledgments. This work is supported by RFBR grant No 00-04-48515; INTAS grant No 97-30950; USA National Institutes of Health grant RO1-RR07801 and GAP awards RBO-685 and RBO-895.

References

1. Lander,E.S.: The new genomics: Global view of biology. Science 274:536, 1996.
2. Sanchez,C., Lachaize,C., Janody,F., et al.: Grasping at molecular interactions and genetic networks in Drosophila melanogaster using FlyNets, an Internet database. Nucleic Acids Research 27:89–94, 1999.
3. Akam,M.: The molecular basis for metameric pattern in the Drosophila embryo. Development 101:1–22, 1987.
4. Spirov,A.V., Timakin,D.L., Reinitz,J. and Kosman,D. Experimental Determination Of Drosophila Embryonic Coordinates By Genetic Algorithms, the Simplex Method, And Their Hybrid. In S. Cagnoni et al., editors, Springer Verlag Lecture Notes in Computer Science, Number 1803, pages 97–106. 2000.
5. Reinitz,J., Kosman,D., Vanario-Alonso,C.E. Sharp,D.: Stripe forming architecture of the gap gene system. Developmental Genetics. 23:11–27, 1998.
6. Kosman,D. and Reinitz,J.: Rapid preparation of a panel of polyclonal antibodies to Drosophila segmentation proteins. Development, Genes, and Evolution 208:290–294, 1998.
7. Kosman,D., Reinitz,J. and Sharp D.H. Automated assay of gene expression at cellular resolution. In Altman,R., Dunker,K., Hunter,L. and Klein,T. editors, Proceedings of the 1998 Pacific Symposium on Biocomputing, Singapore: pages 6–17. World Scientific Press, 1997.
8. Merelo,J.J.: EO Evolutionary Computation Framework. http://geneura.ugr.es/ jmerelo/eo
9. Spirov,A.V., Timakin,D.L., Reinitz,J. and Kosman,D. Using of Genetic Algorithms in Image Processing for Quantitative Atlas of Drosophila Genes Expression. http://www.mssm.edu/molbio/hoxpro/atlas/atlas.html
10. Myasnikova,E., Kosman,D.,Reinitz,J. and Samsonova,M.: Spatio-temporal registration of the expression patterns of *Drosophila* segmentation genes, *Proc. 7-th Int. Conf. on Intelligent Systems for Molecular Biology*, pages 195–201, AAAI Press, 1999.

Selection of Behavior in Social Situations

Application to the Development of Coordinated Movements

Samuel Delepoulle[1,2]*, Philippe Preux[2], and Jean-Claude Darcheville[1]

[1] Unité de Recherche sur l'Évolution des Comportements et des Apprentissages (URECA), UPRES-EA 1059, Université de Lille 3, B.P. 149, 59653 Villeneuve d'Ascq Cedex, France, lastname@univ-lille3.fr

[2] Laboratoire d'Informatique du Littoral (LIL), Université du Littoral Côte d'Opale, UPRES, B.P. 719, 62228 Calais Cedex, France, lastname@lil.univ-littoral.fr

Abstract. The law of effect is a very simple law which relates the probability of emission of a behavior by a living being to the consequences of the emission of this behavior by this living being in the past. As such, this law models very basic learning. This law can be considered as an experimental fact as far as it has been observed for a whole range of living beings including human beings. In this paper, we first show that this general law can be the result of a selection process such as natural selection. Then, we show that the implementation of this law can lead to the design of adaptive systems which can mimic very closely the way a new-born develops coordinated movements. To sum-up, we show that the ability to learn such coordinated movements and exhibit adaptive behaviors can result from a multi-stage process of selection.

1 Introduction

The selection of behaviors by their consequences is of paramount importance in the study of behavior of living beings among behavior psychologists, and ethologists. Initially introduced by Thorndike as the "law of effect" [1,2], this principle has been put to the test, verified, and reported in an innumerable amount of situations and publications. The law of effect holds for all living beings, and it has been investigated for creatures ranging from flies to human beings. The law of effect solely deals with learning during lifetime: hence, it deals with the evolution of behavior of a living being (that is, the evolution of its behavioral repertoire) during its lifetime. We want it to be clear that the law of effect has nothing to do with natural selection proposed by Darwin and successors. Natural selection acts along generations on populations of living beings; selection of behaviors by their consequences and its model, the law of effect, acts along lifetime among populations of behaviors. Given a living being, natural selection has produced innate abilities, while selection of behaviors by

* during this work, Samuel Delepoulle acknowledges the support from "Conseil Régional Nord-Pas de Calais, France", under contract n 97 53 0283

E.J.W. Boers et al. (Eds.) EvoWorkshop 2001, LNCS 2037, pp. 384–393, 2001.

their consequences produces its acquired, or learnt, abilities. In the following, we consider that natural selection is well-known by the reader but we will describe selection of behaviors by their consequences in more details.

The law of effect merely states that when a living being emits a behavior that brings it back favorable consequences, the probability that the living being emits this behavior again in the "same" situation increases. This process leads to a selection of behaviors during the lifetime of an animal. It may seem strange, and even unbearable, to state that human behaviors (at least, some of them) are driven by such a simple law. However, two things should be made clear: it is not claimed that all animal behaviors follow this law; it is also clear that so-called "cognitive" activities are taking place in human beings, as well as in most animals. Anyway, some people argue that cognitive activities may also be explained using the law of effect, but this old and important debate is clearly out of the scope of this paper. The fact is that the evolution of complex behaviors of living beings (including human beings) have successfully been explained using the law of effect [3]. Indeed, the interaction of a set of agents which behavior is driven by the law of effect can show complex patterns of activity, and a continuously adaptive behavior. In this regard, adaptation is considered as learning [4]. A nice thing is that the law of effect is a selectionist law, just like natural selection [5]. However, natural selection works along generations on populations of individuals, while the law of effect works on populations of behaviors during lifetime. Clearly, the law of effect may also have indirect consequences on future generations according to the Baldwin effect. Different authors have studied the influence of learning on evolution (see e.g. [6,7,8]). The interaction between the learning process and the evolution process being important, we come back to this issue in Sec. 4. It is also worth noting that the implementation of the law of effect in computer programs is not obvious although it is simple to express verbally. If that was the case, the problem of creating truly adaptive artefacts would probably be solved to a greater extent.

In this contribution, we aim to set a bridge between different results we have recently obtained. This work is the result of a collaboration between psychologists specialized in the experimental analysis of behavior and computer scientists. The line of thought will give the organization of the sequel of this contribution. First, we show that the law of effect may be acquired by way of genetic selection. Second, we show that agents which behaviors are selected according to the law of effect perform well in a social situation. Then, grounded on a parallel between the law of effect and reinforcement learning, we show that it is possible to model in a very realistic way the development of the reaching arm movement in human beings using reinforcement algorithms. While these results are interesting to scientists studying life, they are also interesting for computer scientists to build new adaptive systems that are deeply inspired by available knowledge on the behavior of living beings, the most adaptive systems we are aware of so far.

2   The Selection of the Law of Effect

In this section, we present briefly the result of computer simulations showing that the law of effect can have been selected among generations by natural selection. As far as we are aware of, this important result has never been demonstrated so clearly up to now. This section summarizes chapter 12 of [9].

In this simulation, a population of agents evolves by way of a genetic algorithm. Each agent interacts with its environment via a set of N input sensors, and N behavior units. The input sensors are sampled at each time step. The emission of a behavior is performed by the behavior units ; at each time step, at most one behavior unit can be active, emitting the corresponding behavior. The activity of an agent is coordinated by a neural network. This neural network is made of C layers of N neurons. A neuron of a layer is never connected to a neuron of the same layer. A neuron can be connected to any neuron of the 2 surrounding layers. Each input sensor is connected to all N neurons of the first layer. Each behavior unit is connected to one neuron of the last layer and feeds these N neurons back. The response of each neuron is characterized by 6 real numbers and a boolean value which indicates whether the neuron is active or not. Due to the inter-connection topology, each neuron is also characterized by $2 \times N$ weights. Each weight itself is characterized by a quadruple. Without going into all details which would require much more space, the parameters of a weight describe how the weight evolves along time and activations. Finally, the whole network is characterized by two numbers A_c and A_p. The activation of the neurons of an agent is made at random: A_c indicates the number of neurons that are activated at each time step, while A_p specifies the number of weight updating steps (or, learning step) that are performed by the network. At each time step, a neuron is drawn at random. So, in general, there are less than A_c different neurons that are activated at each time step, while some of the neurons are activated more than once.

An initial population of 10 agents is formed at random, that is that the characteristics of each neuron are drawn at random. Then, the agents are evaluated and follow the usual genetic algorithm loop of selection, duplication, recombination, and mutation. Recombination is one-point crossover which cuts only between two neurons (instead of between any two bits). Five different mutations are used, each with its own probability: mutation of a weight which means changing the value of a weight of $\pm$ at most 10 %; mutation of a neuron which means resetting all the parameters of a neuron at random; mutation of expression which means toggling the activation bit; mutation of A_c or A_p.

The task that has to be performed by the agents is a discrimination task: two stimuli S1 and S2 are presented alternately; in presence of a certain stimulus, the agent must emit behavior B1; in presence of the other stimulus, the agent must emit behavior B2. Each stimulus is input on a given neuron of the input layer. If the agent behavior is correct with regards to the presented stimulus, then it is reinforced: the evaluation of the agent is equal to the number of reinforcers received over 10 sessions, also called its "score". In each session, a certain association has to be learnt: S1-B1, and S2-B2, or S1-B2, and S2-B1. Then, the next session, the other association may become the correct one. 50 % of the sessions reinforce the S1-B1/S2-B2 associations, while the 50 other % of sessions reinforce the other association. No extra-stimulus indicates which association is reinforced in a given session. Furthermore, it should be mentioned that sessions are not known by the agents; for agents, there is no difference except the fact that they do not receive a reward for the same behavior. Each session is made of 1000 presentations of stimulus. This way of changing the reinforced association selects adaptive agents, not only agents that can learn a given association.

Though this description of agents may seem complicated, it is rather simple and natural and seems to be rather minimal with regards to the task we wish to accomplish.

Simulations have been performed. After 200 generations, approximately 90 % of the agents are able to accomplish the discrimination task.

So, this means that a rather simple neuron architecture can be selected along generations to realize a discrimination task. Thus, discrimination abilities in living beings might be the product of evolution. Being able to discriminate, that means that these agents follow the law of effect: their behaviors are selected by their consequences. So, this means that agents whose behavior are selected by their consequences can be selected by selection along generations, that is, by natural selection.

3 The Law of Effect Selects Social Behaviors

In this section, we show that agents which behaviors are selected by the law of effect outperform by far other agents in a social task, the minimal social situation [10]. This situation is well-known in the social psychology literature. This section summarizes [11].

We do not describe the minimal social situation itself but how we have modeled it to simulate it: this will provide the reader enough information to follow the line of thought. Interested readers are kindly asked to refer to [10] for a thorough description of this situation. We have simulated the realization of this experience by the agents described in the previous section. In this case, the task to be accomplished is as follows: in 50 % of the sessions, if the behavior of agent A is B1, then the agent B receives +1; if A emits B2, then B receives -1; if B emits B1, then A receives +1; if B emits B2, then A receives -1. In the other 50 % of sessions, the consequences of each agent behavior are exchanged: B1 pays -1 to the other agent while B2 pays +1 to the other agent. So, one's behavior only affect his party's reward. The best strategy is for both to emit the behavior that provide +1 to its party. This behavior is a cooperation.

Two sets of simulations have been performed. The only difference between these two sets is the way how the initial population is formed. In the first set S_r, the initial population is formed at random; in the second set S_s, the initial population is formed with agents that have been able to select their behaviors according to their consequences at least some times to times: they are not very good at this task, but they are able to achieve it with statistical significance.

The performance of an agent in the minimal social situation is measured by the number of +1 it receives. Figure 1 plots the evolution along the simulation of these two populations. There is a very clear advantage for agents of S_s which obtain good performances after 50 generations, and already much better performances than S_r after only 20 generations.

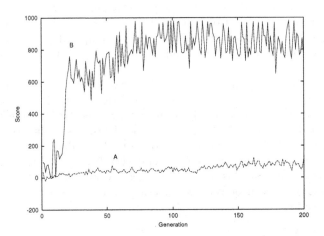

Fig. 1. Evolution of the performance of agents that succeed in the minimal social situation. The curve (A) corresponds to the case where the initial population is formed with agents selects their behaviors according to their consequences that at least some times to times. The curve (B) corresponds to the case where the initial population is formed at random. The difference in performance of the two populations is striking. The best possible performance would be rated 1000.

Thus, in this section, we have shown that agents which select their behaviors according to their consequences can show a striking advantage in a social situation which selects behaviors that lead to cooperation. We emphasize the similarities between the dynamics observed here with the population S_s and with human subjects. So, in this situation, the law of effect selects cooperative behaviors.

4   Social Behaviors Select the Dynamics of the Arm Reaching Movement

Up to now, we have shown that the law of evolution of the behavioral repertoire of a living being during its lifetime can be the product of natural selection. So, natural selection (NS) has induced the behavior selection mechanism (BS), that is, the law of effect which yields the selection of behaviors by their consequences along lifetime; this induction is denoted by thick arrows in figure 2. It is known that behavior during lifetime can modify the genetic material of species by the

Baldwin effect [6]; this feedback is denoted by thin arrows in figure 2. At this stage, it is worth noting that due to the law of effect, the behavior of a living being is modified by its environment, among which, the living beings with which it interacts during its lifetime; this interaction is shown with zigzagged arrows between two NS/BS feedback loops, that is, between two living beings in figure 2. Then, we have a sketch of the way different evolution processes interact to yield living being behaviors. It should be clear that these processes happen at different level (genetic/behavior) and different time scales, as well as the interactions between these processes (see figure 2). NS deals with the adaptation of genomes along generations, while BS deals with the adaptation of behavior during lifetime.

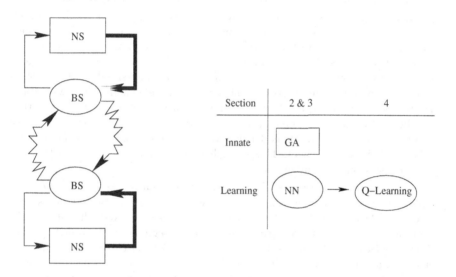

Fig. 2. On the left part, we sketch how the natural selection process (NS) interacts with the behavior selection process (BS), as well as, how the behavior of one living being interacts with that of other living beings (zigzagged arrows): the upper part and lower part NS/BS loops each represents one living being, while the zigzagged arrows represent the interaction between living beings. Clearly, all the environment of a given living being may alter its behavior, but we have only represented the interaction between two beings for the sake of clarity. This sketch also shows the interaction between innate and acquired abilities. It is note worthy that the processes happen on different time scales: NS acts along generations while BS acts along lifetime; likewise, the time it takes for one process to alter an other one is variable, ranging from fractions of a second to generations. The right part indicates how the various processes have been implemented in our work. The shape of the boxes in the left part (a rectangle for NS, an ellipse for BS) refers to the table in the right part: NS is implemented with a genetic algorithm (GA) while BS is implemented by way of neural networks (NN) in Sec. 2 and 3, with Q-learning in Sec. 4.

The two processes are implemented by way of two different algorithms: a genetic algorithm implements natural selection while neural networks and Q-learning implement the behavior selection. Natural selection yields the ability to learn to living beings, while behavior selection lets them acquire, and develop their behavioral repertoire, and learn new abilities during its lifetime.

We have used two algorithms (NN, and Q-learning) to implement the behavior selection process. However, it should be stressed that this is a purely technical point: it is only a matter of efficiency in computations and simplicity in software development: we could have used the neural networks of Sec. 3 in place of Q-learning in Sec. 4. The fact is that we are currently updating the simulator so that this is done that way, using NN instead of Q-Learning.

This having been said, we now describe the interaction (during their lifetime) between two agents which behaviors are selected by their consequences.

Lots of organs and members of living beings are made of cooperating elements: muscles, tendons, ligaments, bones, ... not mentioning neurons which play an important role in the overall organization of movements and behaviors in general. In this section, we show that a combination of cooperative agents whose behaviors are selected by their consequences may simulate the development of the arm reaching movement showing remarkable similarities with those of a human baby facing the "same" situation. Instead of using the agents that have been selected in the simulations reported in the previous sections, we use the Q-learning algorithm to model the law of effect. Indeed, Q-learning models rather well the law of effect. This point has been argued at length in [12] and [13]. This section summarizes [14].

Our goal is to model the reaching movement of an arm. We describe here the 2 dimensional case. An arm is modeled as being composed of two segments joint by an articulation (see fig. 3). One extremity of the arm is fixed (the "shoulder") while the other one can move (the "hand"). Each segment is controlled by a set of two muscles: an agonist muscle and an antagonist muscle. So, there are four muscles. Each muscle can be in either 1 of 50 states which indicate its tension. At each time step, this tension can increase or decrease of 1 unit. Increasing or decreasing its tension are the two possible behaviors of each muscle-agent. Each agent receives three inputs: its current state, and two stimuli that provide very poor visual information. Actually, each agent models a muscle and the motor-neuron associated to it. This gives relevancy to the fact that the agent receives visual information. Finally, the behavior of the agent has a cost in energy. The model, though simple, is rather realistic.

Like the human babies with whom the experiment has been done, the arm receives a reward only when it puts its hand into a certain region of the space. Otherwise, it does not get any reward. So, the system of agents has to reach a balance between the cost of emitted behaviors and the need to put its hand in the reinforcement zone to get its reward. The agent is not initially aware of the fact that it can receive any reward. At the beginning of the simulation and as long as its hand does not come into the reinforcement zone, the agent receives no consequence for its behaviors, except their energetic cost. The visual system indicates whether the hand is rather close of far from this region with an integer 0, 1, 2, or 3 and the relative position of the hand with regards to the region: north-west, north-east, south-west, south-east. A very large majority of points of the space are merely perceived as far, in a certain quadrant (see fig. 3). This visual system has been designed to be rather crude. There are two reasons for that: first, in the new-born, the visual system is also very crude. Second, from a computer scientist point of view, if the visual system was very acute, then, it would be very simple to solve the problem we have assigned the arm by simply moving the hand towards the target using a mere gradient algorithm. As far as the vision does not provide enough information to guide the hand towards the target, such a trick is not possible. This is precisely what we wanted to avoid: putting the solution of the problem into the agent.

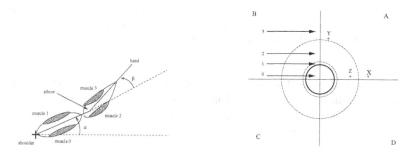

Fig. 3. The leftmost part sketches the architecture of the arm (see text for description). The rightmost part indicates how the arm perceives its environment. The visual field is split into 4 quadrants (A, B, C, and D) and 4 circular zones (0, 1, 2, and 3). Positions X and Z are perceived differently; positions X and Y are perceived as identical though more remote than X to Z. So, this visual system is rather crude and does not help very much the arm to reach the reinforcement zone.

The simulation of this arm shows different stages of development of the movement (see fig. 4) : initially, the hand wanders erratically; after some times, it passes into the reinforcement zone, by chance, but it is unable to remain in it; after the position of the arm has been reset to its initial position, and after many attempts, the arm develops a smoother and smoother direct movement to reach the zone. Then, it is time to change the reinforcement zone. Of course, the arm will first seek the zone in its former position. The thing is that as far as the arm will not receive any reward, it will wander again. Then, it will find the new location. However, it takes less time to reach the new zone than the first one. This fact shows what psychologists call a capacity of generalization. We can go on and on. Each time, the arm develops a smooth and direct movement. The more it has been trained on different positions of the reinforcement, the quicker the arm finds new positions. Finally, we can also reset the reinforcement zone to a former position. Then, the arm finds this former position very quickly. So, the arm has learnt many positions of the reinforcement zone and has developed the ability to reach it by smooth movements.

Fig. 4. The development of the reaching movement: from left to right, the arm first has an erratic movement; after a while, it occasionally comes through the reinforcement zone; the movement becomes smoother and smoother; finally, the arm has a direct and smooth movement. The cross indicates the shoulder of the arm. The circle is the reinforcement zone. In the 3 leftmost plots, the position of the hand is indicated. In the rightmost plot, the whole arm is sketched.

Even though the 2 dimensional arm described here may seem too simple as a system, we have also obtained identical results for a 3 dimensional arm and we are working towards a multi-armed and multi-legged animat, as well as a real robot embedding the same techniques.

5   Discussion

In this contribution, we have summarized some recent works we have done and we have drawn links between these separate results. Our point is three-fold: first, we have shown that the law of effect can result from natural selection: as the law of effect models learning during lifetime of living beings, this first point shows that the way the behavioral repertoire of a living being evolves can have been produced by natural selection along generations since the emergence of life on earth; second, we have shown that agents following the law of effect obtain good performances in a social situation as they come to cooperate; third, using Q-learning as an implementation of the law of effect (that is, Q-learning is considered as an efficient implementation of the law of effect, that is, also the product of natural selection), we have shown that a set of cooperative Q-learners used in a realistic way to simulate an arm has a dynamics which is very similar to that of the arm of a human baby acquiring the same movement. All that work has been done so that artificial agents are built using realistic and well-tested hypotheses with regards to living beings. We think that such a multi-level way of tackling the problem of acquiring new behaviors is highly interesting for two reasons: for the scientists who study living beings, this shows that complex behaviors may be explained using the law of effect, itself being the product of natural selection; for scientists aiming at creating artificial adaptive agents either in software, or in hardware, this work shows that a careful implementation of "natural" laws may give rise to complex self-organized adaptive behaviors.

This work exemplify a multi-stage selectionist process: once the law of effect have been selected and is rather widespread in a population, the cooperation between agents becomes more probable which, in turns, can yield more complex structures among adaptive components.

References

1. E.L. Thorndike. Animal intelligence: An experimental study of the associative process in animals. *Psychology Monographs*, 2, 1898.
2. E.L. Thorndike. *Animal Intelligence: Experimental Studies.* Mac Millan, 1911.
3. J.W. Donahoe and D.C. Palmer. *Learning and Complex Behavior.* Allyn and bacon, 1994.
4. J.E.R. Staddon. *Adaptive behavior and learning.* Cambridge University Press, 1983.
5. B.F. Skinner. Selection by consequences. *Science*, 213:501–514, 1981.
6. J.M. Baldwin. A new factor in evolution. *The american naturalist*, 30, 1896. reprinted in [15, pp. 59-80].
7. G.E. Hinton and S.J. Nowlan. How learning can guide evolution. *Complex Systems*, 1:495–502, 1987. also reproduced in [15], chapter 25, pp. 447-454.
8. D. Parisi and S. Nolfi. The influence of learning on evolution. In *[15]*, pages 419–428.
9. S. Delepoulle. *Coopération entre agents adaptatifs ; étude de la sélection des comportements sociaux, expérimentations et simulations.* PhD thesis, Université de Lille 3, URECA, Villeneuve d'Ascq, October 2000. Thèse de doctorat de Psychologie.
10. J.B. Sidowski, B. Wyckoff, and L. Tabory. The influence of reinforcement and punishment in a minimal social situation. *Journal of Abnormal Social Psychology*, 52:115–119, 1956.
11. S. Delepoulle, Ph. Preux, and J-C. Darcheville. Learning as a consequence of selection, 2001. (submitted).
12. A.G. Barto. Reinforcement learning and adaptive critic methods. In D.A. White and D.A. Sofge, editors, *Handbook of intelligent control: neural, fuzzy, and adaptive approach*, pages 469–491. Van Nostrand Reinhold, 1992.
13. R.S. Sutton and A.G. Barto. *Reinforcement learning: an introduction.* MIT Press, 1998.
14. Ph. Preux, S. Delepoulle, and J-C. Darcheville. Selection of behaviors by their consequences in the human baby, software agents, and robots. In *Proc. Computational Biology, Genome Information Systems and Technology*, March 2001.
15. M. Mitchell and R. Belew, editors. *Adaptive Individuals In Evolving Population Models.* Santa Fe Institute Studies in the Sciences of Complexity. Addison-Wesley Publishing Company, 1996.

Clustering Moving Data with a Modified Immune Algorithm

Emma Hart and Peter Ross

School of Computing, Napier University
Edinburgh EH14 1DJ, Scotland
{emmah,peter}@dcs.napier.ac.uk

Abstract. In this paper we present a prototype of a new model for performing clustering in large, non-static databases. Although many machine learning algorithms for data clustering have been proposed, none appear to specifically address the task of clustering moving data. The model we describe combines features of two existing computational models — that of Artificial Immune Systems (AIS) and Sparse Distributed Memories (SDM). The model is evolved using a coevolutionary genetic algorithm that runs continuously in order to dynamically track clusters in the data. Although the system is very much in its infancy, the experiments conducted so far show that the system is capable of tracking moving clusters in artificial data sets, and also incorporates some memory of past clusters. The results suggest many possible directions for future research.

1 Introduction

As the ability to collect and store vast quantities of data increases, having some facility to intelligently and efficiently analyse that data in order to detect clusters, patterns and meaningful correlations becomes essential. Many algorithms have been proposed to perform some or all of these tasks, however it seems clear that a successful algorithm must address the following key features of larger databases if it is to prove useful in the real-world:

- The data-base is likely to be non-static; data is continually added and deleted
- Trends in the data change over time
- The data may be distributed across several servers
- The data may contain a lot of 'noise'
- A significant proportion of the data may contain missing fields or records

Recently, a growing body of work has shown that the biological immune contains many desirable features which allow it to be used to address some of the characteristics listed above. In (one) extremely simplified view, the immune system (IS) can be considered to be a decentralised self-organising system which operates by producing antibodies which recognise potentially harmful invaders and eliminate them from the body. The recognition takes place via some kind of

E.J.W. Boers et al. (Eds.) EvoWorkshop 2001, LNCS 2037, pp. 394–403, 2001.

sophisticated pattern matching mechanism which allows it to access a content addressable memory of past invaders. The matching mechanism is imprecise — an antibody is stimulated by an antigen if the strength or affinity of the match between the two exceeds some threshold. Any antibody which stimulates an antigen is said to be within the ball of stimulation of the antigen. Moreover, the IS is able to dynamically learn about new substances when it encounters them and add them to its memory. At the same time, any little used information is deleted from the memory, therefore the memory is continually changing. Several of these features have been modelled in a number of very different implementations of artificial immune systems and applied to the problem of clustering data. For example, Potter *et al.* describe a model of an AIS that uses a coevolutionary genetic algorithm (GA) to evolve antibodies to cluster artificial data sets [1] and Congress voting records [2]; Forrest *et. al* [3] describe a GA that uses emergent fitness sharing to find patterns; Hunt *et al.* [4] describe a system named *Jisys* which was used to cluster data for use in mortgage fraud detection and Timmis [5] has adapted this system to successfully cluster the well known but very small benchmark data set containing iris petal sizes. Both the Timmis and Hunt work used a model based on connected networks of antibodies, in which nodes which are connected recognise similar patterns. So far, none of these methods have addressed the question of clustering data in time-varying databases. Although there is no intrinsic barrier to extending either the coevolutionary or network models to deal with non-stationary data, both methods present obstacles. In the network model, there are high computational overheads associated with re-organising large networks as the data changes, which increase as the size of the database increases also. It is also unclear whether the coevolutionary method of evolving clusters is able to cope with extremely large databases, particularly as the antibodies compete to exclusively recognise data, whereas in reality clusters may overlap.

In other work, Smith *et. al* [6] have shown that the immune system can be considered to be representative of the same class of memories as Kanerva's Sparse Distributed memory, [7]. The SDM is a content-addressable memory which was originally proposed as an efficient method for storing a very large number of large binary data patterns using a very small number of physical data addresses, in a manner which allows accurate recall of all the data. An SDM is composed of a set of physical or hard locations, each of which recognises data within a specified distance of itself — this distance is known as the recognition radius of the location. Each location also has an associated set of counters, one for each bit in its length, which it uses to 'vote' on whether a bit recalled from the memory should be set to 1 or 0. An item of data is stored in the memory by distributing it to every location which recognises it — if recognition occurs, then the counters at the recognising locations are updated by either incrementing the counter by 1 if the bit being stored is 1, or decrementing the counter by 1 if the bit being stored is 0. To recall data from the memory, all locations which recognise an address from which recall is being attempted vote by summing their counters at each bit position; a positive sum results in the recalled bit being set to 1, a negative sum in the bit being set to 0. This results in a memory which is particularly robust to noisy data due to its distributed nature and inexact method of storing data. These properties make it an ideal candidate for addressing clustering problems in large databases; For example, we can consider each physical location along with its recognition radius to define a cluster of data; the location itself can be considered to be a concise representation or description of that cluster, and the recognition radius specifies the size of the cluster. Clusters can overlap — indeed, it is this precisely this property which allows all data to be recognised with high precision whilst maintaining a relatively low number of clusters. If no overlap is allowed, then a large number of locations are required to cluster the data, the system becomes overly specific, and hence general trends in the data are lost. In the form described, the SDM is also static and inflexible, however given its powerful and efficient storage and recognition capacities, it is fruitful to adapt it to operate in a dynamic environment. Therefore, the model we now describe combines features of SDM and the type of AIS describe by Potter to produce a system that is dynamic, adaptable and capable of tracking changes in large volumes of data. For simplicity, during the remainder of this paper we use immunological terminology — an antigen is equivalent to a piece of data, an antibody to a description of a cluster, and the ball of stimulation of the antibody defines the size of the cluster.

2   Description of the Proposed Model

The proposed model is shown in figure 1. The basic proposition is to use a coevolutionary GA, running continuously, to find quickly the set of antibodies (and their corresponding balls of stimulation) that best cluster the data currently visible to the system. An *antigen* is represented by a bit string of length L. An *antibody* is also represented by a bit string of length L, and also defines the recognition radius R of the antibody. Each antibody has an associated set of counters, one for each bit, which are used to 'vote' on whether the bit should be set to 1 or 0 as described in the previous section. The accuracy of the SDM formed by the set of antibodies can be determined by attempting to recall each data item stored and comparing the results to the actual data in the database. The coevolutionary GA controls the evolution of k populations of antibodies — each population is attempting to evolve the location and radius of one of the antibodies defining the memory (and therefore the clusters). At any time t, the best antibodies in each population cooperate to form an SDM in which all data visible to the system at this time can be stored and ideally accurately recalled (and hence clustered). The mechanism by which the evolution proceeds is detailed in the next section.

3   Experimental Details

In order to calculate the fitness of an antibody in any population (which only represents a partial solution to the problem), the antibody is added to a serum

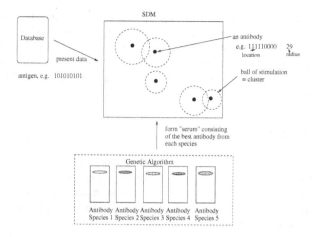

Fig. 1. The proposed model, combining features from an SDM, AIS and coevolutionary genetic algorithm

consisting of itself and the best members of the other populations.[1] The counters of each antibody in the serum are set to 0. All the antigens in the database are then stored in the SDM defined by this serum, and then recall attempted of each antigen. Antigens not recognised by any of the antibody species are allocated to a default cluster which is defined by the antibody '0000...000' with a recognition radius of L. Otherwise, the antigen is assigned a match-score equivalent to the number of correctly recalled bits. The fitness of the antibody is then set to the average value of M. Antibodies which cooperate with other antibodies to more accurately represent the dataset are thus more highly rewarded. Note that antibodies are *not* exclusively competing for antigen — several hard locations in the serum may recognise an antigen and thus collaborate in order produce the recalled data, which should result in a higher recall accuracy than in the system described in [1]. This bears a close analogy to the real immune system in which a cross-reaction between antibodies can occur.

3.1 Control of Number of Species

The number of species is dynamic, that is species are added and deleted from the algorithm as becomes necessary. The rate at which this happens is controlled by 4 parameters; the extinction threshold, e, the learning phase, l, the stagnation phase length, sp and the stagnation level st. If the fitness of the serum composed of the best member of each species does not increase by at least st over spgenerations, then a new species is added to the system, with randomly generated members. Similarly, if the best member of a species does not recognise at

---

[1] In the initial generation of the algorithm, antibodies are chosen at random from populations that have not yet been evaluated when forming the serum

least e antigens from the current antigen population, and the species has been in existence for at least l generations then that species is removed from the system, with the caveat that if the species recognises an antigen that is *not* recognised by any other antibody then the species is allowed to remain. A limit of M species is imposed on the system to prevent it growing too large (and therefore too specialised).

4   Experiments

Two series of experiments were performed in order to investigate the capability of the system. The first series of experiments investigated the ability of the system to track clusters which vary in a random manner with time. The second series was concerned with investigating the performance of the system in an environment in which data appears in cycles, and is designed to test the ability of the system to react more quickly to clusters of data which it has previously encountered.

Generating Data. In all experiments, the system is continually exposed to a set of 100 antigen. The antigen are generated from s schemas. Each schema consists of a string of 64 bits, in which d contiguous bits are set to 1, with the start position of the d bits randomly chosen. All remaining bit positions contain wild-cards. Antigen are generated in equal proportion from each schema by randomly replacing wild-cards with either 0 or 1. In order to generate non-stationary data, the following procedure is followed. 100 antigens are generated at time $t = 0$ from s schema. Every U time-steps, g schemas are chosen at random and replaced by g new randomly generated schema. New antigens are generated from the new schema and replace those antigens generated from the schema being replaced.

Tolerization Period. In each experiment, the system is allowed to undergo a *tolerization* period of T iterations in order to learn the data present at $t = 0$. This is necessary in order to accurately measure the response of the system to data changes from a state in which it has accurately clustered the current data. If this was not present, the system may still be learning the original data when changes occur, and hence we are not measuring the ability of the system to *adapt* to new data from an already stable state. This can be considered similar to the *neonatal* period in humans in which the body is thought to become tolerant of 'normal' proteins, [8]. In all experiments described, T is set to 200.

5   Simple Pattern Tracking

In the first series of experiments, a number of tests were performed for an update rate $U = 50$, varying the parameters s, d, and g. In this paper, due to space constraints, we report the results from experiments with $s = 5, d \in (8, 16, 32)$ and $g \in (1, 2, 3, 4, 5)$. Each experiment was repeated 5 times, and the resulting fitness at each time averaged. In each experiment, the size of each population was set to 50. Initially, 2 populations are created, and a maximum limit of 10 populations is imposed. The populations are evolved using fitness proportionate selection, 2-point crossover, and bit-flip mutation at a rate of $1/L$ per gene. The parameters controlling the dynamics of the evolution were set to $e = 5, l = 10, sp = 5, st = 0.5$.

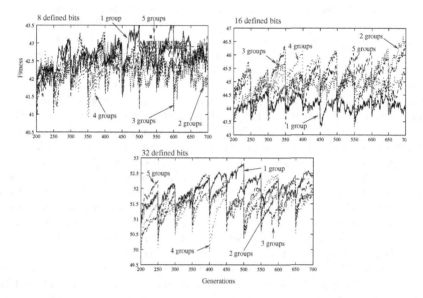

Fig. 2. The figure shows the performance of the proposed system for 3 experiments in which groups of $1 \rightarrow n$ schema are replaced at each update. The vertical axis shows the fitness of the best SDM found, where fitness is equal to the average number of correctly recalled bits across the entire data set.

5.1   Results

Figure 2 shows the results of the experiments for each combination of d and g, i.e. number of defined bits and number of schemas replaced. The results are shown from the end of the tolerization period only so that trends can be more clearly observed. In order to analyse the trends more thoroughly, the magnitude of the average drop in fitness in the system whenever a change in antigen occurs is plotted against the number of schemas replaced. This is shown in figure 3. In all experiments, a single antibody with all bits set to 1 and a radius of L should provide the most general clustering of the system, as this would match all of the possible sequences of d defined bits comprising the data set. This antibody would produce a recall score for the entire data set of 36,40 or 48, for the cases when $d = 8$, $d = 16$ and $d = 32$ respectively. This is calculated by considering that exactly d bits would match perfectly and on average $(l-d)/2$ of the remaining bits would match. This gives a baseline against which to compare the performance of the system.

The first observation to note from Figure 2 is that in all experiments, the fitness of the system immediately before any antigen change always exceeds the baseline fitness described above, therefore the system is clearly performing some clustering. The success of the system varies as the number of antigen replaced at each update changes, and also as the number of bits defining the schema change. The greatest effect on performance of the number of antigens replaced is observed when $d = 32$. This is as expected — if the number of defined bits is large compared with the length of the antigen, then the number of bits that must change in order to recognise different schema may be very large (for example, in the extreme case, if schema consisting of '11111.....#####' is replaced by one described by '#####....11111' then $L/2$ bits may need to change.) Hence we expect to see a large variation in performance as the number of antigen replaced increases. At the other end of the scale, when $d = 8$, and therefore small compared with the length of string, it is likely that common patterns exist in the schema other than those described by the section of defined bits, and therefore introducing new antigen produces much less overall change in the composition of the entire antigen data set, as these spurious patterns can always be generated by chance. In the case where $d = 16$, slightly anomalous results are observed, as the worst fitness values are obtained when $g = 1$, i.e. when only 20% of the antigen population is updated. This requires further investigation, particularly as figure 3 indicates that the drop in fitness between updates increases approximately linearly as the number of schemas replaced increases for $d = 16$ but that the drop is higher than in the case when $d = 32$. Figure 3 shows that although the drop in fitness increases with number of schema replaced, the drop is actually small in proportion to the relative fitness of the system; the largest drop obtained is approximately 1.3, whereas the fitness of the system is generally higher than 41.

6   Investigating the Memory Retention

The second series of experiments aimed to investigate whether the system could retain some memory of past clusters so that if a cluster reappears the system responds to it more rapidly. In this series of experiments, antigen sets were again generated from sets of schema in the following manner:

$c * s$ schema are initially generated in the manner previously described. At any time t, only s of these schema are used to generate the antigen population. A sliding window of size s defines which schemas are used; this window is moves w schemas along the schema list every U generations. The schema list is treated as cyclic and wraps around when the window reaches the end. Thus, if $c = 2$ and $s = 4$, then 8 schemas are initially generated. If w is equal to c, then all antigens are replaced at each update; thus at time $t = 0$, antigens $\{0, 1, 2, 3\}$ define the data set. At time U, antigens $\{4, 5, 6, 7\}$ define the data, at time $2U$, antigens

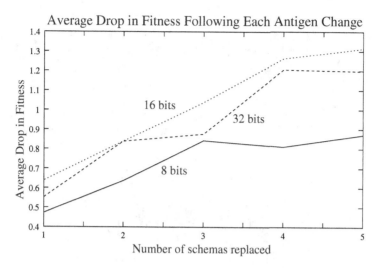

Fig. 3. The figure shows the drop in fitness experienced following a change in antigen for three experiments using schema containing 8,16, or 32 defined bits.

$\{0, 1, 2, 3\}$ again define the data etc. A more incremental update is achieved by setting $w < s$.

The experiments reported have $s = w$ where $s \in (2, 5, 10)$. The update rate U is set to 50 generations, and c is set equal to 2. Again, a tolerization period of T generations is allowed, in which the system learns the first s schemas. Thereafter, the schema set alternates between the two possible schemas (referred to as set A and B in the following discussion) every 50 generations. Figure 4 shows the results of these experiments. In each case, the experiments are compared to an equivalent experiment in which the antigen set is updated from randomly generated schema at each update. Clearly, the experiments which replace antigen with previously encountered antigen outperform the random set, showing that the system must be displaying some kind of memory. Again, the results are only shown from the end of the tolerization period.

To investigate the 'period' of this memory, we analyse the best fitness found for schema set A each time it appears. The experiment described above is repeated for values of c equal to 3, and 4, so that $3 * s$ schemas are generated in the former case, and $4 * s$ in the latter. In each case, the experiments are run for sufficient generations that the schema set A appears 5 times. The best fitness found on each occasion is averaged over the entire experiment and the results are shown in table 1. t-tests applied to each pair of results shows that the only significant difference in values is found between cases $(c = 2, s = 2)$ and $(c = 3, s = 2)$, $(c = 3, s = 2)$ and $(c = 4, s = 2)$, and finally between cases $(c = 3, s = 10)$ and $(c = 4, s = 10)$. Therefore, the system appears relatively robust to the parameter c which controls the period of the memory.

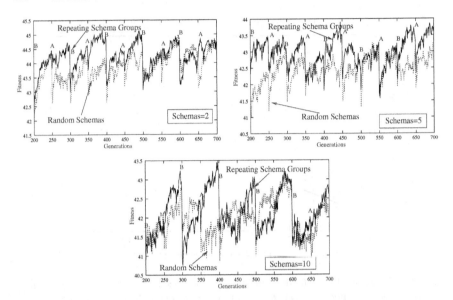

Fig. 4. Comparison of experiments in which new antigen are generated from new, randomly generated, schema to those in which antigen are generated from schema that the system has previously been exposed to

Table 1. The mean and standard deviation (shown in brackets) of the maximum fitness found for schema set A averaged across 5 occurrences of the set

Number of Schemas s	Multiplier for Number of Schema c		
	2	3	4
2	44.63 (0.186)	44.37 (0.274)	44.77 (0.322)
3	43.06 (0.382)	44.1 (0.342)	43.01 (0.405)
4	42.72 (0.266)	42.78 (0.474)	42.79 (0.281)

7   Conclusion

So far, only a very preliminary investigation of the capabilities of the proposed new system have been investigated. However, in light of those experiments reported here, and others to be reported in future publications, we make the following observations;

- The model appears capable of clustering data sets; this has been tested with up to 10 clusters.
- The model satisfactorily copes with moving data; the experiments show that the model tracks both incremental and large changes in data, but performance degrades as the amount of data changing increases.

 – The model exhibits a basic form of memory; when re-exposed to familiar antigen, it reacts more rapidly than to previously unseen antigen.

Clearly, there is much more work to do to fully assess the performance of the model, however the results found so far are promising. The model described seems to provide a sensible method of addressing the difficulties concerned with clustering data in non-stationary databases. Issues that must be addressed in future however include investigating the scalability of the system, the robustness of the system to noise in the data and to the distance between clusters. Although the system produced satisfactory results with arbitrarily chosen parameters, we wish to investigate the sensitivity of the parameter choices. Finally, we intend to compare the model to other methods which potentially could be employed to track non-stationary data.

References

1. M.A. Potter and K.A De Jong. Cooperative coevolution: An architecture for evolving coadapted subcomponents. *Evolutionary Computation*, 8(1):1–29, 2000.
2. M.A Potter and K.A De Jong. The coevolution of antibodies for concept learning. In *Parallel Problem Solving From Nature - PPSN V*, pages 530–540. Springer-Verlag, 1998.
3. S Forrest, B Javornik, R.E Smith, and A.S Perelson. Using genetic algorithms to explore pattern recognition in the immune syste. *Evolutionary Computation*, 1(3):191–211, 1993.
4. D Dasgupta, editor. *Artificial Immune Systems and Their Applications*, chapter Jisys: The Development on An Immune System for Real World Applications, pages 157–184. Springer-Verlag, 1999.
5. J. Timmis and M. Neal. A resource limited artifical immune system for data analysis. In *Expert Systems 2000: International Conference on Knowledge BAsed Systems and Applied Artificial Intelligence*. Springer-Verlag.
6. D.J Smith, S Forrest, and A.S Perelson. *Artificial Immune Systems and Their Applications*, chapter Immunological Memory is Associative, pages 105–112. Springer-Verlag, 1999.
7. P Kanerva. *Sparse Distributed Memory*. MIT Press,Cambridge,MA, 1988.
8. R.E Billingham, L. Brent, and P.B. Medawar. Actively acquired tolerance of foreign cells. *Nature*, 172:603–606, 1953.

Belief Revision by Lamarckian Evolution

Evelina Lamma[1], Luís Moniz Pereira[2], and Fabrizio Riguzzi[1]

[1] Dipartimento di Ingegneria, Università di Ferrara,
Via Saragat 1, 44100 Ferrara, Italy
elamma@ing.unife.it, friguzzi@ing.unife.it
[2] Centro de Inteligência Artificial (CENTRIA), Departamento de Informática,
Faculdade de Ciências e Tecnologia, Universidade Nova de Lisboa, 2825-114
Caparica, Portugal
lmp@di.fct.unl.pt

Abstract. We propose a multi-agent genetic algorithm to accomplish belief revision. The algorithm implements a new evolutionary strategy resulting from a combination of Darwinian and Lamarckian approaches. Besides encompassing the Darwinian operators of selection, mutation and crossover, it comprises a Lamarckian operator that mutates the genes in a chromosome that code for the believed assumptions. These self mutations are performed as a consequence of the chromosome phenotype's experience obtained while solving a belief revision problem. They are directed by a belief revision procedure which relies on tracing the logical derivations leading to inconsistency of belief, so as to remove the latter's support on the gene coded assumptions, by mutating the genes.

1 Introduction

Herein, we propose a genetic algorithm for belief revision that includes, besides Darwin's operators of selection, mutation and crossover [1], a logic based Lamarckian operator as well. This operator differs from Darwinian ones precisely because it modifies a chromosome coding beliefs so that its fitness is improved by experience rather than in random way.

We venture that the combination of Darwinian and Lamarckian operators will be useful not only for standard belief revision problems, but especially for problems where different chromosomes may be exposed to different constraints and environmental observations. In these cases, the Lamarckian and Darwinian operators play different rôles: the Lamarckian one is employed to bring a given chromosome closer to a solution (or even find an exact one) to the current belief revision problem, whereas the Darwinian ones exert the rôle of randomly producing alternative belief chromosomes so as to deal with unencountered situations, by means of exchanging genes amongst them.

We tested this hypothesis on multi-agent joint belief revision problems. In such a distributed setting, agents usually take advantage of each other's knowledge and experience by explicitly communicating messages to that effect. As multiple-population GAs (see [2], for discussion), we allow knowledge and experience to be coded as genes in an agent and consider several sub-populations

E.J.W. Boers et al. (Eds.) EvoWorkshop 2001, LNCS 2037, pp. 404–413, 2001.

which exchange individuals occasionally. In particular, genes are exchanged with those of other agents, not by explicit message passing but through the crossover genetic operator. The new offspring agent chromosomes can be naturally selected according to their gene coded knowledge governing their behaviour.

Crucial to this endeavour, we introduce a logic-based technique for modifying cultural genes, i.e. memes, on the basis of individual agent experience. The technique amounts to a form of belief revision, where a meme codes for an agent's belief or assumption about a piece of knowledge, and which is then diversely modified on the basis of how the present beliefs may be contradicted by observations and laws (expressed as integrity constraints). These self mutations are indeed performed as the outcome of the chromosome phenotype's (i.e., agent's) experience while solving a belief revision problem. They are directed by a belief revision procedure, which relies on tracing the logical derivations leading to inconsistency of belief, so as to remove the latter's support on gene coded assumptions by mutating the memes involved. Each agent possesses a pool of chromosomes containing such diversely modified memes, or alternative assumptions, which cross-fertilize Darwinianly amongst themselves. Such an experience-influenced genetic evolution mechanism is aptly called Lamarckian.

To illustrate how these mechanisms, of individual agent Lamarckian evolution and of Darwinian agent genetics, can jointly lead to improved single agent population behaviour in collaborative problem-solving, we apply them to distributed model-based diagnosis of digital circuits, a natural domain in which belief revision techniques apply [3].

2   Preliminaries

We consider belief revision of first order theories expressed in the language of extended logic programs [4]. For this language, we adopt the Extended Well Founded Semantics ($WFSX$) that extends the well founded semantics (WFS) [5] for normal logic programs to programs extended with explicit negation, besides the implicit or default negation of normal programs.

Extended logic programs are liable to be contradictory because of integrity constraints, either those that are user-defined or those of the form $\bot \leftarrow L, \neg L$ that are implicitely assumed. The *revisables* of a program P are the elements of a chosen subset $Rev(P)$, of the set of all literals L having no rules for them in P , and code revisable beliefs.

3   A Genetic Algorithm for Multi-agent Belief Revision

The algorithm here proposed for belief revision extends the standard genetic algorithm (described for example in [1]) in two ways:

- crossover is performed among chromosomes belonging to different agents,
- a Lamarckian operator called Learn is added in order to bring a chromosome closer to a correct revision by changing the value of revisables.

In two-valued belief revision, each individual hypothesis is described by the truth value of all the revisables. Therefore each hypothesis can be considered as a set containing one literal, either positive or default, for every revisable. A chromosome is obtained by associating a bit to each revisable that has value 1 if the revisable is true and 0 if it is false.

The fitness function that has been used takes the following form:

$$Fitness(h_i) = \frac{n_i}{n} + \frac{f_i}{|h_i|} \times 0.5$$

where n_i is the number of integrity constraints satisfied by hypothesis h_i, n is the total number of integrity constraints, f_i is the number of revisables in h_i that are false, and $|h_i|$ is the total number of revisables. In this way, the fitness function takes into account both the fraction of constraints that are satisfied and the number of revisables whose truth value must be changed to true, preferring hypotheses with a lower number of these. Assuming that the initial value of the revisables is false, this means that minimal revisions are encouraged. The factor 0.5 was chosen in order to give more importance to the accuracy, rather than to the number of unchanged revisables.

The Lamarckian operator *Learn* changes the values of the revisables in a chromosome C so that a bigger number of constraints is satisfied, thus bringing C closer to a solution.

This is done by modifying the belief revision techniques presented in [6]. In particular, in [6] an algorithm for belief revision is presented that is based on the notions of *support sets, hitting sets* and *removal sets*. Intuitively, a support set of a literal is the set of revisable supporting the derivation of the literal. The hitting set of a collection C of sets is formed by the union of one non-empty subset from each $S \in C$. A hitting set is minimal iff no proper subset is a hitting set. A removal set of a literal is a hitting set of all the support sets of the literal. Contradiction in [6] is removed by finding the removal set of $\bot$.

Each agent executes the following algorithm:

GA(*Fitness, max_gen, p, r, m, l*)
    Fitness: a function that assigns an evaluation score to a hypothesis coded as a chromosome, *max_gen*: the maximum number of generations before termination, *p*: the number of individuals in the population, *r*: the fraction of the population to be replaced by Crossover at each step, *m*: the fraction of the population to be mutated, *l*: the fraction of the population that should evolve Lamarckianly.

    Initialize population: $P \leftarrow$ generate p hypotheses at random
    Evaluate: for each h in P, compute $Fitness(h)$
    $gen \leftarrow 0$
    While $gen \leq max\_gen$
    Create a new population P_s:
        Select: Probabilistically select $(1 - r)p$ members of P
            to add to P_s. The probability $Pr(h_i)$ of selecting

hypothesis h_i from P is given by
$$Pr(h_i) = \frac{Fitness(h_i)}{\Sigma_{j=1}^{p} Fitness(h_j)}$$

Crossover:
    For i=1 to rp
        Probabilistically select an hypothesis h_1 from P,
            according to $Pr(h_1)$ given above
        Obtain an hypothesis h_2 from another agent
            chosen at random
        Crossover h_1 with h_2 obtaining h_1'
        Add h_1' to P_s

Mutate: Choose m percent of the members of P_s with
    uniform probability. For each, invert one
    randomly selected bit in its representation

Learn: Choose lp hypotheses from P_s with uniform
    probability and substitute each of them with the
    modified hypotheses returned by the procedure *Learn*

    Update: $P \leftarrow P_s$
Return the hypothesis from P with the highest fitness

The Lamarckian operator *Learn* works in the following way: given a chromosome C, it finds all the support sets for $\perp$ such that they contain literals in C. These support sets are called *Lamarckian support sets* (a formal definition for them is given in [7]). Therefore, it does not find all support sets for $\perp$ but only those that are subsets of C.

Since the Lamarckian support sets for $\perp$ represent only a subset of all the support sets for $\perp$, a hitting set generated from them is not necessarily a contradiction removal set and therefore it does not represent a solution to the belief revision problem. However, it eliminates some of the derivation paths to $\perp$ and, therefore, may increase the number of satisfied constraints, thus improving the fitness, as required by the notion of Lamarckian operator.

In the case of the circuit diagnosis problems in section 4, the support sets procedure becomes simplified in that the occurrences of default negated literals pertain only to revisables.

When computing the support sets, the Lamarckian operator also modifies an extra bit associated with each meme each time the meme is considered in the computation of Lamarckian support sets. This bit indicates whether the meme has been "accessed" by the operator. This is needed for the crossover operator that is described below.

procedure $Learn(C, C')$
    inputs : C: a chromosome translated into a set of revisables
    outputs : C' the revised chromosome
    Find the support sets for $\perp$: $Support\_sets([\perp], C, \{\}, \{\}, SS)$
    Find a hitting set HS: $Hitting\_set(SS, HS)$
    Change the value of the literals in the chromosome C
       that appear as well in HS

procedure $Support\_sets(GL, C, S, SSin, SSout)$:
  inputs : GL: list of goals, C: a chromosome translated into a set of revisables,
  S: the current support set, $SSin$: the current set of support sets

  outputs : $SSout$: a set containing the support sets for the first goal in the
  list

  If GL is empty, then return $SSout = SSin$
  Consider the first literal L of the first goal G of GL
    ($GL = [G|RGL]$ using Prolog notation for lists)
    (1) if G is empty then add the current support set to $SSin$
        and call recursively the algorithm on the rest of GL
        $Support\_sets(RGL, C, \{\}, SSin \cup \{S\}, SSout)$
    (2) if G is not empty ($G = [L|RG]$) then:
        (2a) if L is a revisable and is in C, then add it to S,
            set to 1 L's access bit
            and call the algorithm recursively on the rest of G
            $Support\_sets([RG|RGL], C, S \cup \{L\}, SSin, SSout)$
        (2b) if L is a revisable and it is not in C, or its opposite
            is in C, then set to 1 L's access bit, discard S
            and call the algorithm recursively on the rest of GL
            $Support\_sets(RGL, C, S \cup \{L\}, SSin, SSout)$
        (2c) if it is not a revisable then reduce it with all the rules,
            obtaining the new goals $G_1, ..., G_n$, one for each
            matching rule, add the goals to GL and call
            the algorihtm recursively $Support\_sets([[G_1|RG], ...,$
            $[G_n|RG]|RGL], C, S, SSin, SSout)$
        (2d) if it is not a revisable and there are no rules, then return
            without adding S to SS ($SSout = SSin$)

procedure $hitting\_set(SS, HS)$:
  Pick a literal from every support set in SS
  Add it to HS if it does not lead to contradiction
  (i.e., the literal must not be already present
    in its complemented form).
  If it leads to contradiction pick another literal.

The crossover operator is an extension of a standard uniform crossover operator. The crossover operator produces a new offspring from two parent strings by copying selected bits from each parent. The bit at position i in the offspring is copied from the bit at position i in one of the two parents. The choice of which parent provides the bit for position i is determined by the crossover mask that, in uniform crossover, is generated as a bit string where each bit is chosen at random and independently of the others.

In our setting, one of the parents comes from the agent local population, while the other comes from the population of another agent. However, not all the bits in the chromosome are treated equally. In particular, we distinguish genes from memes: genes are modified only by Darwinian operators, while memes are modified by Darwinian and Lamarckian operators. Genes in the offspring can be copied from both parents, while memes can be copied from the parent coming from another agent only if they have been "accessed" by the other agent as a result of the application of the Lamarckian operator.

In this way, an agent can acquire from another agent only memes that have been checked for consistency. Therefore, the flow of memes is asymmetrical and goes from a "teacher" to a "learner", but not vice versa. In particular, in the asymmetrical crossover operator the mask is generated again as a random bit string and crossover is performed in the following way: if the i-th bit in the mask is 1 and the i-th bit in the other agent's chromosome has been accessed, then the i-th bit of the offspring is copied from the other agent's chromosome, otherwise it is copied from the local agent's chromosome. Simplified versions of this algorithm have also been considered in order to separately test the effectiveness of each of the features added to the standard genetic algorithm. In particular, five algorithms have been considered named in the sequel algorithms 1, 2, 3, 4 and 5. Algorithm 1 is a standard single agent genetic algorithm: crossover is performed only among chromosomes of the same agent and the Lamarckian operator is not used. Algorithm 2 adds to algorithm 1 the use of the Lamarckian operator, with a parameter l (percentage of the population to be mutated Lamarckianly) equal to 0.6. Algorithm 3 is a multi-agent algorithm without the Lamarckian operator, i.e., crossover is performed between chromosomes of different agents but the operator Learn is not applied to them. Algorithm 4 extends algorithm 3 by adding the Lamarckian operator, with a parameter l equal to 0.6. However, it does not distinguish genes from memes, i.e. crossover is always symmetric. Algorithm 5 differs from algorithm 4 because it treats genes and memes differently, exchanging only those memes that have been accessed.

These algorithms have been used in order to experimentally prove the following theses:

1. Lamarckism plus Darwinism outperforms Darwinism alone in the single agent case;
2. the distributed algorithm (with or without the Lamarckian operator) performs better than the non-distributed one, in the same number of generations, because of parallel exploration;
3. Lamarckism plus Darwinism outperforms Darwinism alone in the multi-agent case;
4. the distributed algorithm with the distinction between genes and memes performs better than the one without the distinction.

4   Experiments

The algorithms have been tested on a number of belief revision problems in order to prove the above theses. In particular, we have considered problems of digital circuit diagnosis, as per [3]. A problem of digital circuit diagnosis can be modelled as a belief revision problem by describing it with a logic program consisting of

Table 1. Experiments on digital circuits debugging

Circuit	Algorithm	Fitness	Standard Deviation	Correct solution
voter	1	1.295	0.00634	100 %
	2	1.312	0.01728	100 %
alu4_flat	1	1.193	0.03939	20 %
	2	1.213	0.01765	33 %

four groups of clauses: one that allows to compute the predicted output of each component, one that describes the topology of the circuit, one that describes the observed inputs and outputs, and one that consists of integrity constraints stating that the predicted value for an output of the system cannot be different from the observed value. The revisables are literals of the form ab(Name) which, if true, state that the gate Name is faulty. The representation formalism we use is the one of [3].

If the digital circuit is faulty, one or more of the constraints will be violated. By means of belief revision, the values of the revisables are changed in order to restore consistency.

The system has been tested on some real world problems taken from the ISCAS85 benchmark circuits [8] that has been used as well for testing the belief revision system REVISE [3].[1]

We have considered the voter and alu4_flat circuits: voter has 59 gates and 4 outputs, corresponding respectively to 59 revisables and 8 constraints, while alu4_flat has 100 gates and 8 outputs, corresponding respectively to 100 revisables and 16 constraints.

We have first tested algorithms 1 and 2 on the voter and alu4_flat circuits. The parameters of the genetic algorithms were 30 for the population and 10 for the number of generations. Both algorithms were run 5 times and the resulting maximum fitness averaged. In table 1 the Fitness column shows the value of the fitness function for the best hypothesis after ten generations averaged over the 5 runs together with its standard deviation, while the Correct solution column shows the percentage of times in which a correct solution was found.

From these results it can be seen that thesis 1 is proved, i.e., that the use of a Lamarckian operator improves the fitness of the best hypothesis. Moreover, the algorithm does not heavily depend on the initial population, as shown by the low values for the standard deviation. Finally, the Lamarckian operator does not greatly influence the dependency on the initial population, as can be seen from the fact that in one case (voter) the use of the Lamarckian operator has increased the standard deviation but in the other case (alu4_flat) it has decreased it.

Algorithms 2, 3, 4 and 5 have been tested on the voter circuit. Each algorithm was run 5 times. The parameters that have been used for the runs are:

---

[1] These examples can be found at http://www.soi.city.ac.uk/~msch/revise/.

Table 2. Experiments with algorithms 2, 3, 4 and 5

Circuit	Algorithm	Fitness	Standard Deviation	Correct solution
voter	2	1.319	0.00415	100 %
	3	1.314	0.00928	100 %
	4	1.325	0.03321	100 %
	5	1.392	0.06296	100 %

10 maximum generations, 40 individuals for algorithm 2 (single agent), 10 individuals per agent and 4 agents for algorithms 3, 4 and 5. In algorithms 3, 4 and 5 each agent has the same set of observations and program clauses, while the integrity constraints are distributed among the agents so that each agent knows only the constraints that are related to one same output.

In table 2 we show, for each algorithm, the value of the fitness function for the best hypothesis after ten generations averaged over the five runs.

As can be seen, theses 2, 3 and 4 are also confirmed. If we compare the results of algorithm 1 (table 1) and 3 and those of algorithms 2 and 4 we realize that the cooperation among agents improves the quality of the results with respect to the single agent case in the same number of generations (thesis 2). The fitness increment between algorithms 3 and 4 shows the usefulness of the Lamarckian operator in the multi-agent case (thesis 3). Finally, the fitness increment between algorithms 4 and 5 shows the usefulness of the distinction of memes from genes and of the asymmetrical crossover mechanism among memes (thesis 4). Again, the low values for the standard deviation in all cases show that the result does not heavily depend on the initial population.

5   Related Work

Various authors have investigated the integration of Darwinian and Lamarckian evolution into a genetic algorithm [9,10,11,12]. A Lamarckian operator first translates a genotype into its corresponding phenotype and performs a local search in the phenotype's space. The local optimum that is obtained is then translated back into its corresponding genotype and added to the population for further evolution. [9] has shown that the traditional genetic algorithm performs well for searching widely separated portions of the search space caused by a scattered population, while Lamarckism is more proficient for exploring localized areas of the population that would otherwise be missed by the global search of the genetic algorithm. Therefore, Lamarckism can play an important rôle when the population has converged to areas of local maxima that would not be thoroughly explored by the standard genetic algorithm. The adoption of a Lamarckian operator provides a significant speedup in the performance of the genetic algorithm.

Similarly to the approaches in [9,10,11,12], we adopt a procedure for Lamarckian evolution that first translates the chromosome into its phenotype and then

modifies it in order to improve its fitness. In our case too the Lamarckian operator improves the performance of the genetic algorithm. Differently from [9,10,
11,12], the procedure does not perform a local search but finds an improvement
by tracing logical derivations causally supporting the undesired behaviour.

Our work is also related to coevolutive approaches and distributed GAs (see
[13,14,2]. It can be considered a cooperative coevolutionary approach(see [13,14]
to belief revision since knowledge about the domain problem (and constraints in
particular) are spread among the agents, each of which is ruled by a GA. In this
respect, each species represents a possibly partial solution to the belief revision
problem. While in [13] the complete solutions (to the problem of function optimization, in that paper) are obtained by assembling the representative members
of each of the species present, in our work the solution is obtained by evolution
and exchange between species, and by the application of the crossover operator
to members of two species, the foreigner of which may have already gained in
experience (i.e., it evolved Lamarckianly).

According to the classification given in [2], our approach is a multiple-population coarse-grained GA. Multiple-population (or distributed) GAs consist of
several subpopulations which exchange individuals occasionally by migration.
Rather than migration, we consider instead selection of members of different
sub-population to be merged by crossover. This form of *virtual* migration is synchronous with the application of crossover operator. Furthermore, the topology
we consider is in fully connected, but again the application of crossover chooses
a chromosome from a selected agent.

6   Conclusions and Future Work

We have proposed a novel way of looking at belief revision, which is GA based,
and hence a new application domain for GAs. Since it is still in the initial development stages, and it cannot be expected yet to compete with hard-boiled
methods for belief revision. On the other hand, we believe our method to be
important for situations where classical belief revision methods hardly apply:
Those where environments are non-uniform and time changing. These can be
explored by distributed agents that evolve genetically to accomplish cooperative belief revision, using our approach. Notwithstanding, some type of efficient
hybrid implementation approach might emerge, combining hard-boiled belief revision techniques with the newly introduced GA suplement. Our contribution
has been to get the new approach off the ground.

References

1. T. M. Mitchell. *Machine Learning*. McGraw Hill, 1997.
2. Erick Cantú-Paz. A survey of parallel genetic algorithms.
3. C. V. Damásio, L. M. Pereira, and M. Schroeder. REVISE: Logic programming and
   diagnosis. In *Proceedings of Logic-Programming and Non-Monotonic Reasoning,
   LPNMR'97*, volume 1265 of *LNAI*, Germany, 1997. Springer-Verlag.

4. J. J. Alferes, L. M. Pereira, and T. C. Przymusinski. "Classical" negation in non-monotonic reasoning and logic programming. *Journal of Automated Reasoning*, 20:107–142, 1998.

5. A. Van Gelder, K. A. Ross, and J. S. Schlipf. The well-founded semantics for general logic programs. *Journal of the ACM*, 38(3):620–650, 1991.

6. L. M. Pereira, C. V. Damásio, and J. J. Alferes. Diagnosis and debugging as contradiction removal. In L. M. Pereira and A. Nerode, editors, *Proceedings of the 2nd International Workshop on Logic Programming and Non-monotonic Reasoning*, pages 316–330. MIT Press, 1993.

7. E. Lamma, L. M. Pereira, and F. Riguzzi. Multi-agent logic aided lamarckian learning. Technical Report DEIS-LIA-00-004, Dipartimento di Elettronica, Informatica e Sistemistica, University of Bologna (Italy), 2000. LIA Series no. 44.

8. F. Brglez, P. Pownall, and R. Hum. Accelerated ATPG and fault grading via testability analysis. In *Proceedings of IEEE Int. Symposium on Circuits and Systems*, pages 695–698, 1985. The ISCAS85 benchmark netlist are available via ftp mcnc.mcnc.org.

9. W. E. Hart and R. K. Belew. Optimization with genetic algorithms hybrids that use local search. In R. K. Belew and M. Mitchell, editors, *Adaptive Individuals in Evolving Populations*. Addison Wesley, 1996.

10. D. H. Ackely and M. L. Littman. A case for lamarckian evolution. In C. G. Langton, editor, *Artificial Life III*. Addison Wesley, 1994.

11. Y. Li, K. C. Tan, and M. Gong. Model reduction in control systems by means of global structure evolution and local parameter learning. In D. Dasgupta and Z. Michalewicz, editors, *Evolutionary Algorithms in Engineering Applications*. Springer Verlag, 1996.

12. J. J. Grefenstette. Lamarckian learning in multi-agent environments. In *Proc. 4th Intl. Conference on Genetic Algorithms*. Morgan Kauffman, 1991.

13. M. Potter and K. de Jong. A cooperative coevolutionary approach to function optimization, 1994.

14. Mitchell A. Potter, Kenneth A. De Jong, and John J. Grefenstette. A coevolutionary approach to learning sequential decision rules. In Larry Eshelman, editor, *Proceedings of the Sixth International Conference on Genetic Algorithms*, pages 366–372, San Francisco, CA, 1995. Morgan Kaufmann.

A Study on the Effect of Cooperative Evolution on Concept Learning

Filippo Neri

Marie Curie Fellow at Unilever Research, Port Sunlight, UK
University of Piemonte Orientale, Italy
neri@di.unito.it

Abstract. A preliminary investigation of the results produced by two cooperative learning strategies exploited in the system REGAL is reported. The objective is to produce a more efficient learning system. An extensive description about how to setup a suitable experimental setup is included. It is worthwhile to note that, in principle, these cooperative learning strategies could be applied to a pool of different learning systems.

1 Introduction

Concept learning [1] is the task of finding a rule (in a wide sense) that discriminates between positive and negative instances of a given concept. The relevance of concept learning is well characterized by the variety of its fielded applications like prediction of mutagenetic compounds [2], and management of computer systems and networks [3,4]. Learning concepts means searching large hypothesis spaces. So, the capability to take advantage of effective search becomes a plus. Approaches based on Genetic Algorithms [5,6] proved their potentialities on a variety of concept learning tasks [7,8,9,10].

From these efforts it emerged that the main disadvantage of using GAs, with respect to alternative approaches, stays in their high user waiting time and in their high computational cost. A possible way of reducing GA computational cost is to use distributed computation efficiently: possibly by promoting cooperation among the simultaneous evolving populations. This approach is known as cooperative evolution or co-evolution [11,12,13,14].

In co-evolution, a complex problem is decomposed into simpler subproblems at runtime, then the evolution of several species, each one oriented to a subproblem's solution, is promoted. Periodically, a candidate solution for the problem is assembled from the species' best individuals and evaluated. Finally, the solution evaluation is backpropagated to the existing species through a new problem decomposition that affects their further evolution.

In the past, we investigated how the adoption of cooperative learning into the GA-based system REGAL [13] could produce a more efficient learning system. Research on cooperative learning includes also approaches like: boosting [15] and bagging [16]. These techniques combine a pool of classifiers in order to improve

E.J.W. Boers et al. (Eds.) EvoWorkshop 2001, LNCS 2037, pp. 414–420, 2001.

their separate classification performances. Generally they exploit re-sampling or weighting of the learning instances in order to acquire different classifiers to be combined, and they are independent from the specific used learning method.

The paper organization follows. In Section 2, REGAL and two cooperative learning strategies are briefly described. In Section 3, the experimental context is analyzed. In Section 4, the results are reported. The conclusion ends the work.

2   The System REGAL

REGAL [9,13] learns relational disjunctive concept descriptions in a restricted form of First Order Logic by using cooperative evolution. In REGAL an individual is a conjunctive formula (encoded as a fixed length bitstring) and a subset of the individuals in the populations has to be determined to form a disjunctive description for the target concept. For the scope of this work, we concentrate on REGAL's cooperative architecture as a description of the system's other components have already been published. REGAL's architecture is a network of N processes *GALearners*, coordinated by a *Supervisor* that imposes cooperation among the evolving populations. Metaphorically speaking, each *GALearner* realizes a niche, defined by a subset of the learning instances, where some species live. Each $GALearner_n$ tries to find a description for a subset of the learning instances LS_n by evolving its population. In addition, the *GALearners* may perform migration (exchange) of individuals. The *Supervisor* coordinates the distributed learning activity by periodically assigning different subsets of the learning instances to the *GALearners*. The composition of these subsets depends on the specific cooperative policy used. Two policies of cooperation will be investigated.

2.1   Two Cooperative Learning Strategies

As no a priori information is available on what is a successful assignment of learning instances, we decided to develop two cooperative learning strategies based on different assumptions. First, we analyzed the methods used by well known learning systems (like: AQ [17], C4.5 [18]) to deal with a large set of instances that cannot be covered by a single conjunctive formula. They all exploit a "divide et impera" policy (also known as "learn one conjunct at a time"): learns a description, remove the instances covered by it from the learning set, and restart the learning on the remaining instances. So we decided to implement similar policies as cooperative learning strategies. The first is named Let Seed Expand (LSE) and works as follows: when a learner find a description ψ, remove from its learning set all the instances covered by other already found descriptions and not covered by ψ, and let ψ improve.

An alternative form of cooperation, named Describe Those Still Uncovered (DTSU), forces the learners in dealing as soon as possible with the instances difficult to cover. Essentially, as soon as a promising concept description emerges, the instances not covered by it are included into all the learning sets, whereas each covered instance is inserted into *only one* learning set.

The algorithmic description of both cooperative learning strategies is skipped for space reasons.

3   Empirical Qualitative Evaluation

The effectiveness of any concept learning system is primarily evaluated on the basis of its averaged prediction error estimate. However, in order to provide a closer insight in a system behavior, additional measures may be used, such as, for instance, measures accounting for the structure of the acquired concept description. The comparison of REGAL's performances in terms of its average prediction error has already been analyzed [19]. We are here interested in the qualitative evaluation of how cooperation affects the structure of the found concept descriptions. Consequently, we will study REGAL's behavior with and without a cooperative strategy at work and considering the effect of migration. Given all the previous, setting up a suitable experimental context involves dealing with the following three issues:

1) The selection of what characteristics of concept description should be measured. We chose the following ones: (a) its average prediction error (ε) evaluated on a independent set of instances; (b) its complexity (C); (c) the number of conjuncts (NC) in Concept; (d) the maximum (MXC), average (AVC) and minimum (SMC) number of positive examples covered by any conjunct in Concept; (e) and the user waiting time (T), i.e. cpu time of the slowest learners to complete its task. The complexity (C) of a concept description has been defined as the number of conditions (i.e. its number of constants) to be tested in order to verify it.

2) The selection of the learning problem. In order to be able to compare the learned concept descriptions with respect to reasonable target ones, we chose an applicative domain whose (near to) optimal concept descriptions are *a priori* known. These target concept descriptions are characterized by a null predictive error and by a low complexity value.

3) The selection of a set of operative conditions, including parameters' values, under which to run the learning system.

We now discuss issues 2) and 3) in more details.

3.1   Characteristics of the Selected Application

As applicative domain, we selected a known concept learning dataset: the "Mushrooms"[1] [20] one. This problem is characterized by the absence in its hypothesis spaces of a purely conjunctive concept description and by the existence in its

---

[1] The problem consists in recognizing mushrooms from the Agaricus and Lepiota families as Edible (the firsts) and Poisonous (the seconds). The dataset contains 8124 instances, 4208 of edible mushrooms and 3916 of poisonous ones. Each instance is described by a vector of 22 discrete attributes, each of which can assume from 2

hypothesis spaces of at least a disjunctive concept description. The knowledge about this hypothesis space comes from results appeared in the literature.

From previous experiments, we know that the Mushrooms application admits as good description for the poisonous mushrooms concept that requires 15 conditions to be tested.

Three randomly selected sets of 4000 instances (2000 edible plus 2000 poisonous) have been used as learning sets, while the remaining 4124 instances have been used for testing.

3.2   Choosing Proper Experimental Configurations

In order to run a GA-based system, a set of parameters such as the population size, the number of generations to be accomplished (in short, the generation number), the crossover probability, the mutation rate, etc. have to be fixed [6]. In general, the results obtained by any GA-based system are sensitive to the chosen values. A system is robust to the parameter variation if a little variation in its parameters values corresponds to a little shift in the "quality" of its results. From the point of view of concept learning, we are interested in exploiting a system that is robust with respect to the parameters settings in order to avoid uncontrollable fluctuations in the system outputs. In other words, we would like to use a system whose results only depends on the information provided by the learning instances and not by the system's internal behavior. In the past, we performed series of experiments to understand and set once for ever REGAL's parameters [9] and we determined such set of values for the parameters that we call "classic" one. It is worthwhile to note that the population size and the generation number are problem dependent so exploratory run on the application at hand have to be performed in order to set their values.

In this work, we used our "classic" parameter setting as reported in Table 1. The population size and the generation number were chosen after some exploratory runs which allowed to determine a sufficiently small value.

4   REGAL with or without Using a Cooperative Strategy

The experiments reported in this section aims to study what kind of descriptions are learnt and what computational cost is involved when no cooperation or some cooperation policy is exploited. A set of basic configurations has been selected to act as a baseline. The following configurations, corresponding to the parameter settings appearing in Table 1, have been considered:

CONF1 (16 GAlearners and $\mu = 0.0$) - A basic distributed approach: 16 *GA_Learners*, each one evolving a population of 100 individuals. No cooperative strategy to coordinate the learners. This means that every learner exploits the whole learning set.

---

to more than 6 different values. By defining a predicate for each <attribute, value> pair, the language template for this application could be coded as a bitstring of 126 bits.

Table 1. REGAL's configurations used in this work.

Parameter	Value
Population size	1600
Number of GA learner	16
Crossover probability p_c	0.6
Mutation probability p_m	0.0001
Migration rate μ	0.0 or 0.5
Generation limit	200
Generation gap	0.9
Cooperation	None/LSE/DTSU

CONF2 (16 GA learners and $\mu = 0.5$) - As CONF1 plus migration of individuals among the *GA_learners*.

Plus CONF1 and CONF2 exploiting one cooperative policy.

In Table 2, the results obtained are reported. The leftmost column of the table shows the configuration's identifier. The other columns of the table contains the parameters already described plus the 'Cons & Compl' field that summarizes whether the learned concept description is complete and consistent on the learning set. Finally, the rows, with the value "Target", report the features of the target concept. For each configuration settings three runs have been performed. The reported error rate is an average over the three runs. Instead the other values are the real values of the description found in the experiment with the median error rate.

The experimental findings can be summarized as follows:

A) In CONF1, the maintenance of genetic diversity is mainly deferred to the *locality* of the evolution: each *GA_Learner* only affects the evolution of its population. When migration of individuals occurs (CONF2), genetic diversity across population tends to reduce. Thus letting individuals, describing (part of) their parents' original niches, merge and favoring the appearance of general descriptions. In turn, this biases the learning system toward the discovery of overfit concept description [21] that may decrease the classification performances as observable when passing from CONF1 to CONF2 in the experiments. In addition, migration increases the computational cost of a factor proportional to the number of exchanged individuals. This is due to the double evaluation migrating individuals are subjected to in the leaving and incoming niche. A minor point to be investigated during the system's reimplementation would be to reduce this computational overhead.

Let us evaluate now the contribution of cooperation to REGAL's performances:

1) both forms of cooperation allow to learn good concept descriptions.

2) The effect of migration of individuals is not very evident from the point of view of the error rate but a decrease in the solution complexity is observable when it is used.

3) Quite surprisingly using a cooperative strategy does not significantly increase

the system running cost. The reason may be that the evolving populations tend to converge toward simple descriptions at an earlier generation than when no co-operation is present. In summary, it seems that both cooperative policies perform reasonably well across a variety of system's configurations. Of course, additional study is needed in order to confirm or discard these latter conclusions.

Table 2. REGAL learning the "Poisonous mushrooms" concept.

CoopLS μ		T	C	ND	MXC	SMC	AVC	e[%]	Cons & Compl
CONF1									
None	0.0	76	7	2	1946	1139	1542	2	No
LSE	0.0	72	21	4	1946	62	946	0	Yes
DTSU	0.0	73	63	5	1946	329	1064	0	Yes
CONF2									
None	0.5	97	14	2	1946	1161	1553	4	No
LSE	0.5	103	16	3	1946	414	1089	0	Yes
DTSU	0.5	99	42	4	1946	317	1063	0	Yes
Target	-		15	3	1946	197	1096	0	Yes

5   Conclusion

Some preliminary investigation of two cooperative learning strategies has been reported. We believe that a distributed genetic base learner able to exploit these two cooperative strategies may acquire satisfactory concept descriptions across a wide range of applications. Additional experimentation, required to confirm or discard this claim, is in progress.

References

1. T.G. Dieterich and R.S. Michalski. A comparative review of selected methods for learning from examples. In J.G. Carbonell, R.S. Michalski, and T. Mitchell, editors, *Machine Learning, an Artificial Intelligence Approach*. Morgan Kaufmann, 1983.
2. R. S. King, S. Muggleton, R. A. Lewis, and M. J. E. Sternberg. Theories for mutagenecity: a study in first order and feature based induction. *Artificial Intelligence*, 74, 1995.
3. W. Lee, S. Stolfo, and K. W. Mok. Mining audit data to build intrusion detection models. In *Knowledge discovery in databases 1998*, pages 66–72, Fairfax, VA, 1998.
4. F. Neri. Comparing local search with respect to genetic evolution to detect intrusions in computer networks. In *Congress on Evolutionary Computation 2000*, pages 512–517, IEEE Press, 2000.

5. J. H. Holland. *Adaptation in Natural and Artificial Systems*. The University of Michigan Press, Ann Arbor, Mi, 1975.
6. D. Goldberg. *Genetic Algorithms in Search, Optimization, and Machine Learning*. Addison-Wesley, Reading, Ma, 1989.
7. K. A. De Jong, W. M. Spears, and F. D. Gordon. Using genetic algorithms for concept learning. *Machine Learning*, 13:161–188, 1993.
8. C.Z. Janikow. A knowledge intensive genetic algorithm for supervised learning. *Machine Learning*, 13:198–228, 1993.
9. A. Giordana and F. Neri. Search-intensive concept induction. *Evolutionary Computation*, 3 (4):375–416, 1995.
10. J. Hekanaho. Background knowledge in ga-based concept learning. In *13th International Conference on Machine Learning*, pages 234–242, Bari, Italy, 1996.
11. P. Husbands and F. Mill. A theoretical investigation of a parallel genetic algorithm. In *Fourth International Conference on Genetic Algorithms*, pages 264–270, Fairfax, VA, 1991. Morgan Kaufmann.
12. M. Potter. *The Design and Analysis of a Computational Model of Cooperative Co-evolution*. PhD thesis, Department of Computer Science. George Mason University, VA, 1997.
13. F. Neri. *First Order Logic Concept Learning by means of a Distributed Genetic Algorithm*. PhD thesis, Department of Computer Science. University of Torino, Italy, 1997.
14. J. L. Shapiro. Does data-mod co-evolution improve generalization performances of evolving learners? *Lecture Notes in Computer Science*, LNCS 1498:540–549, 1998.
15. R. E. Schapire. A brief introduction to boosting. pages 1401–1406, 1999.
16. T. Dietterich. An experimental comparison of three methods for constructing ensembles of decision trees: Bagging, boosting, and randomization. *Machine Learning*, 40:139–158, 2000.
17. R. Michalski, I. Mozetic, J. Hong, and N. Lavrac. The multi-purpose incremental learning system AQ15 and its testing application to three medical domains. In *Fifth National Conference on Artificial Intelligence*, pages 1041–1045, Philadelphia, PA, 1986.
18. J. R. Quinlan. *C4.5: Programs for Machine Learning*. Morgan Kaufmann, California, 1993.
19. F. Neri and L. Saitta. Exploring the power of genetic search in learning symbolic classifiers. *IEEE Trans. on Pattern Analysis and Machine Intelligence*, PAMI-18:1135–1142, 1996.
20. J. S. Schlimmer. Concept acquisition through representational adjustement. Technical Report TR 87-19, Dept. of Information and Computer Science, University of Californina, Irvine, CA, 1987.
21. R. Quinlan. Oversearching and layered search in empirical learning. In *International Conference on Machine Learning*, Lake Tahoe, CA, 1995.

The Influence of Learning in the Evolution of Busy Beavers

Francisco B. Pereira[1,2] and Ernesto Costa[2]

[1]Instituto Superior de Engenharia de Coimbra, Quinta da Nora, 3030 Coimbra, Portugal
[2]Centro de Informática e Sistemas da Universidade de Coimbra, Polo II, 3030 Coimbra,
Portugal
{xico, ernesto}@dei.uc.pt

Abstract. The goal of this research is to study how individual learning interacts
with an evolutionary algorithm in its search for good candidates for the Busy
Beaver problem. Two learning models, designed to act as local search
procedures, are proposed. Experimental results show that local search methods
that are able to perform several modifications in the structure of an individual in
each learning step provide an important advantage. Some insight about the role
that evolution and learning play during search is also presented.

1 Introduction

Evolution and learning are the two major forces that promote the adaptation of
individuals to the environment. Evolution, operating at the population level, includes
all mechanisms of genetic changes that occur in organisms over generations. Learning
operates at a different time scale. It gives to each individual the ability to modify its
phenotype during its life in order to increase its adaptation to the environment and,
hence, its chance to be selected for reproduction. In standard evolutionary
computation (EC) optimisation, learning has usually been implemented as local
search algorithms [1], [2], [3], [4]. These methods iteratively test several alternatives
in the neighbourhood of the learning individual trying to discover better solutions. At
the end of the learning process, the quality of an individual will be, not only the
measure of its initial fitness, but also of its ability to improve, which leads to a better
understanding of the fitness landscape. The combination of local search heuristics
with EC techniques is, in some contexts and situations, known as memetic algorithms
[5]. This designation is inspired by Richard Dawkin's concept of meme, which
represents a unit of cultural evolution that can exhibit plastic adaptation.

   In our research we are interested in studying how learning and evolution may be
combined in computer simulations. In this paper we use the Busy Beaver (BB)
problem as the testbed to study the above-mentioned interactions. In 1962, Tibor
Rado proposed this problem in the context of the existence of non-computable
functions [6]. It can be defined as follows: suppose a Turing Machine (TM) with a
two-way infinite tape and a tape alphabet={blank,1}. The question Rado asked was:
what is the maximum number of **1**'s that can be written by a N-state halting TM when
started on a blank tape? This number, which is a function of the number of states, is
denoted by $\Sigma(N)$. A TM that produces $\Sigma(N)$ non-blanks cells is called a Busy Beaver.

E.J.W. Boers et al. (Eds.) EvoWorkshop 2001, LNCS 2037, pp. 421-430, 2001.
© Springer-Verlag Berlin Heidelberg 2001

The BB is considered one of the most interesting theoretical problems and, since its proposal, has attracted the attention of many researchers. Some values for $\Sigma(N)$ and the corresponding TMs are known today for small values of N. As the number of states increases, the problem becomes harder and, for N$\geq$5, there are several candidates that set lower bounds on the value of $\Sigma(N)$. To prove that a particular candidate is the N-state BB we must perform an exhaustive search over the space of all N-state TMs and verify that no other machine produces a higher number of ones. This is extremely complex due to the halting problem.

The search space of the BB problem possesses several characteristics, such as its dimension and its complexity, that make it extremely appealing to the EC field. Several attempts to apply EC techniques were reported in the past few years with different levels of success [7], [8]. In a previous work we proposed two local search algorithms designed to act as learning procedures when seeking for solutions to this problem [2]. Experimental results showed that both of them were unable to improve the performance of the evolutionary algorithm. A partial explanation, proposed in the above mentioned work, suggested that a combination of two factors made those local search methods completely ineffective: the structure of the landscape and the behaviour of the learning procedures. The topology of the search landscape is highly irregular and has a very complex structure. We performed some empirical analysis and verified that, in different areas of the search space, there are small groups of neighbour valid solutions to the BB problem. The size of these groups and the quality of the TMs that compose them varies but, nevertheless, they tend to be surrounded by large low fitness areas composed by invalid solutions. The combination of these factors makes the space very prone to premature convergence. In the beginning of the simulation, EC methods quickly identify one of these areas. Then there is a high evolutionary pressure towards this area and it is almost impossible to escape premature convergence. On the other hand, both of the learning algorithms used in the study act as typical hill-climbers. In each learning step they perform one modification in the current solution, accepting it if it does not lead to a decrease in the fitness. In the described situation, the efforts of learning methods are helpless, since the exploration of very close neighbourhood will be ineffective to escape the basin of attraction of the group of valid TMs. In this paper we introduce a new local search method that is able to perform several modifications in the structure of a TM in each learning step. Our goal is to determine what is the influence that this new situation has in the evolutionary process. In the literature there are some reports of other learning methods that carry out more than one modification in each step to overcome limitations when searching difficult landscapes [9], [10].

The structure of the paper is the following: in the next section we present a formal definition of the BB problem. In section 3 we describe our evolutionary model, including the learning procedures used. Section 4 comprises some experimental details about the simulation. In section 5 we present results of the experiments performed and analyse them. Finally, in section 6, we present some conclusions and suggest directions for future work.

2   The Busy Beaver Problem

A deterministic TM can be specified by a sextuple $(Q, \Pi, \Gamma, \delta, s, f)$, where:

Q is a finite set of states, Π is an alphabet of input symbols, Γ is an alphabet of tape symbols, δ is the transition function, $s \in Q$ is the start state and $f \in Q$ is the final state. The transition function can assume several forms. The most usual one is:

$$\delta: Q \times \Gamma \to Q \times \Gamma \times \{L, R\}$$

where L denotes move the head left and R move right. Machines with a transition function with this format are called 5-tuple TMs. A common variation consists in considering a transition function of the form:

$$\delta: Q \times \Gamma \to Q \times \{\Gamma \cup \{L, R\}\}$$

Machines of this type are known as 4-tuple TMs. When performing a transition, a 5-tuple TM writes a symbol on the tape, moves the head and enters a new state. A 4-tuple TM either writes a new symbol on the tape or moves its head, before entering the new state.

The original definition of the BB [6] considers 5-tuple TMs with N+1 states (N states and an anonymous halt state). The tape alphabet has two symbols, $\Gamma = \{blank, 1\}$, and the input alphabet has one, $\Pi = \{1\}$. The productivity of a TM is defined as the number of **1**'s present, on the initially blank tape, when the machine halts. Machines that do not halt have productivity zero. $\Sigma(N)$ is defined as the maximum productivity that can be achieved by a N-state TM. This TM is called a Busy Beaver. In the 4-tuple variant productivity is defined as the length of the sequence of **1**'s produced by the TM when started on a blank tape, and halting when scanning the leftmost one of the sequence, with the rest of the tape blank. Machines that do not halt, or, that halt on another configuration, have productivity zero [11]. Thus, the machine must halt when reading a **1**, this **1** must be the leftmost of a string of **1**'s and, with the exception of this string the tape must be blank. In our research we focus our attention on the 4-tuple variant.

3   Experimental Model

3.1   Representation

In the experiments reported in this paper we are searching for good candidates for the 4-tuple BB(7). Without loss of generality we consider $Q = \{1, 2, 3, 4, 5, 6, 7, f\}$, set 1 as the initial state and f as the final state. Since $\Gamma = \{blank, 1\}$, the essential information needed to represent a potential solution is reduced to the state transition table. Figure 1 shows a 4-tuple TM with 7 states (plus the halting state f) and its state transition table. To codify the information contained in the table we use an integer string with 28 genes (4 genes per state) with the following format:

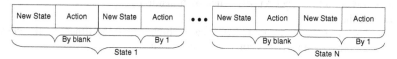

3.2   Simulation and Evaluation

To evaluate an individual we simply decode the information from the chromosome and then simulate the resulting TM. Due to the halting problem we must establish a limit for the maximum number of transitions (MaxT). Machines that don't halt before this limit are considered non-halting TMs. To assign fitness we consider the following factors in decreasing order of importance:

- Halting before reaching the predefined limit for the number of transitions;
- Accordance to the 4-tuple rules [11];
- Productivity;
- Number of used transitions;
- Number of steps made before halting.

We consider all these factors to assign fitness because we intend to explore differences between "bad" individuals. With this fitness function a TM that never leaves state 1 is considered worse than another one that goes through 3 or 4 states, even if both are non-halting machines and have the same productivity. In some preliminary experiments this approach proved to be more effective than using productivity alone as fitness.

δ	By blank		By one	
Q	New State	Action	New State	Action
1	2	1	5	L
2	6	R	2	R
3	2	R	F	1
4	5	R	6	L
5	3	R	7	0
6	4	1	1	L
7	4	1	4	0

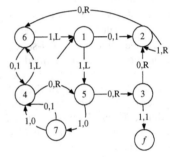

Fig. 1. A seven state 4-tuple TM and its corresponding transition table. The blank symbol is represented by 0.

3.3   Learning Models

After evaluation an individual might be selected for learning. In this research we perform experiments with two different local search procedures.

Random Local Search (RLS).
Given a current TM (machine that was built with the information encoded in the chromosome of the individual selected for learning) perform the following actions:
1. Select one transition T used in the simulation of the current TM.
2. Randomly modify the action performed by transition T[1].
3. Evaluate the resulting TM.
4. If the fitness of the resulting TM is equal or higher than the fitness of the current TM, then the resulting TM becomes the current one.

---

[1] Possible actions for one transition: write blank, write **1**, move left or move right.

5. If the maximum number of learning steps has been equalled stop learning. Otherwise go to 1.

RLS is identical to one the proposals from our previous work [2]. In each learning cycle RLS performs one modification in the structure of the TM accepting it if it does not lead to a decrease in fitness. Changes in the structure of the TM are limited to actions performed by transitions. To ensure that this restriction is not biasing results we performed some additional tests enabling RLS to change either actions or new states and verified that there was no significant difference in the outcomes. Results achieved by experiments using RLS will be useful just to have a comparison measure with results obtained by the new algorithm, Multi Step Learning (MSL). The most important difference between RLS and MSL is that, with the new method, in each learning cycle an individual performs 2 or 3 changes in its structure. An algorithmic description of MSL follows:

Multi Step Learning (MSL).

Given a current TM (machine that was built with the information encoded in the chromosome of the individual selected for learning) perform the following actions:
1. Select one transition T1 used in the simulation of the current TM and that does not lead to the final state.
2. Randomly modify the action performed by transition T1.
3. With a probability of 0.5 randomly modify the state to where transition T1 leads. The final state is not considered as a possibility when selecting the new destiny.
4. Let S be the state to where T1 leads. Select, with equal probability, one transition T2 from state S.
5. Randomly modify the action performed by transition T2.
6. Evaluate the resulting TM.
7. If the fitness of the resulting TM is equal or higher than the fitness of the current TM, then the resulting TM becomes the current one.
8. If the maximum number of learning steps has been equalled stop learning. Otherwise go to 1.

Modifications in the structure of the TM are done in components that are directly connected, starting with the action of one transition, then the destiny state of the transition (it changes with 0.5 probability) and, finally, the action of one of the transitions from this state. With MSL, an individual has the possibility to jump to a point in space that is not so close to its current position. We hope that this might give to the learning individual a higher chance to escape from local optima and allow the evolutionary process to perform better.

In this paper we use a Lamarckian strategy for learning. Lamarckian theory of evolution claims that phenotypic characteristics acquired by individuals during their lifetime are somehow re-encoded in their genes and directly inherited by their descendants. In our experiments, at the end of the learning period, all changes induced in the current TM are coded back to the genotype of the learning individual. Even though Lamarckian theory proved to be wrong in biological systems, the idea has been usefully applied in several experiments in the EC field [12], [13], provided that some special considerations are taken into account. To prevent the tendency of Lamarckian learning to increase the convergence rate of the evolutionary algorithm we use a global parameter, the learning rate (LR), which is defined as the probability of an individual being subject to learning. This way we are able to restrict the number of individuals that learn in each generation and we hope that this might control the tendency to premature convergence.

4   Experimental Settings

The experiments presented concern the search for the 4-tuple BB(7). The settings of the evolutionary algorithm are the following: Number of evaluations: 200,000,000; Population Size: 500; Elitist Strategy; Tournament Selection with tourney size 5; Single point Mutation; Mutation rate: 0.025; Graph Based Crossover; Maximum graph crossover size: 4; Crossover rate: 0.7; MaxT (Maximum number of transitions): 100,000; LR = {0.1, 0.5, 1}.

Graph based crossover was presented in [14]. It was designed to work with individuals with a graph-like structure. The main idea of this operator is the exchange of sub-graphs between individuals. Maximum graph crossover size defines the number of states belonging to each sub-graph. Results presented confirmed that, in this domain, it clearly outperforms classical crossover operators.

During learning the number of steps (i.e., number of learning cycles) performed by an individual is set to 10 and remains fixed for all experiments. Each step counts as one evaluation. The initial population is randomly generated and for every set of parameters we performed 30 runs with the same initial conditions and different random seeds. In the next section we present results from 7 distinct experiments: one where the individuals do not learn (NoLearn) and 3 experiments using each one of the two learning procedures, RLS and MSL. The only difference between the 3 experiments using the same learning method is the LR value. Even though values for different settings were set heuristically we performed some tests with other values and verified that, within a moderate range, there was no significant difference in the outcomes.

5   Results

In graph 1 we present, for all different experiments, the productivity of the best individual of the final generation in each one of the 30 runs. A brief perusal of the results suggests that, with some settings, learning was able to cause a considerable improvement in the search process. In the remainder of this section we try to identify the conditions that were required to obtain such improvements. Before the application of EC techniques to the 4-tuple BB(7) the productivity of the best-known candidate was 37 [15]. We adopt this value as the threshold of minimum quality and focus our attention in runs that were able to find TMs with higher productivity. Assuming this we can see that only experiments with MSL were able to outperform the experiment without learning. A standard evolutionary approach was able to find TMs with productivity > 37 in 10% of the runs (3 out of 30 runs were able to find such machines). Using MSL with LR=0.1 the percentage raises to 20% (6 out of 30 runs). RLS was never able to help the search process. Results for all LRs are of inferior quality than those achieved by experiments with MSL and by the NoLearn experiment. The best experiment with RLS only found two TMs with productivity > 37. This result confirms our previous conclusions. RLS just performs small readjustments in the organisation of the TMs. Given the structure of the landscape it is unable to help search to escape from local optima areas.

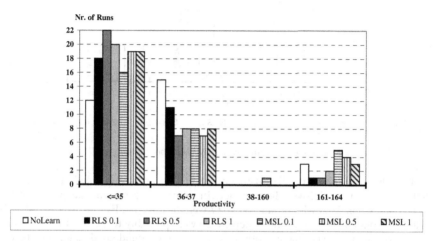

Graph 1. Productivity of the best individual of the final generation for each one of the 30 runs.

Now, we focus our analysis on the results achieved by MSL. It is clear that results obtained by experiments with MSL are clearly better than those obtained by the experiments with RLS. This confirms that MSL is a more effective method to help evolution to escape from the traps that exist in the landscape. When searching for solutions in highly irregular landscapes there is an important advantage if the learning individual is allowed enlarge its neighbourhood region (this region includes all points to where the individual is allowed to jump in just one learning step). There are, however, significant differences in the results achieved by the three experiments that use MSL, suggesting that the ability to perform larger jumps is not a sufficient condition to improve performance. Results from graph 1 show that infrequent learning (LR=0.1) is most beneficial to this situation. When LR=0.1, MSL helps evolution to find 6 TMs with productivity > 37. This number decreases to 4 with LR=0.5 and to 3 with LR=1. It is not surprising that the performance of experiments with MSL decreases as the LR increases. Since it re-encodes all changes back to the genotype, Lamarckian learning is a very strong mechanism and it pushes the search very fast to a local optimum. Given the landscape that we are dealing with, this effect is even magnified and the search process will, with high probability, converge to such a local optimum. Looking at the left columns from graph 1 (runs whose best TMs have productivity≤35) it is possible to verify that this situation happens with higher probability in experiments with LR≥0.5. Convergence occurs nearly in the beginning of the runs and individuals from subsequent generations are not able to escape from this basin of attraction.

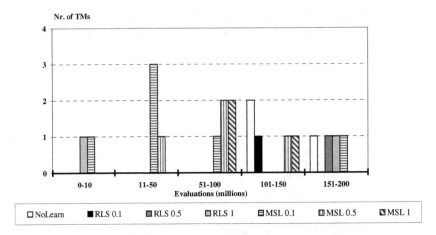

Graph 2. Period in the simulation when TMs with productivity > 37 were found.

In addition to increasing the likelihood of finding promising solutions, MSL also helps evolution to discover them earlier. In graph 2 we present the period in the simulation when TMs with productivity>37 were discovered. Even if we discard TMs that were found before 10 millions steps (they were probably found due to some lucky move) we see that experiments with MSL consistently found good candidates before 100 million evaluations. The experiment without learning only started to find good candidates after this period. This gives credit to the conviction that learning is really helping evolution to find good solutions. Some results collected from the experiments help to clarify what might be happening during the search process. In table 1 we present the contribution of evolution and learning to the discovery of new best individuals. Contribution from evolution includes all new best individuals generated by crossover and/or mutation and contribution from learning includes all new best individuals that result from the application of local search procedures. Values obtained in each one of the experiments are divided over 4 temporal periods. If we look to the results of the line labelled MSL 0.1 (experiment using MSL with LR=0.1) there are two important features that distinguish it from all other experiments with learning. The first one is the number of improvements obtained by evolution. In the period ranging from 1 million to 100 million evaluations, the number of improvements due to evolution is similar to the values achieved by the experiment without learning. Between 10 million and 100 million evaluations this value is even superior (131 vs. 120), which is a remarkable result, especially if we consider that in experiments with LR=0.1 approximately half of the evaluations are spent in the learning process. This result suggests that learning is not preventing evolution from sampling the space. At the contrary, it is helping evolution in this task. The second feature concerns the relative weight of learning in the process of finding new solutions. Values in parenthesis in the columns labelled *Learn* represent the percentage of all improvements from that period that were due to learning. The lowest values are always found on the line labelled MSL 0.1. Evolution is clearly more effective than learning in the process of finding new best solutions. Only in the final period, when the search has become stable, the weight of learning increases to values similar to the ones presented by evolution. This suggests that when an evolutionary algorithm is starting to explore the space, moderate (i.e., low LR) Lamarckian learning might give some clues about the best paths to follow. Nevertheless, it is evolution that is guiding the process. The role of learning is to present some hints that might allow the early discovery of promising areas to explore. However, just like it was shown, this small effect is important to obtain good results. A different scenario occurs if there is too much pressure (LR too high). In this situation learning acts as the primary guiding force of the search process and evolution plays a secondary role. Since learning in this context is, by definition, a local procedure, search will most likely end up in the nearest local optimum.

Table 1. Contributions from evolution (columns labelled Ev.) and learning (columns labelled Learn) to the improvement of the best solution during simulation. For each experiment, results presented are the sum of 30 runs. Results are divided over 4 temporal periods of the simulation. Values in parenthesis in the columns labelled Learn represent the percentage of all improvements from that period that were due to learning.

Evals. (millions)	Periods of the Simulation							
	0 to 1		1 to 10		10 to 100		100 to 200	
	Ev.	Learn	Ev.	Learn	Ev.	Learn	Ev.	Learn
NoLearn	1096		146		120		43	
RLS 0.1	669	302 (31)	110	42 (28)	74	26 (26)	31	22 (41)
RLS 0.5	376	600 (62)	59	103 (64)	50	71 (59)	4	10 (71)
RLS 1	212	768 (78)	44	108 (71)	30	90 (75)	4	14 (78)
MSL 0.1	798	222 (22)	141	50 (26)	131	42 (24)	17	12 (41)
MSL 0.5	365	505 (58)	86	69 (45)	77	67 (46)	12	14 (54)
MSL 1	231	572 (71)	53	91 (63)	70	83 (54)	12	30 (71)

6   Conclusions and Further Work

In this paper we studied the interactions that occur between evolution and learning when searching for good solutions for the BB problem. We presented the results of several experiments performed within a Lamarckian framework and showed that a procedure able to perform several modifications in the structure of the learning individual in each learning step is most beneficial. We also identified some conditions that should be met to maximise the search performance of the evolutionary algorithm:

• Evolution should be the primary force responsible for sampling the landscape.
• A moderate contribution of a learning procedure (with a considerable degree of freedom in what concerns the definition of local neighbourhood) is very important to help evolution in its task.

As future work we will study the same kind of interactions with another learning strategy, known as the Baldwin Effect. We intend to analyse if the conditions identified in the current work are extensible to this learning framework.

Acknowledgments. This work was partially funded by the Portuguese Ministry of Science and Technology, under Program PRAXIS XXI.

References

1. Belew, R. and Mitchell, M. (1996). *Adaptive Individuals in Evolving Populations: Models and Algorithms*, Santa Fe Institute in the Sciences of Complexity, Vol. XXVI, Reading, MA: Addison-Wesley.
2. Pereira, F. B., Machado, P., Costa, E., Cardoso, A., Rodriguez, A. Santana, R., and Soto, M. (2000). Too Busy to Learn. *Proceedings of the Congress on Evolutionary Computation (CEC-2000)*, pp. 720-727.
3. Sasaki, T. and Tokoro, M. (1999). Adaptation under Changing Environments with Various Rates of Inheritance of Acquired Characters: Comparison Between Darwinian and Lamarckian Evolution. In McKay, B., Yao, X., Newton, C. S., Kim, J. H. and Furuhashi, T. (Eds.), *Proceedings of 2nd Asia-Pacific Conference on Simulated Evolution and Learning (SEAL-98)*.
4. Whitley, D., Gordon, S. and Mathias, K. (1994). Lamarckian Evolution, the Baldwin Effect, and Function Optimization. In Davidor, Y. Schwefel, H. P. and Manner, R. (Eds.) *Parallel Problem Solving from Nature (PPSN-III)*, pp. 6-15.
5. Corne, D, Glover, F. and Dorigo, M. (1999). *New Ideas in Optimization*. McGraw-Hill.
6. Rado, T. (1962) On non-computable functions, *The Bell System Technical Journal*, vol. 41, no. 3, pp.877-884.
7. Jones, T. and Rawlins, G. (1993). Reverse Hillclimbing, Genetic Algorithms, and the Busy Beaver Problem. In Forrest, S. (Ed.). *Proceedings of the 5th International Conference on Genetic Algorithms (ICGA-93)*, pp.70-75, San Mateo, CA, Morgan Kaufmann.
8. Machado, P., Pereira, F. B, Cardoso, A., Costa, E. (1999). Busy Beaver – The Influence of Representation, In Poli, R., Nordin, P. Langdon, W. and Fogarty, T. (Eds.). *Proceedings of the Second European Workshop in Genetic Programming (EuroGP-99)*.
9. Bull, L. (1999). On the Baldwin Effect. *Artificial Life*, Vol. 5(3), pp. 241-246.
10. Pereira, F. B. and Costa, E. (1997). The Influence of Learning in the Optimization of Royal Road Functions, *Proceedings of the 3rd International Mendel Conference on Genetic algorithms, Optimization Problems, Fuzzy Logic, Neural Networks and Rough Sets (Mendel'97)*, pp 244-249.
11. Boolos, G., and Jeffrey, R. (1995). *Computability and Logic*, Cambridge University Press.
12. Ackley, D. and Littman, M. (1994). A Case for Lamarckian Evolution. In Langton, C. (Ed.), *Artificial Life III*, pp. 3-10, Addison-Wesley.
13. Imada, A. and Araki, K. (1996). Lamarckian Evolution of Associative Memory. In *Proceedings of the Third International Conference on Evolutionary Computation (ICEC-96)*, pp.676-680.
14. Pereira, F.B., Machado, P., Costa, E. and Cardoso A. (1999). Graph Based Crossover - A Case Study with the Busy Beaver Problem. In Banzhaf, W., Daida, J., Eiben, A. E., Garzon, M. H., Honavar, V., Jakiela, M., & Smith, R. E. (Eds.). *GECCO-99: Proceedings of the Genetic and Evolutionary Computation Conference*, pp. 1149-1155, Morgan Kaufmann.
15. Lally, A., Reineke, J. and Weader, J. (1997). An Abstract Representation of Busy Beaver Candidate Turing Machines, *Technical Report*, Van Gogh Group, Rensselaer Polytechnic Institute.

Automated Solution of a Highly Constrained School Timetabling Problem – Preliminary Results

Marc Bufé, Tim Fischer, Holger Gubbels, Claudius Häcker, Oliver Hasprich, Christian Scheibel, Karsten Weicker*, Nicole Weicker, Michael Wenig, and Christian Wolfangel

University of Stuttgart, Faculty of Computer Science, Germany

Abstract. This work introduces a highly constrained school timetabling problem which was modeled from the requirements of a German high school. The concept for solving the problem uses a hybrid approach. On the one hand an evolutionary algorithm searches the space of all permutations of the events from which a timetable builder generates the school timetables. Those timetables are further optimized by local search using specific mutation operators. Thus, only valid (partial) timetables are generated which fulfill all hard constraints.

1 Introduction

Timetabling problems occur in many companies and educational institutions. Especially in education we can distinguish three different types of problems: the school timetabling problem, the university timetabling problem, and the exam timetabling problem. All three problems have a slightly different focus [1]. Therefore, different techniques are necessary for solving them. In this work we consider the school timetabling problem. The school timetabling problem is a complex real-world problem with many interdependencies. Thus, it can be solved seldom by divide-and-conquer or greedy techniques. Practitioners report that feasible solutions may be found manually within a rather short period of time. However, manual solutions satisfying organizational or didactic requirements may need several person-days work. The timetabling problem has been shown to be NP-complete as soon as unavailabilities of teachers, classes, or rooms are involved [2]. Organizational requirements are of increasing importance since many teachers are only working part-time and inconvenient, flexible time arrangements are necessary for them. Moreover, schools put a lot of thought into didactic issues, e.g. the arrangement of the different subjects within the week. These considerations should be reflected in the timetables.

---

* correspondence address: Karsten Weicker, University of Stuttgart, Breitwiesenstr. 20–22, 70565 Stuttgart, Germany, Email: Karsten.Weicker@informatik.uni-stuttgart.de

E.J.W. Boers et al. (Eds.) EvoWorkshop 2001, LNCS 2037, pp. 431–440, 2001.
© Springer-Verlag Berlin Heidelberg 2001

2   The School Timetabling Problem

When talking about school timetabling problems, two different goals in timetabling must be distinguished. First, the simple form of the problem only requires that a feasible solution is found concerning certain hard constraints. However, this is not enough in practice where, second, additional soft constraints should be fulfilled. The simple timetabling problem involves a set of teachers T, a set of classes C, a set of rooms R, a set of time periods P, and a set of events E. Each event requires a class $c(e) \in C$, a teacher $t(e) \in T$, and a number of weekly hours $hours(e) \in \mathbb{N}$. In order to solve the problem a mapping must be found which assigns each event to a room and a time period for each required hour, i.e.

$$TT : E \rightarrow 2^{P \times R}$$

where $|TT(e)| = hours(e)$. As an example, let e be chemistry for one class, then $TT(e) = \{(M - 1, 101), (W - 3, 102)\}$ assigns the chemistry lectures to the first hour on Mondays and the third hour on Wednesdays, the according rooms have the numbers 101 and 102.

Furthermore, a solution of the timetabling problem must suffice the following hard constraints. In order to simplify the notation in the constraints, the components of an assignment $a = (p, r) \in P \times R$ may be accessed by $p(a) = p$ for the time period and $r(a) = r$ for the room.

– Each teacher participates in at most one event per time slot.

$$\forall_{e \in E} \forall_{e' \in E \setminus \{e\}} \left(t(e) = t(e') \Rightarrow \left(\forall_{a \in TT(e)} \forall_{a' \in TT(e')} p(a) \neq p(a') \right) \right) \quad (1)$$

– Each class participates in at most one event per time slot.

$$\forall_{e \in E} \forall_{e' \in E \setminus \{e\}} \left(c(e) = c(e') \Rightarrow \left(\forall_{a \in TT(e)} \forall_{a' \in TT(e')} p(a) \neq p(a') \right) \right) \quad (2)$$

– Each room is assigned to at most one event per time slot.

$$\forall_{e \in E} \forall_{e' \in E \setminus \{e\}} \forall_{a \in TT(e)} \forall_{a' \in TT(e')} \left(r(a) = r(a') \Rightarrow p(a') \neq p(a') \right) \quad (3)$$

This simple problem is not sufficient for modeling the requirements of a real schools timetable. In the remainder of this section the necessary extensions with additional hard and soft constraints are described.

To begin with, in German schools it is common practice to mix several classes for certain subjects, e.g. for physical education or religious knowledge lessons. As a consequence, more than one class $c(e) \subset C$ is involved in one event. However it is still required that each event must be scheduled to exactly one room. That means that religious knowledge lessons must be split for each age-class into two events. All protestant pupils participate in one event, all catholic pupils in another event. Although both events comprise pupils of the same classes, they are distinct and should be scheduled within the same time periods to guarantee a timetable of high quality. Thus, those events are grouped together in event groups. All other events are called singular events. The singular events are denoted $E_S \subseteq E$, the grouped events $E_G = E \setminus E_S \subseteq E$. For each event $e \in E$, we denote by $[e] \subseteq E$ the events that belong together, i.e. for $e \in E_S$ it holds that $[e] = \{e\}$ and for $e \in E_G$ all events in the group of e are contained in $[e]$. The following hard constraint is affected slightly by this modification.

– Each class participates in at most one event per time slot.

$$\forall_{e \in E} \forall_{e' \in E \setminus [e]} \left((c(e) \cap c(e') \neq \emptyset) \Rightarrow \left(\forall_{a \in TT(e)} \forall_{a' \in TT(e')} p(a) \neq p(a') \right) \right) \tag{2'}$$

In favor of more flexibility in modeling timetabling problems and due to events requiring more than one teacher's attendance, we also allow more than one teacher $t(e) \subset T$ for each event.

– Each teacher participates in at most one event per time slot.

$$\forall_{e \in E} \forall_{e' \in E \setminus \{e\}} \left((t(e) \cap t(e') \neq \emptyset) \Rightarrow \left(\forall_{a \in TT(e)} \forall_{a' \in TT(e')} p(a) \neq p(a') \right) \right) \tag{1'}$$

Moreover, teachers, classes, and rooms are extended by further attributes in order to comprise all necessary constraints. First of all, all three basic elements of the problem may be provided with unavailabilities for certain periods, $unavail(t), unavail(c), unavail(r) \subset P$. The according hard constraints read as follows.

$$\forall_{e \in E} \forall_{a \in TT(e)} \ p(a) \notin unavail(t(e)) \tag{4}$$

$$\forall_{e \in E} \forall_{a \in TT(e)} \ p(a) \notin unavail(c(e)) \tag{5}$$

$$\forall_{e \in E} \forall_{a \in TT(e)} \ p(a) \notin unavail(r(e)) \tag{6}$$

In addition to the unavailabilities, most events have special requirements for the equipment of the room, e.g. a biology lab, speech lab, geography room. Those requirements are summarized in room features F. The features are assigned to the rooms $f(r) \subset F$ and the required features to the events $f(e) \subset F$. The required room features must be considered in the assignment process.

$$\forall_{e \in E} \forall_{a \in TT(e)} \ f(e) \subseteq f(r(a)) \tag{7}$$

All constraints (1'), (2'), (3), ... , (7) must be fulfilled by a solution to be feasible.

The succeeding conditions are soft constraints which are encouraged to be fulfilled. They form the didactic quality of the timetable and consider certain organizational issues a timetable must meet in order to be acceptable.

(S-1) Events should be placed more likely in the morning than in the afternoon, e.g. an average class has at most once or twice a week lessons in the afternoon.

(S-2) Many teachers have part-time jobs which means that they should have at least a minimum number of free days each week.

(S-3) Further requirements for certain events like a minimal and maximal number of double hours, fortnightly or marginal placement should be considered.

(S-4) A uniform distribution of the hours of each event over the week is desired (cf. [3]). In addition, certain event groups or singular events can be marked as associated which means that they should not take place at the same day.

(S-5) There should be no gaps in the morning and afternoon schedules of each class and teacher (cf. [4]).

(S-6) For each time period a teacher must be available to supervise a class in case of an unexpected absence of the regular teacher (cf. [3]).

3   Related Work

Automated timetabling has been an issue within the last 30 years. Schaerf [1] gives an overview of the techniques used for the different problems of school timetabling, university timetabling, and exam timetabling. Junginger [5] provides a list of the early approaches to the school timetabling problem in Germany. A survey on the usage of evolutionary techniques used for timetabling and scheduling can be found in [6] The most available evolutionary approaches to the school timetabling problem use a direct representation of the timetable (e.g. [7,3,4]). Those algorithms have the huge disadvantage that the evolutionary operators may produce infeasible timetables which must be repaired by a genetic repair function. This has very often the effect that the correlation between the parent timetables and the offsprings is very low. Where the mutation operator can be chosen appropriately to minimize these effects, this is not possible for the crossover operator. Thus often the crossover acts more like a macromutation than a recombination. Fernandes et. al. [4] used standard operators where Colorni et. al. [7,3] designed special operators which can only guarantee partial feasibility. As a consequence in [3] it is reported that tabu search outperforms the genetic algorithm in most test cases. Another high-quality tabu search algorithm was presented by Schaerf [8].

Instead of using a direct representation, there are also approaches in university timetabling where a timetable builder is used which generates a feasible timetable from an individual containing several parameters for the timetable builder (e.g. [9]).

4   Concept

In the previous section two different techniques have been pointed out to be very promising in evolutionary timetabling: phenotypic mutations and timetable builders using genotypic operators since in both approaches operators may be defined in such a way that a high correlation between parent and offspring timetables is guaranteed. Furthermore, they can always generate feasible (probably partial) timetables fulfilling all hard constraints. As a consequence a hybrid approach was used to combine the positive aspects of both standard approaches and to minimize their negative drawbacks. The individuals use two different representations of the timetable: a parametric representation and a high level representation. The parametric representation consists of a permutation of the events which is understood as a queue for the timetable builder. The timetable builder is a deterministic algorithm, described in the next section, which creates a timetable from the queue. Its result is stored in the second half of the individual. Note, that this timetable builder might create only partial timetables where certain events remain unplaced. The different levels of this approach are sketched in Figure 1. The timetable builder is an essential technique since it

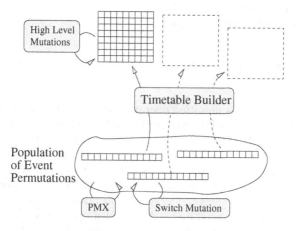

Fig. 1. Hybrid approach using a timetable builder and operators on both levels of representation.

may use many deterministic "intelligent" heuristics to create a timetable. Thus a high quality can already gained by the timetable builder. However, the use of those heuristics narrows the search space critically. Given the huge amount of constraints no timetable builder could be found which actually placed all events as we will discuss in Section 6. As a consequence the phenotypic mutations are necessary to optimize those timetables in order to get all events placed or to fulfill more violated soft constraints.

On the event queues, the evolutionary algorithm uses a mutation swapping two elements and the partial matching crossover (PMX) with the first crossing point fixed at the beginning of the queue to increase the correlation between parents and children. These operators provide the exploration of the search space. Since the timetable builder is deterministic the resulting timetable may be reproduced anytime and slightly changed permutations should also result in related timetables.

```
 1: INPUT: list with all hours of one event
 2: for daytime = morning, afternoon do
 3:   for constraints = all, hardOnly do
 4:     for day = M, W, Th, Tu, F do
 5:       check for each slot ∈ daytime whether there is a room such that
 6:             no unavailability (room, class, teacher) for slot    ∧
 7:             room, class, and teacher are disposable    ∧
 8:             constraints = all ⇒ soft constraints are fulfilled
 9:       if such a room and slot exists then
10:         assign first hour in list to slot
11:       end if
12:     end for
13:   end for
14: end for
15: if not all hours in list are placed then
16:   undo all assignments of the event
17: end if
```

Fig. 2. Placing algorithm within the timetable builder.

Furthermore, various mutation operators are defined on the complex timetables. They respect the feasibility of the timetable and apply little changes to the timetable. The result of this mutation has no effect on the permutation used by the timetable builder. This local optimization of the timetables should encourage a high exploitation of interesting solutions in the search space. Currently three different types of mutation operators are used. The first operator unplaces an event or a part of an event. The second operator tries to place an unplaced event or a partially unplaced event. A the third operator combines both operations by moving an event from one time slot to a different time slot. Also all three operators try to place the unplaced events after their primary operation.

Since the algorithm uses uniform parental selection and only replaces the worst 40 percent of the population in each generation those operators are able to perform an extensive parallel hill-climbing search on the better individuals.

5   Details

Different kinds of constraints need different constraint handling within the evolutionary algorithm. The timetable builder considers the hard constraints (1'), (2'), (3), ... , (7), i.e. all solutions generated from the permutation of events fulfill those requirements – if this is not possible a partial timetable is created and certain events remain unplaced. Also the timetable builder takes care of the soft constraints (S-2) and (S-3) directly. Constraints (S-1) and (S-4) are tried to be fulfilled indirectly by the timetable builder. The high-level mutation operators can only consider the hard constraints (1'), (2'), (3), ... , (7) where soft constraints may be hurt. All soft constraints are considered within the fitness function.

Table 1. Considered constraints in the fitness function.

	level of constraint	averaged value	standard deviation
(S-1)	classes C	$E_{S\text{-}1}[TT] = \frac{1}{|C|} \sum_{c \in C}$ violation for c	$\sqrt{V_{S\text{-}1}[TT]}$
(S-2)	teachers T	$E_{S\text{-}3}[TT] = \frac{1}{|T|} \sum_{t \in T}$ violation for t	$\sqrt{V_{S\text{-}3}[TT]}$
(S-3)	events E	$E_{S\text{-}4}[TT] = \frac{1}{|E|} \sum_{e \in E}$ violation for e	$\sqrt{V_{S\text{-}4}[TT]}$
(S-4)	events E	$E_{S\text{-}5}[TT] = \frac{1}{|E|} \sum_{e \in E}$ violation for e	$\sqrt{V_{S\text{-}5}[TT]}$
(S-5)	T and C	$E_{S\text{-}6}[TT] = \frac{1}{|T \cup C|} \sum_{x \in T \cup C}$ violation for x	$\sqrt{V_{S\text{-}6}[TT]}$
(S-6)	periods P	$E_{S\text{-}7}[TT] = \frac{1}{|P|} \sum_{p \in P}$ violation for p	$\sqrt{V_{S\text{-}7}[TT]}$
(P)	timetable	$E_P[TT] = \frac{|unplacedevents|}{|allevents|}$	—
(Q)	timetable	$E_Q[TT] = \frac{|partiallyplacedevents|}{|allevents|}$	—

The timetable builder consists of three phases. In the first phase, free days are assigned to the part time teachers (cf. S-2) using a round-robin method. This method guarantees that the free days are distributed equally over the week. The strategy which days are assigned to which teacher depends on the event queue in the individual. The second phase, splits up the events into single or paired lecture hours based on the preferences contained in the data (cf. S-3). Then, in the third phase, the placing algorithm in Figure 2 is applied to all events. This algorithm guarantees that first all hours are tried to be placed in the morning (in the first instance under consideration of all constraints, then only meeting the hard constraints). If this fails they are placed in the afternoon. This ensures that constraint (S-1) is fulfilled as good as possible. Also the uniform distribution of the hours of each event over the week (S-4) is encouraged by the order of the days for trying to find a valid time slot.

To define a fitness function the degree of constraint violation has to be measured. Since the different constraints are defined on different levels within the timetable we define first all those measures in Table 1. Then, the fitness function is composed of three components, the number of placed events f, the soft constraints g, and the standard deviation h concerning the soft constraints:

$$f(TT) = \kappa_1 E_P[TT] + \kappa_2 E_Q[TT]$$

$$g(TT) = \sum_{i=1}^{6} \gamma_i E_{S\text{-}i}[TT]$$

$$h(TT) = \sum_{i=1}^{6} \rho_i \sqrt{V_{S\text{-}i}[TT]}$$

438   Marc Bufé et al.

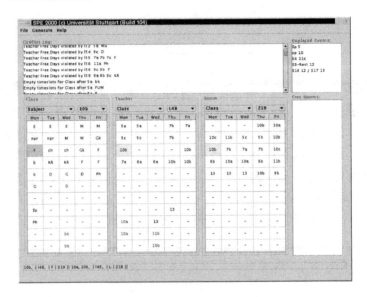

Fig. 3. Graphical user interface displaying the timetables of a class, a teacher, and a room from left to right. Conflicts are displayed in the top section and at the right unplaced events and possible free rooms for an event are shown.

where κ_i, γ_i, and ρ_i are adjustable weights. The standard deviation is considered to guarantee a fair distribution of the constraint violations. The fitness function is defined as the length of the vector $(f(TT), g(TT), h(TT))$ which has to be minimized.

6   Results

The presented concept is implemented in Java (cf. Figure 3). It allows not only the generation of a timetable using the evolutionary algorithm, also manual optimizations and further local search are possible.

The timetabling data is provided by a German high school and consists of 61 teachers, 23 classes, 49 rooms, and 351 events. The high school comprises 9 age-classes – the first 7 age-classes are split into three parallel classes, the last 2 age-classes are taught using individual selections of courses. Thus each of the latter age-classes is considered as one class where only fractions take part in various parallel courses. The rooms contain many special rooms for the different subjects.

Several optimization runs have been executed where always 4000 generations are computed using a population size of 20. Because of restrictions caused by the Java programming language the rather small population size is necessary. Each optimization run needs approximately 12 hours computation time. We have carried out experiments using genotype operations only (A), experiments starting with genotype only operations where after 1200 generations the genotype

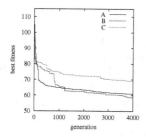

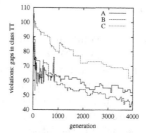

A: only genotype operations
B: first genotype, then additional phenotype
C: phenotype mutation only

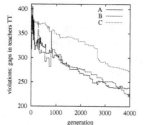

	A	B	C
Best fitness	61.0	58.6	69.0
Unplaced events	1	0	0
Partially placed events	0	1	1
Gaps in class' timetables	52	45	63
Gaps in teachers' timetables	238	219	272
Teachers free day	22	25	28
Double hours	26	28	33

Fig. 4. Experimental results.

mutation is replaced by the phenotype mutations (B), and experiments using phenotypic mutation only (C).

Due to the expensive fitness function tuning the algorithm is a tedious (and ongoing) task. As a consequence, we could not produce enough results to get statistical confidence for any of the above experimental setups. Thus, we provide the best result for each of the setups.

The experiments show that the phenotype mutations starting after 1200 generations help to reach a better fitness compared to the experiments using genotype operators only. Primary reason for this effect is the improved ability of the phenotype operators to move single hours and thus create partially placed events. This also leads to better scores concerning the gaps in individual timetables. The algorithm using only phenotype mutation is also able to place almost all events but without the timetables of high quality as starting points created by the genotype operators it cannot reach an overall quality comparable to the hybrid approach. Other experiments with slightly changed setups tend to similar results.

These results indicate that the algorithm is an encouraging approach. However, the results are still not good enough to be used in daily school practice. To improve the algorithm in the future, the phenotypic operators are planned to get more intelligence in choosing time slots for placing events and in resolving constraint violations in the timetable.

7   Conclusion

This work presents an evolutionary timetabling algorithm for highly constrained school timetabling problems. Where most previous work uses an algorithm either on the high-level representation of a timetable or on a low-level encoding (together with a timetable builder) this work chooses a hybrid approach where both techniques are combined. The algorithm respects all hard constraints, i.e. no infeasible solutions are generated, – additional soft constraints may be weighted and are considered within the fitness function. In particular the algorithm is designed in such a way that there is a high correlation between the individuals before and after the operators' application. This enables an effective search where on the one hand the recombination preserves characteristics of parent candidate solutions and on the other hand the mutation on the high-level representation allows the use of very sophisticated heuristics for the fine-tuning of timetables. These heuristics will be included in the near future. The consideration of standard deviations within the fitness function guarantees a fair distribution of soft constraint violations over all events, classes, and teachers.

References

1. Andrea Schaerf. A survey of automated timetabling. *Artificial Intelligence Review*, 13(2):87–127, 1999.
2. S. Even, A. Itai, and A. Shamir. On the complexity of timetabling and multicommodity flow problems. *SIAM Journal of Computation*, 5(4):691–703, 1976.
3. Alberto Colorni, Marco Dorigo, and Vittorio Maniezzo. Metaheuristics for high-school timetabling. *Computational Optimization and Applications*, 9(3):277–298, 1998.
4. Carlos Fernandes, João Paulo Caldeira, Fernando Melicio, and Agostinho Rosa. High school weekly timetabling by evolutionary algorithms. In *ACM SAC 99*, pages 344–350, New York, 1999. ACM.
5. Werner Junginger. Timetabling in germany – a survey. *Interfaces*, 16(4):66–74, 1986.
6. Emma Hart and David Corne. The state of the art in evolutionary approaches to timetabling and scheduling. EvoStim – The EVONET Working Group on Evolutionary Scheduling and timetabling, 1998.
7. Alberto Colorni, Marco Dorigo, and Vittorio Maniezzo. Genetic algorithms: A new approach to the time-table problem. In M. Akgül, editor, *Combinatorial Optimization*, pages 235–239. Springer, Berlin, 1990.
8. Andrea Schaerf. Tabu search techniques for large high-school timetabling problems. Technical Report CS-R9611, CWI, Amsterdam, NL, 1996.
9. Ben Paechter, R. C. Rankin, Andrew Cumming, and Terence C. Fogarty. Timetabling the classes of an entire university with an evolutionary algorithm. In Agoston E. Eiben, Thomas Bäck, Marc Schoenauer, and Hans-Paul Schwefel, editors, *Parallel Problem Solving from Nature – PPSN V*, pages 865–874, Berlin, 1998. Springer. Lecture Notes in Computer Science 1498.

Design of Iterated Local Search Algorithms
An Example Application to the Single Machine Total Weighted Tardiness Problem

Matthijs den Besten[1], Thomas Stützle[1], and Marco Dorigo[2]

[1] Darmstadt University of Technology, Intellectics Group,
Alexanderstr. 10, 64283 Darmstadt, Germany
[2] Université Libre de Bruxelles, IRIDIA,
Avenue Franklin Roosevelt 50, CP 194/6, 1050 Brussels, Belgium

Abstract. In this article we investigate the application of iterated local search (ILS) to the single machine total weighted tardiness problem. Our research is inspired by the recently proposed iterated dynasearch approach, which was shown to be a very effective ILS algorithm for this problem. In this paper we systematically configure an ILS algorithms by optimizing the single procedures part of ILS and optimizing their interaction. We come up with a highly effective ILS approach, which outperforms our implementation of the iterated dynasearch algorithm on the hardest benchmark instances.

1 Introduction

In the single machine total weighted tardiness problem (SMTWTP) n jobs have to be sequentially processed on a single machine. Each job j has a processing time p_j, a weight w_j, and a due date d_j associated, and the jobs become available for processing at time zero. The tardiness of a job j is defined as $T_j = \max\{0, C_j - d_j\}$, where C_j is the completion time of job j in the current job sequence. The goal is to find a job sequence which minimizes the sum of the weighted tardiness given by $\sum_{i=1}^{n} w_i \cdot T_i$.

The SMTWTP is an $\mathcal{NP}$-hard [9] scheduling problem and instances with more than 50 jobs can often not be solved to optimality with state-of-the-art branch & bound algorithms [1,4]. Therefore, several heuristic methods have been proposed for its solution. These include simple construction heuristics like the Earliest Due Date or the Apparent Urgency heuristics (see [17] for an overview) and metaheuristics like simulated annealing [14,17], tabu search [4], genetic algorithms [4], ant colony optimization (ACO) [5,15], and iterated local search (ILS) [3].

ILS appears to be a very promising approach for solving the SMTWTP, because the ILS algorithm by Congram, Potts, and de Velde, called *iterated dynasearch* [3], has shown so far, together with the recent ACO algorithm due to den Besten, Stützle, and Dorigo[5], the best performance results for the SMTWTP. Despite the very good performance of iterated dynasearch, it is not

E.J.W. Boers et al. (Eds.) EvoWorkshop 2001, LNCS 2037, pp. 441–451, 2001.
© Springer-Verlag Berlin Heidelberg 2001

Algorithm 1 Algorithmic outline of iterated local search.

1: $s_0 = \bullet \bullet\bullet\bullet\bullet\bullet\bullet\bullet\bullet\ \bullet\bullet\bullet\ \bullet\bullet\ \bullet\ \bullet\bullet\ \bullet\bullet\bullet\ \bullet$
2: $s^* = \bullet\ \bullet\bullet\bullet\bullet\ \bullet\bullet\bullet\bullet\ (s_0)$
3: **repeat**
4:     $s' = \bullet\ \bullet\bullet\bullet\bullet\ \bullet\bullet\ \bullet\bullet\bullet\ \bullet\ (s^*, history)$
5:     $s^{*\prime} = \bullet\ \bullet\bullet\bullet\bullet\ \bullet\bullet\ \bullet\bullet\bullet\bullet\ (s')$
6:     $s^* = \bullet\ \bullet\bullet\bullet\bullet\ \bullet\bullet\ \bullet\bullet\bullet\ \bullet\bullet\bullet\ \bullet\bullet\bullet\ \bullet\ (s^*, s^{*\prime}, history)$
7: **until** termination criterion met

---

very clear, whether this ILS algorithm has been designed in a best possible way. Therefore, this paper examines the systematic, experimentally driven configuration of an ILS algorithm for the SMTWTP. In particular, we first optimize the single ILS components and derive in this way a highly effective algorithm. This step-by-step methodology can also be used as a guideline for the development of ILS algorithms for other combinatorial optimization problems.

The paper is structured as follows. Section 2 introduces ILS and Section 3 studies the influence of the single procedures which are part of an ILS algorithm on its performance and gives experimental results with the final ILS algorithm. We end with some concluding remarks in Section 4.

2   Iterated Local Search

The underlying idea of ILS [2,13,12] is that of building a random walk in $\mathcal{S}^*$, the space of local optima defined by the output of a given local search algorithm. Four basic "ingredients" are needed to derive an ILS algorithm: a procedure $\bullet\ \bullet\bullet\bullet$ $\bullet\bullet\bullet\bullet\bullet\ \bullet\bullet\bullet\ \bullet\bullet\ \bullet\bullet\bullet\bullet$, which returns some initial solution, a local search procedure $\bullet\bullet\bullet\bullet\bullet\ \bullet\bullet\bullet\bullet$, a scheme of how to perturb a solution, implemented by a procedure $\bullet\bullet\bullet\bullet\bullet\ \bullet\bullet\bullet\bullet$, and an $\bullet\ \bullet\bullet\bullet\bullet\bullet\bullet\bullet\bullet\bullet\ \bullet\bullet\bullet\bullet\bullet$, which decides from which solution the search is continued. An algorithmic outline for ILS is given in Algorithm 1. The particular walk in $\mathcal{S}^*$ followed by the ILS algorithm can also be depend on the search history, which is indicated by *history* in $\bullet\ \bullet\bullet\bullet\bullet\bullet\ \bullet\bullet\bullet\bullet\ \bullet$ and $\bullet\ \bullet\bullet\bullet\bullet\bullet\bullet\bullet\bullet\bullet\ \bullet\bullet\bullet\bullet$ $\bullet\bullet\bullet\ \bullet$.

The effectiveness of the walk in $\mathcal{S}^*$ depends on the definition the four component procedures of ILS: The effectiveness of the local search algorithm is of major importance, because it strongly influences the final solution quality of ILS and its overall computation time. The perturbations should allow the ILS to effectively escape local optima but at the same time avoid the disadvantages of random restart (hence, not be too strong). The acceptance criterion, together with the perturbation, strongly influence the type of walk in $\mathcal{S}^*$ and can be used to control the balance between intensification and diversification of the search. The initial solution will be mainly be important in the initial part of the search.

The configuration problem in ILS is to find a best possible choice for the four components such that best overall performance is achieved. Because of the interactions among the components, this is a difficult problem and it has to be solved in a heuristic way. Here, we do this by considering at each step only the influence of one single component, keeping the others at some fixed, "reasonable" choices. These are that (i) as the initial solution we use the best construction heuristic, (ii) the acceptance criterion forces the cost to decrease (this means the perturbation is always applied to the best solution found so far), and (iii) the perturbation uses a number of random moves in a given neighborhood. First, we will optimize the choice of •••••••••• by investigating different local search algorithms. Once found a good local search, we reconsider the choices for the solution perturbation and the acceptance criterion in that order.

3   Iterated Local Search for the SMTWTP

3.1   Local Search

Local search for the SMTWTP starts from some initial sequence and repeatedly tries to improve the current sequence by replacing it with neighboring solutions. The simplest local search algorithm, iterative descent, repeatedly replaces the current sequence π with a better sequence found in the neighborhood of π and stops at the first local minimum encountered. For the SMTWTP we considered the following two neighborhood structures:

(1) exchanges of jobs placed at the ith and the jth position, $i \neq j$ (*interchange*)
(2) removal of the job at the ith position and insertion in the jth position (*insert*)

To allow for a fast evaluation of moves in these neighborhoods, the data structures proposed in [3] were implemented. The neighborhood structure is critical for the performance of the local search. Often, with more complex neighborhoods than the two presented above better solutions may be found. An example of a more complex neighborhood is the one used in dynasearch [3]. Dynasearch uses dynamic programming to find a best move which is composed of a set of independent interchange moves; each such move exchanges the jobs at positions i and j, $j \neq i$. Two interchange moves are independent if they do not overlap, that is if for two moves involving positions i, j and k, l we have that $\min\{i, j\} \geq \max\{k, l\}$ or vice versa. This neighborhood is of exponential size and dynasearch explores this neighborhood in polynomial time, to be more exact in $\mathcal{O}(n^3)$. In [3] very good performance with dynasearch has been reported.

To achieve further improvements of the solution quality, we considered the application of a variable neighborhood descent (VND) [16]. In our VND we concatenate iterative descent algorithms using two different neighborhoods; such an approach was also proposed in [18] for the permutation flow shop problem. VND exploits the observation that a local optimum with respect to one neighborhood structure need not be a local optimum for the other one. In fact, the variable neighborhood search (VNS) metaheuristic [16] systematically applies the idea of changing neighborhoods in the search. There are two possible ways of concatenating the two neighborhoods; these will be denoted in the following as *interchange+insert* and *insert+interchange*, depending on which neighborhood is searched first. Additionally, we also considered replacing the

Table 1. Comparison of the local search effectiveness for the SMTWTP. Results on the 100 job instances without local search and using the interchange, the insert, and the VND variants. We give the average percentage deviation from the best known solutions (Δ_{avg}), the number of best-known solutions found (n_{opt}), and the average CPU time in seconds (t_{avg}) averaged over the 125 benchmark instances.

start	no local search			insert			interchange			dyna interchange		
	Δ_{avg}	n_{opt}	t_{avg}	Δ_{avg}	n_{opt}	t_{avg}	Δ_{avg}	n_{opt}	t_{avg}	Δ_{avg}	n_{opt}	t_{avg}
EDD	135	24	0.0004	1.19	38	0.29	2.09	26	0.23	1.25	26	0.41
MDD	62	24	0.0007	1.31	36	0.32	1.03	33	0.16	1.02	33	0.27
AU	62	20	0.0018	0.56	39	0.11	0.81	33	0.05	0.90	30	0.12

start	dynasearch+insert			insert + interchange			interchange+insert			insert+dynasearch		
	Δ_{avg}	n_{opt}	t_{avg}	Δ_{avg}	n_{opt}	t_{avg}	Δ_{avg}	n_{opt}	t_{avg}	Δ_{avg}	n_{opt}	t_{avg}
EDD	0.30	45	0.46	0.46	49	0.30	0.52	42	0.28	0.46	50	0.34
MDD	0.39	46	0.33	0.42	42	0.33	0.37	44	0.20	0.42	42	0.40
AU	0.67	46	0.16	0.34	48	0.12	0.63	50	0.08	0.34	48	0.15

interchange local search algorithm with dynasearch, yielding two more variants, namely *dyna+insert* and *insert+dyna*.

We evaluated the proposed local search algorithms using a benchmark set of randomly generated instances, available via ORLIB at `http://www.ms.ic.ac.-uk/info.html`. There are three sets of instances with 40, 50, and 100 jobs. While for the 40 and 50 job instances the optimal solutions are known, the 100 job instances are still unsolved and only the best known solutions are available. The instances are generated by drawing the processing time p_j for each job j randomly according to a uniform distribution of integers between 1 and 100 and assigning it a weight w_j randomly drawn from a uniform distribution over the integers between 1 and 10. The due dates are randomly drawn integers from the interval $[(1 - TF - RDD/2) \cdot \sum p_i, (1 - TF + RDD/2) \cdot \sum p_i]$, where TF, the tardiness factor, and RDD, the relative due date, are two parameters. There are five instances for each pair of TF and RDD from the set $\{0.2, 0.4, 0.6, 0.8, 1.0\}$. This makes three sets of 125 instances each. The tardiness factor and the relative due dates determine critically the difficulty of solving the instances. For example, we found that most of the instances with TF = 0.2 are solved after one single application of the best local search procedures, while for larger TF, the instances were much harder to solve. All the experiments were run on a 700MHz Pentium III CPU with 512 MB RAM. Programs were written in C++ and run under Suse Linux 6.1.

The computational results are given in Table 1 for three different construction heuristics and all the described local search variants. Some of the benchmark instances are very easily solved, as indicated by the large number of best-known solutions found by the local search algorithms alone. In general, the best performance is obtained by the VND local search algorithms, which yield significantly

Table 2. Comparison of ILS algorithms using different choices for local search. Results on the 100 job instances with one trial per instance of 10 secs. We give the average percentage deviation from the best known solutions (Δ_{avg}), the number of best-known solutions found (n_{opt}), and the average CPU time in seconds to find the best solution in a trial (t_{avg}) averaged over the 125 benchmark instances.

start	ILS-inter+insert			ILS-dyna+insert			ILS-insert+inter			ILS-insert+dyna			ILS-dynasearch		
	Δ_{avg}	n_{opt}	t_{avg}	Δ_{avg}	n_{opt}	t_{avg}	Δ_{avg}	n_{opt}	t_{avg}	Δ_{avg}	n_{opt}	t_{avg}	Δ_{avg}	n_{opt}	t_{avg}
AU	0.0006	116	1.95	0.0019	109	2.62	0.0057	107	2.79	0.0046	107	2.97	0.0848	79	5.02

better solution quality at only a slight increase in computation time compared to the single neighborhood local search.[1] The dynasearch local search seems not to give significantly better solution quality than the interchange algorithm.[2]

Based on the results reported in Table 1, we run one trial on each instance for 10 seconds with the basic ILS algorithm using the VND variants for the local search and then measured the final solution quality and the number of best known solutions found. Before discussing the results, let us identify the perturbation applied in the ILS algorithm: We used either six random interchange moves or six random insert moves, depending on which neighborhood is used in the next local search. The idea is that the perturbation should be complementary to the particular local search and it should be difficult for the local search to undo the perturbation. For example, when applying ILS using the *interchange+insert* local search, we use random insert moves for the perturbation, because they are complementary to the following interchange local search.

The results of these latter experiments are reported in Table 2. In general, VNDs *interchange+insert* and *dyna+insert* appear to perform best when used inside an ILS algorithm. The *insert+interchange* VND performs only slightly worse; significantly worse results are obtained with the ILS-dynasearch. The results with ILS-dynasearch obtained with our re-implementation of dynasearch also appear to be worse than those presented in [3], especially when taking into account computation time (the experiments in [3] were run on a much slower computer). We verified that our dynasearch implementation works properly from a solution quality point of view, but it appears to be slower than the interchange local search, while in [3] interchange was slower than dynasearch. Therefore, we conjecture, that in particular the dynamic programming algorithm used in dynasearch to examine the neighborhood could still be speed up and our results with dynasearch should be taken as preliminary. Additionally, the use of C++ and some of its features like inheritance and templates may make our code significantly slower.

---

[1] The single results are slightly different to those published in [5], because of minor changes in the local search implementation.

[2] This fact has also noted in [3], where it was argued that the main advantage with dynasearch comes from a repetitive application of dynasearch in an ILS.

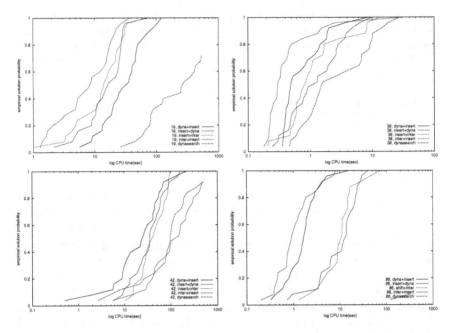

Fig. 1. We compare different choices for the local search with RTDs. The x-axis gives the logarithm of the CPU time, the y-axis the cumulative empirical solution probability (the more to the left a curve is located, the better performs the algorithm). RTDs are given for the five local searches tested in Table 2. The number in the caption gives the instances number; for example instance 42 is the 42nd instance of the 125 available one from ORLIB.

In a second experiment we analyzed the ILS run-time behavior by using run-time distributions (RTDs). RTDs give the cumulative empirically observed probability of finding an optimal solution (or a solution within a specific solution quality bound) as a function of the CPU time [7,20]. Here, for each instance 25 runs have been performed. In total we examined the run-time behavior of ILS algorithms with different choices for •••••••• on 10 instances which were known to be relatively hard; in particular they are not solved by applying one single local search. Figure 1 presents only results for four instances, the behavior on the others was similar. The RTDs show, that no single local search algorithm gives the best behavior on all instances. The best performance is obtained when applying the *interchange+insert* and *dyna+insert* VND (an exception is, for example, instance 42); the ILS with dynasearch is significantly worse for most instances. Because our interchange implementation is faster than our dynasearch implementation, we use the *interchange+insert* VND for the following experiments.

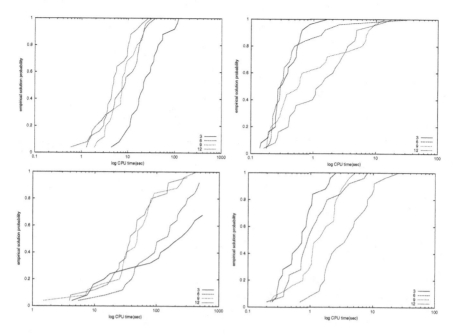

Fig. 2. We compare different choices for the perturbation with RTDs. The x-axis gives the logarithm of the CPU time, the y-axis the cumulative empirical solution probability. RTDs are given for the four different perturbation strengths for pure random perturbations. Results are given for instances 19 (top left), 38 (top right), 42 (bottom left) and 86 (bottom right).

3.2   Perturbation

Once fixed the choice for the local search, we closer examined the role of the solution perturbation. We addressed two important issues:

Perturbation strength: We will refer to the *strength* of a perturbation as the number of solution components which are modified. In the SMTWTP this is the number of jobs directly affected by a perturbation. Different choices for the perturbation strength, from three to twelve in steps of three, were examined.

Nature of perturbations: As said before, the perturbations should be complementary to the local search. Here, in addition we examined a variant, in which the random perturbations were required to involve only late jobs.

Again, we examined the different choices for the perturbation strength using RTDs, which are given in Figure 2 (results with perturbation focusing on late jobs are not given to keep the figures as clear as possible). As a first result we can observe that no single perturbation strength is best among all instances. Additionally, we found that a focus on late jobs in the perturbation does not

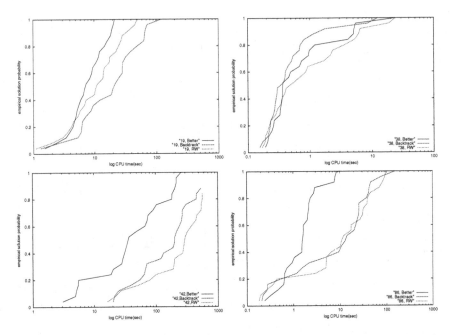

Fig. 3. We compare different acceptance criteria with RTDs. The x-axis gives the logarithm of the CPU time, the y-axis the cumulative empirical solution probability. RTDs are given for the three different acceptance criteria. The number in the caption gives the instances number.

significantly improve performance. Therefore, we settled for the following choice of the perturbation: (i) we do not focus on late jobs, because a simpler choice gave similar performance and (ii) we varied the perturbation strength in a range from three to twelve random moves. This latter variation was done in a scheme analogously to that introduced in the basic VNS HanMla99:mic.

3.3   Acceptance Criterion

A natural choice for the acceptance criterion is to force the cost to decrease by accepting an $s^{*\prime}$ if its cost is less than that of s^* (we refer to this acceptance criterion as Better in the following). Such a choice leads to a very strong intensification of the search and may lead to bad behavior for long run-times, when diversification of the search becomes more important. Diversification of the search is extremely favored if every $s^{*\prime}$ is accepted as the new solution, resulting in a random walk in $\mathcal{S}^*$. We call this acceptance criterion RW. In [3], a Backtrack acceptance criterion is proposed, which is a combination of the Better and the RW: For β iterations RW is used. If no improved solution is found, the ILS continues again from the best solution seen so far.

The results of the RTD-based analysis of the acceptance criteria with RTDs is plotted in Figure 3. We found that with respect to the acceptance criteria

Table 3. We give some basic statistics on the distribution of the computation times to solve instances of the three problem sets. We indicate the number of jobs (n), the average time (averaged over the 125 instances) to solve the benchmark set (t_{avg}) and its standard deviation (σ_t), the average time to solve the easiest and the hardest instance (t_{min} and t_{max}, respectively), and the quantils of the average time to solve a given percentage of the instances. Q_x indicates the average time to solve $x\%$ of the benchmark instances.

n	t_{avg}	σ_t	t_{min}	t_{max}	Q_{25}	Q_{50}	Q_{75}	Q_{90}
100	5.75	14.50	0.0076	105.50	0.052	0.98	5.14	13.12
50	0.20	0.86	0.0017	8.71	0.0064	0.018	0.118	0.28
40	0.040	0.13	0.0011	1.23	0.0031	0.0072	0.033	0.062

the results were somewhat clearer compared to the findings on the other two components: For almost all instances ILS with the `Better` acceptance criterion showed best behavior. Hence, this acceptance criterion was also chosen for our final ILS algorithm.

3.4   Experimental Results

The final ILS algorithm has the following shape: (i) it uses the AU construction heuristic to generate the initial solution, (ii) it uses the *interchange+insert* VND local search, (iii) it varies the perturbation strength between three to twelve random insert moves, and (iv) it accepts only better solutions in the random walk in $\mathcal{S}^*$. We conducted some experiments with this ILS algorithm on all 40, 50, and 100 job SMTWTP instances available from ORLIB. On each instance 25 trials were performed with a large computation time limit, which was enough that each instance could be solved to the best-known solutions, which we conjecture to be optimal, in each single trial. Of the 125 instances with 100 jobs (those with 40 or 50 jobs are very easily solved, see Table 3), only 15 took an average time to optimal larger than 10 seconds; only 7 of these longer than 20 seconds. The large majority of the benchmark instances was either solved with a single local search or within very few seconds.

Our ILS algorithm also compares very favourably to our earlier ACO algorithm presented in [5]. This shows that ILS may be an easily adaptable alternative to other, often more complex metaheuristics, showing an excellent performance after some straightforward optimizations. A more detailed investigation of different metaheuristics applied to the SMTWTP and a detailed search space analysis of the SMTWTP are the next steps we will take.

4   Conclusion

The results of this research can be summarized as follows:

1. The independent optimization of the single components of an ILS algorithm for the SMTWTP has led to a high performing ILS algorithm for the SMTWTP.

2. VND [16] leads to a very powerful local search for the SMTWTP.
3. For the SMTWTP the optimization of the ILS algorithm leads to improved performance. Yet, the improvement due to these optimizations is not as spectacular as observed for other problems [6,19,21], possibly due to the powerful local search.
4. The SMTWTP instances from ORLIB do not pose a challenge for state-of-the-art algorithms.

Clearly, these conclusions do raise further questions. The optimization of ILS for the SMTWTP was certainly rather straightforward and heuristic. We conjecture that for harder problems, this process will be much more important: it will need more iterations through the choices for the single components and statistical methods of experimental design will become more important. The excellent performance of VND for the SMTWTP naturally lends to the question whether VND can improve the efficiency of local search also for other scheduling problems. Preliminary results suggest, that the answer strongly depends on the particular problem. For example, in [18] encouraging results have been reported with a VND for the flow shop problem, but some experiments with an ILS algorithm [19] suggest that this improvement does not carry over to a significant improvement of ILS.

Future work will include tests of the ILS algorithms on larger instances, and an extension of our approach to other single machine scheduling problems. The results of this paper and the very good performance of a variety of ILS algorithms on several classes of scheduling problems [3,8,11,10,19] suggest that ILS is a very appropriate metaheuristic to obtained very high quality solutions in scheduling applications.

Acknowledgments. Marco Dorigo acknowledges support from the Belgian FNRS, of which he is a Senior Research Associate. This work was partially supported by the "Metaheuristics Network", a Research Training Network funded by the Improving Human Potential programme of the CEC, grant HPRN-CT-1999-00106. The information provided is the sole responsibility of the authors and does not reflect the Community's opinion. The Community is not responsible for any use that might be made of data appearing in this publication.

References

1. T. S. Abdul-Razaq, C. N. Potts, and L. N. Van Wassenhove. A survey of algorithms for the single machine total weighted tardiness scheduling problem. *Discrete Applied Mathematics*, 26:235–253, 1990.
2. E. B. Baum. Towards practical "neural" computation for combinatorial optimization problems. In J. Denker, editor, *Neural Networks for Computing*, pages 53–64, 1986. AIP conference proceedings.
3. R. K. Congram, C. N. Potts, and S. L. Van de Velde. An iterated dynasearch algorithm for the single–machine total weighted tardiness scheduling problem. IN-FORMS Journal on Computing, to appear, 2000.

4. H. A. J. Crauwels, C. N. Potts, and L. N. Van Wassenhove. Local search heuristics for the single machine total weighted tardiness scheduling problem. *INFORMS Journal on Computing*, 10(3):341–350, 1998.

5. M. den Besten, T. Stützle, and M. Dorigo. Ant colony optimization for the total weighted tardiness problem. In M. Schoenauer et al., editor, *Proceedings of PPSN-VI*, volume 1917 of *LNCS*, pages 611–620. Springer Verlag, Berlin, Germany, 2000.

6. I. Hong, A. B. Kahng, and B. R. Moon. Improved large-step Markov chain variants for the symmetric TSP. *Journal of Heuristics*, 3(1):63–81, 1997.

7. H. H. Hoos and T. Stützle. Evaluating Las Vegas algorithms — pitfalls and remedies. In *Proceedings of the 14th Conference on Uncertainty in Artificial Intelligence 1998*, pages 238–245. Morgan Kaufmann Publishers, 1998.

8. S. Kreipl. A large step random walk for minimizing total weighted tardiness in a job shop. *Journal of Scheduling*, 3(3):125–138, 2000.

9. J. K. Lenstra, A. H. G. Rinnooy Kan, and P. Brucker. Complexity of machine scheduling problem. In P. L. Hammer et al., editor, *Studies in Integer Programming*, volume 1 of *Annals of Discrete Mathematics*, pages 343–362. North-Holland, Amsterdam, NL, 1977.

10. H. R. Lourenço. Job-shop scheduling: Computational study of local search and large-step optimization methods. *European Journal of Operational Research*, 83:347–364, 1995.

11. H. R. Lourenço and M. Zwijnenburg. Combining the large-step optimization with tabu-search: Application to the job-shop scheduling problem. In I.H. Osman and J.P. Kelly, editors, *Meta-Heuristics: Theory & Applications*, pages 219–236. Kluwer, 1996.

12. O. Martin and S.W. Otto. Combining Simulated Annealing with Local Search Heuristics. *Annals of Operations Research*, 63:57–75, 1996.

13. O. Martin, S.W. Otto, and E.W. Felten. Large-Step Markov Chains for the Traveling Salesman Problem. *Complex Systems*, 5(3):299–326, 1991.

14. H. Matsuo, C. J. Suh, and R. S. Sullivan. A controlled search simulated annealing method for the single machine weighted tardiness problem. Working paper 87-12-2, Department of Management, University of Texas at Austin, TX, 1987.

15. D. Merkle and M. Middendorf. An ant algorithm with a new pheromone evaluation rule for total tardiness problems. In S. Cagnoni et al., editor, *Proceedings of EvoWorkshops 2000*, volume 1803 of *LNCS*, pages 287–296. Springer Verlag, Berlin, Germany, 2000.

16. N. Mladenović and P. Hansen. Variable Neighborhood Search. *Computers & Operations Research*, 24:1097–1100, 1997.

17. C. N. Potts and L. N. Van Wassenhove. Single machine tardiness sequencing heuristics. *IIE Transactions*, 23:346–354, 1991.

18. C. R. Reeves. Landscapes, operators and heuristic search. To appear in *Annals of Operations Research*, 2000.

19. T. Stützle. Applying iterated local search to the permutation flow shop problem. Technical Report AIDA–98–04, FG Intellektik, TU Darmstadt, August 1998.

20. T. Stützle. *Local Search Algorithms for Combinatorial Problems — Analysis, Improvements, and New Applications*. PhD thesis, Darmstadt University of Technology, Department of Computer Science, 1998.

21. Thomas Stützle. Iterated local search for the quadratic assignment problem. Technical Report AIDA–99–03, FG Intellektik, FB Informatik, TU Darmstadt, March 1999.

An Evolutionary Algorithm for Solving the School Time-Tabling Problem

Calogero Di Stefano[1] and Andrea G. B. Tettamanzi[2]

[1] Genetica S.r.l.
Via San Dionigi 15, I-20139 Milan, Italy
distefano@genetica-soft.com
[2] Università degli Studi di Milano
Dipartimento di Tecnologie dell'Informazione
Via Bramante 65, I-26013 Crema (CR), Italy
tettaman@dsi.unimi.it

Abstract. This paper describes an evolutionary algorithm for school time-tabling, demonstrated through applications to the Italian school system. Heuristics have been found and perfected which offer good generalization capabilities. A particular attention has been devoted to problem formulation, also in terms of fuzzy logic, as well as to testing different genetic operators and parameter settings. This work has obtained results of remarkable practical relevance on real-world problem instances illustrated in the paper, and eventually gave rise to a successful commercial product.

1 Introduction

The time-table problem (TTP) is the planning of a number of meetings (e.g., exams, lessons, matches) involving a group of people (e.g., students, teachers, athlets) for a given period and requiring given resources (e.g., rooms, laboratories, sports facilities) according with their availability and respecting some other constraints.

The TTP is an exciting challenge for computational intelligence and operations research, essentially because it is NP-complete [1]. School time-tabling is often even more complicated by the details of a given real situation and real-world problem instances often involve constraints escaping an exact representation, such as constraints related to user personal preferences. Advanced search techniques exploit various heuristics in order to rule out from the search space those regions where an optimum is not expected to exist. One of these heuristics, namely evolutionary algorithms, have been successfully applied to various types of TTP [2,3,4] and are among the most promising techniques available to date for solving this kind of problem.

This paper is organized as follows: Section 2 presents the data and the constraints that make up a problem instance and the form of a solution, in order to identify its characterizing aspects. Section 3 describes the evolutionary algorithm implemented. Finally, Section 4 demonstrated the experiments carried out with

E.J.W. Boers et al. (Eds.) EvoWorkshop 2001, LNCS 2037, pp. 452–462, 2001.

the algorithm on some real-world problem instances, and Section 5 discusses the results.

2   The Problem

2.1   Instance Representation

School time-tabling is generally based on a significant volume of data and on different types of constraints. A problem instance is identified by the following entities and relationships:

- rooms, defined by their type, capacity and location;
- subjects, identified by the room type they require and by a priority with respect to other subjects;
- teachers, characterized by the subjects they teach and by their fuzzy availability matrix, defining the degree to which each teacher is available in each period (see Figure 1);

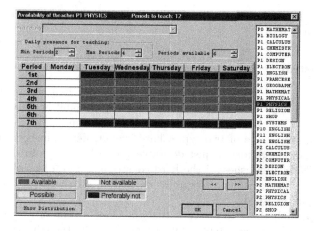

Fig. 1. An example of a fuzzy availability matrix for a teacher.

- classes (i.e., groups of students, including the degenerate case of single students), assigned to a given location (e.g., building) and having their availability matrix, such as the one in Figure 2;
- lessons, meaning the relation $\langle t, s, c, l \rangle$, where t is a teacher, s is the subject, c is the class attending the lesson and l is the duration in periods. More than one teacher and more than one class can participate in a lesson, in which case we speak of *grouping*.

As the reader will have noticed, some constraints of the school TTP are most naturally expressed in fuzzy terms [5].

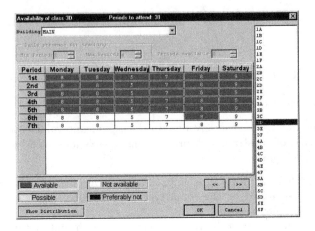

Fig. 2. An availability matrix for a class.

A candidate solution to this TTP is an assignment of a starting period to every lesson.

2.2 Constraints

A candidate solution can be considered satisfactory if it meets the teachers' and students' requirements, and if it respects the availability of resources. These properties are formalized through a number of constraints.

Constraints can be of two types: hard or soft, depending on whether their satisfaction is mandatory or just desirable. Accordingly, a solution will be said to be *feasible* if it satisfies all the hard constraints, and a feasible solution will be said to be more or less *acceptable* depending on the degree to which soft constrains are satisfied.

An important thing to notice is that the classification of each constraint as hard or soft is to some extent arbitrary, and may be thought as being part of the problem instance definition process.

For the school TTP, based on a thorough analysis of the Italian system, the following set of general constraints was identified.

- Constraints on rooms: in each period, the number of rooms used per type cannot be larger than the number of available rooms of that type (*bounded rooms*); rooms of some types are not available for a certain time before or after use for preparation or tidying up (*room availability*).
- Constraints on subjects: some subjects are better taught before others, or earlier in the day when students are fresh, or lessons of some other subjects should not take place after one another, e.g., similar subjects (*priority*).
- Constraints on teachers: a teacher cannot teach more than one lesson at a time (*physical availability*); a teacher must teach a given number of lessons

to each class as provided for by the programme for every subject (*contractual availability*); a teacher should be assigned lessons only in the periods he or she has marked as available (*subjective availability*); if the school has more than one building or location, appropriate time must be allowed for a teacher to travel between buildings when required (*displacement*); finally, teachers prefer that their lessons be concentrated in time, minimizing gaps or isolated lessons (*concentration*).

- Constraints on classes: if the school has more than one location, each class belongs to just one location and classes participating in a lesson must be from the same location (*co-location*); a lesson cannot take place when one of the participating classes is not at school (*presence*); while at school, a class should always be attending a lesson, except for breaks (*no gaps*)—this is a strong requirement in the Italian system, but may not be so important in other systems.

- Constraints on lessons: some lessons must stick to a pre-assigned schedule for organizational reasons (*pre-assignment*); lessons of the same subject should be uniformly distributed over the week (*distribution*); the total weight of lessons for each day should be as uniform as possible over the week (*uniformity*); lesson weight (i.e., priority) should be decreasing within a day (*decreasing burden*); multiple lessons on the same subject, with the same classes on the same day should be scheduled sequentially or, put another way, only a single lesson on a subject is allowed per day for a class (*sequentiality*); subsequent lessons on the same subject must take place in the same room (*locality*); different lessons involving the same teacher or the same class cannot overlap (*non-overlap*).

- Other organizational constraints: for instance, on each day, classes take their lunch break at different hours in order to avoid overcrowding of the lunch room (*break balance*);

The search for an acceptable timetable strongly depends on how constraints are classified. The case where all constraints are soft brings forth an exceedingly broad search space, whereas the opposite case where all constraints are hard is practically meaningless, for it gives rise to a very sparse if not even empty search space.

For a standard problem instance relevant to a typical Italian secondary school, a possible partition of the constraints described above into hard and soft constraints might be the one reported in Table 1.

3   The Evolutionary Algorithm

The approach adopted to solve the problem described in Section 2 is a quite standard elitist evolutionary algorithm, except for the perturbation operator, which alternates between an *intelligent mutation* operator and an *improvement* operator, whose details are given below.

Table 1. A possible classification of constraints for the Italian school system.

hard constraints		soft constraints	
Constraint	Entity	Constraint	Entity
bounded rooms	rooms	room availability	rooms
physical availability	teachers	priority	subjects
contractual availability	teachers	subjective availability	teachers
co-location	classes	displacement	teachers
presence	classes	concentration	teachers
pre-assignment	lessons	distribution	lessons
sequentiality	lessons	decreasing burden	lessons
no gaps	classes	locality	lessons
non overlap	lessons	uniformity	lessons

3.1   Genetic Representation

A *direct* representation has been chosen, along the lines of [4,3,6,7], even though recent work [8] would suggest indirect representation to work much better.

However, given the use of the improvement operator, which is a Lamarckian local search operator, one might argue that our method should not really be classed as having a direct representation. If this operator were always applied then one could possibly describe our algorithm as having an indirect representation with the chromosome providing a parameterisation to the *improvement* operator.

So, in the end, our approach might show the advantages both of an indirect representation and of a direct representation. Moreover, such representation lends itself to specializations of the kinds required by a many-faceted reality like the Italian school system.

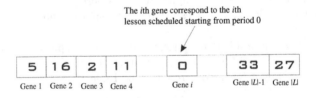

Fig. 3. Representation scheme of a genotype.

An individual in the population is a vector of integers, as depicted in Figure 3, having the same size as the number of lessons to be scheduled. Each integer encodes the starting period of the associated lesson.

The values each gene can take up depends on data like the length of the relevant lesson in periods, the number of days in the school week, the number of periods in each day, and so on.

Other advantages of this direct representation are:

1. a simple procedure for calculating the penalties to assign to the violation of each constraint;
2. easy design of operators which preserve feasibility, thus allowing significant savings of computational power;
3. last but not least, a particularly efficient decoding function which, given a genotype, reconstructs the corresponding timetable.

3.2   Initialization

The population is seeded with genotypes generated by a randomly driven greedy algorithm, which places the lessons one by one by selecting at random a starting period among those that would not violate any constraint. This incremental procedure goes on until the set of "open" starting periods becomes empty, and then places all remaining lessons at random.

The main advantage of this seeding heuristics is that it gives the evolutionary algorithm a jump start, by producing much better timetables than a completely random initialization would do—practically at the same computational cost.

3.3   Fitness

The fitness function employed by our evolutionary algorithm for school timetabling is of the form

$$f(g) = \frac{1}{1 + \sum_i \alpha_i h_i} + \frac{\gamma}{1 + \sum_j \beta_j s_j}, \tag{1}$$

where $h_i \in \mathbb{R}^+$ is the penalty associated to the violation of the ith hard constraint, whose relative weight is $\alpha_i \in \mathbb{R}^+$, and $s_j \in \mathbb{R}^+$ is the penalty associated to the degree of violation of the jth soft constraint, whose relative weight is $\beta_j \in \mathbb{R}^+$; finally, $\gamma \in \{0, 1\}$ is zero until all hard constraints are satisfied, and one since.

The weights attached to each hard and soft constraint allow the algorithm to be fine-tuned to better direct search. In practice, the weights were chosen according to the relative importance of constraints in the problem instances treated. It should be stressed that the results are *very* sensitive to the choice of the weights, because soft constraints correspond to conflicting criteria.

Satisfaction of soft constraints begins only after at least one feasible solution has been found, and γ is switched to one: this device does not let evolution trade off slight violations of hard constraints with high satisfaction of soft constraints.

3.4   Selection

The individuals are chosen for reproduction using tournament selection [9], with tournament size $k \geq 2$, which is a parameter of the algorithm. The implementation of this selection scheme enforces elitism by introducing at least one copy of the best individual into the next generation, without perturbation. Apart from this detail, the algorithm uses a generational replacement reproduction strategy.

3.5   Recombination

Among the cross-over operators described in the literature for an integer representation like the one adopted by our algorithm (see e.g. [10]), the one that proved to be best suited for the TTP, based on experiments on the three sample problem instances discussed in Section 4, was uniform cross-over, whose mechanism is illustrated in Figure 4: this operator treats the integer genes atomically. Recombination is performed on each pair of individuals with probability p_{cross}, which is a parameter of the algorithm.

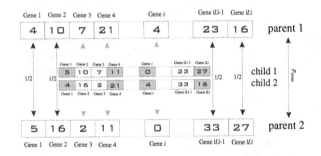

Fig. 4. Illustration of uniform crossover on the genetic representation of a school timetable.

3.6   Perturbation

Typically, mutation in EAs is a blind operator, in the sense that it produces random perturbations, regardless of the fact that they improve or worsen the solution represented by the affected individual.

In contrast with that, the two perturbation operators employed in combination by our algorithm are rather of the Lamarckian type, i.e., based on the law of use and disuse which, if wrong when referred to natural genetics, sometimes leads to more efficient search in artificial evolutionary systems.

The two mutually exclusive perturbation operators are:

- *intelligent mutation*, applied with probability p_{mut}, and

– *improvement*, applied with probability p_{improve}.

At each generation, the decision whether to enable the intelligent mutation or the improvement operator is made by tossing a biased coin. The bias p_{pert}, in favor of intelligent mutation, is a parameter of the algorithm.

Intelligent Mutation. This operator, while retaining a random nature, is directed towards performing changes that do not decrease the fitness of the individual. In particular, if the operator affects the ith gene, it will indirectly affect all the other lessons involving the same class, teacher or room. The choice of the "action range" of this operator is random and the variation suffered by the ith gene is such as to make fitness increase. In practice, intelligent mutation aims at reducing constraint violation.

Improvement. This operator, which is the perturbing alternative to intelligent mutation, restructures an individual to a major extent.

Restructuring commences by randomly selecting a gene (i.e., a lesson) and concentrates on the partial timetables for the relevant class, teacher and room.

First of all, it aims at making periods for which no lesson has been scheduled contiguous (i.e., compacting "gaps" in the class, teacher, and room timetable). Then it allocates the contiguous free space thus claimed to the the lesson corresponding to the selected gene.

Application of this operator generally results in a fitness increase and preserves the feasibility of a solution. Furthermore, this operator is responsible for injecting substantially novel genetic material into the population.

4   Experiments and Results

Experiments have been carried out on a constructed problem instance, designed to test most aspects of the problem. The instance is a realistic model of a generic Italian "industrial technical high-school", which we will call "G. Marconi" for convenience, summarized in Table 2. A total of 942 lessons are to be scheduled.

Further experiments have been carried out on two real state lycées located in Milan, namely the G. Carducci classic state lycée and the Leonardo da Vinci scientific state lycée, both enrolling ca. 1,000 students.

Data for the G. Carducci lycée refer to the school year 1997/1998, with 924 lessons to schedule. Data for the Leonardo da Vinci lycée refer to school year 1996/1997, with 1003 lessons to schedule. Table 3 summarizes information on these two schools.

The parameter settings for the evolutionary algorithm reported in Table 4, are the ones that consistently gave the best results on the three problem instances used for experiments.

A typical run for these problem instances takes 40 minutes on average to come up with a feasible solution and $5\frac{1}{2}$ hours to devise an acceptable solution on a 500 MHz PC with 64 Mbyte RAM and the Microsoft NT operating system.

Table 2. Summary of the G. Marconi model industrial technical high-school.

Resources	Main Building	Branch Building	Remarks
classes	12	18	6 sections of 5 classes
teachers	35	27	3 Physical Education teachers work in both buildings
subjects	10	13	6 subjects are taught in both buildings
lessons	558	384	no lesson requires class grouping
labs	8	4	
gyms	2	1	the figures refer to the number classes which can use the gym at one time

Table 3. Summary of the G. Carducci and Leonardo da Vinci lycées.

Resources	G. Carducci	Leonardo da Vinci
classes	33	35
teachers	60	71
subjects	13	23
lessons	924	1003
labs	1	3
gyms	2	2

Table 4. Optimal parameter settings for the evolutionary algorithm.

Parameter	Value
Population size	100
p_{cross}	0.3
p_{pert}	0.5
p_{mut}	0.02
$p_{improve}$	0.02
Tournament size k	30
α_i, β_j	1 for all i, j

Figure 5 plots the fitness and the penalties associated to constraint violations during a run of the algorithm on the Leonardo lycée. The first feasible solution was found at generation 2,001, and the run was stopped at generation 10,200 when the degree of satisfaction of soft constraints did not look likely to increase anymore.

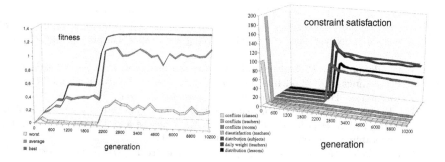

Fig. 5. Fitness graph and evolution of constraint satisfaction during a run for the Leonardo lycée.

5   Discussion

The algorithm described in Section 3 cannot be considered a "pure" evolutionary algorithm; instead, its efficiency and effectiveness rely critically on the hybridization with two heuristics based on local search, implemented by the intelligent mutation and improvement operators.

Even without fine tuning of the algorithm parameters and of the relative constraint weights α_i and β_j, the quality of the results obtained, which cannot be shown here for lack of space, was remarkable, all the more so because no simplifying assumption or constraint elimination had to be made to obtain them.

The timetables obtained were submitted to the timetable committee of the Leonardo da Vinci scientific state lycée, which has been in charge for years of assembling the school timetable at every opening of the school year. Their expert assessment of the results was highly positive and their implementation did not require any manual adjustment.

The practical relevance of the results obtained led Genetica S.r.l. to package the approach decribed in this paper into a full-fledged commercial software product, EvoSchool, which has been successfully launched on the Italian market. Adaptations for other countries are also under way. EvoSchool, version 2.0, is freely available for evaluation at the URL

    http://www.genetica-soft.com/eng/evoschool.html

The demo comes with the datasets for the three examples described in Section 4.

References

1. S. Even, A. Itai, A. Shamir. *On the Complexity of Timetable and Multicommodity Flow Problems.* Siam Journal of Computing, Vol. 5, No.4, December 1976, 691-703.

2. W. Erben. *A Grouping Genetic Algorithm for Graph Colouring and Exam Timetabling*. Proceedings of the Third International Conference on the Practice and Theory of Automated Timetabling, Constance, Germany, August 16–18, 2000.

3. P. Adamidis and P. Arakapis. *Weekly lecture timetabling with genetic algorithms*. Proceedings of the 2nd International Conference on the Practice and Theory of Automated Timetabling, University of Toronto, Canada, 1997.

4. J.P. Caldeira and A.C Rosa. *School timetabling using genetic search*. Proceedings of the 2nd International Conference on the Practice and Theory of Automated Timetabling, University of Toronto, Canada, 1997.

5. L.A. Zadeh. *Fuzzy Sets and Applications: Selected Papers*. John Wiley & Sons, New York, 1987.

6. A.M. Barham and J.B. Westwood. *A Simple Heuristic to Facilitate Course Timetabling*. J. Opnl. Res. Soc. 29, 1055-1060.

7. D. Corne,P. Ross, H. Fang. *Evolutionary Timetabling: Practice, Prospects and Work in Progress*. Presented at the UK Planning and Scheduling SIG Workshop, (Strathclyde, UK, September 1994), organised by P Prosser.

8. B. Paechter, R. C. Rankin, A. Cumming. *Improving a Lecture Timetabling System for University Wide Use*. Practice and Theory of Automated Timetabling II, Springer-Verlag, LNCS 1408, Berlin, 1998.

9. A. Brindle. *Genetic algorithms for function optimization*. Technical Report TR81-2, Department of Computer Science, University of Alberta, Edmonton, 1981.

10. Z. Michalewicz. *Genetic Algorithms + Data Structures = Evolution Programs*. Springer-Verlag, Berlin, 1992.

11. D.E. Goldberg. *Genetic Algorithms in Search, Optimization and Machine Learning*. Addison-Wesley, Reading, MA, 1989.

Optimizing Employee Schedules by a Hybrid Genetic Algorithm

Matthias Gröbner[1] and Peter Wilke[2]

[1] Lehrstuhl für Programmiersprachen
Universität Erlangen-Nürnberg
Martensstrasse 3, 91058 Erlangen, Germany
email: Groebner@informatik.uni-erlangen.de
[2] Centre for Intelligent Information Processing Systems
The University of Western Australia
Nedlands, 6907, Australia
email: wilke@ee.uwa.edu.au

Abstract. Creating an employee schedule means taking into account many heavy constraints like employee contracts or minimal staffing levels on the one hand and many global, difficult to formalize constraints like aspects of fairness on the other hand. Optimisation is quite difficult especially when fix rostering schemata cannot be used, e.g. because of frequently varying staffing levels. In this paper we present how real-life employee scheduling problems can be solved by applying a Hybrid Genetic Algorithm that uses problem specific knowledge. First we briefly describe the given problem domain, then the idea and implementation of the Genetic Algorithm is presented. Finally we show some application results and the outlook.

1 Introduction

Rosters (employee schedules) are used in a broad range of economic sectors such as in industrial production, on the health sector or other service sectors and therefore their construction is a quite common task. Special planning software has been developed to support e.g. the supervisor who used to create a roster by hand for a certain period, typically one week or one month. Most employee planning algorithms are based on heuristical methods or artificial intelligence techniques as branch and bound search or constraint logic programming. Implementations of these methods are limited in most cases to one concrete real-world problem domain. Off the shelf software has not been developed because the environment and constraints to be fulfilled turn out to be very specific and unique even if the optimisation task looks similar for different applications. When the problem description has to be changed slightly, e.g. with respect to extensions and changes to get out to most of the available resources, elementary parts of the algorithm have to be redesigned. In addition most of these techniques cannot deal with global or soft constraints that can be found in those real-world problems [1].

E.J.W. Boers et al. (Eds.) EvoWorkshop 2001, LNCS 2037, pp. 463–472, 2001.
© Springer-Verlag Berlin Heidelberg 2001

So our goal was to develop a method that facilitates companies in a number of economical sectors to create rosters automatically and optimise them with respect to a certain number of global and local constraints. The key idea is to establish an algorithm which covers the most important constraints of a broad variety of employee scheduling tasks and can easily be adopted to the specific circumstances. With this approach we accept the fact that this algorithm will be only suitable for most of the case while it isn't appropriate for some.

Because Genetic Algorithms have been successfully applied to some other similar timetabling and scheduling problems [2] [3] [4] [5] [6] we decided to base our toolkit on a Genetic Algorithm.

A less detailled presentation of this work is published in [7].

2   The Data

Fortunately data from several real world employee scheduling tasks was available to us. To make comparison easier these data sets were standardised as follows:

- All datasets describe work around the clock which is divided in shifts.
- In each shift different positions are to be filled, e.g. supervisor, engineer, driver, secretary etc. Each position requires one or more specific functions in arbitrary combination to be performed and therefore can only be filled by a worker qualified for these functions.
- For each position within a shift there are requirements regarding the number of staff able to perform to fill this position (staffing levels). These requirements are noted according to the following scheme: minimal requirement / target requirement / maximal requirement (see figure 1)
- There is a record for each staff member indicating his qualifications for the functions to be performed and as a result for which positions he qualifies.
- The employees may have different workloads during a week.

In figure 2 a scheduled roster is shown in part.

3   Constraints

Different constraints have to be considered when planning a roster. For this we distinguish between constraints that have to be fulfilled under all circumstances (hard constraints) and constraints that should be fulfilled if possible (soft constraints). Examples for hard constraints are:

- Each employee works at most one shift per day.
- Enforcing the minimal required breaks between two working shifts.
- Not to exceed or fall below the employee's monthly target working hours (within a tolerance limit).
- Not to exceed the maximum number of working days in uninterrupted sequence.

Shifts/Pos	Friday, Mar. 1 Min/Target/Max			Saturday, Mar 2. Min/Target/Max			Sunday, Mar. 3 Min/Target/Max			Monday, Mar. 4 Min/Target/Max		
Shift Early 1												
Assistant	1/	1/	2	0/	0/	0	0/	0/	0	1/	1/	2
Surgery 1	1/	3/	3	1/	1/	2	1/	1/	2	1/	3/	3
Surgery 2	1/	3/	3	1/	1/	2	1/	1/	2	1/	3/	3
Chief	1/	1/	2	0/	0/	0	0/	0/	0	1/	1/	2
Shift Early 2												
Assistant	0/	1/	2	0/	0/	0	0/	0/	0	0/	1/	2
Surgery 1	0/	2/	3	0/	1/	2	0/	1/	2	0/	2/	3
Surgery 2	0/	2/	3	0/	1/	2	0/	1/	2	0/	2/	3
Chief	0/	0/	0	0/	0/	0	0/	0/	0	0/	0/	0
Late Shift												
Assistant	1/	1/	4	0/	0/	0	0/	0/	0	1/	1/	4
Surgery 1	2/	2/	4	1/	1/	2	1/	1/	2	2/	2/	4
Surgery 2	2/	2/	4	1/	1/	2	1/	1/	2	2/	2/	4
Chief	1/	1/	1	0/	0/	0	0/	0/	0	1/	1/	1
Night Shift												
Assistant	0/	0/	0	0/	0/	0	0/	0/	0	0/	0/	0
Surgery 1	1/	1/	4	1/	1/	2	1/	1/	2	1/	1/	4
Surgery 2	1/	1/	4	1/	1/	2	1/	1/	2	1/	1/	4
Chief	0/	0/	0	0/	0/	0	0/	0/	0	0/	0/	0

Fig. 1. Example of a staffing level schedule for a hospital (surgery).

- To keep the assigned number of staff within the lower and upper limit of the staffing level for all separate positions within a shift.

  Among the soft constraints are:

- Keep the number of assigned staff members close to the target staffing level.
- The working and holiday blocks of the employees should be as compact as possible, i.e. employees should work in sequence as many days as possible respectively have as many days off as possible. Singular working days or holidays should be avoided.
- Homogenous shift patterns within each employee's working block, e.g. not always alternate between night shift and early shift.
- Aspects of fairness. The shifts, especially the night and weekend shifts, should be distributed uniformly among the employees.

4   The Algorithm

The core of our algorithm is a standard Genetic Algorithm [8] with tournament selection and two-point crossover with three individuals. This kind of tourna-

```
January      1  2  3  4  5  6  7  8  9 10 11 12 13 14 15 16 17 18 19 20 21 22 23 24 25 ...
Weekday     Th Fr Sa Su Mo Tu We Th Fr Sa Su Mo Tu We Th Fr Sa Su Mo Tu We Th Fr Sa Su ...


Shift Early 1 (E1)
SURGERY1     3  2  1  1  2  1  3  2  1  1  1  1  2  2  3  3  1  1  2  1  2  1  1  1  1
SURGERY2     1  2  1  1  2  1  2  2  2  1  1  2  1  1  2  1  1  1  2  2  1  2  2  1  1
ASSISTANT    1  1        1        1  1  1        1  1  1  1  1        1  1  1  1  1
CHIEF        1  1        1        1  1  1        1  1  1  1  1        1  1  1  1  1


Shift Early 2 (E2)
SURGERY1     2  2  1  1  1  1  2  2  1  1  1  1  1  2  2  2  1  0  1  1  1  1  1  0
SURGERY2     1  1  0  0  0  1  0  1  0  0  1  1  1  0  1  1  1  1  0  1  1  2  1  0  1
ASSISTANT    1  1        0        0  1  1        1  1  1  1  1        1  1  1  1  1


Late Shift (LS)
SURGERY1     2  2  1  1  2  1  2  2  2  1  1  2  2  2  2  2  1  1  2  2  2  2  2  1  1
SURGERY2     2  2  1  1  2  1  2  2  2  1  1  2  2  2  2  2  1  1  2  2  2  2  2  1  1
ASSISTANT    1  1        1        1  1  1        1  1  1  1  1        1  1  1  1  1
CHIEF        1  1        1        1  1  1        1  1  1  1  1        1  1  1  1  1


Night Shift
SURGERY1     1  1  1  1  1  1  1  1  1  1  1  1  1  1  1  1  1  1  1  1  1  1  1  1  1
SURGERY2     1  1  1  1  1  1  1  1  1  1  1  1  1  1  1  1  1  1  1  1  1  1  1  1  1


January      1  2  3  4  5  6  7  8  9 10 11 12 13 14 15 16 17 18 19 20 21 22 23 24 25 ...
Weekday     Th Fr Sa Su Mo Tu We Th Fr Sa Su Mo Tu We Th Fr Sa Su Mo Tu We Th Fr Sa Su ...


Adam,T.      LS LS LS LS        E2 E2 E2              E1 E1 E1 E1           NS NS NS LS     -1.5
Alexander,P. NS NS NS NS        E2 E2        LS LS LS LS LS E1        LS LS           NS NS  +1,1
Huber,E.     E1 E1        LS LS LS NS NS NS        E2 E2 E2 E2 E2 E2        LS LS            -0.8
Beisser,H.   E1 E1        LS        E1 E1 E1        LS LS LS E1 E1        E1 E1 E1 E1 E1     +0.5
Cäsar,J.     E1 E1 E1 E1 E1        LS LS        E1 NS NS LS LS LS        E2 E2 E2 E2        -3.1
Escher,H.    E1 E1        NS NS NS NS NS        LS LS LS E1        LS LS        E1 E1 E1     +0.3
Huber,G.        LS NS NS NS NS NS        E2 E2        LS LS LS        E2 E2 E2 E1           +2.3
Kunz,R.         LS LS        E1 E1 E1 E1        NS NS NS NS NS        LS E2 E2              -0.2
Biber,A.     LS        E2 E2 E2 LS           LS LS LS LS LS        E1 E1 E1 E1 E1          -2.1
... (further employees)
```

Fig. 2. Example of a roster computed by the Hybrid Genetic Algorithm. The upper part shows the number of employees assigned to the positions of each shift. The part below shows the shifts the employees have been assigned to and the difference to their target monthly working hours.

ment selection caused better fitness values than tournament selection with two individuals and one-point-crossover only. The individuals created that way are added to the population if their fitness value is better than the fitness value of the worst individual of the population, which is removed then.

4.1   Coding

Direct coding has been chosen to represent the monthly roster in a chromosome, i.e. the genes represent the sequence of workplaces of the functions of each shift. Thus we have a gene for each possible assignment of an employee to a function (see figure 4). The consequence of this coding method is a comparable long gene string. If the assignment of more employees than provided by the maximum staffing level should be permitted, additional genes have to be inserted.

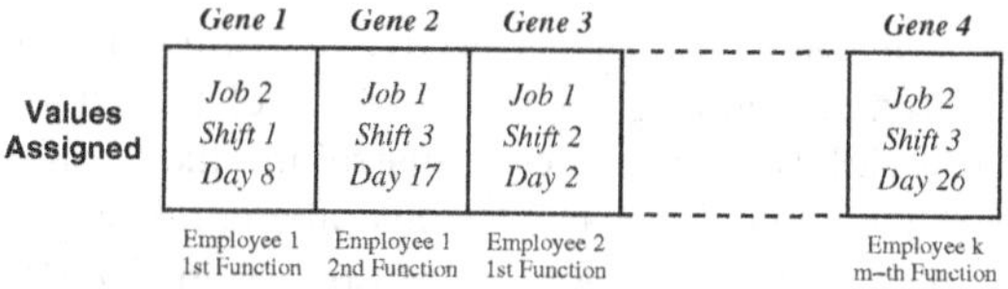

	Gene 1	Gene 2	Gene 3		Gene 4
Values Assigned	Job 2 Shift 1 Day 8	Job 1 Shift 3 Day 17	Job 1 Shift 2 Day 2	- - - - -	Job 2 Shift 3 Day 26
	Employee 1 1st Function	Employee 1 2nd Function	Employee 2 1st Function		Employee k m–th Function

Fig. 3. Coding of the roster from the point of view of the employees.

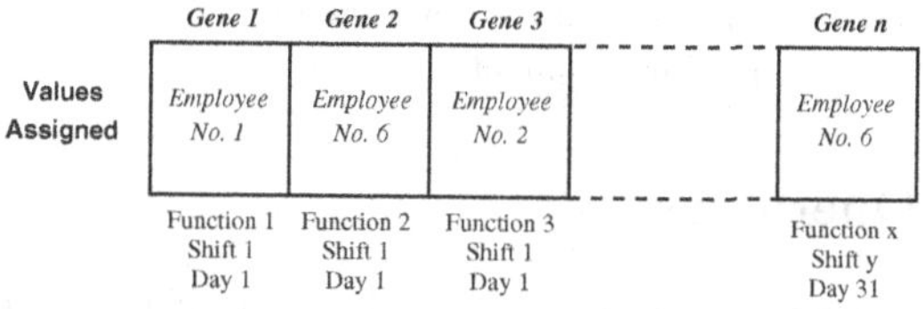

	Gene 1	Gene 2	Gene 3		Gene n
Values Assigned	Employee No. 1	Employee No. 6	Employee No. 2	- - - - -	Employee No. 6
	Function 1 Shift 1 Day 1	Function 2 Shift 1 Day 1	Function 3 Shift 1 Day 1		Function x Shift y Day 31

Fig. 4. Chosen coding of the roster as a pattern of shifts.

The alleles represent the employees respectively the "empty" employee. But there are only assignments of employees allowed if the employees are able to fulfill the required function of the position.

Using the above encoding scheme yields assignments of the employees to the functions of the shifts which are directly apparent and time consuming decoding algorithms are avoided. The drawback is that several hard constraints are not enforced, i.e. the genotype represents invalid solutions, e.g. the repeated assignment of the same employee to one day. However, a larger part of the search space is searched by this method as if certain assignments would be forbidden by the coding scheme. For this reason direct representation has been preferred to implicit representation [3]. Of course these violations have to be sorted out in a later stage of the algorithm.

4.2 Mutation

The chosen coding scheme induces that the mutation operation randomly changes the assigned employee. In doing so the initially defined mutation rate is reduced linear to two changes at the end. This is done according to the idea of Evolution Strategies [9] where the mutation rate is reduced to ensure that the algorithm converges towards an optimum.

4.3 Fitness Function

The fitness of an individual is computed by assigning penalty costs (penalty points) for the violation of constraints. These penalty costs are specified by the user according to the importance of the constraint fulfillment. As mentioned above the violation of most hard constraints is allowed with the chosen coding scheme. To sort things out violations result in lower fitness scores (higher penalty costs) and therefore increase the evolutionary pressure on these individuals.

Penalty costs are assigned for each occurrence of a constraint violation. The penalty is calculated depending on the severity of the violation. E.g. for the penalisation of the violation of fairness constraints like "uniform distribution of weekend shifts among the employees" the average number of weekend shifts has to be calculated first and the penalty costs are assigned representing deviation of each employee's number of weekend shifts from the average value.

The quality of a chromosome is expressed by its penalty costs. The better individual is the one with less penalty points which results in a higher probability for reproduction by tournament selection.

5   Repair Operators

The Genetic Algorithm described so far yielded relative good rosters, but still an unacceptable high number of hard constraints remain violated. To improve the quality and speed of the algorithm it was necessary to introduce problem specific knowledge resulting in an accelerated convergence. The idea is to apply repair operators which outperform the simple penalisation of constraint violations. These new operators are applied to the individuals after selection, recombination and mutation, but before fitness calculation.

The disadvantage which comes with this is that the search behaviour becomes more locally. So the parameters of the algorithm have to be chosen carefully so that local search isn't introduced too early restricting the search space in an inappropriate fashion. So our repair operators are applied best if their impact is gradually increased. This parameter is to be fine tuned during the experiment.

The repair operators carry out the following modifications:

Cancelling of assignments. When an employee has been assigned more than once a day all assignments can be cancelled except one.

Selected assignment. An employee can be assigned to days to those he has not yet been assigned if his current monthly working hours are less than his target working hours.

Selected cancelling of an assignment and reassignment of the employee. Sometimes it is more favourable with respect to the staffing levels of the shifts to cancel the assignment of an employee and to reassign him to another function within a shift of the same day.

Interchanging of the assignments of two employees to yield a more homogenous sequence of shifts for the employees.

The improvements of the roster found by the repair operators are subsequently re-decoded to the genotype. This is establishes a kind of Lamarckian evolution.

6   Results

The Genetic Algorithm presented here has been tested on several real world databases and yielded feasible solutions of the problem in nearly all cases i.e.

a roster without violations of hard constraints. In figure 5 different real world hospital databases with different configurations are shown. The databases differ in the number of employees to be scheduled, the employees' weekly working hours and the number of shifts and functions. The task was to compute a roster for one month in advance. The corresponding test run results showing the remaining violations of hard constraints can be seen in figure 6.

Database	No. Employees	hours per week	No. Shifts	No. Functions	Comp. Time
Internal Med. 1	22	all 38.5	4	4	3 min
Internal Med. 2	20	38.5, 19.25	3	5	5 min
Geriatrics	24	38.5, 20, 15	2	7	5 min

Fig. 5. Three different input databases for the Hybrid Genetic Algorithm.

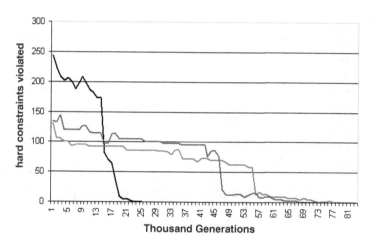

Fig. 6. The Hybrid Genetic Algorithm applied to three different hospital databases. The curves show the number of remaining violations of hard constraints.

An adapted version of the algorithm has been integrated in the employee planning software SP-EXPERT© [1] and is used by several customers since then.

In figure 5 computing times are listed, too. The algorithm required at most 5 minutes on a 600 MHz Pentium to get feasible solutions for several test databases with less than 25 employees. Longer running times with more generations do not necessary result in an improvement of solution quality as the example in figure 7 shows. Results get even worse with too long running times because the repair operators are applied to a more resticted search space then.

[1] Contact: Astrum GmbH, Am Wolfsmantel 2, D-91058 Erlangen, Germany or www.astrum.de

Computing time	Final fitness value	Computing time	Final fitness value
0.5 min	-721	8 min	-1701
1 min	-702	10 min	-1284
2 min	-1009	15 min	-1106
5 min	-1862	20 min	-1324

Fig. 7. Impact of running time to the final result of the Genetic Algorithm (average value of several test runs). Smaller fitness values indicate better solutions.

The presented technique shows that Genetic Algorithms are suitable to automatically create rosters. But it seems to be necessary to use problem specific knowledge — in our example repair operators — to yield finally feasible rosters. Figure 8 shows the effect of the repair operators and proves that no feasible solutions can be found even after a large number of generations when renouncing the repair operators (figure 9).

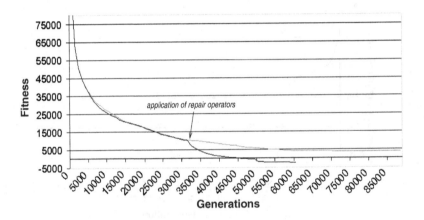

Fig. 8. Comparison of two test runs with and without repair operators. The figure shows the different trends of the fitness values (penalty costs).

Furthermore the presented Genetic Algorithm verifies that Lamarckian and Elitist strategies can speed up the search process even if there exists the danger to converge in local optima [6]. We tried to avoid this by introducing the repair operators smoothly during the run of the evolution process.

It hasn't escaped our attention that the chosen fitness calculation is not completely without problems because there is a large variety of incompatible constraints. The question is how the algorithm "knows" that a solution with a better fitness value really has less hard constraints violated. We solved this problem in our test databases by appropriate parameter values for the penalty used by the fitness function.

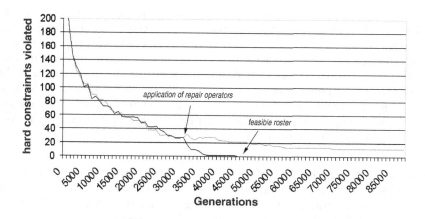

Fig. 9. Comparison of two test runs with and without repair operators. The figure shows the number of remaining violations of hard constraints.

7   Conclusion and Outlook

Our tests showed that the presented Genetic Algorithm is flexible with respect to the integration of future constraints. The only thing to do is to extend the fitness function to calculate the appropriate penalty costs. Normally the convergence is slowed by this and sometimes some soft constraints are not satisfied in the final solution. In some cases the repair operators have to be adapted to take the new constraints in account. Otherwise the repairs would not be able to improve the solution found so far. Two problems are connected with the repair operators. The first problem is the choice of the penalty costs for the fitness function. Often it is rather difficult for users to express the different constraint violations in penalty costs. A rating by linguistic formulations ("very important", "doesn't matter", "not so important") would make sense. Thus a combination of the presented algorithm and fuzzy logic would be interesting. Another problem is the control of the impact of the repair operators. If these operators are introduced too early the search space turns locally, if they are introduced too late the repair may not be complete to yield valid solutions. In our tests the optimal point in time to begin with the repair operators has been determined heuristically.

Further research will be directed towards finding an automated mechanisms to control the repair operator's impact on the search space.

The disadvantage of the Genetic Algorithm presented is that coding scheme and repair operators have to be adapted for each new scheduling problem. But it looks promising to develop a more general version of this problem specific method to be able to integrate new constraints easier and to apply this method to new similar optimisation problems with little effort. For this we will try to generalize the problem representation and the allowed constraints to cover most such scheduling problems to be able to apply more general repair operators.

As a first step towards this goal we are testing the application of our Genetic Algorithm to similar new scheduling problems like Employee Scheduling for Call Centers, student timetabling for University courses or school timetabling. In the Call Center case the difference is that e.g. employees do not work in shifts but with variable working times around the clock where for the University timetable student's course preferences play a more important role. So the Genetic Algorithm has to be changed if necessary with respect to the coding scheme and the concrete implementation of the repair operators. First tests verify the good results yielded with the rostering case.

Acknowledgement. This work has been partly supported by a grant from the Lehrstuhl fuer Programmiersprachen, Institut fuer Informatik, Friedrich-Alexander-Universitaet Erlangen-Nuernberg, Martensstrasse 3, 91058 Erlangen, Germany. We would like to thank them for their support.

References

1. A. Meisels and N. Lusternik, "Experiments on Networks of Employee Timetabling Problems", in *Proceedings of the Second International Conference on the Practice and Theory of Automated Timetabling, ed.s E. Burke and M. Carter*, pp. 215-228, Springer, 1997.
2. E. Burke and D. Elliman and R. Weare, "Specialised Recombinative Operators for Timetabling Problems", in *Proceedings of the AISB (AI and Simulated Behaviour) Workshop on Evolutionary Computing*, pp. 75-85, Heidelberg, Springer, 1995.
3. D. Corne and P. Ross and H.-L. Fang, "Evolutionary Timetabling: Practice, Prospects and Work in Progress", in *Proceedings of the UK Planning and Scheduling SIG Workshop, ed. P. Prosser*, University of Strathclyde, 1994.
4. C. Fernandes and J. P. Caldeira and F. Melicio and A. Rosa, "High School Weekly Timetabling by Evolutionary Algorithms", in *Proceedings of 14th Annual Acm Symposium On Applied Computing*, San Antonio, Texas, 1999.
5. D. Mattfeld, "Scalable Search Spaces for Scheduling Problems", in *Proceedings of GECCO99, ed.s W. Banzhaf et al*, pp. 1616-1621, Morgan Kaufmann, 1999.
6. R. Weare and E. Burke and D. Elliman, "A Hybrid Genetic Algorithm for Highly Constrained Timetabling Problems", in *Proceedings of the Sixth International Conference on Genetic Algorithms, ed. L. J. Eshelman*, pp. 605-610, Pittsburg, Morgan Kaufmann, 1995.
7. M. Gröbner and P. Wilke, "Rostering with a Hybrid Genetic Algorithm", to be published in *Proceedings of Fifth International Conference on Artificial Neural Networks and Genetic Algorithms*, Springer, 2001.
8. D. E. Goldberg, "Genetic Algorithms in Search, Optimization and Machine Learning", Addison-Wesley, 1989.
9. I. Rechenberg, "Evolutionsstrategie: Optimierung technischer Systeme nach Prinzipien der biologischen Evolution", Fromann-Holzboog, 1973.
10. M. Gröbner, "Optimierung der Einsatzplanung für Personal im Schichtdienst", *Master Thesis*, Universität Erlangen-Nürnberg, October 1998.

A Genetic Algorithm for the Capacitated Arc Routing Problem and Its Extensions

Philippe Lacomme, Christian Prins, and Wahiba Ramdane-Chérif

University of Technology of Troyes, Laboratory for Industrial Systems Optimization
12, Rue Marie Curie, BP 2060, F-10010 Troyes Cedex (France)
{Lacomme, Prins, Ramdane}@univ-troyes.fr

Abstract. The NP-hard Capacitated Arc Routing Problem (CARP) allows to model urban waste collection or road gritting, for instance. Exact algorithms are still limited to small problems and metaheuristics are required for large scale instances. The paper presents the first genetic algorithm (GA) published for the CARP. This hybrid GA tackles realistic extensions like mixed graphs or prohibited turns. It displays excellent results and outperforms the best metaheuristics published when applied to two standard sets of benchmarks: the average deviations to lower bounds are 0.24 % and 0.69 % respectively, a majority of instances are solved to optimality, and eight best known solutions are improved.

1 Introduction

The *Capacitated Arc Routing Problem* (*CARP*) is defined in the literature on an undirected network $G = (V,E)$ with a set V of n nodes and a set E of m edges. A fleet of identical vehicles of capacity Q is based at a depot node s. A subset R of edges require service by a vehicle. All edges can be traversed any number of times. Each edge (i, j) has a traversal cost $c_{ij} \geq 0$ and a demand $r_{ij} \geq 0$. The CARP consists of determining a set of vehicle trips of minimum total cost, such that each trip starts and ends at the depot, each required edge is serviced by one single trip, and the total demand handled by any vehicle does not exceed Q. The cost of a trip comprises the costs of its serviced edges and of its intermediate connecting paths.

Many applications occur in road networks: urban waste collection, snow plowing, sweeping, gritting, etc. Demands are amounts to be collected along the streets (urban waste) or delivered (salt for ice clearance). Costs are distances or travel times. The undirected version concern roads that can be serviced during one traversal and in any direction, which happens in low-traffic suburban areas. In the directed version of the CARP, each arc represents one street (or one side of street) with a mandatory direction for service. Both versions are NP-hard, even in the special case where one trip is sufficient (*Rural Postman Problem*).

E.J.W. Boers et al. (Eds.) EvoWorkshop 2001, LNCS 2037, pp. 473-483, 2001.
© Springer-Verlag Berlin Heidelberg 2001

Golden and Wong [1], Benavent *et al.* [2] and Belenguer and Benavent [3] have investigated integer linear programming formulations and lower bounds for the CARP. Since exact algorithms like the branch-and-bound method of Hirabayashi *et al.* [4] are still limited to small instances (30 edges), larger instances must be tackled in practice by heuristics. Good greedy heuristics include Path-Scanning from Golden and Wong [1], Construct-Strike as improved by Pearn [5], Augment-Insert from Pearn [6], Augment-Merge from Golden *et al.* [7] and Ulusoy's tour splitting method [8]. Concerning metaheuristics, Li [9] applied simulated annealing and tabu search to a road gritting problem, and Eglese [10] designed a simulated annealing approach for a multi-depot gritting problem with side constraints. The most efficient metaheuristic published so far is a sophisticated tabu search method from Hertz *et al.* [11].

The context of municipal waste collection is adopted in the following, for making the problem more concrete and talkative. We first present in section 2 our extension of the basic CARP with its data structures. Section 3 describes the basic components of our GA, which is tested in section 4 on two standard sets of benchmarks.

2   Extended CARP and Data Structures

Compared with the basic CARP in introduction, we already tackle three extensions. First, we add a set U of directed arcs to get a mixed graph $G = (V,U,E)$ combining edges and arcs. m becomes the number of *links* (arcs or edges). We call *tasks* the t links in R (usually defined by a non-zero amount of waste). Second, we define for each task (i, j) a collecting cost $w_{ij} \geq 0$ distinct from the true traversal cost c_{ij} (without collecting). Third, we handle prohibited turns, *e.g.* undesirable U-turns.

In memory, G is converted into an entirely directed internal graph $H = (V, A)$. A contains $nia = |U|+2.|E|+1$ *internal arcs*: one per arc of U, two opposite arcs per edge of E, and one dummy loop on the depot. We drop the nodes to use arc indexes only. Each internal arc u is defined by a demand $r(u)$, a traversal cost $c(u)$, a collecting cost $w(u)$ if u is required, a pointer $inv(u)$ to the inverse arc when u codes one edge, and a list $Succ(u)$ of successor arcs to which it is allowed to turn. Turn prohibitions become transparent with this data structure. Shortest path costs are pre-computed in a matrix D, $nia \times nia$. For any pair of arcs (u,v), $D(u,v)$ is the traversal cost of a shortest path from u to v (*not included* to ease arc insertions or deletions in trips), taking turn prohibitions into account. A modified Dijsktra's algorithm with arc labels can compute D in $O(m^3)$, or in $O(m^2.\log m)$ with a heap structure. Since $m \approx 4n$ in real road networks, these complexities reduce to $O(n^3)$ and $O(n^2.\log n)$ in practice.

In any solution, each trip is stored as a sequence of tasks (internal arc numbers, in the order of service) with a total cost and a total load. Shortest paths between tasks are assumed. They are not stored, since their cost can be read in $O(1)$ from D. The cost of a trip cumulates the collecting costs of its tasks and the traversal costs of its intermediate paths. The total cost of a solution is the sum of all trip costs.

3 Genetic Algorithms for the Extended CARP

This section describes the basic components of the hybrid GA: chromosomes and fitness, population, reproduction step (selection of parents and crossover), mutation by local search, replacement method, and stopping criteria.

3.1 Chromosomes and Fitness

Most GAs for the travelling salesman problem (TSP) use *permutation chromosomes*. For the CARP, such a chromosome could be viewed as the order in which a vehicle must collect all the t tasks, assuming the same vehicle performs all trips in turn. This encoding is appealing because *there always exists one optimal sequence*. However, a given sequence may be split into trips in many ways, and trip delimiters seem inevitable to avoid ambiguous chromosomes. This raises several problems, for instance defining trip limits in the children generated by a crossover.

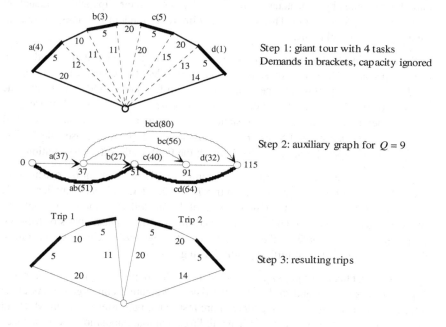

Fig. 1. Main steps of Ulusoy's algorithm

Ulusoy's algorithm provides an elegant solution for keeping simple permutation chromosomes. Usually, this algorithm is a CARP heuristic better explained by figure 1. First, capacity is ignored to build one giant tour T covering all tasks (a, b, c and d in the figure). Second, an auxiliary graph is built in which each arc denotes a subsequence of T that can be done by one trip. Each arc is weighted by the cost of this trip. A shortest path in this graph shows where to split T into trips and gives the cost of the corresponding solution. Third, the solution is built with one trip per arc of the path. This algorithm is heuristic, because the giant tour giving an optimal CARP solution is difficult to obtain. However, *it splits any given tour optimally.* Our GA use step 2 to evaluate the chromosomes, considered as giant tours. The trips are never expressed, except to output solutions when the GA stops. The fitness is simply the total cost of the underlying CARP solution, as returned by step 2.

3.2 Population Structure and Initial Population

The population is an array P of nc chromosomes, *sorted in decreasing cost order.* It is initialized with random solutions and three good heuristic solutions. Random ones are computed by a sequential heuristic that randomly selects at each iteration one task not yet collected, plus its collecting direction if it is an edge. The current trip ends when one more task would violate truck capacity. The good solutions are given by Ulusoy's heuristic (see 3.1), Path-Scanning and Augment-Merge. All solutions are converted into chromosomes by concatenating the lists of tasks of their trips and evaluated by step 2 of Ulusoy's algorithm. This significantly improves random solutions but seldom the good heuristic solutions.

Path-Scanning builds trips one by one, starting from the depot, and extends each trip task per task. The extension step at a node i considers two cases: if no free task is incident to i, the trip moves to the nearest node with free tasks incident to it, else the next task (i, j) to be collected is selected using five criteria: 1) maximize distance from j to the depot, 2) minimize this distance, 3) maximize waste density r_{ij} / c_{ij}, 4) minimize waste density and 5) use rule 1 if truck load $\leq Q / 2$, else use rule 2. Five solutions are computed (one for each criterion); the final result is the best solution.

Augment-Merge builds one trip per task and sorts these t trips in decreasing cost order. An *Augment phase* compares each trip T_i, $i=1,2,...,t-1$, with each smaller trip T_j, $j=i+1,i+2,...,t$. If T_i traverses the unique task u of T_j, can collect u, and if this improves total cost, then T_i collects u and T_j is discarded. In the *Merge phase*, pairs of trips are merged in descending order of the saving produced, subject to the capacity constraints. This process stops when no further positive saving remains.

Clones (identical solutions) are forbidden in the population, to have a better dispersion of solutions and to diminish the risk of premature convergence. As clone detection is time consuming, we adopt a more restrictive but faster policy in which all solutions must have distinct costs. When building the initial population, we then check that each inserted solution has a new cost. We try up to *mnt* times (50 for instance) to generate each random solution. If all attempts generate a duplicate cost, the population P is truncated, but this occurs only when nc is too large.

3.3   Reproduction Step and Extended OX Crossover

Parents are chosen by *binary tournament selection*. We first randomly select two solutions from the population. We then select from these two the least cost solution to be the first parent P1. This procedure is repeated to get the second parent P2. The two parents undergo an extended version of the classical *order crossover* (OX). The reproduction step ends by randomly keeping only one child C and by discarding the other. This policy works slightly better than keeping two children or the best one.

For two parents P1 and P2 of length t, the classical OX crossover draws two random subscripts p and q with $1 \leq p \leq q \leq t$. To build child C1, it copies the string P1[p]-P1[q] into C1[p]-C1[q]. Finally, it scans P2 in a circular way from $q+1$ (mod t) and copies each element not yet taken to fill C1 circularly, starting from $q+1$ (mod t) too. The roles of P1 and P2 are exchanged to get the other child C2. OX must be extended for the CARP and our data structure. Each chromosome contains all t tasks, but an edge can appear as one internal arc u or as its inverse $inv(u)$. Therefore, when copying u from a parent, we must check whether u and $inv(u)$ are not already taken. As explained in 3.1, the children are evaluated using step 2 of Ulusoy's algorithm.

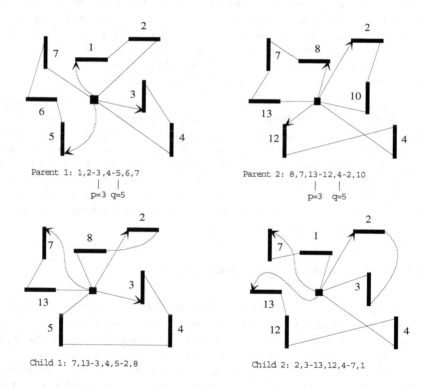

Fig. 2. An example of two parents with children generated by the OX crossover

Figure 1 gives two parent chromosomes with children obtained by the extended OX crossover, for a CARP with 7 edge-tasks. The trip limits are imaginary and result in practice from Ulusoy's algorithm, whose execution cannot be detailed here. They are shown as dashes in the chromosomes only to see the link with the graphical solutions. Recall that such limits are not stored by the GA. Thick lines are required edges. Thin lines are shortest paths in the actual network. Edges as traversed in parent 1 correspond to internal arcs 1-7. For each internal arc $u \leq 7$, the opposite arc is here $inv(u) = u + 7$, e.g., the leftmost edge give internal arcs 6 and 13.

3.4   Local Search as Mutation Operator

We mutate with a fixed rate *pm* the child C produced by the crossover. The mutation operator is a *local search LS*, giving a *hybrid GA*. It is clear nowadays that hybrid GAs are better than Holland's basic GA and can even outperform other metaheuristics. Before applying LS, the child is converted into a set of trips, using steps 2-3 of Ulusoy's algorithm. Each iteration of LS scans in $O(t^2)$ all possible ways of moving a task (in one trip or to another trip) and all possible permutations of two tasks (in one trip or between two distinct trips). The collecting direction of an edge-task u may be inverted during this process, *i.e.* we try to reinsert u or $inv(u)$. We also inspect two kinds of *2-opt moves* by removing two shortest paths (possibly empty) u-v and x-y, in the same trip or in distinct trips (u, v, x and y denote a task or the depot loop). We replace them by shortest paths u-y and v-x or by paths u-x and v-y. 2-opt moves are not always possible in mixed networks because they may invert the trip order.

At each iteration, the first improving move is executed. The process is repeated until no further saving can be found. Some trips may become empty and are removed at the end. The child C is kept. A new chromosome S is rebuilt from the final trips (by concatenating their tasks) and re-evaluated with step 2 of Ulusoy's algorithm. Quite often, this slightly improves LS by shifting some trip limits. In [11], Hertz use sophisticated improvement procedures for the basic CARP, like one called *Shorten*. Testing shows that adding such procedures in LS has no significant effect on the overall GA. We took simpler neighborhoods for speed and also because they remain valid for our extended CARP model.

3.5   Replacement Method and Stopping Criteria

The replacement is *incremental*. One chromosome $P(k)$ is randomly selected below the median: $k \leq int(nc/2)$. After a mutation by LS, we try to replace $P(k)$ by S if no clone is created. In case of clone, or when LS is not applied, we replace $P(k)$ by C, avoiding cost duplication. If this fails too, the current GA iteration is said *unproductive* and discarded. This method keeps the best solution $P(nc)$ and allows a child to reproduce immediately. The GA stops after a maximum number of iterations *mni*, after a maximal number of unproductive iterations *mnui*, or when it reaches a lower bound *LB* known for some instances (in that case *P(nc)* is optimal).

3.6 Summary – General Structure

Before starting the algorithm, we fix *nc, mnt, mni, mnui* and the mutation rate *pm*. A chromosome is encoded in *P* as a record with two fields *Seq* (sequence of tasks) and *Cost* (fitness). This gives the following general structure:

```
//Initial population
Put in P the solutions of Path-Scanning, Augment-Merge and Ulusoy
k := 3
Repeat
  k := k + 1
  Randomly generate P(k) with a new cost (mnt attempts)
Until (k = nc) or (failure)
If failure then nc := k - 1
Sort P in decreasing cost order
//Main iteration
ni, nui := 0
Repeat
    Select P1 and P2 by binary tournament
    Apply extended OX crossover and select one child C
    Evaluate C using Ulusoy's algorithm
    Choose k randomly in [1,int(nc/2)]
    If Random < pm then
       Mutate C by local search giving S
       If S.Cost not in P\{P(k)} then C := S
    EndIf
    If C.Cost not in P\{P(k)} then //Productive iteration
       ni := ni + 1
       If S.Cost < P(nc).Cost then nui := 0 Else nui := nui + 1
       P(k) := S
       Shift P(k) to resort P
    EndIf
Until (ni = mni) or (nui = mnui) or (P(nc).Cost = LB).
```

4 Computational Evaluation

The GA is implemented in Delphi on a 500 MHz PC under Windows 95 and compared with the best method for the CARP, the tabu search Carpet from Hertz *et al.* [11]. These tests are done on two standard sets of undirected instances in which all edges are required. The first set (table 1) contains 23 instances from DeArmon [12] with 7 to 27 nodes and 11 to 55 edges. The second set (table 2) contains 34 harder instances from Belenguer and Benavent [3], with 24 to 50 nodes and 34 to 97 edges. All these files can be obtained at *ftp://matheron.estadi.uv.es/pub/CARP*.

In both tables, *Prob* gives the instance number and *n,m* the numbers of nodes and of edges. *LB* is a lower bound from Belenguer *et al.* [3]. *TS* is the result of Carpet with the parameter setting yielding the best results on average (same setting for all instances). *Best* gives the best solution published, generally obtained by Carpet with various settings of parameters. *GA* is the solution of our hybrid GA, with a unique parameter setting to ease comparison with Carpet. *Dev* is the deviation of GA to Carpet in %. *New best* is the new best solution when running the GA with several settings. Boldface indicates instances for which the GA improves Carpet. New best solutions are in boldface italics. Asterisks denote new optimal solutions. Our results are obtained with a small population of $nc = 30$ solutions. The local search is applied with a rate $pm = 0.1$. The GA performs a first phase stopping after $mni = 20000$ productive crossovers, $mnui = 6000$ crossovers without improving the best solution, or when *LB* is reached. If *LB* is not reached, it restarts for 10 short phases with $mni = mnui = 2000$ and pm pushed up to 0.2. nc is small to avoid losing too much time in unproductive crossovers (the rejection rate is 10 to 20% for $nc = 30$).

Table 1. Results of the GA on De Armon's instances

Prob	n,m	LB	Best	TS	GA	Var %	New best
1	12,22	316	316	316	316	0.00	316
2	12,26	339	339	339	339	0.00	339
3	12,22	275	275	275	275	0.00	275
4	11,19	287	287	287	287	0.00	287
5	13,26	377	377	377	377	0.00	377
6	12,22	298	298	298	298	0.00	298
7	12,22	325	325	325	325	0.00	325
8	27,46	344	348	352	**350**	-0.57	348
9	27,51	303	311	317	***303****	**-4.42**	***303****
10	12,25	275	275	275	275	0.00	275
11	22,45	395	395	395	395	0.00	395
12	13,23	448	458	458	458	0.00	458
13	10,28	536	544	544	***540***	-0.74	***538***
14	7,21	100	100	100	100	0.00	100
15	7,21	58	58	58	58	0.00	58
16	8,28	127	127	127	127	0.00	127
17	8,28	91	91	91	91	0.00	91
18	9,36	164	164	164	164	0.00	164
19	11,11	55	55	55	55	0.00	55
20	11,22	121	121	121	121	0.00	121
21	11,33	156	156	156	156	0.00	156
22	11,44	200	200	200	200	0.00	200
23	11,55	233	233	235	235	0.00	235

The hybrid GA is *very* efficient: on all instances, it is at least as good as Carpet. On the 23 DeArmon's instances, it outperforms Carpet 3 times, improves 2 best known solutions with one to optimality, and reaches *LB* 19 times. The average and worst deviations to *LB* are roughly divided by 2 compared to Carpet and become respectively 0.24 % and 2.23 %. For the 34 harder instances from Belenguer et al., Carpet is improved upon 16 times and the best solution 5 times, with one new optimal solution. Optimality is proven on 23 instances. Again, the average and worst deviations to *LB* are halved and become 0.69 % and 4.26 %. *Table 3* summarizes this comparison with Carpet and gives some other indicators. CPU times are difficult to compare: Carpet was tested by Hertz on a Silicon Graphics Indigo 2 at 195 MHz, while we use a Pentium III PC at 500 MHz.

Table 2. Results of the GA on the instances from Belenguer and Benavent

Prob	n,m	LB	Best	TS	GA	Var %	New best
1A	24,39	173	173	173	173	0.00	173
1B	24,39	173	173	173	173	0.00	173
1C	24,39	235	245	245	245	0.00	245
2A	24,34	227	227	227	227	0.00	227
2B	24,34	259	259	260	259	-0.38	259
2C	24,34	455	457	494	462	-6.48	457
3A	24,35	81	81	81	81	0.00	81
3B	24,35	87	87	87	87	0.00	87
3C	24,35	137	138	138	138	0.00	138
4A	41,69	400	400	400	400	0.00	400
4B	41,69	412	412	416	412	-0.96	412
4C	41,69	428	430	453	428*	-5.52	428*
4D	41,69	520	546	556	541	-0.92	530
5A	34,65	423	423	423	423	0.00	423
5B	34,65	446	446	448	446	-0.45	446
5C	34,65	469	474	476	474	-0.42	474
5D	34,65	571	593	607	581	-4.28	581
6A	31,50	223	223	223	223	0.00	223
6B	31,50	231	233	241	233	-3.32	233
6C	31,50	311	317	329	317	-3.65	317
7A	40,66	279	279	279	279	0.00	279
7B	40,66	283	283	283	283	0.00	283
7C	40,66	333	334	343	334	-2.62	334
8A	30,63	386	386	386	386	0.00	386
8B	30,63	395	395	401	395	-1.50	395
8C	30,63	517	528	533	533	0.00	527
9A	50,92	323	323	323	323	0.00	323
9B	50,92	326	326	329	326	-0.91	326
9C	50,92	332	332	332	332	0.00	332
9D	50,92	382	399	409	391	-4.40	391
10A	50,97	428	428	428	428	0.00	428
10B	50,97	436	436	436	436	0.00	436
10C	50,97	446	446	451	446	-1.11	446
10D	50,97	524	536	544	535	-1.65	530

6   Conclusion

This paper presents the first GA ever published for the CARP. On two standard sets of benchmarks, it improves eight best known solutions and its average deviation to the lower bound is very small, less than 0.7 %. On all tested instances, despite simple neighborhoods in its local search, this hybrid GA is as least as good as the best algorithm published, a powerful tabu search. This shows that, contrary to a still widespread opinion, hybrid GAs can outperform other metaheuristics. Moreover, thanks to our data structure, our GA can already handle an extended CARP, with a mixed graph, collecting and traversal costs, and prohibited turns or turn penalties. Our goal is now to design large instances of that type and to investigate new useful extensions, like heterogeneous fleets, multitrips, selective collection of several kinds of waste, several dumping sites, periodic problems over several days or weeks, sectors with different priorities, and time windows on the arcs.

Table 3. Overall comparison between Carpet and the GA on the two data sets

Algorithm	DeArmon instances		Belenguer's instances	
	Carpet	GA	Carpet	GA
Avg. dev. to LB%	0.5	0.24	1.9	0.69
Worst dev. to LB %	4.62	2.23	8.57	4.26
Avg. dev. to Best %	0.2	-0.08	1.1	-0.10
Proven optima	18	19	16	23
Solutions < Carpet	0	3	0	16
Solutions = Carpet	23	20	34	18
Solutions > Carpet	0	0	0	0
Solutions < Best	0	2	0	5
Solutions = Best	20	19	17	27
Solutions > Best	3	2	17	2
Avg. CPU time (s)	49	21	346	120

References

1. B.L. Golden, R.T.Wong, Capacitated arc routing problems, *Networks*, 11, 305-315, 1981.
2. E. Benavent, V. Campos, A. Corberan, E. Mota, The capacitated arc routing problem: lower bounds, *Networks*, 22, 669-690, 1992.
3. J.M. Belenguer, E. Benavent, *A cutting plane algorithm for the capacitated arc routing problem*, Research Report, Dept. of Statistics and OR, Univ. of Valencia (Spain), 1997.
4. R. Hirabayashi, Y. Saruwatari, N. Nishida, Tour construction algorithm for the capacitated arc routing problem, *Asia-Pacific Journal of Oper. Res.*, 9, 155-175, 1992.
5. W.L. Pearn, Approximate solutions for the capacitated arc routing problem, *Computers and Operations Research*, 16(6), 589-600, 1989.
6. W.L. Pearn, Augment-insert algorithms for the capacitated arc routing problem, *Computers and Operations Research*, 18(2), 189-198, 1991.
7. B.L. Golden, J.S DeArmon, E.K. Baker, Computational experiments with algorithms for a class of routing problems, *Computers and Operation Research*, 10(1), 47-59, 1983.
8. G. Ulusoy, The Fleet Size and Mixed Problem for Capacitated Arc Routing, *European Journal of Operational Research*, 22, 329-337, 1985.
9. L.Y.O. Li, Vehicle routing for winter gritting. Ph.D. dissertation, Department of OR and OM, Lancaster University, Lancaster, UK, 1992.

10. R.W. Eglese, Routing winter gritting vehicles, *DAM*, 48(3), 231-244, 1994.
11. A. Hertz, G. Laporte, M. Mittaz, A Tabu Search Heuristic for the Capacitated Arc Routing Problem, *Operations Research*, 48(1), 129-135, 2000.
12. J.S. DeArmon, *A Comparison of Heuristics for the Capacitated Chinese Postman Problem*, Master's Thesis, The University of Maryland at College Park, MD, USA, 1981.

A New Approach to Solve Permutation Scheduling Problems with Ant Colony Optimization

Daniel Merkle and Martin Middendorf

Institute for Applied Computer Science and Formal Description Methods
University of Karlsruhe, Germany
{merkle,middendorf}@aifb.uni-karlsruhe.de

Abstract. A new approach for solving permutation scheduling problems with Ant Colony Optimization is proposed in this paper. The approach assumes that no precedence constraints between the jobs have to be fulfilled. It is tested with an ant algorithm for the Single Machine Total Weighted Deviation Problem. The new approach uses ants that allocate the places in the schedule not sequentially, as the standard approach, but in random order. This leads to a better utilization of the pheromone information. It is shown that adequate combinations between the standard approach which can profit from list scheduling heuristics and the new approach perform particularly well.

1   Introduction

The Ant Colony Optimization (ACO) metaheuristic has recently been applied to several scheduling problems like the Job-Shop problem [1,2,3], the Flow-Shop problem [4], the Single Machine Total Tardiness problem (SMTTP) and its weighted variant the SMTWTP [5,6,7], and the Resource Constrained Project Scheduling problem [8]. In ACO several generations of artificial ants search for good solutions. Information exchange between the ants is based on principles of communicative behavior found in real ant colonies (for an introduction and overview see [9]). Every ant builds up a solution step by step going through several probabilistic decisions until a solution is found. Ants that found a good solution mark their paths through the decision space by putting some amount of pheromone on the edges of the path. The following ants of the next generation are attracted by the pheromone so that they will search in the solution space near good solutions. In addition to the pheromone values the ants will usually be guided by some problem specific heuristic for evaluating the possible decisions.

It has already been shown that ACO can solve permutation scheduling problems like SMTWTP [5,6,7] and Flow-Shop [4] very successfully. A comparison between ACO and other heuristics on a set of benchmark problems from [10] for the SMTWTP was done in [6]. ACO was able to find for all 125 test instances with 100 jobs the best known solutions. This was significantly better

E.J.W. Boers et al. (Eds.) EvoWorkshop 2001, LNCS 2037, pp. 484–494, 2001.

than the best known Tabu Search method for SMTWTP. Only iterated dynasearch reached a similar performance as ACO.

There is a general approach to solve permutation scheduling problems like
SMTWTP and Flow-Shop with ACO. Starting with the first place of the schedule
every ant constructs a solution by deciding iteratively which job is at the next
place. For every place i in the schedule and every job j there is pheromone
information τ_{ij} about the desirability to put job j at place i. This approach is
natural since for many permutation scheduling problems there exists good list
scheduling heuristics which can be used by the ants in addition to the pheromone
information. All ant algorithms that have been proposed so far for the SMTTP,
SMTWTP, and the flow-shop problem follow this approach (cmp. [5,6,7,4]).

In this paper we identify a disadvantage of the standard approach to solve
permutation scheduling problems with ACO. We propose a method to combine
the advantages of the standard approach with a new approach that circumvents
the disadvantage of the standard approach. Since the ACO algorithm of [6] that
uses the standard approach was already able to find the best solutions for all
of the large benchmark instances for the SMTWTP in the OR-Library [10] we
tested our method on a somewhat more difficult problem where in addition to
the weighted tardiness costs also weighted earliness cost have to be considered.
This problem is called the Single Machine Total Weighted Deviation Problem
(SMTWDP).

The paper is organized as follows. The definition of the SMTWDP is given in
Section 2. The new approach is described in Section 3. In Section 4 we describe
the ant algorithm for the SMTWDP. Some variants and further aspects of the ant
algorithm are considered in Section 5. The choice of the parameter values of the
algorithms that are used in the test runs and the test instances are described in
Section 6. Results are reported in Section 7 and a conclusion is given in Section 8.

2   The Single Machine Total Weighted Deviation Problem

The Single Machine Total Weighted Deviation Problem (SMTWDP) is to find
a one machine schedule for a given set of jobs that minimizes the sum of the
earliness and tardiness of the jobs. Formally, SMTWDP is to find for n jobs,
where job j, $1 \leq j \leq n$ has a processing time p_j, a due date d_j, and two
weights h_j, w_j, a non-preemptive one machine schedule that minimizes $D = \sum_{j=1}^{n} (h_j \cdot \max\{0, d_j - C_j\} + w_j \cdot \max\{0, C_j - d_j\})$ where C_j is the completion
time of job j. D is called the total weighted deviation of the schedule. Observe
that $h_j \cdot \max\{0, d_j - C_j\}$ is the weighted earliness of a job and $w_j \cdot \max\{0, C_j - d_j\}$
is its weighted tardiness. It is known that SMTWDP is NP-hard in the strong
sense even when all weights are the same [11]. Note, that the SMTWTP problem
easier since it can be solved in pseudopolynomial time when all weights are equal
[11].

3   The New Approach

The standard approach that ants build up a schedule by always extending an already fixed prefix can often profit from existing list scheduling heuristics. In the following we identify a disadvantage of this approach.

The general principle of ant algorithms is that the pheromone information reflects the outcomes of the decisions that have been made by former ants that found good solutions. Due to pheromone evaporation older generations of ants have a smaller influence on the pheromone values than newer ones. The ants of the actual generation should use this pheromone information in an adequate way. Hence their decisions should be made according to the probability distribution that is determined by the relative size of the pheromone values corresponding to the possible outcomes of the decision. Note, that this probability distribution might be modified by heuristic information.

The following observation shows a general problem for ant algorithms to follow this guideline. While the probability distribution for the first decision of an ant correctly reflects the relative size of the pheromone values this is not necessarily true for the following decisions. This is illustrated by an example. Assume that an ant decides with respect to the relative amount of pheromone and no additional heuristic information is considered. Let there be $n = 3$ jobs and the pheromone matrix given by

$$|\tau_{ij}|_{i,j \in [1:3]} = \begin{vmatrix} \frac{1}{2} & \frac{1}{3} & \frac{1}{6} \\ \frac{1}{6} & \frac{1}{3} & \frac{1}{2} \\ \frac{1}{3} & \frac{1}{3} & \frac{1}{3} \end{vmatrix}$$

Then, for the first place in the schedule an ant chooses job j, $j \in [1 : 3]$ with probability $p_{1j} = \tau_{1j}$. But for the second place the probabilities are different from the relative amount of pheromone in the row because the decision for the first place has to be considered. In particular, $p_{21} = p_{12}(\tau_{21}/(\tau_{21} + \tau_{23})) + p_{13}(\tau_{21}/(\tau_{21} + \tau_{22})) = 25/180 < 1/6$. Similarly, $p_{22} = 56/180 < 1/3$ and $p_{23} = 99/180 > 1/2$.

In the following we propose a method to cope with the above mentioned problem. The new approach is applicable to ant algorithms that solve permutation scheduling problems where no precedence constraints between the jobs have to be considered. The general idea is that every ant determines in a random order over the places which job is assigned to the next place. The advantage is that every place has the same chance to be the first that is assigned a job. Therefore, in the average the decisions of the ants will better reflect the information that is contained in the pheromone matrix. It is likely that this will improve the optimization behavior compared to an ant algorithm following the standard approach.

But there is a problem with this approach. An ant that assigns jobs to the places in a random order can usually not profit from heuristics that are based on list scheduling. Since there exist good list scheduling heuristics for many permutation scheduling problems and heuristic information is in general important for the optimization behavior of ant algorithms it is not clear whether the new approach will be better than a standard ant algorithm that uses a powerful list scheduling heuristic.

Therefore, we propose to combine both approaches. Some ants should decide according to the standard sequential order and use the (list scheduling) heuristic while the other ants make decisions according to a random order without using heuristic information. But some care has to be taken when two types of ants work together. When both types of ants are in the same generation it might often be the case that ants of one type are better than the others. Then, only the ants of one type will have a chance to update the pheromone matrix and the other ants are useless (we assume here that only the better ants are allowed to update the pheromone information). We circumvent this problem by using ants of different type in different generations so that competition occurs only between ants of the same type.

4   The Ant Algorithm for SMTWDP

The ant algorithm for the SMTWDP works as follows. In every generation each of m ants constructs one solution for the SMTWDP. We use two types of ants. The so called sequential ants select the jobs in the order in which they will appear in the schedule. Whereas the so called random ants select the jobs according to some random order in which they will appear in the schedule. For the selection of a job both types of ants use the pheromone information. The sequential ants use also some heuristic information. The heuristic information, denoted by η_{ij}, and the pheromone information, denoted by τ_{ij}, are indicators of how good it seems to have job j at place i of the schedule. The heuristic value is generated by some problem dependent heuristic whereas the pheromone information stems from former ants that have found good solutions. The next job is chosen according to the probability distribution over the set of unscheduled jobs S determined by

$$p_{ij} = \frac{\tau_{ij} \cdot \eta_{ij}}{\sum_{h \in S} \tau_{ih} \cdot \eta_{ih}} \quad \text{or} \quad p_{ij} = \frac{\tau_{ij}}{\sum_{h \in S} \tau_{ih}}$$

for the sequential ants respectively the random ants.

The heuristic values η_{ij} are computed according a heuristic that has been obtained by modifying a heuristic proposed in [12]. The idea is that a sequential ant choses the next job from the set of jobs that already exceed their due date (with respect to the sum of the processing times of all jobs that are scheduled so far) or will exceed it when they are scheduled next (if there exists such a job). For every of these jobs the costs will become higher when it is scheduled later. From these jobs the shorter ones and those with a high tardiness weight should be scheduled first. Hence, for $d_j \leq T + p_j$, where T is the sum of the processing times of all jobs that have already been scheduled, the heuristic value is

$$\eta_{ij} = \frac{w_j}{p_j}$$

Some of the jobs which will not exceed their due date when they are scheduled next might exceed it when they are scheduled after some other job that is scheduled next. To be able to consider those jobs we give the heuristic some foresight. All those jobs that have a processing time which is smaller than the average processing time and where the due date exceeds $T + p_j$ by at most $(\bar{p} - p_j) \cdot \frac{w_j}{w_j + h_j}$ are given some positive heuristic value. Observe that the threshold becomes larger for jobs which have tardiness costs that are relatively high compared to the earliness costs. The heuristic values then fell linearly with increasing due date from $\frac{w_j}{p_j}$ to zero in the interval $[T + p_j, T + p_j + (\bar{p} - p_j) \cdot \frac{w_j}{w_j + h_j}]$. Hence if $d_j \in [T + p_j, T + p_j + \max\{0, \bar{p} - p_j\} \cdot \frac{w_j}{w_j + h_j}]$ then

$$\eta_{ij} = \left(1 - \frac{d_j - T - p_j}{\max\{0, \bar{p} - p_j\} \cdot \frac{w_j}{w_j + h_j}}\right) \cdot \frac{w_j}{p_j}$$

When each remaining job j has a due date $d_j > T + p_j + (\bar{p} - p_j) \cdot \frac{w_j}{w_j + h_j}$ then the heuristic value η_{ij} equals

$$\eta_{ij} = \frac{p_j}{h_j}$$

so that jobs with a long processing time and small earliness weight will be preferred.

After all m ants of the generation have constructed a solution the ant that found the best solution in that generation is allowed to update the pheromone matrix. But before that some of the old pheromone is evaporated according to

$$\tau_{ij} = (1 - \rho) \cdot \tau_{ij}$$

The reason for this is that old pheromone should not have a too strong influence on the future. Then, for every job j in the schedule of the best solution found in the generation some amount of pheromone is added to element (ij) of the pheromone matrix where i is the place of job j in the schedule. The amount of pheromone added is $1/D$ where D is the total deviation of the schedule, i.e.,

$$\tau_{ij} = \tau_{ij} + \frac{1}{D}$$

The algorithm stops when a certain number of generations has been done. We tested different modes of alternation between generations of sequential ants and generations of random ants.

5   Additional Aspects and Variants

Some variants of the ant algorithm that was described in the last section which concern the pheromone evaluation and the type of ants used are described in this section.

5.1   Pheromone Summation Rule

An alternative way to use the pheromone information was proposed in [7] for the SMTWTP. Since the SMTWTP is the variant of the SMTWDP where all weights h_j are zero, i.e. only the tardiness of a job counts, we use this pheromone evaluation method — called pheromone summation evaluation in [7] — here also.

The following problem occurs when using the relative pheromone values directly as the probability to choose the next job. Assume that by chance an ant chooses to put some job h at place i of the schedule that has a low pheromone value τ_{ih} (instead of a job j that has a high pheromone value τ_{ij}). Then in order to have a high chance to still end up with a good solution it will likely be necessary for the ant to place job j not too late in the schedule when j has a small due date. To handle this problem it was proposed in [7] to let a pheromone value τ_{ij} also influence later decisions when choosing a job for some place $l > i$. A simple way to guaranty this influence is to use the sum of all pheromone values for every job from the first row of the matrix up to row i when deciding about the job for place i. Then a sequential ant chooses the next job for place i in the schedule according to the probability distribution over S determined by

$$p_{ij} = \frac{\sum_{k=1}^{i} \tau_{kj} \cdot \eta_{ij}}{\sum_{h \in S} \left(\sum_{k=1}^{i} \tau_{kh} \cdot \eta_{ih} \right)} \tag{1}$$

5.2   Backward Ants

In [13] it was proposed to use (sequential) forward and (sequential) backward ants for solving the Shortest Common Supersequence problem. Here we study which kind of ants, forward ants or backward ants, are better for the different types of problem instances. Backward ants construct solutions by assigning jobs to the places in reverse order, i.e. they first decide which job is the last in the schedule. Clearly, for the sequential backward ants the heuristic they use has to be modified accordingly. This was done as follows.

A sequential backward ant starts with choosing the last job that finishes at time $\sum_{j=1}^{n} p_j$. It always choses the next job from the set of jobs that already exceed their due date (if there exists such a job). Of these jobs the shorter ones and those with a high earliness weight should be scheduled first. Hence, for $d_j \geq T$, where T is the sum of the processing times of all remaining jobs that have not been scheduled so far, the heuristic value is

$$\eta_{ij} = \frac{h_j}{p_j}$$

Similar as for the sequential forward ants some of the jobs which will not exceed their due date when they are scheduled next will also be considered. All those jobs that have a processing time which is smaller than the average and where the due date is not before $T - (\overline{p} - p_j) \cdot \frac{h_j}{w_j + h_j}$ are given some positive heuristic value. If $d_j \in \left[T - \max\{0, \overline{p} - p_j\} \cdot \frac{h_j}{w_j + h_j}, T \right]$ then

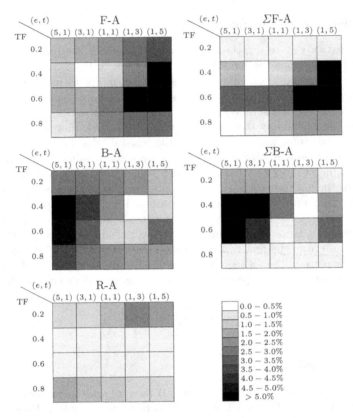

Fig. 1. Results for F-A, B-A, ΣF-A, ΣB-A, and R-A. For each combination of TF and (e, t) values: average loss in solution quality compared to the best performing variant for that combination of TF and (e, t) values.

$$\eta_{ij} = \left(1 - \frac{T - d_j}{\max\left\{0, \overline{p} - p_j\right\} \cdot \frac{h_j}{w_j + h_j}}\right) \cdot \frac{h_j}{p_j}$$

6   Test Instances and Parameters

We tested the different variants of the algorithm on instances for the SMTWDP of size 100 jobs. These instances were generated as follows: for each job $j \in [1, 100]$ an integer processing time p_j is taken randomly from the interval $[10, 100]$, the earliness weight h_j is taken randomly from the interval $[1e, 2e]$, the tardiness weight w_j is taken randomly from the interval $[1t, 2t]$, and an integer due date d_j is taken randomly from the interval

$$d_j \in \left[\sum_{j=1}^{100} p_j \cdot (1 - TF - \frac{RDD}{2}), \sum_{j=1}^{100} p_j \cdot (1 - TF + \frac{RDD}{2}) \right]$$

Note, that this rule was also used for creating the benchmark instances for the SMTWTP that can be found in the OR-Library [10]. The parameters e and t allow to control the average influence of the earliness and tardiness weights for a problem instance. The value RDD (relative range of due dates) determines the length of the interval from which the due dates were taken. TF (tardiness factor) determines the relative position of the center of this interval between 0 and the sum of the processing times $\sum_{j=1}^{100} p_j$. The values for TF are chosen from the set $\{0.2, 0.4, 0.6, 0.8\}$. RDD was set to 0.4, i.e. the due dates cover a range of 40% of the computation interval. For (e, t) we tested the combinations $(5, 1)$, $(3, 1)$, $(1, 1)$, $(1, 3)$ and $(1, 5)$. For each combination of TF and (e, t) we use a set of 15 test instances. The parameter ρ was set 0.01 and the number of ants in every generation was $m = 10$. Every run was stopped after 20000 generations.

We use the following notations for the different versions of the ant algorithm. F-A: only sequential forward ants are used as in Section 4. B-A: only sequential backward ants are used as in 5.2. The corresponding versions where the ants use the pheromone summation rule as in 5.1 are called ΣF-A respectively ΣB-A. R-A: only random ants are used. Combinations between F-A, B-A, ΣF-A and ΣB-A with the R-A where exactly the even generations work according to R-A are denoted by FR-A, BR-A, ΣFR-A and ΣBR-A.

7 Experimental Results

A comparison between F-A, B-A, ΣF-A, ΣB-A, and R-A is shown in Figure 1. In general R-A performed quite well. It was best in about half of the cases where TF=0.4 and TF=0.6, i.e. the due dates lie in the middle of the scheduling interval. ΣF-A performed best for TF=0.2 where the due dates are more at the beginning of the scheduling interval. The opposite is true for ΣB-A. For relative high earliness weights (large e values) ΣF-A performs better than for relative high tardiness weights (large t values). Again, the opposite is true for ΣB-A. Analogous remarks hold for F-A and B-A which performed, in general, worse than their counterparts with the pheromone summation evaluation. Our results coincide with an observation of den Besten et al. [6] that for the SMTWTP the instances with high TF values are more difficult than those with smaller TF values. We can conclude here that the preferred working direction of the ants should depend on the type of the problem instances.

Comparing the ant algorithms that use two types of ants (FR-A, BR-A, ΣFR-A, ΣBR-A) to their single type counterparts (F-A, B-A, ΣF-A, ΣB-A) our test results show that the two type variants are in every case better than the corresponding single type variant. A comparison between the algorithms using two types of ants and algorithm R-A is given in Figure 2. For TF=0.2 and TF=0.8 ΣFR-A and ΣBR-A are the best variants. It is interesting that in this

cases the sequential ants using the pheromone summation evaluation can profit
from the combination with the random ants. For TF=0.4 and TF=0.6 algorithm
FR-A is the best when the earliness weights are relatively large compared to the
tardiness weights. For the opposite case where earliness weights are relatively
small compared to the tardiness weights BR-A is the best variant.

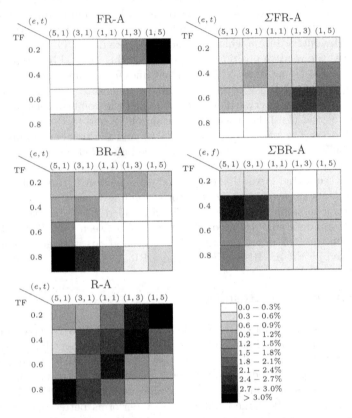

Fig. 2. Results for FR-A, BR-A, ΣFR-A, ΣBR-A, and R-A. For each combination of
TF and (e,t) values: average loss in solution quality compared to the best performing
variant for that combination of TF and (e,t) values.

The effect of the relative influence of the sequential ants and the random ants
was tested for TF=0.4 and $e = t = 1$. We changed the relative rate of sequential generations and random generations in the algorithm. With FR-(i,j)-A we
denote the algorithm where i generations of sequential forward ants alternate
with j generations of random ants. Other combinations are denoted analogously.
Result are shown in figure 3 (Note, that FR-$(0,\infty)$-A is the same as R-A). For
the test case we see that a similar influence of sequential generations and random

generations gives the best results for F-(x, y)-A. For ΣF-(x, y)-A more random influence is better.

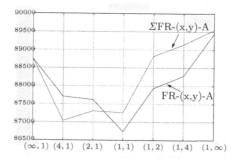

Fig. 3. Results for FR-(x, y)-A and ΣFR-(x, y)-A for TF=0.4, $e = t = 1$

Finally we give the absolute deviation values of the best algorithm versions in table 7. The table clearly shows that the two type ant algorithms perform best.

Table 1. Best results

	(5,1)	(3,1)	(1,1)	(3,1)	(5,1)
TF=0.2	ΣBR-A	ΣBR-A	ΣBR-A	ΣBR-A	ΣBR-A
	739122	443851	148453	149288	149834
TF=0.4	FR-A	FR-A	FR-A	FR-A	BR-A
	349575	218472	86738	124358	160203
TF=0.6	FR-A	BR-A	BR-A	BR-A	BR-A
	152589	121241	87874	226185	363407
TF=0.8	ΣFR-A	ΣFR-A	ΣBR-A	ΣBR-A	ΣBR-A
	155756	155734	154640	462516	770590

8   Conclusion

We have proposed a new approach for solving permutation scheduling problems with Ant Colony Optimization. It was shown for the Single Machine Total Weighted Deviation Problem (SMTWDP) that a combination between generations of ants that use the standard method of always extending a prefix of the schedule with generation of ants that allocate the places in the schedule in random order leads to significant improvements. Further we studied the influence of the type of problem instances on the best working direction (forward or backward) that the ants should follow.

References

1. A. Colorni, M. Dorigo, V. Maniezzo, and M. Trubian. Ant system for job-shop scheduling. *JORBEL - Belgian Journal of Operations Research, Statistics and Computer Science*, 34:39–53, 1994.
2. M. Dorigo, V. Maniezzo, and A. Colorni. The ant system: Optimization by a colony of cooperating agents. *IEEE Trans. Systems, Man, and Cybernetics – Part B*, 26:29–41, 1996.
3. S. van der Zwaan and C. Marques. Ant colony optimisation for job shop scheduling. In *Proceedings of the Third Workshop on Genetic Algorithms and Artificial Life (GAAL 99)*, 1999.
4. T. Stützle. An ant approach for the flow shop problem. In *Proceedings of the 6th European Congress on Intelligent Techniques & Soft Computing (EUFIT '98)*, volume 3, pages 1560–1564. Verlag Mainz, Aachen, 1998.
5. A. Bauer, B. Bullnheimer, R.F. Hartl, and C. Strauss. An ant colony optimization approach for the single machine total tardiness problem. In *Proceedings of the 1999 Congress on Evolutionary Computation (CEC99), 6-9 July Washington D.C., USA*, pages 1445–1450, 1999.
6. M.L den Besten, T. Stützle, and M. Dorigo. Ant colony optimization for the total weighted tardiness problem. In M. Schoenauer et al., editor, *Parallel Problem Solving from Nature: 6th international conference*, volume 1917 of *Lecture Notes on Computer Science*, pages 611–620, Berlin, September 2000. Springer Verlag.
7. D. Merkle and M. Middendorf. An ant algorithm with a new pheromone evaluation rule for total tardiness problems. In *Proceeding of the EvoWorkshops 2000*, number 1803 in Lecture Notes in Computer Science, pages 287–296. Springer Verlag, 2000.
8. D. Merkle, M. Middendorf, and H. Schmeck. Ant colony optimization for resource-constrained project scheduling. In D. Whitley et al., editor, *Proceedings of the Genetic and Evolutionary Computation Conference (GECCO-2000)*, pages 893–900, Las Vegas, Nevada, USA, 10-12 July 2000. Morgan Kaufmann.
9. M. Dorigo and G. Di Caro. The ant colony optimization meta-heuristic. In D. Corne, M. Dorigo, and F. Glover, editors, *New Ideas in Optimization*, pages 11–32. McGraw-Hill, 1999.
10. http://mscmga.ms.ic.ac.uk/jeb/orlib/wtinfo.html.
11. M.R. Garey and D.S. Johnson. *Computers and Intractibility: A Guide to the Theory of NP-Completeness*. W. H. Freeman and Company, New York, 1979.
12. P.S. Ow and T.E. Morton. The single machine early/tardy problem. *Management Science*, 35:177–191, 1998.
13. R. Michels and M. Middendorf. An ant system for the shortest common supersequence problem. In D. Corne, M. Dorigo, and F. Glover, editors, *New Ideas in Optimization*, pages 51–61. McGraw-Hill, 1999.

Street-Based Routing Using an Evolutionary Algorithm

Neil Urquhart, Ben Paechter, and Kenneth Chisholm

School Of Computing
Napier University
219 Colinton Rd
Edinburgh
EH14 1DJ
n.urquhart@napier.ac.uk

Abstract. Much research has been carried out into solving routing problems using both Evolutionary Techniques and other methods. In this paper the authors investigate the usage of an Evolutionary Algorithms to solve the Street-Based Routing Problem (SBRP). The SBRP is a subset of the Travelling Salesman Problem that deals specifically with a street-based environment. The paper also compares two possible strategies for evolving networks of routes. This paper may be considered introduction to the particular problem, and opens the way for future research into this area.

1 A Description of the Problem

Much previous research into routing has concentrated on solving the Travelling Salesman Problem (TSP) [11][12]. The objective of the TSP is to produce the optimum (shortest) route around a set of delivery points, visiting each point only once and returning to the start. Most previous attempts at solving this problem using evolutionary algorithms have used a permutation-based representation.

If the pattern of deliveries to households within an urban area is examined, it will become immediately apparent that significant adjacent clusters of delivery points occur as they are grouped into streets. It follows that all delivery points within a street will normally be serviced in sequence, before moving on to the next street. Some considerable advantage may be drawn from grouping delivery points into street sections and solving the problem by ordering street sections rather than individual delivery points. For practical purposes a street section is defined as all the delivery points on one side of a street between two junctions. Thus most named streets within a town are divided into several street sections. The authors have named this technique Street Based Routing (SBR).

The street based routing problem (SBRP) may be considered to be a sub-problem of the Travelling Salesman Problem. It is the authors' belief that when the proposed representation is applied to street-based routing problems the majority of excluded solutions will be inefficient solutions. For example, given a problem that involves delivery to 471 delivery points a permutation-based approach would yield 471! = 1.499×10^{1056} potential solutions. By grouping the houses into short sections, and

E.J.W. Boers et al. (Eds.) EvoWorkshop 2001, LNCS 2037, pp. 495-504, 2001.

then using a permutation of sections, the reduction in search space is significant. For instance if the delivery points are grouped into 113 sections the search space is now 113!=2.23x10^184.

Within each street section, one of three pre-defined delivery patterns may be applied to obtain the order in which the delivery points within the section are to be serviced. Within each street section the possible patterns of delivery have been identified, as:

1. Deliver to all the households on one side of the street, then cross over and deliver to the opposite side, ending up at the start point (see fig1).
2. Traverse the street from end to end, delivering to both sides, crossing over as required, finishing at the opposite end of the street from the starting point (see fig 2).
3. Deliver to all of the households on one side, then deliver to the opposite side at a later stage in the route (see fig 3).

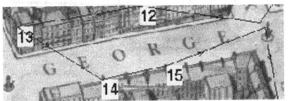

Fig. 1. Deliveries pattern 1; deliver to both sides of the street, crossing at the last house.

Fig. 2. Deliveries pattern 2; deliver to both sides of the street, crossing as required.

Fig. 3. Deliveries pattern 3; the two sides of the road receive deliveries at separate points within the round.

2   An Evolutionary Algorithm for Solving the SBRP

2.1   Introduction

Evolutionary algorithms have been utilised to solve a wide range of tasks, including scheduling and routing tasks [2][3][4][5][6][7][10][12][13][14]. It is assumed that the reader has a reasonable understanding of Evolutionary Techniques (if the reader requires any background information [9] is a useful reference work).

2.2   The Initial Algorithm

In the initial attempt at constructing an algorithm, utilised a direct representation, each chromosome consisted of a permutation of street-section genes. Within each gene the direction of traversal and the delivery pattern used had to be specified explicitly. The direction and delivery patterns were altered by mutations. A recombination operator was used to alter the ordering of the delivery sections. This initial algorithm produced very disappointing results, mostly due to inappropriate delivery patterns and streets being traversed in the 'wrong' direction.

2.3   Improvements to the Initial Algorithm

A memetic algorithm was constructed, this utilised an indirect representation in the form of 'memes'. The genotype only holds the order in which the street sections are to be considered. No information is held within the genotype as to in what direction the section is to be traversed or which delivery pattern is to be used. These genotypes may then be decoded into phenotypes (i.e. a complete route) by applying a simple set of rules to determine the appropriate delivery pattern. Previous research in a variety of areas has suggested that the addition of local search to an evolutionary algorithm can improve its performance. [10][6].

The mutation operator randomly selects a sub-string of genes from within the chromosome and moves this sub-string to another point selected at random within the gene. The recombination operator is based on uniform crossover, but makes use of a global precedence operator, in this case it is the total length of the route. When two parents have been selected, the first gene from one parent is copied to the child. Subsequent genes are copied from either parent, based on selection of the gene that results in the smallest increase in walk length. In some cases the choice will be restricted by the requirement for the child chromosome to contain a valid permutation.

The algorithm used employed a steady state population incorporating elitism and tournament selection and replacement. Initial experiments showed that a population size of 100, a recombination rate of 0.3 and a mutation rate of 0.2 were reasonable parameters. Having built a route based on the above methods, the fitness value is the length of the route under consideration.

2.4   Initial Experimentation

The initial experiments were carried out on a delivery network representing the centre of Edinburgh, this data set has 471 delivery points, distributed amongst 113 street sections and a total "mileage" of 2346 units.

Some experimentation was carried out to determine sizes for the population and the selection tournaments. Populations of 500 and 250 were tried and tournament sizes of 20 and 50 were compared also. The initial results are shown in table 1. An example of a route produced using local search is shown in fig 4.

2.5 Integrating a Heuristic with the EA

Some recent research [8][10][2][3] has concentrated on integrating heuristics and EAs. To improve the results obtained with algorithm as initially constructed, a simple nearest neighbour, re-ordering heuristic has been added. Because our algorithm has a computationally expensive evaluation function (due to it having to build the routes) a hill climbing heuristic which would require a large number of evaluations was not felt to be suitable.

Although the heuristic dramatically improves inefficient routes there are however situations where the next street in the optimum solution is not necessarily the closest street, in this situation the heuristic can actually lengthen the route. The heuristic only applied to individuals in the population every 1,000 evaluations. When the heuristic is applied to an individual the modified individual is only accepted back into the population if the heuristic has improved the fitness else the original unmodified phenotype is retained. Thus the EA may make subtle adjustments to the route without the heuristic removing them. See fig 5.

3   Constructing a Delivery Network

3. Introduction

Construction of a single delivery route is an interesting academic exercise, but the real-life problem involves the construction of multiple delivery routes. A maximum number of deliveries is specified for the routes; this is a hard constraint and thus cannot be contravened. It is also desirable to ensure that the routes are as close to equal length as is possible. Some similarity exists with the classic graph-partitioning problem, which researchers have attempted to solve using a mixture of evolutionary algorithms and heuristics [1][16]. The individual routes are constructed using the SBRP methods outlined in section 2.

3.2 Two Approaches to Building Networks

The authors have opted to investigate two approaches to building delivery networks using Evolutionary Algorithms.

Table 1. Distances averages over 10 runs ,each for 250,000 evaluations.

Run	P=500 T =5	P=250 T = 5	P=500 T=20	P=250 T=20
Average	2281.4	2164.6	1973.5	2037.6
Deviation	407.098459	198.9417559	176.639149	192.7913092

Fig. 4. The best result produced using the hybrid algorithm (reproduced with kind permission of James Brown Designs).

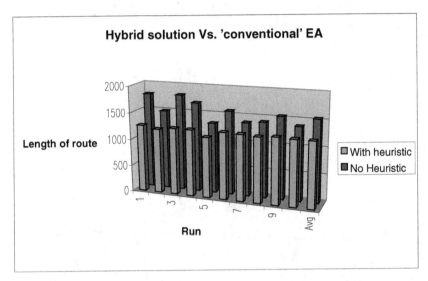

Fig. 5. A comparison of the hybrid and conventional heuristic.

The first approach modifies the routing algorithm to allow a single chromosome to contain multiple routes. Within the chromosome a series of marker genes define where one route stops and the next commences. Thus the routes may be evolved and evaluated together. This approach is referred to as the 'Single Chromosome' approach.

A second approach that may be taken allows the construction of the delivery network to take place in two distinct stages, firstly a grouping algorithm is utilised to divide the streets into groups. Each group of streets contains no more than the maximum number of deliveries and should be as geographically adjacent as possible. The groups of streets are then passed individually to the SBRP routing algorithm, allowing the each route to be constructed and optimised. This approach is known as 'Group and Build'.

The grouping stage ensures that the route builder algorithm commences with data that does not contain more the maximum number of deliveries for each route, it also ensures that the streets are geographically adjacent. The route builder algorithms can fully optimise a route for the streets allocated to it by the grouping algorithm. This version of the architecture would appear to overcome one of the disadvantages of the first architecture in that it allows co-evolution of the routes without interference.

3.3   Description of the Grouping EA Used for the 'Group and Build' Approach

The user initially enters the maximum no of deliveries permissible on each route, the minimum number of routes (n) can then be calculated as the total deliveries divided by the maximum allowed per route, and rounded up to the nearest whole number.

The initial population is seeded by randomly allocating each street section to a group. A mutation consists of randomly reallocating a street to another route. The recombination works by selecting a route from parent 1 this is then copied to the child. The remaining routes are copied from parent 2. The first route is preserved without being altered, the other routes may be changed slightly to prevent them from 'overwriting' the initial set of genes copied from parent 1.

The fitness function is as follows; for each group calculate the deviation between number of delivery points and the target, if the route has too many deliveries (ie more deliveries than the user has specified as a maximum) then multiply the deviation by a constant (4) and add a penalty (100). This penalises groups that contain too many delivery points. For each group calculate the average distance between each street end within that group. The closer the streets are clustered the smaller this value. The final value is the total deviation from the desired number of delivery points, added to the average distance. The deviation is multiplied by a weight (usually 2) to force the EA to quickly discount those individuals whose routes break this hard constraint.

If after a set number of evaluations one or more routes contain more than the *max* number of deliveries, the algorithm is executed again, but with the number of routes incremented.

4.2 Experiments and Results Obtained

The division of the 'Edinburgh' data set (471 deliveries) into 5 routes, each with a maximum of 100 deliveries has been attempted using both the Group And Build-EA (GAB-EA) and the single chromosome-EA(SC-EA) approaches. The results are shown in tables 2 and 3. The SC-EA was allowed to run for 250,000 evaluations, whilst the first stage of GAB-EA ran for 50,000 evaluations and each of the routes were evolved over 40,000 evaluations, thus both networks were created using the same number of evaluations.

Regarding the ability of each algorithm to satisfy the hard constraint that no single route contains more than 100 deliveries, table 3 shows that the GAB-EA satisfies this constraint 100% of the time. An inspection of table 5 reveals that the SC-EA breaks this constraint 22 times (44% of the total routes constructed). A perusal of tables 4 and 6 reveals that the average length of routes constructed with the GAB-EA is 384 units compared to 749 units (GAB-EA produced routes are on average 49% shorter than those produced with the with the single-EA).

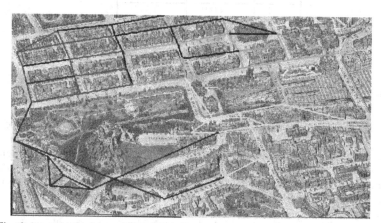

Fig. 6. A grouping of streets (by colour) after running the grouping EA. (Reproduced with kind permission of James Brown Designs.)

Using the GAB-EA approach gives a much superior result to utilising the SC-EA approach. The reasons for this include the fact that the GAB-EA attempts to satisfy the maximum delivery constraint prior to beginning the routing process. Because both the grouping and the routing process are taking place concurrently within the SC-EA it may be deduced that it is unable to evolve satisfactory routes. If we consider that each group of streets that makes up a route has its own section of the fitness landscape, then the addition and removal of streets from a particular route alters the nature of the landscape. In this manner the SC-EA is unable to evolve routes due to the continuously changing nature of the landscape sections being considered. The GAB-EA however, divides the landscape into sections when grouping the routes and then does not alter the section boundaries while building the routes.

Table 2. The number of deliveries per route when constructing a network of 5 routes (averaged over 10 runs) using the Multi-EA.

Route	Avg Distance (multi-EA)	Avg deliveries (single EA)
1	96.1	156.9
2	95.6	129.3
3	94.7	82.6
4	89.7	70.6
5	94.9	31.6
StdDev	2.57682	49.43071

Table 3. The length of routes when constructing a network of 5 routes (averaged over 10 runs).

Route	Avg length (Multi EA)	Avg length (single EA)
1	362.5	869.7
2	392.5	825.4
3	377.4	733.9
4	431.9	656.1
5	359.2	660.2
Avg	384.7	749.1

5   Conclusions and Future Work

The concept of using the SBR representation for particular problems has been presented and discussed in this paper. It can be proven wthat the search space for an SBR-based routing problem is smaller than one based on a more 'traditional' routing problem. Future work intends to prove that for problems exhibiting specific characteristics in their grouping of delivery points and streets the smaller search space of solutions allowable by SBR contains the optimum solutions. Thus this search space may be navigated more quickly to find the desired solution.

The representation was altered to an indirect representation by the elimination of the StartAt portion of the gene, the system using the simple heuristic described earlier to set the direction for each street. As previous work on EAs has shown, moving from a direct to an indirect representation reduces the search space and thus produces faster and higher quality results.

The research is currently concentrating on comparing the SBR approach outlined above to other approaches to the TSP, it is hoped to prove that for specific types of problem the SBR can outperform more traditional heuristics. Future research will look at the application of hybrid solutions (heuristics and evolutionary algorithms) to the SBR problem.

References

1. A Hybrid Genetic Algorithm for Multiway Graph Partitioning. So-Jin Kang, Byung-Ro Moon. Proceedings of the Genetic and Evolutionary Computation Conference 2000. Eds D. Whitley, D Goldberg, E Cantu-Paz, Lee Spector, Ian Parmee, Hans-Georg Beyer. Morgan Kaufman Publishers 2000

2. A Memetic Algorithm With Self-Adaptive Local Search: TSP as a case study. Natalio Krasnogor, Jim Smith. *Proceedings of the Genetic and Evolutionary Computation Conference 2000. Eds D. Whitley, D Goldberg, E Cantu-Paz, Lee Spector, Ian Parmee, Hans-Georg Beyer. Morgan Kaufman Publishers 2000.*

3. A Comparison of Nature Inspired Heuristics on the Traveling Salesman Problem. Thomas Stutzle, Andreas Grun, Sebastian Linke, Marco Ruttger. *Parallel Problem Solving from Nature VI. Eds Marc Schoenauer, Kalyanmoy Deb, Guenter Rudolph, Xin Yao, Evelyne Lutton, Juan Julian Merelo, Hans-Paul Schwefel Eds.* Pub Springer-Verlag 2000.

4. Evolving Schedule Graphs for the Vehicle Routing Problem with Time Windows. H Timucin Ozdemir, Chilukuri K. Mohan. *Congress on Evolutionary Computation 2000.* Pub IEEE 2000.

5. Scheduling Chicken Catching – An Investigation Into The Success Of A Genetic Algorithm On A Real World Scheduling Problem. Hart E, Ross P, Nelson J. Annals Of Operations Research 92 Baltzer Science Publishers 1999.

6. A Genetic Algorithm for Job-shop problems with various schedule criteria. Hsio-Lan Fang, David Corne, Peter Ross. *Evolutionary Computing, AISB Workshop Brighton UK April 1996 Ed. Terence C. Fogarty* Pub: Springer-Verlag 1996

7. Extensions to a Memetic Timetabling System. Paechter B, Norman M, Luchian H. *Practice and theory of Automated Timetabling*, Burke and Ross Eds. Springer Verlag 1996.

8. New Genetic Local Search Operators for the Traveling Salesman Problem. Bernd Freisleben and Peter Merz. Parallel Problem Solving from Nature - PPSN IV Eds: Hans-Michael Voigt, Werner Ebeling Ingo Rechenberg, Hans-Paul Schwefel Springer Verlag 1996..

9. Genetic Algorithms + Data Structures = Evolution Programs (Third, Revised and Extended Edition). Michalewicz Z. Springer-Verlag 1996.

10. Two Solutions to the General Timetable Problem Using Evolutionary Methods. Peachter B, Cumming A, Luchian H, Petruic. *Proceedings of the First IEEE Conference on Evolutionary Computionary Computation 1994.*

11. A Comparison Study of Genetic Codings for the Travelling Salesman Problem. Tamaki H, Kita H, Shimizu N, Maekawa K, Nishikawa Y. *Proceedings of the First IEEE Conference on Evolutionary Computionary Computation 1994.*

12. A new Genetic Approach for the Travelling Salesman Problem. Bui T, Moon B. *Proceedings of the First IEEE Conference on Evolutionary Computionary Computation 1994.*

13. Vehicle Routing with Time Deadlines using Genetic and Local Algorithms. Thangiah S, Vinayagamoorthy R, Gubbi A. *Proceedings of the Fifth International Conference on Genetic Algorithms* Forrest S Ed. Morgan Kaufmann 1993.
14. Multiple Vehicle Routing with Time and Capacity Constraints using Genetic Algorithms.
15. *Proceedings of the Fifth International Conference on Genetic Algorithms* Forrest S Ed. Morgan Kaufmann, 1993.
16. Intelligent Structural Operators for the K-way Graph partitioning Algorithm. Gregor von Laszewski. *4th International Conference on Genetic Algorithms. Morgan Kaufmann 1991.*

Investigation of Different Seeding Strategies in a Genetic Planner

C. Henrik Westerberg and John Levine

AI Applications Institute, University of Edinburgh,
80 South Bridge, Edinburgh, EH1 1HN
{carlw,johnl}@dai.ed.ac.uk
http://www.dai.ed.ac.uk/homes/carlw/

Abstract. Planning is a difficult and fundamental problem of AI. An alternative solution to traditional planning techniques is to apply Genetic Programming. As a program is similar to a plan a Genetic Planner can be constructed that evolves plans to the plan solution. One of the stages of the Genetic Programming algorithm is the initial population seeding stage. We present five alternatives to simple random selection based on simple search. We found that some of these strategies did improve the initial population, and the efficiency of the Genetic Planner over simple random selection of actions.

1   Introduction

This paper presents an investigation into different seeding strategies for initial population formation. This work is performed in the context of Genetic Programming (GP) applied to classical planning. An already implemented Genetic Planner [1,2] was altered to use the different seeding strategies. We then experimented with the different strategies to determine their effect on the GP algorithm in terms of CPU time used to solve the problem, the quality of the initial population, and the number of generations required to solve the problem.

There are many areas in the GP algorithm which can be improved, such as population seeding, crossover, and the fitness function. We started at the beginning and wanted to improve the initial population seeding stage. The efficiency of the GP algorithm could be improved by increasing the quality of the initial population. This can be done by adopting simple rules or doing some limited search to decide which non-terminal should come next in the candidate. An easy example, from timetabling, would be add the easy time-course relationships into the initial population. We used simple search strategies like Depth First Search (DFS) and Best First Search (BFS)to improve the quality of the initial population.

This paper will first discuss the available work on Genetic Planning and seeding. The subsequent section gives a quick overview of planning and the types of planning problems the Genetic Planner caters for. This is followed by a description of how the GP algorithm was implemented a description of some specific

E.J.W. Boers et al. (Eds.) EvoWorkshop 2001, LNCS 2037, pp. 505–514, 2001.
© Springer-Verlag Berlin Heidelberg 2001

algorithmic features. The next section describes the different seeding strategies which were implemented and results from experiments done using those strategies. Finally the last section discusses some conclusions and possibilities for further work.

2   Current Work

There are three papers that relate closest to the idea of domain independent planning by evolutionary means. Spector looked at using GP for simple planning problems in the blocks domain and also using GP to come up with general which work on any problem [3]. Muslea looked at the effectiveness of using GP for domain independent planning [4]. Westerberg summarises some work done studying the effectiveness of using GP on classical planning domains [1,2].

There is also other related work to Genetic Planning. Handley looked at GP applied to robot control planning [5]. Aler has looked at generating heuristics for planning [6] and generating control rules for planning problems [7]. Finally, Koza looked at generating a general plan for a blocks stacking problem [8].

Work on population seeding for evolutionary methods is equally sparse. One example of a positive result from improving the seeding algorithm is Langon's work [9]. In Langdon's work the author starts by seeding the population with perfect individuals and then asks the GP algorithm to generalise them. The author compared this method with a random seeding strategy. The population based on perfect individuals was able to generalise better than the random method. This is an indication of the fact that there are benefits to improving the initial population seeding method rather than resorting to a simple random method.

3   Style of Planning

Planning is the task of deciding on which sequence of actions will best achieve some goal or set of goals. In general a planning system is given the world, represented by a set of facts, a set of descriptions of actions the planner can choose from, and goals that need to become true or remain true. Given this information the planner must then find a plan that when executed will achieve the goal or set of goals. There is interaction between actions in that some actions have to occur before others. For example if someone wants to buy some bread they would have to go to a shop before trying to purchase the bread. There are a large variety of planning problems depending on how the actions, goals, and world state are described.

The Genetic Planner concentrates on a style of planning called Classical Planning. One form of classical planning is called STRIPS planning [10]. This style of planning gives us the rules on how to set out a planning domain. The world state is made up of a list of facts that are true about the world. Using the Briefcase Domain as an example, also used by Muslea [4], facts look like: at(papers,london), at(mybriefcase,london). STRIPS planning also tells us how we should design our operators. For example, one of the planning operators from the Briefcase Domain is putin:

```
Operator: putin(Object,Briefcase)
Preconditions: at(Object,Location),
               at(Briefcase,Location)
Add List: in-briefcase(Object,Briefcase)
Delete List: at(Object,Location)
```

We give the operator's name and what types of objects the operator takes. The preconditions are those facts that need to be present in the world state before the action can execute. If the action executes the facts in the add list get added to the world state and the those facts in the delete list are deleted from the world state. Those facts which slowly get added and deleted from the knowledge base should hopefully push the world state towards the goal state.

Given the following initial state and goal state we can come up with a plan using our operators to transform the initial state to the goal state.

```
Initial State                        Goal State
at(mypapers,london)                  at(mypaper,paris)
at(mybriefcase,london)               at(mybriefcase,paris)
```

The plan shown in Fig.1 solves the above problem:

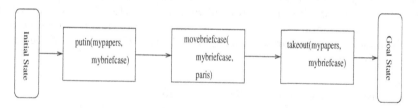

Fig. 1. An example plan/chromosome

The first step is to put the papers in the briefcase, the second step is to move the briefcase to paris. The last step is to take the papers out of the briefcase. The task of the Genetic Planner is to solve this style of planning problems. Finding solutions to these problems is quite difficult but we are more often interested in finding an optimal plan. For a domain independent system this is known to be NP-hard for most cases [11].

4 Genetic Planning

Genetic Planning is the application of GP to planning. An important difference between this and normal GP is that Genetic Planning works on plans as opposed to programs. Programs and plans are analogous as they can both be considered as an ordered set of instructions. The Genetic Planner starts by producing an initial population of candidates according to some strategy. The Genetic Planner then evolves the population of plans using survival of the fittest in order to produce a plan that will achieve some set of goals.

The Genetic Planner, reported here uses a generational algorithm with tournament selection, a domain independent fitness function, 1-point crossover, reproduction, and an addition mutation.

To highlight the other major differences between Genetic Planning and GP we will consider three specific algorithmic details.

4.1   Representing a Candidate/Chromosome

We decided to represent candidates using a linear genome rather than a tree genome [12]. This decision resulted in a considerable speed up and simplification of the system [1]. The linear genome is a linear list of planning operators and their arguments, much like Fig. 1. Each planning operator and associated arguments makes up an atomic action. Each candidate is made up of these atomic actions. In a GP context the planning operators from a particular problem domain are the non-terminals and the objects those actions take are the terminals. Each planning operator also takes arguments of a particular type. For example the putin operator from the Briefcase Domain takes an object and a briefcase. Those terminals that belong in each of these sets is already predefined so that no non-working actions are created due to having arguments of the wrong type in the operator. Also crossover in the system works at the action level so that actions are always being swapped with other actions. Crossover points can never occur in the middle of the arguments to an action. The atomic structure of the actions is thus preserved.

4.2   Creating a Candidate/Chromosome

The first strategy we implemented for creating new candidates for the initial population was the random method (Strategy 0). First the candidate is constrained to a specific length. For example, we may want to seed the population with candidates of length ten actions. For each action one of the available non-terminals is selected, and arguments for that non-terminal are selected. The non-terminal and the arguments are selected at random and the arguments are selected from sets of the correct type. The non-terminal and its arguments make up a single planning action. The strategy then adds the created action to the candidate and continues creating new actions until the maximum length of the candidate is reached.

4.3   Finding the Fitness

A difference between GP and Genetic Planning is that GP operates on a multitude of fitness cases and in Genetic Planning there is only one fitness case for each problem. Before the fitness function can assess a particular candidate the candidate must first be simulated. The simulation function starts with the initial state of the problem and sequentially starts applying the actions in the candidate to the initial state. For each action the preconditions of that action are checked against the current state. If the preconditions are satisfied that action is allowed to execute by updating the world state and the next action is considered. If the preconditions are not present in the current state then that action is ignored and the simulator moves to the next action. The current state is carried over to the next action untouched. The resulting state from applying all the actions in the candidate is then used as input for the fitness function. The fitness function works by comparing the input state with the goal state.

The fitness value in the Genetic Planner is made up of two parts. The first and main part considers how many facts from the goal state are present in the resulting state. This forms a proportion of number of achieved goals over number of actual goals. A small part of the fitness value comes from the number of working actions in the candidate. Currently 80% of the fitness value is given by the correctness of the plan, and the remaining 20% is given by the proportion of working actions in the plan.

5   Seeding Strategies

We now present the different initial population seeding strategies we considered. The intuition that prompted this work was to provide the Genetic Planner with a higher quality initial population to evolve from and therefore reducing the evolution time. As the order of actions is important in planning, we should have a seeding strategy that takes this ordering into account somehow, but the random strategy ignores this completely. Hence, we developed some seeding strategies based on different search methods. Each of the strategies works at the candidate level, and each candidate is set to a particular initial length.

5.1   Strategy 1: All Action Method

The first seeding strategy is a modification of the random one. In this case all possible actions for a particular domain-problem pairing are created and stored in a list. For a new candidate the strategy then selects actions randomly from the list and adds those actions to the new candidate. Each time an action is selected it is removed from the possible list of actions and random selection takes place on this smaller list. This continues until the list is empty, at this stage the list is reset. This method ensures that all the actions are present in the initial population and are present with almost equal diversity.

5.2   Strategy 2: DFS

The second new seeding strategy was one based on DFS. In this case the strategy starts with the initial state and the set of all actions. The strategy then searches for an action that can execute from the current state. Once an action is then added to the candidate. The action is applied to the current state to produce a new state. The strategy then proceeds to work from this new state. The strategy continues adding actions until the maximum initial size for plans is reached.

5.3   Strategy 3: DFS + Random Actions

A variation to the second seeding strategy is to force it to occasionally select a random action. In this instance at a particular state the strategy is either going to select an action at random from the set of all actions or the strategy is going to search for an action which executes. The probability of selecting either is set by a parameter. This strategy was added in case Strategy 2 produces initial populations which are not diverse enough.

5.4   Strategy 4: DFS from Random States + Random Actions

A second variation to the DFS strategy would be to alter the initial state the strategy did DFS from. This strategy works by creating a random initial state and then working like strategy number 3 from that initial state. Once a candidate has been produced a new initial state is created and the strategy proceeds. This would increase the variation of the initial population but would also give strings of applicable actions from all over the search space.

5.5   Strategy 5: BFS + Random Actions

The final strategy is based on Best First Search. This strategy starts with the initial state and the set of all possible actions. Using the initial state as input BFS strategy attempts to find the best action to apply from that state. Therefore it applies all actions to the state and records the fitness for each of the states. The action that does best is then added to the candidate. The state is updated by applying that action and the search continues. If more than one action return the same best fitness then an action is chosen randomly from the set of best actions. The final alteration to make is that like Strategy 2, this strategy occasionally picks random actions rather than the best action.

6   Experiments

Each experiment was conducted using the parameters described in Table 1, unless otherwise indicated.

One large experiment was done on all the Blocks World problems available to the planner for the Blocks World Domain. We reproduce results for the four hardest problems called Large A to Large D. The problem specifications were taken from the BlackBox distribution [13]. We used Spector's puton and newtower actions thereby shrinking the plan length by half when compared with plans

Table 1. Experimental Setup

Parameter	Setting
Termination	Maximum number of generations is 100
Pop Size	2000
Initial Length	15-30 actions
Tournament Size	4
Max plan size	200 actions
Genetic op. prop	80% crossover and 20% reproduction
Addition Mutation	Applied 10% of the time and adds 8 random actions to random positions
Seeding Strategy	All, with 50% random actions for 3,4, and 5
Fitness	As described in Section 4.3.

created using the BlackBox distribution. This is due to the Blocks World Domain implemented for BlackBox explicitly modelling the robot arm, thus the their domain containing four planning operators.

Each problem is an extension of the predecessor by adding more blocks to the problem. Large A contains 9 blocks with a shortest plan of 6 actions, up to Large D, which contains 19 blocks with a shortest plan of 18 actions. For each of the problems, each strategy was used, and for each strategy there were ten runs.

6.1   Results

The following four tables report the results of the experimentation. Table 2 shows the average fitness of the initial populations and average value of the best individual in parenthesis. Table 3 shows the average number of generations used to solve each problem. The number in parenthesis for Table 3 indicates the number of failures. No parentheses indicates that there were no failures. This average also includes those runs that went to a hundred generations and failed to find a solution. A dash indicates that none of the ten runs were successful. Table 4 shows the average CPU time in seconds for that particular problem. This time includes population initialisation time and at the moment there is no accurate means to separate the two. This time was computed by dividing the total CPU time taken for the ten runs divided by the number of successful runs. Table 5 shows the results of varying the proportion of randomly selected actions for the Large B problem and using 10 runs for each proportion.

7   Discussion

The results from using Strategy 0 are the benchmark results. The first refinement, Strategy 1, reduced the amount of time and generations needed, but had little impact on the last two problems. An interesting point is that though initial

Table 2. Average and Best Fitness of Initial Populations

Strategy	Large A	Large B	Large C	Large D
0	32.1 (60.7)	26.1 (49.6)	10.0 (29.7)	9.8 (22.6)
1	26.4 (60.7)	21.9 (46.1)	9.4 (18.2)	8.0 (16.8)
2	26.3 (62.8)	23.8 (53.8)	13.3 (42.0)	11.6 (28.5)
3	27.4 (61.5)	24.8 (52.5)	12.1 (36.6)	11.0 (27.1)
4	26.9 (57.5)	22.1 (46.3)	9.4 (22.6)	8.2 (18.9)
5	57.2 (100.0)	51.3 (90.1)	30.5 (72.7)	29.3 (74.5)

Table 3. Generations used to Solve Problem

Strategy	Large A	Large B	Large C	Large D
0	7.9	16.3	80.8 (5)	-
1	6.7	15.4	76.7 (6)	-
2	4.6	9.4	44.7	89.5 (7)
3	5.0	9.8	51.5 (2)	-
4	7.2	16.0	75.0 (3)	-
5	0.0	2.9	8.1	5.5

Table 4. Time Taken to Solve Problem

Strategy	Large A	Large B	Large C	Large D
0	51.8	197.4	3862.4	-
1	41.9	168.8	4492.9	-
2	53.3	112.1	1064.7	8834.3
3	43.9	97.0	1484.5	-
4	64.8	194.8	2614.8	-
5	302.9	557.5	1459.8	3176.6

Table 5. Varying the Proportion of Random Actions for Strategy 5, for Large B

Proportion Random	Average Generations
0.1	20.6
0.2	13.4
0.3	5.2
0.4	0.0
0.5	2.0
0.6	3.9
0.7	3.8
0.8	6.1
0.9	11.8

population looks slightly worse, Strategy 1 does better than Strategy 0. Strategy 2, using DFS, allowed for significantly less time and generations when compared with Strategy 0. Also notable is that Large D was solvable using this strategy. Strategy 3 performs slightly worse than Strategy 2. Both 2 and 3 produced initial populations with good fitness values. Strategy 4 resulted in being equivalent to Strategy 0 in terms of number of generations, time taken, and quality of initial population.

Perhaps the most interesting result is Strategy 5, based on BFS. Using this seeding strategy, most of the work goes into seeding the population, about 80% of the CPU time. This strategy used the fewest generations to solve each of the problems. Although it took longer to solve the simpler problems, the rate at which the time taken to solve a particular problem is increasing at a slower rate than the other seeding strategy methods. This last strategy also produced by far the best individuals for the initial population.

Table 5 confirms that a 50/50 proportion of random actions and searched actions was a good idea. In fact, having nearly no random actions gave performance which was worse than the random strategy. But having initial candidates made of 40% random actions and 60% best actions gave the least number of generations for Large B.

8   Conclusions

We believed that the efficiency of the Genetic Planner can be improved by improving the quality of the initial population. We also had some intuitions about what features a good initial population should have. There should be many working actions in the initial populations. The order of the actions is important in planning so it's important that this is reflected in the initial population. Also the initial population should be diverse in that it includes all the actions. We feel that some of these intuitions have been weakened. The strategies that worked best, strategies 2 and 5, produced the least diverse populations but performed well. This is probably because the order of the actions is more useful than having a diverse population. Strategy 4 would satisfy all three requirements, but performed only as well as a random method. Perhaps this is because the ordering was completely lost. However the final table indicates that some randomness in the initial population is clearly beneficial. When this combined with an initialisation strategy that gives the best ordering of actions, the strategy produces the best result for Large B. Clearly, however more statistical methods and experimental work has to be carried out before any of these conclusions can be confirmed.

Acknowledgements. The first author would like to thank the EPSRC for support via a quota studentship.

References

1. Westerberg, C.H.: An investigation into the use of using Genetic Programming to solve Classical Planning Problems. 4th Year CS/AI Project, Division of Informatics, University of Edinburgh, 2000
2. Westerberg, C.H., Levine, J.: "GenPlan": Combining Genetic Programming and Planning. Proceedings for the UK Planning and Scheduling Special Interest Group 2000, Milton Keynes, UK, 2000
3. Spector, L.: Genetic Programming and AI Planning Systems. Proceedings of Twelfth National Conference on Artificial Intelligence, Seattle Washington USA AAAI Press/MIT Press, 1994
4. Muslea, I.: SINERGY: A Linear Planner Based on Genetic Programming. Proceedings of the 4th European Conference on Planning, 1997
5. Handley, S.G.: The automatic generations of plans for a mobile robot via genetic programming with automatically defined functions. In Advances in Genetic Programming, MIT Press, pages 391-407, 1994
6. Aler, R., Borrajo, D., Isasi, P.: GP Fitness Functions to Evolve Heuristics for Planning. Proceeding of the GECCO 2000 Conference, EvoPlan Workshop (2000)
7. Aler, R., Borrajo, D., Isasi, P.: Genetic Programming and deductive-inductive learning: A multistrategy approach. Proceedings of the Fifteenth International Conference on Machine Learning, ICML'98, Madison Wisconsin, (1998) 10-18
8. Koza, J.R.: Genetic Programming. The MIT Press, Cambridge MA (1992)
9. Langdon, W.B., Nordin, J.P.: Seeding Genetic Programming Populations. Proceedings of the Third European Conference on Genetic Programming, EURO GP 2000, Edinburgh (2000)
10. Fikes, R.E., Nilsson, N.: STRIPS: A New Approach to the Application of Theorem Proving to Problem Solving. Artificial Intelligence 2 (1971) 189-208
11. Bylander, T.: The computational complexity of propositional STRIPS planning. Artificial Intelligence, 69(1-2):165-204, (1994)
12. Banzhaf, W., Nordin, P., Keller, R.E., Francone, F.D.: Genetic Programming An Introduction. Morgan Kaufmann Publishers, San Francisco CA (1998)
13. Blackbox: http://www.cs.washington.edu/homes/kautz/blackbox/index.html
14. Tate, A., Hendler, J., Drummond, M.: A Review of AI Planning Techniques. Readings in Planning. Morgan Kaufmann Publishers, San Mateo CA (1990)
15. Weld, S.: An Introduction to least commitment planning. AI Magazine 15(4), (1994)

Author Index

Lecture Notes in Computer Science

For information about Vols. 1–1937
please contact your bookseller or Springer-Verlag

Vol. 1976: T. Okamoto (Ed.), Advances in Cryptology – ASIACRYPT 2000. Proceedings, 2000. XII, 630 pages. 2000.

Vol. 1977: B. Roy, E. Okamoto (Eds.), Progress in Cryptology – INDOCRYPT 2000. Proceedings, 2000. X, 295 pages. 2000.

Vol. 1978: B. Schneier (Ed.), Fast Software Encryption. Proceedings, 2000. VIII, 315 pages. 2001.

Vol. 1979: S. Moss, P. Davidsson (Eds.), Multi-Agent-Based Simulation. Proceedings, 2000. VIII, 267 pages. 2001. (Subseries LNAI).

Vol. 1983: K.S. Leung, L.-W. Chan, H. Meng (Eds.), Intelligent Data Engineering and Automated Learning – IDEAL 2000. Proceedings, 2000. XVI, 573 pages. 2000.

Vol. 1984: J. Marks (Ed.), Graph Drawing. Proceedings, 2001. XII, 419 pages. 2001.

Vol. 1985: J. Davidson, S.L. Min (Eds.), Languages, Compilers, and Tools for Embedded Systems. Proceedings, 2000. VIII, 221 pages. 2001.

Vol. 1987: K.-L. Tan, M.J. Franklin, J. C.-S. Lui (Eds.), Mobile Data Management. Proceedings, 2001. XIII, 289 pages. 2001.

Vol. 1988: L. Vulkov, J. Waśniewski, P. Yalamov (Eds.), Numerical Analysis and Its Applications. Proceedings, 2000. XIII, 782 pages. 2001.

Vol. 1989: M. Ajmone Marsan, A. Bianco (Eds.), Quality of Service in Multiservice IP Networks. Proceedings, 2001. XII, 440 pages. 2001.

Vol. 1990: I.V. Ramakrishnan (Ed.), Practical Aspects of Declarative Languages. Proceedings, 2001. VIII, 353 pages. 2001.

Vol. 1991: F. Dignum, C. Sierra (Eds.), Agent Mediated Electronic Commerce. VIII, 241 pages. 2001. (Subseries LNAI).

Vol. 1992: K. Kim (Ed.), Public Key Cryptography. Proceedings, 2001. XI, 423 pages. 2001.

Vol. 1993: E. Zitzler, K. Deb, L. Thiele, C.A.Coello Coello, D. Corne (Eds.), Evolutionary Multi-Criterion Optimization. Proceedings, 2001. XIII, 712 pages. 2001.

Vol. 1995: M. Sloman, J. Lobo, E.C. Lupu (Eds.), Policies for Distributed Systems and Networks. Proceedings, 2001. X, 263 pages. 2001.

Vol. 1997: D. Suciu, G. Vossen (Eds.), The World Wide Web and Databases. Proceedings, 2000. XII, 275 pages. 2001.

Vol. 1998: R. Klette, S. Peleg, G. Sommer (Eds.), Robot Vision. Proceedings, 2001. IX, 285 pages. 2001.

Vol. 1999: W. Emmerich, S. Tai (Eds.), Engineering Distributed Objects. Proceedings, 2000. VIII, 271 pages. 2001.

Vol. 2000: R. Wilhelm (Ed.), Informatics: 10 Years Back, 10 Years Ahead. IX, 369 pages. 2001.

Vol. 2003: F. Dignum, U. Cortés (Eds.), Agent Mediated Electronic Commerce III. XII, 193 pages. 2001. (Subseries LNAI).

Vol. 2004: A. Gelbukh (Ed.), Computational Linguistics and Intelligent Text Processing. Proceedings, 2001. XII, 528 pages. 2001.

Vol. 2006: R. Dunke, A. Abran (Eds.), New Approaches in Software Measurement. Proceedings, 2000. VIII, 245 pages. 2001.

Vol. 2007: J.F. Roddick, K. Hornsby (Eds.), Temporal, Spatial, and Spatio-Temporal Data Mining. Proceedings, 2000. VII, 165 pages. 2001. (Subseries LNAI).

Vol. 2009: H. Federrath (Ed.), Designing Privacy Enhancing Technologies. Proceedings, 2000. X, 231 pages. 2001.

Vol. 2010: A. Ferreira, H. Reichel (Eds.), STACS 2001. Proceedings, 2001. XV, 576 pages. 2001.

Vol. 2013: S. Singh, N. Murshed, W. Kropatsch (Eds.), Advances in Pattern Recognition – ICAPR 2001. Proceedings, 2001. XIV, 476 pages. 2001.

Vol. 2015: D. Won (Ed.), Information Security and Cryptology – ICISC 2000. Proceedings, 2000. X, 261 pages. 2001.

Vol. 2018: M. Pollefeys, L. Van Gool, A. Zisserman, A. Fitzgibbon (Eds.), 3D Structure from Images – SMILE 2000. Proceedings, 2000. X, 243 pages. 2001.

Vol. 2020: D. Naccache (Ed.), Topics in Cryptology – CT-RSA 2001. Proceedings, 2001. XII, 473 pages. 2001

Vol. 2021: J. N. Oliveira, P. Zave (Eds.), FME 2001: Formal Methods for Increasing Software Productivity. Proceedings, 2001. XIII, 629 pages. 2001.

Vol. 2022: A. Romanovsky, C. Dony, J. Lindskov Knudsen, A. Tripathi (Eds.), Advances in Exception Handling Techniques. XII, 289 pages. 2001

Vol. 2024: H. Kuchen, K. Ueda (Eds.), Functional and Logic Programming. Proceedings, 2001. X, 391 pages. 2001.

Vol. 2026: F. Müller (Ed.), High-Level Parallel Programming Models and Supportive Environments. Proceedings, 2001. IX, 137 pages. 2001.

Vol. 2027: R. Wilhelm (Ed.), Compiler Construction. Proceedings, 2001. XI, 371 pages. 2001.

Vol. 2028: D. Sands (Ed.), Programming Languages and Systems. Proceedings, 2001. XIII, 433 pages. 2001.

Vol. 2029: H. Hussmann (Ed.), Fundamental Approaches to Software Engineering. Proceedings, 2001. XIII, 349 pages. 2001.

Vol. 2030: F. Honsell, M. Miculan (Eds.), Foundations of Software Science and Computation Structures. Proceedings, 2001. XII, 413 pages. 2001.

Vol. 2031: T. Margaria, W. Yi (Eds.), Tools and Algorithms for the Construction and Analysis of Systems. Proceedings, 2001. XIV, 588 pages. 2001.

Vol. 2034: M.D. Di Benedetto, A. Sangiovanni-Vincentelli (Eds.), Hybrid Systems: Computation and Control. Proceedings, 2001. XIV, 516 pages. 2001.

Vol. 2035: D. Cheung, G.J. Williams, Q. Li (Eds.), Advances in Knowledge Discovery and Data Mining – PAKDD 2001. Proceedings, 2001. XVIII, 596 pages. 2001. (Subseries LNAI).

Vol. 2037: E.J.W. Boers et al. (Eds.), Applications of Evolutionary Computing. Proceedings, 2001. XIII, 516 pages. 2001.

Vol. 2038: J. Miller, M. Tomassini, P.L. Lanzi, C. Ryan, A.G.B. Tettamanzi, W.B. Langdon (Eds.), Genetic Programming. Proceedings, 2001. XI, 384 pages. 2001.